The Nature of Scientific Evidence

THE NATURE OF
SCIENTIFIC EVIDENCE

Statistical, Philosophical, and Empirical Considerations

EDITED BY MARK L. TAPER AND SUBHASH R. LELE

The University of Chicago Press / Chicago and London

MARK L. TAPER is associate professor in the Department of Ecology at Montana State University.
SUBHASH R. LELE is professor in the Department of Mathematical and Statistical Sciences at the
University of Alberta.

The University of Chicago Press, Chicago 60637
The University of Chicago Press, Ltd., London
© 2004 by The University of Chicago
All rights reserved. Published 2004
Printed in the United States of America

13 12 11 10 09 08 07 06 05 04   1 2 3 4 5

ISBN: 0-226-78955-1 (cloth)
ISBN: 0-226-78957-8 (paper)

Library of Congress Cataloging-in-Publication Data
The nature of scientific evidence : statistical, philosophical, and empirical considerations / edited by
    Mark L. Taper and Subhash R. Lele.
        p.   cm.
    Includes index.
    ISBN 0-226-78955-1 (cloth : alk. paper)—ISBN 0-226-78957-8 (pbk. : alk. paper)
    1. Science—Statistical methods.   2. Science—Methodology.   I. Taper, Mark L., 1952–.
    II. Lele, Subhash
    Q180.55.S7N37 2004
    507′.2—dc22
                                                                                    2004003548

⊗ The paper used in this publication meets the minimum requirements of the American
National Standard for Information Sciences–Permanence of Paper for Printed Library
Materials, ANSI Z39.48-1992.

MARK L. TAPER DEDICATES THIS BOOK

> *To his parents Phyllis and Bernard Taper,*
>
> *To his wife Ann Patricia Thompson,*
>
> *And to his daughter Oona Alwynne Taper, who,*
> *at the age of three, ceased to ask "Why?"*
> *and began to ask "How do you know dat?"*

SUBHASH R. LELE DEDICATES THIS BOOK

> *To Mr. D. V. Sathe and the late Dr. V. V. Pendse*
> *for inspiring him to be a scientist with a social conscience.*

Contents

Foreword

Real knowledge is to know the extent of one's ignorance.
—Confucius

How is new knowledge created? In a philosopher's view, knowledge is infallible and the instrument of acquiring knowledge is deductive reasoning. For a scientist, all knowledge is fallible. Available knowledge (or theory) is acceptable if it explains, within certain limits, a large body of natural phenomena. A scientist strives to acquire better knowledge or theory through experimentation and interpretation of results. From a statistical point of view, knowledge derived from observed data is uncertain, but is usable if the amount of uncertainty is known. Statistical methodology is concerned with quantification of uncertainty and optimum utilization of available knowledge, however meager. The quest for knowledge involves all these aspects. This book, edited by Mark L. Taper and Subhash R. Lele, examines, in particular, the nature of scientific evidence from statistical, philosophical, and empirical considerations.

No scientist would dream of baldly saying, "You should believe this because I believe this." These days a forensic attitude in which data are seen as clues and hypotheses are seen as suspects pervades science. This attitude is reflected in the titles of such popular books as *The Ecological Detective* (Hilborn and Mangel, 1997), *The Statistical Sleuth* (Ramsey and Schafer, 2001), and *Statistics and Truth: Putting Chance to Work* (Rao, 1997).

But if evidence is central to science and we need to learn more about it, what discipline should we look to for answers about scientific evidence? Scientific evidence is not exclusively a philosophical question, neither is it an exclusively statistical or scientific question. It lies at the juncture of three disciplines: philosophy, statistics, and science. Statistical inference is the foundation of modern scientific thinking and has been since the days of

Galton, Fisher, and Pearson, who developed statistics as an aid to research workers in all fields of enquiry in interpreting experimental data. The influence of philosophy on science has also been profound. The thinking of such philosophers as Aristotle, Bacon, Whewell, Popper, and Kuhn has become so integrated into the scientific culture that many young scientists are unaware of the extent that fundamental scientific thought processes have been dictated by philosophers.

The three disciplines have a long history of pairwise interactions. Well-developed products are often passed across disciplines. For example, the writings of Popper and Kuhn, although originating in philosophy, have had tremendous impact on the practice of science. Scientific problems have inspired statisticians to develop many basic tools of statistical inference for testing given hypotheses and building models for prediction. Statisticians do work with scientists on a daily basis, but mostly on a technical level, not on the development of the foundations of scientific thought process. In the past, there have been efforts to connect statistics and philosophy at a fundamental level. Examples of this are the *Foundations of Statistical Inference* conference organized by Godambe in 1970, which approached the union from a statistical viewpoint (Godambe and Sprott, 1971), and Hacking (1965), which came at it from philosophy.

This book brings all three disciplines together for an intimate discussion of foundational problems on the nature of scientific evidence. While many voices are heard, overall the book embraces an evidential viewpoint. Evidence plays a fundamental role in the scientific process. Evidence may be utilized in the decision-making process, evidence may be utilized in modifying the prior beliefs; however, it is *evidence* that is the keystone. An important characterization of evidence, extending the thoughts of Platt (1964), is that evidence is comparative. One cannot discuss evidence for a single hypothesis; one has to have alternative explanations with which a given hypothesis is compared. This is a viewpoint that has not been taken very seriously by academic statisticians, philosophers, or scientists with the exceptions of probably Hacking, Edwards (1992), and Royall (1997).

This volume is not an end in itself. It reveals an important problem, sketches out a potential solution, but makes only modest offerings in the way of practical tools. However, while not an end, this book may very well be a beginning. The evidential approach discussed here has the potential to profoundly influence the practices of empirical science, statistics, and philosophy. This book is perhaps best viewed as forwarding a challenge for further development of the concept of evidence and practical techniques based on it. As with all other approaches in science, evidential statistics will stand or fall on its proven utility. I applaud the editors of this book for bringing together such a luminous collection of statisticians, philosophers, and empirical scientists to discuss an important and timely subject. I urge the respective intellectual communities to consider this viewpoint seriously.

C. R. Rao
Eberly Emeritus Professor of Statistics
Pennsylvania State University

REFERENCES

Edwards, A. W. F. 1992. *Likelihood* (expanded ed.). Baltimore: John Hopkins University Press.

Godambe, V. P., and D. A. Sprott, eds. 1971. *Foundations of Statistical Inference.* Toronto: Holt, Rinehart, and Winston.

Hacking, I. 1965. *Logic of Statistical Inference.* Cambridge: Cambridge University Press.

Hilborn, R., and M. Mangel. 1997. *The Ecological Detective: Confronting Models with Data.* Princeton: Princeton University Press.

Platt, J. R. 1964. Strong Inference. *Science* 146:347–353.

Ramsey, F. L., and D. W. Schafer. 2001. *The Statistical Sleuth: A Course in Method of Data Analysis.* Belmont, CA: Duxbury.

Rao, C. R. 1997. *Statistics and Truth: Putting Chance to Work.* Singapore: World Scientific.

Royall, R. M. 1997. *Statistical Evidence: A Likelihood Paradigm.* London: Chapman and Hall.

Preface

Scientists strive to understand the operation of natural processes, creating and continually refining realistic and useful descriptions of nature. But which of many possible descriptions most closely approximate reality? To help in this evaluation, scientists collect data, both experimental and observational. The objective and quantitative interpretation of data as evidence for one model or hypothesis over another is a fundamental part of the scientific process. Many working scientists feel that the existing schools of statistical inference do not meet all their needs. It is safe to say that to some extent, scientists have had to twist their thinking to fit their analyses into existing paradigms. We believe that a reexamination and extension of statistical theory is in order, reflecting the needs of practicing scientists.

Much of the statistical analysis conducted presently in science uses methods that are designed fundamentally as decision-making tools, for example Fisher's significance tests or Neyman-Pearson tests (strictly speaking, the significance test was not designed to be a decision-making tool; Fisher apparently suggested its use for evidential purposes). In science, these methods are used to "decide" between hypotheses. Unfortunately, these methods are also being used to make statements about the strength of evidence, a task for which they are not directly designed. A common alternative to the frequentist approach of statistical analysis is subjective Bayesian analysis. However, subjective Bayesian analysis contains elements of personal belief unacceptable to many scientists. Although in recent years, attempts have been made to create "objective" priors, definition of what "objective" means is still unclear. Frequentist analysis has troubling subjective elements as well. For example, in multiple-comparison situations, P-values depend on the intent of the investigator regarding hypotheses to be tested and thus have strong subjective elements.

Richard Royall's *Statistical Evidence: A Likelihood Paradigm* (1997) clarified the situation by distinguishing three fundamental questions necessary for statistical inference: (1) What do I *believe,* now that I have these data? (2) What should I *do,* now that I have these data? (3) How should I interpret these data as *evidence* regarding one hypothesis versus the other competing hypothesis? Bayesian methodology addresses question (1) quite naturally. Question (2) requires a decision rule; standard "frequentist" (hypothesis-testing/confidence interval) approaches can provide such rules, as can Bayesian methods. Question (3) is more slippery. As discussed by Royall (1997), attempting to answer question 3 with either Bayesian or frequentist statistics can lead to serious problems.

Royall's book inspired a workshop and a symposium hosted in August 1998 by the Ecological Statistics section of the Ecological Society of America. These were so successful and so intriguing that they were followed by a second symposium and an extended working group funded by the National Center for Ecological Analysis and Synthesis (NCEAS). We believe that at the heart of scientific investigation is the determination of what constitutes *evidence* for a model or a hypothesis over other models or hypotheses. The purpose of this volume is first to demonstrate the need for a statistical viewpoint, which focuses on quantifying evidence, and second, through a dialogue among practicing scientists, statisticians, and philosophers, to outline how such a new statistical approach might be constructed.

In this volume, we have brought together eminent ecologists (both theoretical and empirical), statisticians, and philosophers to discuss these issues and to explore new directions. We feel this book will appeal to all three disciplines and more broadly to scientists in general. Although all the scientists contributing to this volume are ecologists, we do not believe that interest in the book will be limited to this discipline. Ecology is used principally as an exemplar science. *The Nature of Scientific Evidence* should speak to all sciences where uncertainty plays an important role, to all statisticians interested in scientific applications, and to philosophers of science interested in the formalities of the role of evidence in science.

This volume is directed toward students, as well as established scientists, statisticians, and philosophers. It can serve a particularly important educational role. Statistics is an increasingly important tool in science and in particular in ecology. As a consequence there has been an increase in the sophistication of ecologists' application of statistical techniques. Recently there has been an increase in the number of texts addressing new and different statistical approaches, reflecting the progress in the field of statistics over the last four decades. There are texts on the application of Monte Carlo techniques, meta-analysis, and general statistical approaches. But this increase in technical sophistication has outpaced the attention generally paid in graduate education in the sciences to the philosophical foundations of statistical and scientific thinking. This book is intended to address that need.

This volume is not synoptic. Rather than extensively review and rehash existing literature, we hope to stimulate thought and research in a new field. We, the editors, as scientists and statisticians, have developed strong opinions as to the profitable directions for this field. However, as editors, we have aggressively sought contributors whose expertise and opinions differ from ours.

We have structured this volume so that it could be digested in a typical one-semester course. The core of the volume is composed of fifteen main chapters (chapters 2 through 16)—one chapter for each week of a standard semester. Chapter 16 provides a synthesis of the volume, while chapter 1 serves as a simple introduction to basic ideas in statistics. It discusses the general structure of the scientific process and how one translates a scientific hypothesis into a statistical hypothesis. It provides a brief overview of Fisher's significance testing (P-values), Neyman-Pearson testing, Bayesian testing, and the concept of likelihood. The main purpose of this chapter is pedagogical and brings everyone to a common footing.

All chapters, except the introduction and synthesis, are followed by commentaries from recognized authorities, which in turn are followed by rebuttals from the chapter authors. We've chosen this format so as to investigate deeply the implications of a few key ideas rather than to touch lightly on many. This approach should help readers understand papers from outside their own fields. Another advantage of this approach is that it allows presentation of thoughts that might be at odds with the ideas in the main chapters, thus providing a balance in the presentation. Many of the chapters have benefited from this format by forcing the authors to think through their ideas with additional care and clarity.

The book is divided into five main parts: "Scientific Process," "Logics of Evidence," "Realities of Nature," "Science, Opinion, and Evidence," and "Models, Realities, and Evidence." Each part is prefaced by an overview of the questions the chapters struggle with. Despite being a collection of independent chapters written by practitioners of very different fields, the finished volume has considerable coherence. Most of the contributors have interacted through two workshops and two symposia. They are well aware of the thinking of the other contributors. Furthermore, contributed chapters are supported by the introduction and synthesis chapters. These are constructed to give continuity to the entire volume.

We acknowledge the support of the National Center for Ecological Analysis and Synthesis (NCEAS) for the Evidence Project Working Group. Montana State University at Bozeman and the University of Alberta at Edmonton contributed to publication costs. This work was partly supported by a National Science and Engineering Research Council (NSERC) grant to Subhash R. Lele. We thank the participants of a graduate seminar conducted at Montana State University for many useful discussions.

We are extremely grateful to Christie Henry, Michael Koplow, and the other editorial staff at the University of Chicago Press for their continued encouragement and support, not to mention many wonderful dinners and discussions on science in general. Without their help, this project would never have seen the light of day. We are also thankful to the reviewers for patiently reading this long and difficult manuscript and providing insightful comments. Intellectually, our debts are too great and too numerous to number, but Mark Taper particularly acknowledges the faculty of the Rudolf Steiner School of New York City for introducing him to the pleasures of thinking broadly as well as deeply.

REFERENCE

Royall, R. M. 1997. *Statistical Evidence: A Likelihood Paradigm*. London: Chapman and Hall.

PART 1

Mark L. Taper and Subhash R. Lele

OVERVIEW

This section sets the stage for the discussion of science, statistics, and evidence. Its main goal is to review and critically appraise basic ideas in statistics, philosophy of science, and empirical studies as they relate to scientific evidence.

Chapter 1, by Nicholas Lewin-Koh, Mark Taper, and Subhash Lele, is a simple introduction to basic ideas in statistics. It discusses the general structure of the scientific process from a statistical viewpoint, focusing on how one translates a scientific hypothesis into a statistical hypothesis. It then provides a brief overview of Fisherian significance testing (P-values), Neyman-Pearson testing, Bayesian testing, and the concept of likelihood. The purpose of this chapter is pedagogical. Its goal is to introduce all readers to existing basic concepts of statistical inference, and thus to set everyone on a common footing for reading the rest of the book. Because the authors do not introduce any new concepts, there is no commentary on this chapter.

Brian Maurer (chapter 2) argues that science is a legitimate method of learning about the real world. He discusses the ideas of deductive and inductive science and the implications of this dialectic for ecological science in particular and science in general. He claims that both methodologies are useful, but for different purposes and at different stages of scientific maturity of a topic. Maurer discusses the differences between "logical empiricism" and "naturalistic realism." He claims that there are no laws of nature but that nature nevertheless behaves in a lawlike fashion. This resonates well with the dictum "All models are approximations." This sentiment is repeated frequently throughout the rest of the volume. One can use deductive or inductive logic to discover the lawlike structure of natural processes.

Samuel Scheiner (chapter 3) explores the concept of evidence from a scientific standpoint. What is important is not the type of observation, but whether it matches the question at hand. This is what decides whether an observation counts toward evidence in a given instance. Evidence comes from scientifically guided empirical observations combined with background information, logic, and scientific expertise. Observations may come from both manipulative and observational experiments. Experiments may be either theory driven experiments, designed to test a particular theory, or poke-at-it experiments, designed just to look at something potentially interesting. Because all theories of global import deal with phenomena, mechanisms, and processes at many different scales, all kinds of evidence must be used in building and testing these theories. Scheiner's thesis is that scientific theories are built upon the consilience of the evidence. Maurer and Scheiner express several common themes. One is the importance of consilience in scientific process. Another is that evidence is evidence, whether data were collected before theory was proposed or after. This idea needs to be examined in light of the frequentist ideas of multiple testing and joint versus individual confidence intervals.

1   A Brief Tour of Statistical Concepts

Nicholas Lewin-Koh, Mark L. Taper, and
Subhash R. Lele

ABSTRACT

This chapter serves as a tutorial, introducing key concepts of statistical inference. We describe the language and most basic procedures of Fisherian *P*-value tests, Neyman-Pearson tests, Bayesian tests, and the ratio of likelihoods as measures of strength of evidence. We demonstrate each method with an examination of a simple but important scientific question, Fisher's thesis of equal sex ratios. Even within the confines of this simple problem, these methods point toward very different conclusions.

SCIENCE AND HYPOTHESES

In the seventeenth century, Francis Bacon proposed what is still regarded as the cornerstone of science, the scientific method. He discussed the role of proposing alternative explanations and conducting tests to choose among them as a path to the truth. Bacon saw the practice of doing science as a process of exclusions and affirmations, where trial and error would lead to a conclusion. This has been the principal framework under which science has been conducted ever since. Bacon saw science as an inductive process, meaning that explanation moves from the particular to the more general (Susser, 1986).

Nicholas Lewin-Koh thanks his wife Sock-Cheng for the frequent conversations and advice that helped to shape his understanding of statistical concepts, and Paul Marriot for comments, which helped to clarify explanations and the general readability. Mary Towner and Prasanta Bandyopadhyay gave helpful comments on an earlier version of this manuscript.

Karl Popper in the twentieth century argued entirely differently. Popper (1959) argued that science progresses through deduction, meaning that we proceed from the general to the specific. Popper proposed the doctrine of falsification, which defines what is acceptable as a scientific hypothesis: if a statement cannot be falsified, then it is not a scientific proposition.

Both Popper and Bacon promote the idea that we learn through trial and error. Popper's deductive approach stipulates that, through reasoning from logical premises, we can make predictions about how systems should behave, and the hallmark of a scientific statement is that it can be tested.

> I do not think that we can ever seriously reduce by elimination the number of [the] competing theories, since this number remains infinite. What we do—or should do—is hold on, for the time being, to the most improbable of the surviving theories or, more precisely, to the one that can be most severely tested. We tentatively "accept" this theory—but only in the sense that we select it as worthy to be subjected to further criticism, and to the severest tests we can design. (Popper, 1959, p. 419)

Popper's views have been criticized extensively, and it is generally agreed that confirmation of theories plays an important role (Sober, 1999; Lloyd, 1987). What is not controversial is that theories, models, and hypotheses need to be probed to assess their correctness. This creates a need for an objective set of methodologies for assessing the validity of hypotheses. If experimental outcomes were not subject to variation and experiments were controlled to the point where measurement error and natural variation were negligible, then hypothesis testing would be a relatively simple endeavor. However, as any practicing scientist knows, this is not the case; measurements are not always reliable, and nature is not uniform. In this context, scientists have a need for tools to assess the reliability of experimental results and to measure the evidence that the outcome of an experiment provides toward accepting or rejecting a hypothesis. To this end, statistical methods have been developed and applied, as criteria to evaluate hypotheses in the face of incomplete and imperfect data.

There are a variety of statistical approaches to hypothesis validation. As background for this book, we will briefly introduce without advocacy the ones most influential in modern science. We will relate these methods to the general framework of scientific practice that we have outlined. In the interests of concision, many practical considerations and fine details will be glossed over.

HYPOTHESES AND MODELS

Hypotheses play a key role in scientific endeavors. A hypothesis stems from theory in that a good theory suggests some explicit statements that can, in some sense, be tested (Pickett, Kolasa, and Jones, 1994; Sober, 1999). However, we must differentiate between a *scientific* hypothesis and a *statistical* hypothesis and understand the relationship of both to statistical models.

We usually start a study or set of experiments within the context of a body of knowledge and thought built from previous studies, observations, and experience. From this context arises a set of questions. We try to organize these questions into a coherent order with explanatory properties. To do this, we construct a model of how the system works. This is a *scientific model* or scientific hypothesis. To be scientifically testable, this model must be connected to observable quantities.

The expression of the scientific hypothesis in the form of a model leads to a statistical hypothesis and a corresponding statistical model. A statistical model is an explicit quantitative model of potential observations that includes a description of the uncertainty of those observations due to natural variation,[1] to errors in measurement, or to incomplete information, such as when observations are a sample from a larger or abstract population. A statistical hypothesis is a statement about the attributes of a statistical model whose validity can be assessed by comparison with real observations. If a statistical model is found to be inadequate to describe the data, the implication is that the scientific model it represents is also inadequate. Thus, a scientist may compare scientific hypotheses through the analysis of statistical models and data. Figure 1.1 shows these relationships diagrammatically. It is important to realize that, in this book, when we talk of testing hypotheses and evidence, we are referring to the statistical hypotheses, and model selection refers to the selection of statistical models. The scientific inference derived from any statistical analysis can be no better than the adequacy of the models employed.

In the next section, we demonstrate the process of transformation of a scientific hypothesis into a statistical hypothesis with a simple example (see Pickett, Kolasa, and Jones, 1994, for a more complete discussion of translation and transformation in ecological theory).

---

1. Natural variation occurs when entities within a class are not homogeneous, i.e., when the differences among entities are not extreme enough to class them separately but cause measurements of an attribute to vary within the class.

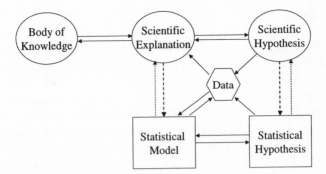

FIGURE 1.1 The relationships between statistical model and scientific explanation: The solid lines are direct translation paths, while the dotted lines are transformation paths. In this model of science, we translate an existing body of knowledge into an explanation and derive hypotheses from the explanations. An explanation is transformed into a statistical model, and hypotheses are generated and tested from the model.

SIMPLE BIOLOGICAL QUESTION

Fisher (1958, 158–60) showed that, in a sexually reproducing population with random mating and equal parental investment in the sexes, natural selection should favor equal proportions of male and female offspring. When these assumptions hold, most studies have shown that there are indeed equal proportions of the sexes in natural populations.

This problem of the sex proportions gives us a framework to demonstrate some key concepts in statistics. What does the statement "the ratio of investment between males and females should be equal for sexually reproducing species" really mean? Certainly, one does not expect that a litter of 3 offspring should have 1.5 males and 1.5 females. Instead, what is being specified is the probability that a given offspring will be male or female is equal for the entire population. If we combine this idea with the ancillary assumption that the sexes of individuals within a litter are independent, then a statistical model describing the proportion of sexes is that the number of males in a litter of a given size is binomially distributed.[2]

2. The binomial distribution arises as the sum of independent trials where each trial can be either a "success" or a "failure" (also known as Bernoulli trials). If we imagine a sequence of independent coin flips with probability θ of landing heads, if we flip the coin n times, then the probability of x successes (defined in this case as landing heads) is $\binom{n}{x}\theta^x(1-\theta)^{n-x}$ for $x = 1, 2, \ldots, n$. The term $\binom{n}{x}$ is the binomial coefficient, which counts the number of possible sequences of n events with x successes.

A statistical hypothesis is a statement about the parameters of a distribution. If θ is the probability of an offspring being born male, then $\theta = .5$, $\theta = .7$, $\theta > .5$, and $\theta \neq .5$ are possible statistical hypotheses. The scientific hypothesis of equal parental investment now corresponds to the statistical hypothesis $\theta = .5$.

To validate or test a statistical hypothesis corresponding to a scientific hypothesis, an investigator must gather real-world data. We utilize a classic data set from the literature on sex ratios in pig litters. The data comes from pig births registered in the British National Duroc-Jersey Pig Record, vol. 67 (Parks, 1932). There are 7929 male and 8304 female births recorded for a total of 16233 births.

How can one investigate whether these data are consistent with Fisher's scientific hypothesis of a 50% proportion of males? In the following sections, we will use these data to illustrate the mechanics of several different statistical approaches to this problem that have been very influential in science. However, before we can do this, we need to introduce a few critical statistical concepts.

THE SAMPLE SPACE, RANDOM VARIABLES, AND THE PARAMETER SPACE

A basic concept essential for understanding statistical inference is the notion of a sample space. The sample space S is the space of all possible outcomes that can occur when an experiment/observation is made. As a simple demonstration, consider our sex proportion example. For a family with three offspring, there are eight ($2^3 = 8$) possible birth sequences. Thus, the sample space is: $S = \{\{F, F, F\}, \{F, F, M\}, \{F, M, F\}, \{M, F, F\}, \{F, M, M\}, \{M, F, M\}, \{M, M, F\}, \{M, M, M\}\}$, where F indicates a female offspring and M a male.

However, our interest is in the proportion of males (or females) in the population, meaning that order is not important. Let X be the number of male piglets in a litter. Then $X(\{F, M, F\}) = 1$ and $X(\{M, M, F\}) = 2$, and so on. Corresponding to each possible litter in our sample space, we can assign a specific number to X. Such an X is called a random variable. For the case of a family size of three, the possible values of X are 0, 1, 2, or 3. The set of possible values of X is often called the range or support of the random variable.

Assuming that the probability of an offspring being male (or female) is one-half, $\theta = \frac{1}{2}$, sexes of littermates are independent, and θ is constant for all litters, then the probability of any specific litter of 3 being entirely com-posed of males is $\frac{1}{2} \times \frac{1}{2} \times \frac{1}{2} = \frac{1}{8}$. Because only one birth sequence can oc-cur at a time, the probability that the random variable X takes on a specific value is equal to the sum of the probabilities of all birth sequences with that number of males. Thus, as only a single birth sequence will produce all fe-males, $P(X = 0) = \frac{1}{8}$, similarly only a single sequence will produce all males and $P(X = 3) = \frac{1}{8}$. However, there are three possible sequences with one male and also three sequences with two males; thus $P(X = 1) = \frac{3}{8}$ and $P(X = 2) = \frac{3}{8}$. The set of probabilities for each possible value of a random variable is known as the probability distribution of the random variable. If the random variable is discrete, as in this example, we refer to the distribu-tion as a probability mass function (pmf) and if the random variable has continuous support we refer to the distribution as a probability density function (pdf).

The probability θ of a piglet being male need not be $1/2$. It could con-ceivably be any number between 0 and 1. For a fixed value of θ, the proba-bility distribution is given by $\{(1 - \theta)^3, 3(1 - \theta)^2\theta, 3(1 - \theta)\theta^2, \theta^3\}$. We can further generalize this to the probability distribution for any θ and any lit-ter size as

$$\left\{ \binom{n}{0}(1 - \theta)^n(\theta)^0, \binom{n}{1}(1 - \theta)^{n-1}(\theta)^1, \ldots, \binom{n}{n}(1 - \theta)^0(\theta)^n \right\}.$$

This is called the binomial distribution with parameters θ and n and is de-noted by Bin(n, θ). Often the number of observations, n, is considered known and fixed, and is thus not a parameter.

Finally, the parameter space Θ is defined as the collection of all possible values that the parameter θ could conceivably take. In this case where X is a random variable from a binomial distribution with parameter θ, where $0 < \theta < 1$, then Θ is the set $\{\theta : 0 < \theta < 1\}$.

With this background in hand, we proceed to the problem of testing sta-tistical hypotheses.

THE MECHANICS OF TESTING STATISTICAL HYPOTHESES

There are several basic approaches to testing statistical hypotheses. In this section, we give a gentle introduction to those approaches and couch them in the framework we have presented so far. We also give examples of how these approaches work using our sex proportions example. The basic ap-

meant to provide an exhaustive description of all approaches taken by scientists. Neither are they mutually exclusive; indeed, any given scientist probably employs elements of both models in his or her daily activities or at different times in his or her career. Rather, the reader should view these models as end points along a continuum of possibilities, realizing that individual scientists or research programs will fall somewhere in between these extremes.

My purpose in discussing specific ways of doing science is to examine the structure of the products of those activities. If there are many ways of doing science, is it necessary that the resulting products of different scientific strategies lead to different bodies of knowledge that cannot be integrated in any meaningful sense? Alternatively, will all scientific knowledge be capable of being conceptually unified across the wide range of spatial and temporal scales represented by the phenomena studied by scientists in different disciplines? Which of these alternatives eventually turns out to be true has some rather profound implications for the way society in general, and scientists in particular, should view the program of scientific investigation. If scientific knowledge is integrative, then each piece is important to the entire whole, and scientific research of all kinds will be necessary to elaborate that whole. If science is not integrative, then each field should be valued separately, as each will contribute something very different to society. I return to this discussion at the close of this chapter.

A DEFINITION OF THE SCIENTIFIC PROCESS

Attempts by philosophers to define what it is scientists do have changed dramatically in the last century. It is necessary to contrast some of these viewpoints, since individual scientists often see themselves somewhere within a broad spectrum of philosophies of science. I will argue that among the alternatives, a philosophical approach called "naturalistic realism" (Giere, 1999) is closest to the actual mind-set that most scientists take. This viewpoint, as will be seen, avoids many of the unsupportable assumptions made by other philosophies of science about the nature of the real world and the way that scientists do science. According to my understanding of this philosophy, it advocates a program of establishing models of reality that then are compared to the real world by scientific investigators to determine their utility. The utility of a model lies in its ability to produce accurate conceptual statements about specific natural systems. The major assumptions naturalistic realism makes are: (1) measurements taken by scientists are taken on real properties of the world, and though such measurements may be constructions made from the nonobjective nature of human experience, there still exist fundamental or intrinsic properties that the measurements represent; and (2) there are no supernatural or other a priori properties of the world that cannot potentially be measured by direct human experience (with appropriate technological extensions).

Science as a way of understanding the world has received intense scrutiny recently (Giere, 1999). This scrutiny has come as a reaction to a set of assumptions made about science that are unrealistic, having been derived from the very different cultural milieu we now refer to as "the Enlightenment." Two major assumptions are required to establish the Enlightenment perspective: (1) there exist immutable and unchanging laws that determine the nature of the world that scientists attempt to understand; and, (2) there is an objective method for discovering these truths that is based on the set of rules that define the process of logical thinking. The program outlined by philosophers that incorporates these assumptions has been called "logical empiricism."

Although the postulates of natural realism and logical empiricism may seem similar, there are deep differences. The two basic assumptions of logical empiricism have faced serious challenges. The existence of universal laws that transcend human experience has been challenged by showing how scientific thought is contingent upon social circumstance (Kuhn, 1970). In its extreme, this line of reasoning states that scientists discover only what they expect to, based on their life experiences within a relatively narrow set of cultural boundaries. The existence of an objective set of laws is irrelevant to the form that such "discoveries" take. Similar phenomena experienced in different cultural contexts lead to different scientific "laws." Although there is clearly an important role for the cultural experiences of individual scientists in what and how they study nature, nonetheless, definitive statements can be made about common phenomena that do transcend cultural experience (see, for example, Hull's 1988 discussion of the changes in biological systematics in the past several decades). To assume otherwise is to fall into the solipsist trap (Deutsch, 1997). The justification for accepting this ability of scientists to develop knowledge about the world that is independent of culture is in the products that have come from application of scientific principles. For example, the gravitational constant measured by a Chinese scientist will be the same as that measure by an American scientist. Furthermore, systematists steeped in modern "Western" culture who study Neotropical animal assemblages develop classifications of species that differ very little from those of peoples raised in Stone Age cultures.

The process of logical thought as the sole and sufficient program for discovering and systematizing knowledge about the real world is also open to some scrutiny. A major problem for this program is the demonstration by Gödel that in any system of logical thought, there are true statements that exist but cannot be deduced from a closed set of fundamental postulates (Hofstadter, 1979). Another way of viewing this result is to say that in any system of logic, there is an infinite set of fundamental postulates (Ruelle, 1991). If this is so, then one should be able to prove anything. All one needs to do is to find a new fundamental postulate if those already in use cannot be used to construct the appropriate proof. Such a state of affairs is troubling indeed if one is trying to discover a finite set of fundamental laws that govern the universe!

Naturalistic realism avoids these pitfalls by relaxing both assumptions made by logical empiricism (Giere, 1999). Instead of assuming the existence of a finite set of laws that govern the behavior of the universe, one need only assume that the universe works in a lawlike manner, so that there are actual phenomena that need to be explained. The nature of the universe may be extremely complex; indeed, much of it might be unfathomable (Barrow, 1998). All that is necessary for science to be successful is that a sufficiently large number of real phenomena exist that *are* understandable within the context of human experience. Relaxing the second assumption of logical empiricism, one need not assume that an objective, transcendental method to describe nature exists. All that is needed is to assume that scientists can build models that approximate nature sufficiently to allow further progress in understanding particular phenomena. If this is so, then as scientific knowledge accumulates, the number of phenomena that are described and understood increases. Scientists need not worry about describing and understanding everything about the universe.

It is important to briefly consider what is meant by "understanding" in the last sentence. As will be seen in a moment, the exact sense in which one uses the word varies profoundly from one scientific strategy to the next. I will use the broad description offered by Pickett, Kolasa, and Jones (1994, 28): "Understanding is an objectively determined, empirical match between some set of confirmable, observable phenomena in the natural world and a conceptual construct." By this definition, there are a number of ways to determine the level of understanding one has about a system. These include, but are not limited to: (1) a priori prediction of unobserved phenomena;

(2) logical consistency between experimental outcomes and statements made by the conceptual construct; and (3) applicability of a conceptual construct to a relatively wide variety of phenomena.

INDUCTIVE AND DEDUCTIVE SCIENCE

Science, then, is the activity of building models that come sufficiently close to mimicking nature that they are useful in developing an understanding about how nature works. To this end, there are a variety of specific activities that might constitute science. The strategy of doing science is the strategy of evaluating how well a conception of a natural phenomenon actually describes it. This evaluation depends not only on the present conceptualization of the phenomenon and the description of it obtained from empirical investigation, but also on how much understanding exists regarding related phenomena. I assume that a continuum of ways to conduct such evaluations exists; I begin by contrasting the two ends of the spectrum.

The two modes of scientific practice that I contrast are *inductive science* and *deductive science*. These modes are recognized by the general goals that scientists employing them are intent on achieving, and not so much by the particular rules of logic that they employ. That is, inductive and deductive science are strategies for achieving scientific understanding, not specific tactics that a scientist employs in any particular investigation. Scheiner (1993) discusses induction and deduction from a tactical viewpoint, while Holling (1998) outlines strategic approaches to ecological science similar to those discussed here. I do not intend to imply that the alternative models discussed here represent formal philosophies of science. During everyday practice of science, individual scientists are rarely conscious of how their work elaborates specific scientific philosophies. Rather, they adhere to an implicit set of rules that define what kinds of questions are interesting to ask and which data and methods are appropriate to answer those questions.

Inductive Science
Inductive science is equivalent to the search for patterns in nature. The fundamental starting place for inductive science is observation (see Scheiner, 1993, for a discussion of tactical uses of induction). Observations are the outcome of experiments. Here I use the term "experiment" to refer to both manipulative and observational studies. Given a set of observations, patterns are sought within them. Once a pattern is demonstrated to exist, usually by some sort of statistical argument, tentative explanations for the pat-

terns are sought. These explanations are post hoc, dealing only with the patterns seen in the data and how they might be consistent with or relate to other established patterns. The adequacy of an explanation is determined by how well it can account for observed patterns.

Documentation of observed patterns and the explanations derived from them stimulate the collection of new data. Explanations obtained from previous data are labeled as hypotheses, and testing such hypotheses tends to be a matter of showing that a new data set demonstrates the same patterns seen in other data sets. This pattern-detection approach to hypothesis testing leads to two possible outcomes. If patterns are repeated among many data sets, explanations are expanded into generalizations. Generalizations provide the basis for the construction of synthetic theories. These theories are inductive in the sense that they are the product of many observations accumulated across numerous experiments. Such theories provide general explanations for a wide variety of data. The other possibility is that new observations fail to conform to expectations obtained from patterns seen in previous experiments. If this is true, then new explanations are sought. New explanations often refute or limit previous explanations for similar data.

The end product of inductive science is a synthetic construct of ideas based on accumulated evidence from many individual experiments. These synthetic explanations are structured inductively based on many specific pieces of information that correspond to individual empirical patterns. Induction is the appropriate mode of reasoning because each pattern alone is relatively uninformative. It is only when the patterns are assembled together that their logical relationship becomes evident. The logical relationship arises because the patterns all share a common structure or inherent similarity. Such similarities imply the same general cause is responsible for each individual observed pattern.

Inductive science is attractive to many scientists because it follows the general mechanism by which humans learn things. Just as repetition of a difficult task makes that task easier, repetition of a pattern among a large number of data sets can make a mechanistic explanation more obvious. Inductive science makes use of human intuition, where explanations may initially arise subconsciously, only later to be consciously described in rigorous language.

Deductive Science
Deductive science begins at a very different starting point than inductive science. In inductive science, the data are the basis upon which explanations are built. In deductive science, the process begins with a proposed explana-

tion for some natural phenomenon (again, see Scheiner, 1993, for the tactical uses of deduction). The phenomenon being examined does not even need to have been observed; it need only exist as a physical possibility (this sets deductive science apart from mathematics because the latter is not constrained by the assumption that a concept is physically realizable). From the first principles and assumptions governing the phenomenon and how it is to be measured, a formal hypothesis is deduced. First principles come from a general theory that is proposed to govern the phenomenon of interest. The formal hypothesis makes specific qualitative and/or quantitative statements that refer to the outcome of a planned experiment.

Rather than exploring for patterns, the purpose of data collection in deductive science is to provide for a "strong test" of the formal hypothesis (Mayo, 1996). This is accomplished by developing an experimental design that maximizes the chance of falsifying the hypothesis should it in fact be false. Ideally the experimental design will eliminate all possible explanations for the phenomenon save one. If the remaining unfalsified hypothesis is the formal hypothesis deduced from theory, then the hypothesis has withstood a strong test, and more confidence is placed in the theory. Over time, theories are constantly exposed to strong tests of their predictions via hypothetico-deductive experimental designs. Those theories that consistently produce falsified predictions are eventually discarded, and other, more robust theories sought.

Deductive science is inherently more difficult to do than inductive science because it requires a detailed knowledge of theories proposed to explain phenomena in order to generate appropriate hypotheses. With a clear understanding of the theoretical significance of a hypothesis, insight into how to design rigorous, strong tests of it is also required. Outcomes of experiments are not very useful unless they are related directly to an existing theory. Good experiments are conducted with a clear statement of the theory that they are testing and how the outcomes of the experiments will provide a clear test of competing hypotheses.

A CASE STUDY: DEVELOPMENT OF MODERN COMMUNITY ECOLOGY

As an example of the roles played by inductive and deductive science in the development of a field of inquiry, consider the development of modern community ecology. Modern community ecology developed from theoretical explorations of early twentieth-century scientists attempting to extend evolutionary theory to cover interactions among organisms (Kingsland, 1995).

In the late 1950s students of G. Evelyn Hutchinson, particularly Robert MacArthur, attempted to meld the theoretical insights made by scientists of the first half of the twentieth century with mathematical models proposed to elaborate patterns observed in natural systems. MacArthur's dissertation (MacArthur, 1958) was an archetype of such thinking. In it, he elaborated patterns in foraging behavior and population dynamics of several species of song birds of the genus *Dendroica,* showing how they could be explained by models incorporating theoretical statements about interactions among competing species. Later, MacArthur (1972,1) argued that "to do science is to search for repeated patterns." MacArthur's approach, and that followed by many of his contemporaries, was inductive: identify real patterns and seek to explain them in a consistent and coherent manner. Much of the research in community ecology that followed MacArthur adopted a similar approach (e.g., see Cody and Diamond, 1975).

In the early 1980s, community ecology entered a period of intense debate (Strong et al., 1984; Diamond and Case, 1986). Part of the debate dealt with the problem of differentiating between patterns generated by ecological processes and those that were artifacts of simple stochastic sampling processes. For example, a point of contention was whether the identities of species found together on different islands in an archipelago were determined by competitive exclusion of some species by others or by chance events. Another component focused on the limits of inferring processes from patterns. As a consequence of this debate, community ecologists began to focus efforts on designing manipulative experiments that could clearly document ecological processes (Diamond and Case, 1986; Hairston, 1989; Underwood, 1997; Resetarits and Bernardo, 1998). Community ecology expanded to incorporate deductive methodologies in order to circumvent problems that arose from limitations imposed by inductive science when applied to complex systems. Modern community ecology progresses by the interplay of inductive and deductive approaches.

STATISTICAL PRACTICE IN INDUCTIVE AND DEDUCTIVE SCIENCE

Statistical reasoning figures very differently in the two models of science outlined in the preceding section. This is primarily due to the differing goals of each model. Statistical techniques useful in pattern detection are often not appropriate for analysis of designed experiments and vice versa. Here I discuss how the use of statistics in inductive science differs from that in deductive science.

The goal of inductive science is to uncover patterns in nature. To this end, a wide variety of statistical techniques are used. Generally, the pattern detection goal lends itself more readily to a use of estimation as opposed to hypothesis testing. If a pattern is thought to exist, the goal is to use data to place some measure of confidence in the existence of the pattern. For many standard applications, what is of interest is not knowing precisely what the outcome of a statistical hypothesis test is, but rather to estimate the range of values that a statistical parameter might take on in order to have some indication of what the pattern looks like. Thus, confidence regions that give some indication of the location and reliability of the best parameter estimates provide more information than rejection regions based on strictly defined null hypotheses. Exploratory statistical methods are particularly applicable in inductive science.

Inductive science can make fruitful use of hypothesis tests when the null hypothesis can be framed in terms of "no pattern" or "random pattern." Rejecting a null hypothesis of "no pattern" lends credibility to the existence of some real pattern in the data, although it doesn't necessarily specify exactly what that pattern is. Rejecting a null hypothesis of "random pattern" suggests that the data has structure that cannot be described by a simple stochastic process. This implies that there is a specific cause-effect process that incorporates known or putative causative factors.

Model fitting is a particularly strong form of statistical inference that can be profitably used in inductive inference. Generally, there may be several different models that correspond to a set of potential causative factors that may have generated a particular data set. Using a model selection criterion, the inductive scientist attempts to identify a specific model that appears to be most likely to have generated the data. By selecting among competing models, the inductive scientist hopes to find an explanation that is best supported by the data. Exposing the same suite of models to a variety of data sets allows the formation of tentative generalizations. In particular, if a specific model corresponding to a specific explanation provides the best description for the most data sets, it increases the confidence of the researcher that the explanation is a general one.

Bayesian statistics provide another strong tool for inference of pattern. They allow the incorporation of previous information regarding a parameter in the evaluation of a particular pattern. The strength of the Bayesian approach is that it provides a way to show that a pattern is consistent among many data sets. Here again, estimation is of more interest than hypothesis testing. By weighting observed data by results obtained from previous experiments, Bayesian methods produce "conservative" estimates. These estimates are conservative in the sense that the researcher expects that a pattern observed in one data set will be generated by the same process that generated other data, so the pattern actually detected by Bayesian estimates will be weighted by previous instances where the same process was obtained.

The use of Bayesian estimation methods has pitfalls as a technique for pattern detection. In particular, the researcher should provide more than simple subjective arguments to justify the expectation that a particular process has operated unchanged to produce several different data sets. This requires a reliable understanding of the process of interest. But in an immature science, this is precisely what is lacking. Research is focused on detecting evidence for the operation of a particular process. Thus, there is a danger that using Bayesian estimation methods may lead to circular reasoning or arbitrariness.

Statistical tests are much more useful in deductive science than in inductive science. Since the purpose of deductive science is to provide strong tests of hypotheses deduced from existing theories (Mayo, 1996, 2004), well-defined statistical hypotheses are more straightforward to construct. When the null hypothesis of a statistical test is a restatement of the prediction made from the deduced hypothesis, rejection using standard Neyman-Pearson procedures unequivocally removes the deduced hypothesis as an explanation for the outcome of the experiment. When a null hypothesis excludes the prediction made by the deduced hypothesis, the rejection strengthens the evidence in favor of the deduced hypothesis. In this case, power calculations can help experimenters evaluate the adequacy of the statistical test.

Bayesian statistical tests are inappropriate in deductive science because they obfuscate the strong test designed for the theoretical hypothesis by bringing to bear information not explicitly contained in the experiment itself. The objective of the experiment is to maximize the chance of getting a result that is more consistent with one hypothesis than another. Such a result can provide a strong logical test of the theoretical explanation. Observed data should stand independently and make as clear a statement about the validity of the proposed explanation as possible. Previous information is useful in the experimental design phase of the research, not during the hypothesis-testing phase.

DIFFERENCES BETWEEN INDUCTIVE AND DEDUCTIVE SCIENCE

In deductive science, theories gain acceptance by consistently producing predictions that survive strong experimental tests. In a sense, a theory that survives does so because of the accumulation of evidence, but the evidence is obtained in a very different way from that obtained in inductive science. Inductive and deductive approaches to science are not mutually exclusive pathways to scientific knowledge. Inductive science can be used to accumulate sufficient evidence to allow the formulation of theories that can be tested by deductive methods. Novel results or the rejection of an existing theory by deductive experiments may lead to new avenues of inquiry that initially may require an inductive approach to outline the issues in the field sufficiently to allow the formulation of appropriate theories. Hence, deductive and inductive science can be complementary approaches useful at different times in the development of a particular field of scientific inquiry.

Statistical inference, rather than representing a single well-defined set of techniques applicable in specific, relatively uniform conditions, should instead be viewed as a collection of tools each of which has applicability in different situations. Exploratory research in inductive science requires one set of tools while confirmatory research in deductive science makes use of different tools. Of course, the sets of tools used by different kinds of science overlap, and when they do, there is sometimes confusion on what the limitations and uses of the tools should be. Consider, for example, analysis of variance. The technique was developed to allow inference about causation from well-designed experiments where researchers control specific factors about which they were attempting to draw inferences. Hence, it is a valuable tool in deductive science, where results from analyses of variance have relatively unambiguous interpretations. However, there is nothing inherent in the statistical technique that prevents it from being used to model observational studies where the "levels" of a factor correspond to naturally occurring differences among different statistical populations. Interpretations of statistical tests in these situations are more ambiguous than in controlled experiments and are often better viewed as model fitting than as a rigorous analysis of causation. Thus, analysis of variance is also useful as a tool for inductive science, where it can be used to identify patterns. However, the interpretations drawn are of a very different nature than those that arise from use of analysis of variance of controlled experiments. Scientists who fail to make such distinctions can invoke explanations that go beyond the scope of their ability to draw inferences from their data.

I conclude that inductive science and deductive approaches to science are different, but synergistic and not mutually exclusive, ways by which progress is made in the process of scientific inquiry. Inductive science tends to occur in relatively young fields of inquiry where there are relatively few generalities and guidelines that define what is of interest to study. Deductive science is more suited for fields of inquiry where generalities are well established and can provide sound guidelines for asking appropriate questions. Inductive science tends to rely more on observational studies and statistical techniques that allow the identification of repeated patterns that can be synthesized into generalities. Deductive science tends to rely more on carefully designed experiments accompanied by statistical techniques that can be used as strong evidence of causality. Individual scientists may use either approach, and most fields of scientific inquiry employ both simultaneously. The use and interpretation of methods of statistical inference depend on the type of science being done. Some statistical methods can be useful in one kind of science but not in another. Other statistical methods can be used in either type of science, but require different assumptions and imply different things about data depending on whether an investigation is primarily inductive or deductive. Wide, sweeping generalizations about the appropriateness of any one method of statistical inference over others are unwarranted.

THE STRUCTURE OF SCIENTIFIC KNOWLEDGE

Returning to the definition of science proposed above, it should now be evident that there are many ways to build models about nature and obtain understanding from them. The contrasting characteristics of inductive and deductive science often can be seen in individual investigations and research programs. This can make discussions of the nature of science and scientific evidence confusing, since people often argue from different points along the continuum of perspectives between the extremes of pure inductive science and pure deductive science. If there is so much diversity in the methods and goals of different scientists, what is the structure of scientific knowledge? Is scientific knowledge a single, integrated body of knowledge, or merely a conglomeration of the findings of different people doing many different things? Does science produce understanding that is integrative, or are the products of different scientific research programs independent of one another?

Wilson (1998) argued persuasively that all scientific understanding accumulated thus far by the human race should be viewed as a single, coherent body of knowledge, where what is known in one field of study is not contradicted by that obtained in other fields. Following Whewell (1858) he termed this self-consistency of scientific knowledge "consilience." Wilson further argued that knowledge from fields other than science should be consilient with scientific knowledge. Clearly, such a view is consistent with a logical empiricist view of science, but can it also be consistent with the less restrictive viewpoint outlined here? There are two questions that need to be addressed. First, is there a logical *necessity* that science produce a consilient body of knowledge? Second, does science, in fact, produce consilient knowledge?

The answer to the first question is a metaphysical one, and one we have implicitly assumed cannot be answered. Scientific naturalism trivially answers the first question negatively, because it is beyond the scope of naturalism to consider such questions. More concretely, a scientific naturalist might argue that the answer is irrelevant to the conduct of science. Better yet, naturalism suggests proceeding as if scientific knowledge is consilient, but always remaining open to the possibility that eventually consilience will be shown to be wrong. Either way, scientific understanding depends only on the ability to evaluate the degree to which a scientific concept actually represents something real about nature. There is no prerequisite that such understanding be consilient with previous understanding. Conversely, lack of consilience for a particular scientific result does not necessarily imply that the concept must fit poorly with data. In fact the opposite is true: it is quite informative if a well-supported conceptualization *is not* consilient with previous knowledge. When this happens, it often points out gaps in previous concepts or a lack of appropriate scientific data.

This brings us to the second question: has science accumulated a consilient body of knowledge so far? The answer to this question is purely empirical, and in being so must always be considered to be tentative. There may at some future time arise a set of data that will be found to be not consilient with the whole of previous science. But putative instances where such observations were thought to exist (considering only situations where the data themselves were not faulty) simply illustrated flaws in conceptualizations rather than fundamental inconsistencies. Quantum mechanics is a good case in point. Classical Newtonian mechanics suffices to explain nearly all physical phenomena at the perceptual scale of human senses, but fails at very small scales (and for that matter, very large scales as well). Quantum mechanics adds new elaborations to the basic model of the universe to explain anomalies that arise at the small scale. And so, it seems, does every new piece of scientific information. Operationally, then, consilience can be considered a tentative assumption that need not reflect reality. It is useful insofar as it allows science to continue to accumulate understanding regarding the nature of the universe.

If we proceed, then, with consilience as a "meta-hypothesis" regarding the products of scientific investigations, it should be clear that all kinds of scientific investigations should be conducted in order to assess its validity. In particular, instances where inconsistencies arise between different disciplines should receive attention. It is at the interface of very different lines of scientific inquiry that the chances of rejecting the consilience hypothesis will be greatest. Yet it is at these same interfaces that one is likely to find very different scientific strategies being employed on different sides of the interface. This makes it particularly challenging to demonstrate nonconsilience, since it is necessary to account for the possible differences in perspective that might result from different scientific strategies. However one views such instances, all scientific fields of inquiry should continue to be given equal status, because each has the potential to add further insights into the overall structure of scientific knowledge.

CONCLUSIONS

On the surface one might think that modern science is a heterogeneous set of activities employing a wide variety of strategies and using very different sets of techniques to pose and answer questions. It is not even a logical necessity that one body of science produce statements that are meaningful within another scientific discipline. Certainly within the context of increased specialization and technical sophistication, divergence in the goals, philosophies, and languages of different branches of science may seem of little importance.

However, in a society with limited resources and increasing demands on scientific disciplines of all sorts, rivalry among disciplines for those resources will inevitably arise. If science is viewed as a single enterprise with the overarching goal of testing the meta-hypothesis of consilience, then the differences among disciplines can be considered things that need to be explained rather than indications that fundamentally different processes of knowledge gathering and evidential evaluation are going on in different disciplines. Each discipline, in this view, has much to contribute to understanding the whole, and no discipline has any logical or practical priority. No discipline can explain everything.

Commentary

Prasanta S. Bandyopadhyay and John G. Bennett

INTRODUCTION

Influenced by the new development of mathematical logic in the early part of the century, logical empiricists and others adopted *deductive methods* as the standard of rigor to understand scientific methods and theories. Impressive amounts of early literature on hypothetico-deductive account of confirmation theories bear witness to this.[1] Gradually the deductive model of approaching science became the received view in philosophy of science, winning the hearts of even some hard-core scientists.[2] Lately there is, however, a growing awareness among working scientists that the received view is grossly oversimplistic. Brian Maurer is one scientist-ecologist who has challenged this view.[3]

Maurer asserts the need for a *plurality of methods* to model scientific inquiry. He argues that depending on the maturity of the disciplines, working scientists are more prone to exploit one method rather than the other. For him, if the discipline is physics, for example, then its methods are more deductive in nature as contrasted with less mature disciplines like some parts of ecology, in which scientists are more often likely to use inductive methods. For the development of a specific discipline, according to him, the use of one method is crucial and as a result preferable to that of the other. Hence, for him the received view is oversimplistic in its preoccupation solely with deductive methods.

We agree with Maurer's insightful comments on the nature of scientific methodology; we need to understand both what he calls inductive science

---

We wish to thank Gordon Brittan Jr., Robert J. Boik, and the editors of the book for helpful comments regarding our paper.

1. C. Hempel's ravens paradox (Hempel, 1945), W. V. Quine's underdetermination problem (Quine, 1961), and N. Goodman's grue paradox (Goodman, 1955) are in point. Those paradoxes/problems are related directly or indirectly to how one construes deductivism. There are, however, two possible objections to our approach. We defer our discussion of those paradoxes along with the objections until we define deductivism in a later part of the commentary.

2. P. Medawar and J. Eccles are two of them.

3. Other ecologists who have joined this crusade are Pickett, Kolasa, and Jones (1994). See also Scheiner (1993).

and what he calls deductive science if we would like to understand how science is being practiced. However, Maurer's terms, "inductive science" and "deductive science" could be misunderstood. We first address this possible misunderstanding and then suggest why Maurer might think that an exclusive preoccupation with a deductive view of scientific evidence fails to provide a correct account of evidence.

TWO KINDS OF ARGUMENTS[4]

In logic, the terms "inductive" and "deductive" refer to types of inference or argument. The concept of validity is essential to differentiate deductive arguments from inductive arguments. An argument is deductively valid if and only if it is logically impossible for its premises to be true while its conclusion is false. An inductive argument by definition is one that is not a deductive argument. This means that it is logically possible for its premises to be true while allowing its conclusion to be false. Here is one example of each. "DE1," "IN1," and so on represent types of deductive and inductive arguments. Here, "p," "q," and "r" stand for sentences; "H," "M," and "L" stand for predicates and properties of entities that are represented by "x," "y," "z," and "w."

DE1
1. If p, then q.
2. p.
Therefore, q.

IN1
1. x has H and x has M.
2. y has H and y has M .
3. z has H and z has M.
4. w has H and w has M.
Therefore, all cases of H are cases of M.

DE1 is a deductively valid argument. That is, if the premises are true, then the conclusion is true. In IN1, by contrast, even though premises are true, the conclusion may not be true. One way to look at the distinction be-

4. Readers are here invited to compare our discussion with Zellner's (1996) discussion on the same topic.

tween deductive and inductive arguments is to consider what the computer scientist calls the *monotonic* property of deductive arguments (see Pearl, 1991). According to that property, if an argument is valid, then any argument that has the same conclusion and includes among its premises all the premises of the first argument is also valid. Consider DE2, which is same as DE1 except that it has an additional premise 3 that says that r is true.

DE2
1. If p then q.
2. p.
3. r.
Therefore, q.

If DE1 is deductively valid, so is DE2. This is because deductive arguments are monotonic in nature, and that alone distinguishes deductive arguments from inductive arguments. Consider IN2, which is the same as IN1 except that it adds an extra premise 5 to the former. Premise 5 says that x has both H and L.

IN2
1. x has H and x has M.
2. y has H and y has M .
3. z has H and z has M.
4. w has H and w has M.
5. x has H and x has L.
Therefore, all cases of H are cases of M.

Notice that IN1 is a strong inductive argument in the sense that its premises provide strong support for its conclusion. Suppose now that "H" represents the property of being a human, "M" represents the property of being mortal, and "L" represents the quality of living for 6,000 years. Then premise 1 in IN1 and IN2 says that x is a human being and x is mortal, while premise 5 in IN2 says that x is a human being and she survives 6,000 years. In this situation, adding premise 5 in IN2 weakens the strength of the inductive argument. After all, we think that a human being could survive at most 200 years. Although premise 5 does not entail the falsity of premise 1, the fact alone raises skepticism about the truth of premise 1, thereby weakening the strength of the conclusion that *all* human beings are mortal. So even though IN1 is an example of a strong inductive argument, IN2 is not. Adding a new premise to an existing strongly supported inductive argument may weaken its strength. Inductive arguments share this nonmonotonic property. Consider DE3, which is the same as DE1, except that it now has an additional premise 3, which says that p is false.

DE3
1. If p then q.
2. p.
3. It is false that p.
Therefore, q.

According to the monotonic property, if DE1 is valid, so is DE3. However, DE3 contains a pair of inconsistent sentences, p and not p. Although DE3 contains a pair of inconsistent sentences, which means one of them must be false, DE3 remains valid. There is no such analogue in inductive logic. If we take inductive logic to behave like conditional probability, then the conditional probabilistic sentence corresponding to DE3 is $P(q|p$ and not $p)$, which is undefined.[5] The probability of p and not p is zero.

Now consider an instance of DE1.

1. If Bozeman is a university town, then $2 + 2 = 4$.
2. Bozeman is a university town.
Therefore, $2 + 2 = 4$.

Although this is an instance of DE1, hence valid, there is apparently no relationship between the antecedent (Bozeman is a university town) and the consequent $(2 + 2 = 4)$ of that argument.

Contrast here this instance of DE1 with DE3. We have noticed that if DE1 is valid then so is DE3, even though DE3 contains an inconsistent pair of premises. Some might wonder whether there is something fishy going on in DE3. If DE3 is a valid argument, then we could derive anything, literally anything, from DE3. So what good is the notion of validity if DE3 is regarded as valid?

One way to dispel this worry is to distinguish the validity of an argument

---

5. There are some systems of nonmonotonic logic; Kyburg's (1974) is one of them that could get around that problem. One thing to notice, however, is that although the above conditional probability is undefined, for a Bayesian the probability of the proposition that p and not p entail the conclusion still has probability 1. The conclusion is a deductive consequence of its premises. We thank Kyburg for making us aware of this Bayesian point, although he is a staunch anti-Bayesian.

from its soundness. An argument is sound if and only if it is a deductively valid argument with true premises. Based on this distinction, one might say that DE3 is a deductively valid argument, but it is not a sound argument. Hence, it is not an instance of a good/strong argument. For an argument to be good/strong, according to this point of view, it must be sound. We, however, contend that if the purpose of scientific inference, and for that matter, any confirmation theory, is to go beyond what is already known to be true, then soundness won't serve our purpose. We have to impose a different constraint on what should be counted as good/strong arguments. Here is why.

Going back to our contrast between DE1 and DE3, we indicated that DE3 does not pass for a sound argument but that DE1 does. The premises of DE1 are true and also it is a valid argument. If we were to consider soundness to be the mark of a good/strong argument, then we should have considered DE1 a good/strong argument. We don't, however, think DE1 is in accord with our intuition of what a good/strong argument should look like. We define a "good/strong argument" as an argument that proves its conclusion. According to us, an argument is good/strong if and only if it is sound and we are justified in believing that its premises are true. In the case of DE3, we have no reason to believe an inconsistent set of premises. Hence, DE3 does not count as a good/sound argument for us.

In the case of DE1, we seldom have any reason to believe the sort of premises of that argument unless we have reason either to believe the antecedent (Bozeman is a university town) false or the consequent $(2 + 2 = 4)$ true. But in neither of these cases will we be in a position to get new knowledge from the argument in question. The most important basis of scientific inference is induction where we go beyond our given knowledge to gather new knowledge although that leaves room for uncertainty and error. Although the above argument, which is an instance of DE1, is valid, there is apparently no evidential relationship between the antecedent and the consequent of that argument. We take "evidential relation" between premises of an argument and its conclusion to mean that premises in an argument must provide some support, weak or strong, for the truth of its conclusion. In this instance of DE1, the premises don't provide either kind of support for its conclusion. In this case, the conclusion is a necessary truth. If one premise of an inductive argument is false, then this will weaken the strength of that argument. In addition, unlike this instance of DE1, if the truth of the premises of an inductive argument has nothing do with the truth of its conclusion, then we can't really make any inference about that conclusion. Under this scenario, in inductive/probabilistic logic the conclusion and the premises will be regarded as probabilistically independent of one another.

From these examples one key feature about deductive arguments emerges. That is, once an argument is found to be valid, it will always remain so due to its monotonic property. This will be so even though its premises could be false or there is no evidential relationship between its premises and the conclusion, provided one maintains the original premises that led to its validity in the first place. So far, we have shown that neither monotonicity nor soundness of deductive arguments is sufficient to make those arguments proofs. And an argument's ability to provide a proof is related to the evidential relation between the premises of that argument and its conclusion. If we are interested in the evidential relation between the premises of an argument and its conclusion, then we are interested in an epistemological investigation. Confirmation theory confronts questions that are epistemological in nature.

We are now in a position to explain why Maurer's terms could be misleading and to forestall any misunderstanding that might stem from them. Both what Maurer calls inductive science and what he calls deductive science rely on both inductive and deductive inference. All statistical inference, whatever methods are used, is inductive inference. Deduction, on the other hand, plays a role in rejecting generalizations: when data do not conform to a generalization, the former falsifies the latter as a matter of deductive logic. We do not think Maurer would disagree with what we say here, as he recognizes that his terms are not intended to imply that any particular forms of inference are used exclusively in the two sorts of methods he describes.

Any confirmation theory, whether it is primarily a deductive confirmation theory or an inductivistic/probabilistic conformation theory, is interested in providing answers to questions like "what is evidence?," "how and when is a theory confirmed by its evidence?," "could two mutually incompatible theories be equally supported by evidence?," and so on. In one sense, deductive arguments don't help us get a deeper understanding about that epistemological investigation, because deductive logic is interested in the validity or soundness of an argument, which is nonepistemic in character. However, empiricists influenced by deductive logic have a clever way of addressing epistemological questions. In next section we discuss their replies.[6]

6. Although our discussion revolves round a distinction between two kinds of arguments, it should not give the impression that there is no relationship between them. There are some important results that show interrelations between propositional logic (a branch of deductive logic) and probability logic. Statements such as "if p is a deductive consequence of q, then the probability of q can't exceed that of p," "the probability of a tautology is one," and "the probability of a contradiction is zero" show some of those relations. Vigorous interdiscipli-

DEDUCTIVISM

Maurer's emphasis on the plurality of methods represents a more recent attitude toward science in contrast to the logical positivist's emphasis on what we call *deductivism*. Here we will both consider and explain the reasons for rejecting deductivism, which consists of (i) all and only observable consequences are evidence for the theory (hereafter D_1), and (ii) two theories having exactly the same observable consequences (i.e., evidence) are equally confirmed/disconfirmed by any particular total body of evidence (hereafter D_2). D_1 provides an answer to the question *what counts* as *evidence,* while D_2 answers *how* well are theories *confirmed.* For deductivists, observable consequences are all that matters in theory confirmation.

Deductivism is not correct. Larry Laudan and Jarrett Leplin have argued that it mistakenly identifies observable consequences with supporting evidence for a theory (Laudan and Leplin, 1991; Leplin, 1997).[7] They maintain that observable consequences are neither necessary nor sufficient for evidence of a theory. We give examples that are similar to theirs to explain their position. Although their objection is directed to other issues, it sheds some light on deductivism.

Consider a hypothesis: The earth and moon have an identical number of stones. An observable consequence for this hypothesis is: either the earth does not have five stones, or the moon has five stones. Assume that the earth doesn't have 5 stones. The observable consequence expressed in the disjunctive statement is still true. However, evidence or observation that the earth doesn't have five stones fails to be supporting evidence for the hypothesis. Therefore, an observable consequence fails to be sufficient for supporting evidence of the hypothesis.

Consider a hypothesis that a given coin is approximately fair, and the observation that in 100,000 random tosses the coin came up heads 50,003 times. Surely, the observation is good evidence for the hypothesis, but it is certainly not a deductive consequence of the hypothesis. Therefore, observable consequences are not necessary for having supporting evidence for the hypothesis.

Deductivism is false because it mistakenly equates a theory's observable consequences with its supporting evidence. The simple examples above,

---

nary researches are under way to explore this relationship further (Adams, 1998). For an elementary but solid introduction to deductive/inductive arguments, see Skyrms (2000).

7. For a critique of their position, see Kukla (1998).

which could be multiplied easily, show that deductivism is mistaken in its characterization of evidence.

CONCLUSION

In these comments we have tried to supplement Maurer's discussion in two ways. First, we tried to forestall any misunderstandings that could arise from Maurer's choice of terminology. We did this by explaining the logical distinction between induction and deduction and distinguishing it from Maurer's distinction between inductive science and deductive science. Second, we supplemented Maurer's account by providing a general characterization of evidence common to all scientific work within deductivism. The sense in which we supplemented his account is by discussing why he might think that such a deductivistic account of evidence failed to provide a viable account of scientific evidence. His possible reasoning, we argued, was that such an account failed to capture the evidential relation between premises of an argument and its conclusion. So, if the deductivistic account of scientific evidence is not the correct account, then the pertinent question is, what is the correct account of evidence? Several authors in this volume (especially Mayo [chapter 4], Royall [chapter 5], Forster and Sober [chapter 6], Boik [comment to chapter 6], and Lele [chapters 7 and 13]) have taken this question seriously, and proposed distinct answers to the question.

2.2	Commentary

Mark L. Wilson

Professor Maurer's thought-provoking summary of how the science of ecology can be viewed as either inductive or deductive raises questions for the study of more complex ecological processes and for related scientific domains in which knowledge is intended to be used to forecast or modify nature. His characterization describes how the inductive approach to ecological inquiry employs accumulated observations to characterize reappearing patterns, thereby allowing inference of likely underlying process. In this way, repeated observations of associations that should not have occurred by chance are considered as evidence supporting an explanation for that association. Deduction, on the other hand, begins with a proposed explanation for natural phenomena, which in turn is used to derive a hypothesis based on underlying mechanisms or principles. Observations, then, are made in an attempt to falsify or reject that hypothesis. Maurer correctly notes that these two "strategies for achieving scientific understanding" are not "mutually exclusive pathways to scientific knowledge." They are presented, however, as the fundamental alternative approaches that most ecologists use in their research.

Maurer's thesis addresses the design of observations and statistical analysis of data used to infer or deduce pattern. He justly recognizes that the statistical tools of "exploratory research in inductive science" differ from those for "confirmatory research in deductive science." Much of the practice of research ecologists corresponds to either of the two approaches that Maurer outlines. For example, observations of natural historical events, descriptions of temporal or spatial trends, and comparisons of different regions all could be considered part of the inductive-inferential approach. On the other hand, most studies involving experimentation, whether manipulative or natural (Scheiner and Gurevitch, 2001), would be considered deductive. Ecologists, however, also practice their science through investigations that focus directly on the behavior of underlying processes rather than analyses of resulting patterns. It is well recognized that the complexity of most ecological systems of interest derives from multiple links among many variables, nonlinear effects that may have contradictory influences, and even changes in direction of effects beyond. Indeed, theories of chaotic behavior suggest that knowledge of the underlying processes of many complex systems is necessary to explain the observed patterns (Perry, Smith, and Woiwodeds, 2000). Otherwise, outcomes of experiments or observations involving such complex systems may appear as statistically nonreproducible or unpredictable (Legendre and Legendre, 1998). In such systems, similar patterns may arise from different processes, just as the same processes may produce diverse outcomes that are rarely repeated.

This possibility raises a number of questions concerning how we ask less traditional questions in the ecological sciences, particularly where large-scale or long-term effects are the objects of study. Regional studies of biodiversity and prolonged investigations of multiyear responses are becoming increasingly prominent elements of ecological science. How should we categorize, for example, investigations that use empirical-statistical models to project likely future changes? The statistical properties of such extrapolations typically assume constancy of conditions or parameter values, when in fact they are likely to vary considerably, thus confounding traditional statistical approaches. Similarly, what should we do with process-based simulation models, which are now an important part of ecological research, as they seem difficult to group into either of the methodological-statistical frameworks described above? More generally, how might forecasts or predictions from such process-based models rigorously be compared to the actual events that eventually occur? When shall we feel satisfied that the predictions conform sufficiently to our hypotheses of underlying process? A careful examination of how statistical properties of forecast results relate to the biological processes of concern is needed, but usually there are few true replications, and each case may be unique.

In domains of inquiry closely related to ecology, knowledge and methods are being borrowed and adapted with the intent of addressing real-world problems. Environmental pollution, water resource management, human health, and agricultural sustainability are but a few matters that drive such research. These "applied" scientific efforts are not aimed solely at generating new insights; rather, they are intended to be used to modify nature. Thus, manipulative experiments often are designed to explore temporal or spatial behavior rather than the illumination of cause, and natural space- or time-pattern analysis also may be employed in an attempt to anticipate future events. Such observations and experiments and the resulting interpretations seem to be neither inductive nor deductive in the sense that Maurer has described. These types of ecological science generate new information and eventually greater understanding, yet the large-scale and urgent goal of application may suggest different study designs and methods of data analysis. Again, how should these kinds of applied ecology research efforts be viewed?

Two examples of such applied ecology studies, both aimed at altering natural processes given a possibly limited knowledge base, involve infectious disease epidemiology and natural resource management research. Both raise the question of what statistical and observational designs are appropriate for analyses intended to simulate process rather than simply characterize pattern. Should we require the same design and statistical "standards" for what constitutes evidence in the sense that Maurer has described, given that logistical problems may be insurmountable and the applied goals urgent? When getting it right can literally be a matter of life or death, should our science consider the deductive and inductive design and methods relevant and useful? On the other hand, expedient decision making and action based on a faulty evidentiary process may do more harm that good.

Ecological research pertinent to natural resource management (e.g., biodiversity, meta-population dynamics, climate change impacts) presents special problems of scientific design and analysis. Many diverse variables, long time periods, and large geographic areas are involved, which means that simple experiments are useful solely to provide information on parts of a more complex dynamic (Bissonette, 1997; Chichilnisky, Heal, and Vercelli, 1998; National Research Council, 2001). Rigorous tests of hypotheses in a standard deductive framework are impossible for such problems. Rather, most investigations involve a sort of inductive approach, even though it is also impossible to repeat observations many times so as to produce a post hoc explanation. Furthermore, the urgency of producing knowledge that can be used to intervene or alter the course of processes severely limits the amount of study that is reasonable. In practice, applied environmental research of this kind must depend on some sort of reasonable guess that evaluates parts of dynamic systems, awareness of uncertainty, and consideration of the costs of being wrong. Might there be another, perhaps better organized (statistical?), method for agreeing on the current state of knowledge? What level of ecological understanding is scientifically sufficient for action with this applied information?

Another realm of ecological research that poses similarly difficult methodological issues involves the dynamics of infectious agents. Infectious disease epidemiology is undergoing a radical change from a period when the observed patterns of exposure and risk represented the focus of description. Increasingly, new methods of analysis focus on the transmission processes that create these patterns (Anderson and May, 1991; Kaplan and Brandeau, 1994; Koopman and Lynch, 1999). Thus, simulation models that produce results of expected patterns of disease, given predominant behaviors, are being used to analyze when and where infection incidence should be increased or diminished. Because there are typically no contemporary data to compare with what will be observed in the future, how shall we classify such ecological investigations? Here, the population ecology of contact, transmission, infection, and disease, as well as continued transmission, have been viewed as part of a larger ecological dynamic. Because simulations typically are not evaluated with regard to true properties, and because there are few comparisons other than new simulations, how might such eco-epidemiological simulation experiments be classified and evaluated?

This commentary is intended more to ask questions than to answer them. The techniques of ecological research, including the design of studies, the data-gathering methods employed, and the approaches to data analysis, are in transition. Application of traditional approaches to ecology may not be able to capture the manner in which questions have changed and how some of the science is evolving. Without being either "physics" or a "meta-physics," how can certain kinds of evidence be determined to represent sufficient knowledge on which intervention should be based?

2.3 Rejoinder

Brian A. Maurer

The object of inquiry of most scientific investigations is some part of a large, extremely complex universe. In my essay, I focused primarily on the way that scientists go about describing that part of the universe they study. Both commentaries have made important observations and clarifications regarding parts of my argument. Here, I briefly revisit their comments and my original essay with the assumption that the universe is exceedingly complex.

The first point I wish to make concerns the nature of logic itself. Bandyopadhyay and Bennett have suggested more precise definitions for the terms "deduction" and "induction," which I used liberally in my essay. I concur with their definitions and note that deduction confers much more stringent conditions on the structure of a logical statement than does induction. Scientific theories and the hypotheses derived from them are essentially complicated strings of logical statements tied together to present an overall description of an idea that is thought to capture the essence of some physical phenomenon. Thus, any theory will incorporate a number of different kinds of statements that may be individually classified as either an induction or a deduction. My definition of deductive science implied that theories and hypotheses in such a science would have a higher frequency of deductive statements in them than theories and hypotheses from an inductive science. This would make the logical structure of theories in a deductive science appear to be more rigorous or precise than similar theories from an inductive science.

The question I wish to pose here is whether the long, complicated strings of logical statements that form the substance of theories are adequate to describe the complexity of the universe or any arbitrary subset of it. The answer to this question has implications for whether the distinctions between induction and deduction, and between inductive science and deductive science, are important distinctions to make, or merely conveniences to help conceptualize the otherwise unfathomable complexity of the scientific process.

To answer the question posed in the previous paragraph, I here reiterate a premise made in my essay in a slightly different form. I will assume that logic and mathematics are products of the unique physiological construction of the human mind and have no existence beyond that. This is simply restating the relaxed assumption regarding the method of describing nature that frees science from the limitations of logical empiricism discussed in more detail in my essay. Another way of stating this assumption is to view the human mind as a computational device[8] that belongs to a general class of computational devices that includes Turing machines, digital computers, etc. The question can then be reformulated in terms of computations (Barrow, 1998): do limits to computations exist, and are such limits applicable to the computations needed to describe the universe or to arbitrary subsets of it?

The answer to the first part of the question is that yes, there are some rather profound limits to computations. Such limitations are exemplified by the computations necessary to solve a certain subset of nonlinear equations often referred to as chaotic attractors. Nonlinear equations whose solutions are chaotic attractors cannot be computed in the following sense: there are an infinite number of nonoverlapping trajectories that can be mapped onto the chaotic attractor, each of which differs by an infinitesimally small amount. Such mathematical solutions can be characterized by certain quantities that define aspects of their mathematical behavior, or the geometric properties of their attractor, but they cannot be completely specified by any string of logical statements no matter how long.

The second part of the computational question posed above is more difficult to answer. Chaotic attractors and other such mathematical constructions may simply be oddities of mathematics that cannot be applied to reality. Do parts of the universe exist that we are sure cannot be computable? Regardless of whether or not chaotic attractors exist in nature, the answer to this question is also yes. This becomes obvious when we consider certain kinds of scientific problems (Barrow, 1998). Basically, any portion of the universe that contains a large number of smaller units, such as a protein

8. I note here that there are some very important objections to this view of human consciousness (Penrose, 1989). Many of these objections, however, can be effectively countered (Dennett, 1995).

molecule, can be found to have so many different potential states that enumerating them all becomes a computational impossibility. Consider, for example, the problem of deciding how a protein will fold. An arbitrary protein made up of a hundred amino acids can take on so many possible three-dimensional configurations that finding the actual configuration that minimizes the energy required to maintain its structure is impossible. There are many such incomputable problems (Barrow, 1998).

Logic, we can conclude, cannot fully capture the complexity of the universe. In fact, there may be many phenomena within the universe that cannot be adequately represented by the strings of inductions and deductions that make up the structure of a scientific theory. There needs to be an additional property of the universe that is necessary to render it at least partially representable in terms of the logical statements that human minds can produce. That property is that the universe must be divisible into a number of classes, wherein each class is composed of a large number of nearly identical subsets. If this is true, then a theory can be used to describe the class, and not be concerned about each member of the class. If this is the case, then the theory provides a convenient simplification of an otherwise potentially bewildering number of possibilities. The question is, do such classes exist in the universe? The answer to this question appears to be yes (Layzer, 1990). In fact, work in theoretical physics seems to imply that all physical phenomena in the universe can ultimately be attributed to a very small class of fundamental forces. This class may be so small as to contain only a single member. In other words, it may be possible to construct a "theory of everything" (TOE).

Upon first glance, the possible existence of a single TOE seems to imply an underlying simplicity to the universe. However, it is important to consider the logical structure of this TOE. Such a theory essentially must unite several subtheories about the universe that are encapsulated in quantum mechanics and general relativity. These theories are themselves *very complex* in their logical structure. The mathematics needed to write these theories are among the most complicated mathematical ideas of which we are currently aware. In fact, recent explorations of theories that can potentially unite quantum mechanics and general relativity indicate that such theories are so complicated that they have yet to be fully described (Greene, 1999). The implication is that the more broadly applicable a theory becomes, the more complicated its logical structure must be. This calls into question the old principle of Occam's razor, that states the "best" theories among alternatives are the simplest ones. Bandyopadhyay and Bennett's Bayesian approach to scientific epistemology implies that simple theories should generally be assigned higher prior probabilities than more complicated ones. Although this may be a useful approach when considering a relatively small set of phenomena, it may break down as the number of phenomena being considered increases.

To summarize, complexity places severe limits on the ability of scientific theories to represent reality. First, logical strings can become sufficiently large that they are essentially incomputable. In terms of human minds this means that theories can become so complicated that human minds may not be able to understand them. Second, the phenomena that theories attempt to describe may be too complex to be described by any theory that has practical comprehensibility. Even if the universe is divisible into classes of physical phenomena that can be represented by a small number of theoretical statements, these statements become more complex as the number of phenomena they attempt to represent increases. We must conclude, then, that induction and deduction are no more than tools that can be applied to descriptions of a complex universe, but these tools appear to be limited in their applicability. There is no simple solution to this problem. Any attempt to construct a usable tool for the evaluation of scientific evidence must take this problem into account. I do not believe either frequentist or Bayesian approaches to statistics will fully overcome the problems posed by the need to describe a complex universe in a manner fathomable by human minds.

These problems of representation of complex systems are echoed in Wilson's comments about ecology. Ecological systems are exceedingly complex, and their behavior may be so complicated as to frustrate attempts to describe them using any single set of statistical or mathematical tools. This is particularly unfortunate, because ecological problems are among the most serious that face the human race. It is conceivable that human impacts on complex ecological systems are so great that they may threaten the long-term sustainability of human populations on the earth. Yet the solutions to these problems approach the limits of science that I discussed above. Wilson's litany of questions about ecological science and its application to ecological problems reflect a concern about such limits.

Despite the looming problem of complexity, there are, I think, approaches that have yet to be fully explored that may help reduce the problem to a more manageable size. In particular, the strategy of developing complementary descriptions of complex systems has the potential to provide new ways of conceptualizing and measuring such systems. Complementarity is a principle that is required to fully comprehend the theoretical and empirical developments of quantum mechanics. The idea that matter and energy are composed of discrete units (quanta) that demonstrate different properties depending upon how they are measured is best understood when we view these properties as complementary descriptions of the same phenomenon (Nadeau and Kafatos, 1999). Treating them otherwise leads to contradictory and unsatisfactory conclusions such as the idea that a measurement of a physical process causes it to happen.

Complementarity need not be confined to the microscopic world of quantum physics (Nadeau and Kafatos, 1999). In fact, the structure of statistical mechanics can be interpreted as an application of the complementarity principle. The general idea is that if both the microscopic and macroscopic descriptions of a system are assumed to be correct, then the quantities of the macroscopic system can be interpreted as deterministic outcomes of small-scale randomness confined by large-scale constraints. Such ideas are readily applicable to ecological systems (Brown, 1995; Maurer, 1999) and may provide a way to make large-scale ecological problems more tractable.

Science in general, and ecological science in particular, is faced with the awesome task of making a vast, complex universe understandable to the human minds it contains, and to assist those minds in ensuring their perpetuation into the future. The universe is more complex than human minds because it contains them and all of their cultural products, which include, among other things, science. In a sense, science is a recursive activity of the universe as it looks at itself through human eyes and attempts to understand itself through human minds. Yet the nature of human minds places rather severe constraints on this recursion. Individual scientists must come to grips with these limitations and attempt to determine the boundaries they place on what questions can be answered and which problems can be solved using the scientific method.

REFERENCES

Adams, E. 1998. *A Primer of Probability Logic.* Stanford, CA: CSLI.
Anderson, R. M., and R. M. May. 1991. *Infectious Diseases of Humans: Dynamics and Control.* New York: Oxford University Press.
Barrow, J. D. 1998. *Impossibility.* Oxford: Oxford University Press.
Bissonette, J. A., ed. 1997. *Wildlife and Landscape Ecology: Effects of Pattern and Scale.* New York: Springer.
Boik, R. J. 2004. Commentary on Forster and Sober. Chapter 6.2 in Taper, M. L., and S. R. Lele, eds., *The Nature of Scientific Evidence: Statistical, Philosophical, and Empirical Considerations.* Chicago: University of Chicago Press.

Brown, J. H. 1995. *Macroecology*. Chicago: University of Chicago Press.

Chichilnisky, G., G. Heal, and A. Vercelli, eds. 1998. *Sustainability: Dynamics and Uncertainty*. Boston: Kluwer.

Cody, M. L., and J. M. Diamond, eds. 1975. *Ecology and Evolution of Communities*. Cambridge: Harvard University Press.

Dennett, D. C. 1995. *Darwin's Dangerous Idea*. New York: Simon and Schuster.

Deutsch, D. 1997. *The Fabric of Reality*. New York: Allen Lane.

Diamond, J., and T. J. Case, eds. 1986. *Community Ecology*. New York: Harper and Row.

Earman, J. 1993. Underdetermination, Realism, and Reason. In French, P., T. E. Uehling, and H. K. Wettstein, eds., *Philosophy of Science*. Minneapolis: University of Minnesota Press.

Forster, M. R., and E. Sober. 2004. Why Likelihood? Chapter 6 in Taper, M. L., and S. R. Lele, eds. *The Nature of Scientific Evidence: Statistical, Philosophical, and Empirical Considerations*. Chicago: University of Chicago Press.

Giere, R. N. 1999. *Science without Laws*. Chicago: University of Chicago Press.

Goodman, N. 1955. *Fact, Fiction, and Forecast*. Cambridge: Harvard University Press.

Greene, B. 1999. *The Elegant Universe*. New York: Vintage.

Hairston, N. G., Sr. 1989. *Ecological Experiments: Purpose, Design, and Execution*. Cambridge: Cambridge University Press.

Hempel, C. 1945. Studies in the Logic of Confirmation. *Mind* 54: 1–26, 97–121. Reprinted in Hempel, *Aspects of Scientific Explanation and Other Essays* (New York: Free Press, 1965).

Hofstadter, D. R. 1979. *Gödel, Escher, Bach: An Eternal Golden Braid*. New York: Vintage.

Holling, C. Z. 1998. Two Cultures of Ecology. *Cons. Ecol.* [on line] 2:4. *http://www.consecol.org/vol2/iss2/art4*.

Hull, D. L. 1988. *Science as a Process*. Chicago: University of Chicago Press.

Kaplan, E. H., and M. L. Brandeau. 1994. *Modeling the AIDS Epidemic: Planning, Policy, and Prediction*. New York: Raven.

Kingsland, S. E. 1995. *Modeling Nature*. 2nd ed. Chicago: University of Chicago Press.

Koopman, J. S., and J. W. Lynch. 1999. Individual Causal Models and Population System Models in Epidemiology. *Am. J. Public Health* 89:1170–1174.

Kuhn, T. S. 1970. *The Structure of Scientific Revolutions*. 2nd ed. Chicago: University of Chicago Press.

Kukla, A. 1998. *Studies in Scientific Realism*. New York: Oxford University Press.

Kyburg, H. 1974. *The Logical Foundations of Statistical Inference*. Boston: Reidel.

Laudan, L., and J. Leplin. 1991. Empirical Equivalence and Underdetermination. *J. Phil.* 88:449–472.

Layzer, D. 1990. *Cosmogenesis*. New York: Oxford University Press.

Legendre, P., and L. Legendre. 1998. *Numerical Ecology*. Amsterdam: Elsevier.

Lele, S. R. 2004a. Elicit Data, Not Prior: On Using Expert Opinion in Ecological Studies. Chapter 13 in Taper, M. L., and S. R. Lele, eds., *The Nature of Scientific Evi-*

dence: Empirical, Statistical, and Philosophical Considerations. Chicago: University of Chicago Press.

Lele, S. R. 2004b. Evidence Functions and the Optimality of the Law of Likelihood. Chapter 7 in Taper, M. L., and S. R. Lele, eds., *The Nature of Scientific Evidence: Statistical, Philosophical, and Empirical Considerations.* Chicago: University of Chicago Press.

Leplin, J. 1997. *A Novel Defense of Realism.* New York: Oxford University Press.

MacArthur, R. H. 1958. Population Ecology of Some Warblers of Northeastern Coniferous Forests. *Ecology* 39 : 599–619.

MacArthur, R. H. 1972. *Geographical Ecology: Patterns in the Distribution of Species.* New York: Harper and Row.

Maher, P. 1999. Inductive Logic and the Ravens Paradox. *Phil. Sci.* 66 : 50–70.

Maurer, B. A. 1999. *Untangling Ecological Complexity: The Macroscopic Perspective.* Chicago: University of Chicago Press.

Mayo, D. G. 1996. *Error and the Growth of Experimental Knowledge.* Chicago: University of Chicago Press.

Mayo, D. G. 2004. An Error-Statistical Philosophy of Evidence. Chapter 4 in Taper, M. L., and S. R. Lele, eds., *The Nature of Scientific Evidence: Empirical, Statistical, and Philosophical Considerations.* Chicago: University of Chicago Press.

Nadeau, R., and M. Kafatos. 1999. *The Non-local Universe.* Oxford: Oxford University Press.

National Research Council, Committee on Climate, Ecosystems, Infectious Disease, and Human Health. 2001. *Under the Weather: Climate, Ecosystems, and Infectious Disease.* Washington: National Academy Press.

Pearl, J. 1991. *Probabilistic Reasoning in Intelligent Systems.* San Mateo, CA: Morgan Kaufmann.

Penrose, R. 1989. *The Emperor's New Mind.* New York: Penguin.

Perry, J. N., R. H. Smith, and I. P. Woiwodeds. 2000. *Chaos in Real Data: The Analysis of Non-linear Dynamics from Short Ecological Time Series.* Boston: Kluwer.

Pickett, S. T. A., J. Kolasa, and C. G. Jones. 1994. *Ecological Understanding.* San Diego: Academic Press.

Quine, W. V. [1953] 1961. Two Dogmas of Empiricism. In *From a Logical Point of View.* Cambridge: Harvard University Press.

Resetarits, W. J., and J. Bernardo, eds. 1998. *Experimental Ecology.* Oxford: Oxford University Press.

Royall, R. M. 2004. The Likelihood Paradigm for Statistical Evidence. Chapter 5 in Taper, M. L., and S. R. Lele, eds., *The Nature of Scientific Evidence: Empirical, Statistical, and Philosophical Considerations.* Chicago: University of Chicago Press.

Ruelle, D. 1991. *Chance and Chaos.* Princeton: Princeton University Press.

Scheiner, S. M. 1993. Introduction: Theories, Hypotheses, and Statistics. In Scheiner, S. M., and J. Gurevitch, eds., *Design and Analysis of Ecological Experiments.* New York: Chapman and Hall.

Scheiner, S. M., and J. Gurevitch, eds. 2001. *Design and Analysis of Ecological Experiments.* 2nd ed. New York: Oxford University Press.

Skyrms, B. 2000. *Choice and Chance.* 4th ed. Belmont, CA: Wadsworth.

Strong, D. R., Jr., D. Simberloff, L. G. Abele, and A. B. Thistle, eds. 1984. *Ecological Communities: Conceptual Issues and the Evidence.* Princeton: Princeton University Press.

Underwood, A. J. 1997. *Experiments in Ecology.* Cambridge: Cambridge University Press.

Whewell, W. 1858. *Novum Organon Renovatum.* London.

Wilson, E. O. 1998. *Consilience: The Unity of Knowledge.* New York: Knopf.

Zellner, A. 1996. *An Introduction to Bayesian Inference in Econometrics.* New York: Wiley.

3 Experiments, Observations, and Other Kinds of Evidence

Samuel M. Scheiner

The game's afoot, Watson! —Sherlock Holmes

INTRODUCTION

How do we come to conclusions in science? What sorts of evidence do we use and how do we use them? These issues form the heart of this book. In this chapter, I focus on the question of the spectrum of types of data and evidence that we use, particularly on the use of experimental vs. observational data. My major contention is that all types of evidence play a role; that scientific decisions are based on the consilience of the evidence, the bringing together of different, even disparate, kinds of evidence (Whewell, 1858). Scientists are like detectives (Hilborn and Mangel, 1997). Each datum is a clue, a piece of evidence, with conclusions built on the totality of the evidence: all observations plus any background information plus the logical relationship among all of this information.

Before I address the nature of evidence, as a statistical consultant I need to address the nature of the questions we are trying to answer. When some-

Many of the examples used here on productivity-diversity relationships are based on a workshop conducted at the National Center for Ecological Analysis and Synthesis, a Center funded by the National Science Foundation (Grant DEB-9421535), the University of California at Santa Barbara, and the State of California. I note, in particular, the efforts of Gary Mittelbach and Chris Steiner in the large-scale literature review. The analysis of the Wisconsin data was primarily carried out by my student, Sharon Jones.

This material is based on work done while serving at the National Science Foundation. The views expressed in this paper do not necessarily reflect those of the National Science Foundation or of the United States Government.

one comes into my office and asks, "What statistical test should I use?" my answer is always "What question are you asking?" Data are one piece of a bigger puzzle, one tool in our toolbox. Data combine with past empirical observations, logic, and our expert judgment to form our evidence. This evidence informs us about what questions are important to ask, how to weigh the various kinds of evidence, and how to judge the results.

Despite what we sometimes learn, science is rarely about a single test of a hypothesis upon which a theory will stand or fall. It is not a simple matter of Popperian falsification (Popper, 1959). See Mayo (2004 [chapter 4, this volume]) for a much more thorough explication of these issues. Suffice it here to say that theory construction and testing is a complex process that involves testing assumptions, methods, and data at many different levels. All of these tests become bits of evidence to add to the whole. Advancement of knowledge takes place one small step at a time. Even a seemingly revolutionary advance such as Darwin's theory of evolution by natural selection was actually just the next logical step in a process that had been under way for over a half century (Ruse, 1999).

In ecology, we add to this complex process by the nature of many of the questions that we ask. Some questions concern theory construction, such as "what is the nature of competition?" That is, some questions are about how ecological processes operate. Layered on those questions are ones like "is competition or predation more important in structuring freshwater aquatic communities?"

Note that neither question deals with issues of truth versus falsity in the larger sense of theory testing. Rather, they are about the relative importance of various mechanisms and processes (Quinn and Dunham, 1983). In the examples that I sketch out below, dealing with the relationship between productivity and diversity, many mechanisms and processes affect this relationship. As is often the case, almost all of the proposed mechanisms have been shown to occur somewhere. Many of the theories that have been put forward are at least plausible. Our job as scientists is to decide between the possible, the plausible, and the consequential. Doing so requires utilizing all possible types of evidence.

I explicate this process by exploring a particular ecological issue: diversity-productivity relationships. After laying out the scientific issues, I consider the types of sources of empirical observations and how we should weigh data that come from manipulated experiments vs. observational experiments, an especially important issue in ecology. Finally, I touch on the relationship between evidence and theory in the context of a priori vs. post hoc explanations.

FRAMEWORK FOR DISCUSSION

It is always best to make philosophical presentations as grounded in reality as possible by putting them in the framework of an actual issue. A useful controversy, in this regard, is the relationship between productivity and species diversity and the mechanisms responsible for that relationship. As with many ecological issues, we have questions concerning both pattern and process.

The pattern question is seemingly simple and straightforward: "how does diversity vary relative to productivity?" Immediately, we run into the problems of what we mean by diversity and productivity and how we measure them. For current purposes, I will keep this simple. By diversity I mean the number of vascular plant species in a specified area. I am restricting this discussion to vascular plants because patterns at different trophic levels require different methods of measurement and different explanations. By productivity, I mean net primary productivity (NPP), the amount of energy captured by the plants over a (typically) one-year period. I will put off specifics of how diversity and productivity are measured until the next section, when I can put the measurements into the context of particular types of experiments and observations.

As with almost all scientific questions, the pattern issue arises within a larger context. During the eighteenth and nineteenth centuries, the observation emerged that there were large-scale patterns of species diversity. Voyages of discovery brought back information on flora and fauna from around the world, especially from the tropics. Naturalists quickly realized that species diversity was much higher in the tropics than in temperate regions (Wallace, 1878).

One explanation for this latitudinal gradient in diversity was a latitudinal gradient in productivity (Connell and Orias, 1964; Currie and Paquin, 1987; Adams and Woodward, 1989; Currie, 1991). One early theory provided a mechanistic basis for this relationship (Preston, 1962a,b; Wright, 1983). Under this theory, (1) as productivity increases, an area can support more individuals; (2) there are a minimal number of individuals necessary for a population to be self-sustaining; and (3) the more individuals there are in an area, the more species will be above the minimal viable population size. This explanation was readily accepted because it was consistent with other ecological mechanisms.

Other patterns were also discovered. Throughout the twentieth century, ecologists measured diversity and productivity on a variety of scales, from distances of meters to whole continents and beyond. Many people started

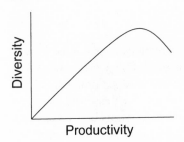

FIGURE 3.1 Many sets of species show a negative quadratic relationship between productivity and diversity with the highest diversity found at intermediate levels of productivity. This relationship is generally known as the hump-shaped productivity-diversity curve.

to find a hump-shaped relationship; at high productivity levels, diversity was lower than at intermediate productivities (figure 3.1). These patterns sparked the interest of ecologists because they seemed to contradict the theory presented above. As these observations accumulated in the 1970s and 1980s, the hump-shaped relationship was soon crowned as *the* relationship (Tilman and Pacala, 1993; Huston, 1994; Rosenzweig, 1995).

Accumulating alongside the observed patterns were additional theories. Many were created specifically to address the hump-shaped pattern. Others were created in different contexts, but co-opted into the productivity-diversity debate because they were also capable of explaining the hump-shaped pattern (see Rosenzweig, 1995, for a detailed analysis). By recent count, some two dozen theories have been put forth to explain the productivity-diversity relationship (Scheiner and Jones, 2002). Some of these theories are mutually exclusive, but many others are not. For example, intertaxon competition predicts hump-shaped relationships at regional scales because any given clade has a limited environmental range to which it can adapt (Tilman and Pacala, 1993; Rosenzweig, 1995). In contrast, speciation-extinction equilibrium predicts the same pattern because the maximal difference between speciation and extinction occurs at intermediate productivities (Vander-Meulen, Hudson, and Scheiner, 2001). These two theories make identical predictions, based on different mechanisms that may both operate at the same time. In other cases, we have theories that use similar mechanisms in different ways at different scales. For example, the resource competition model (Tilman, 1982) is also based on the mechanism of competition, like the intertaxon competition theory. This theory predicts a hump-shaped relationship at local scales because more species can coexist where resource quantity and resource heterogeneity are greatest, which is at intermediate productivities. Again, both theories may be accurate descriptions of the world. This multiplicity of mechanisms and scales means that nonexclusivity is typical of ecological issues.

Complicating the issue further is deciding which is cause and which is effect. Does productivity drive diversity, or vice-versa? Some theories suggest that communities with higher diversity should be more productive because different plant species will exploit different sets of resources (Naeem et al., 1994). This last issue becomes especially important when discussing what types of evidence can be used and in what ways.

USING EVIDENCE TO ADDRESS THE RELATIONSHIP
BETWEEN PRODUCTIVITY AND DIVERSITY

Types of Experiments

All sciences, but ecology in particular, rely on a range of types of data or empirical observations. Those observations may occur as part of a designed experiment or as part of a survey of the natural world. This is not a simple dichotomy, but a continuum. At one end are experiments in which all extraneous factors are held constant, as far as possible, and the scientist varies one or more factors of interest in a controlled fashion. At the other end are observational surveys with no manipulation. Intermediate are field experiments in which some factors are manipulated, some are controlled, and others vary naturally. All observations coming from this variety of situations potentially have epistemological import. The weight to be given each type when coming to scientific conclusions depends not on the type of evidence, but on the relationship between the type of evidence and the question being addressed. As we shall see, all scientific observations are made in the context of some theoretical framework. In this sense, all observations are a consequence of experiments.

I am using the term "experiment" in its broadest sense. There are several types of experiments. We typically think of an experiment as a test of a prediction. The most usual sense of an experiment is the manipulative experiment, including manipulation in laboratories, environmental chambers, and greenhouses, the creation of seminatural environments, and the imposition of experimental treatments on natural environments.

A second type is the observational experiment. This form of experiment is especially important in ecology because the scale of phenomena addressed or ethical considerations often preclude manipulative experiments. For example, the theory may predict that prairies will be more species rich than forests. One would test this hypothesis by measuring (observing) species numbers in each of those habitats. Intermediate between manipulative and observational experiments are natural experiments in which one observes the outcome of some naturally occurring perturbation (Diamond, 1986).

(Some [e.g., Hairston, 1989] lump natural experiments in with observational experiments.)

But not all experimental observations need be in the service of a particular prediction. I refer to these as "poke-at-it experiments," where we do some sort of manipulation just to see what will happen. Such experiments blend with strict observational studies of nature. Theory guides these experiments in the sense that it tells us which experiments are likely to lead to more interesting answers than others. Sometimes, these guidelines will be very general. For example, our general theoretical framework tells us that patterns of species diversity provide interesting information. We know very little about patterns of diversity in Indonesia. Therefore, interesting information will be obtained by observing patterns of species diversity in Indonesia. We may have general expectations of high diversity in lowland tropical forests and lower diversity in montane tropical forests, but we have no particular predictions in mind that we are testing.

What Is Evidence?
An experiment may be defined as a scientifically guided set of empirical observations. Scientific guidance includes analyses of the quality of those observations plus expert decisions on the relevance of an observation to the question at hand. These observations are one type of evidence. Scientifically guided observations are embedded in some sort of theoretical framework. This is what constructivist philosophers mean by theory-laden observation (Kuhn, 1962). I, and realist philosophers, take issue with the constructivist claim that being theory laden means that the observation cannot be used to test the theory.[1] Scientific guidance also means that how we decide what is important to look at is based on scientific judgment. Our poke-at-it experiments are more than random stabs at the universe. They are likely to produce interesting and useful information because a general theoretical framework guides them. Note that a scientifically guided empirical observation need not be made by a scientist, or even within the context of a scientific study, as long as its accuracy can somehow be verified. That is, observations can be considered scientifically guided even after the fact by subjecting the observations to the same level of scrutiny and skepticism that we would accord observations made within the context of a planned experiment.

The totality of the evidence then is all of the observations plus background information plus the logical relationship among all of this information. Assembling this totality includes decisions about how to weigh parts

1. See Laudan (1990) for a thorough deconstruction of this claim.

of the evidence. Some observations made in the course of an experiment may be discarded because they are found to be faulty (e.g., because of equipment malfunction). Not all background information will be relevant to a particular question. Here, expert knowledge is critical in choosing the relevant information, including observations made in other contexts. Scientists may disagree about which background information is relevant. These disagreements are resolved among scientists through decisions about the logical relationships among parts of the evidence plus agreements about what additional observations may be needed. For simplicity, I am ignoring sociological aspects of the resolution process; see Hull (1988) and Mayo (1996) for discussions of this issue. See Pickett, Kolasa, and Jones (1994, box 3.2, p. 59) for a comprehensive description of the components of a scientific theory including how various types of evidence fit into that framework.

Manipulative Studies

How has this range of types of experimental observations been used to address the issue of the productivity-diversity relationship? First, let us consider traditional manipulative experiments. Manipulative experiments are very useful to address issues of direction of causation, although as I discuss below, they are not the only type of experiment that can be used to derive causation. Each type has its strengths and weaknesses.

Experiments have been done manipulating productivity to explore the effect on diversity and manipulation of diversity to explore the effect on productivity. The latter experiments can get complex because species number gets mixed with species identity (Aarssen, 1997). The ideal experiment would consist of all possible combinations of species taken one at a time, two at a time, and so forth. Unfortunately, the ideal experiment is impractical because the number of experimental treatments (species combinations) rises factorially with the total number of species. Instead, experimenters sample a subset of the species and combinations, usually stratified by functional group. A recent experiment tackled this issue by replicating combinations of species and species numbers (from 1 to 32) among eight sites across Europe (Hector et al., 1999). They concluded that diversity does affect productivity, because plots with more species had greater aboveground biomass in all sites, despite differences in details of the response. They also concluded that niche complementarity or positive species interactions—rather than a simple sampling effect—might be the mechanisms responsible for the results, based on the effects of number of functions groups on productivity and patterns of overyielding. Note that these are plausibility arguments based on consistency of the results with the hypothesized mechanisms rather than a direct test of the mechanisms. Thus, besides the data from the experiments themselves, they also used background information and logic to assemble their evidence.

In manipulating productivity, a common experimental protocol is to water or fertilize a set of plots and see what happens to diversity. The usual result is that diversity decreases following fertilization (Gough et al., 2000). But how should we interpret such results? By studying which species typically win in such experiments, they tell us something about how competition works in controlling diversity. The limitation is that these experiments are typically done at small time scales (one to a few years). Local diversity may be controlled by regional diversity, the available species pool (Pärtel et al., 1996; Caley and Schluter, 1997). Given sufficient time for new species to arrive at the fertilized plot, species diversity might increase instead of decrease. One solution to this problem is to speed up migration rates by sowing seeds into plots. But how do you define the species pool? Is it all of the species in the surrounding field, all of the species in nearby fields, or all of the species within a five-kilometer radius? Moreover, if diversity patterns are in part determined by coevolutionary processes, one might need to use coadapted sets of species or even populations.

The common issue here is the problem of extrapolation. When interpreting the results from an experiment done in a laboratory or greenhouse, you must assume that interactions with nonmanipulated factors will not overwhelm the results. For example, in the greenhouse, one could study the effects of soil fertility levels on competition between clover and orchard grass and discover that orchard grass always wins. In the field, however, at low nitrogen levels, a third player enters the game, rhizobium. These bacteria form a symbiosis with the clover, giving it the competitive advantage. This three-way interaction is well studied and understood. Ecologists know under what conditions the results of the competition experiment can be extrapolated and when they cannot. Additional information plays an important role in deciding how to weigh the results of the greenhouse experiment.

Interactions and Extrapolation

Even field experiments face the extrapolation problem. The European diversity experiment dealt with this issue by replicating the experiment across a broad geographic scale. The experimenters then performed an analysis of covariance to test for interactions expressed as differences among sites in the slope of the diversity-productivity relationship. Because they failed to find interactions, they can be confident that their results are generalizable,

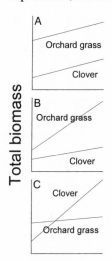

Total biomass

Nitrogen level

FIGURE 3.2 Three hypothetical experiments of the effects of nitrogen level on competition between orchard grass and clover. (A) There are no interactions between extent of competition and nitrogen level, and thus the results can be confidently extrapolated. (B) The interaction is of the noncrossing type, with no change in rank relationship. Extrapolations, especially toward lower nitrogen levels, are suspect. (C) The addition of rhizobium changes the rank-order relationships; the interaction is of the crossing type. Extrapolations, especially across ecological scales, must be done very carefully.

that productivity increases with diversity in a broad range of herbaceous communities.

Interactions can take two general forms, one of which creates more problems with extrapolations. In figure 3.2, we consider a hypothetical experiment examining the effects of nitrogen fertilization on growth of clover and orchard grass. The species are being grown together in pots and thus they are competing with each other. In figure 3.2A, the lines are parallel. Orchard grass always outcompetes clover, and we might confidently extrapolate these results to other fertilizer levels. In figure 3.2B, the lines are no longer parallel. There is an interaction. While orchard grass still outcompetes clover under all observed conditions, we might be more skeptical that our results can be extrapolated, especially toward low nitrogen levels. In figure 3.2C, we have added a new species to the mix, rhizobium. Now the lines cross and clover outcompetes orchard grass under some conditions. We are now sure that we cannot simply extrapolate our results. Thus, crossing interactions, ones with changes in rank-order relationships, create more problems than noncrossing interactions.

Another way to consider this issue is as one of scale. The greater the scale of the extrapolation, the lower is its reliability, because our extrapolation assumes a noncrossing interaction. Scale can be both spatiotemporal and ecological. In our grass-clover-rhizobium example, we are changing the ecological scale, the membership of the ecological community. Often, the ecological scale may be more important than spatiotemporal scale in creating interactions.

Thus, laboratory experiments do not necessarily carry greater weight than field experiments. While laboratory experiments are more powerful for exploring the mechanistic bases of ecological processes, field experiments have greater potential to be generalizable because, by incorporating more potentially interacting species, they can encompass a greater ecological scale.

Any extrapolation, with or without such interactions, assumes that the observed trend in the data (e.g., linear, quadratic, logistic) holds beyond the observed range. But even with this assumption, the uncertainty in our predictions increases as we move away from the core of our data. If our extrapolation is based on a mathematical model (e.g., a regression), we can formalize this uncertainty as a confidence band. This confidence band is an example of how we combine data with logical inference to give weight to a piece of evidence. The further we go from the center of our data, the less weight we would give to our extrapolation as the confidence band widens. Other kinds of logical evidence can be used in a similar fashion. For example, if an extrapolation of a species-area curve predicted substantially more than the known number of species in the region, we would discount that extrapolation.

Observational Studies

Observational data avoid some of those problems while raising new issues. What kinds of observational studies are there? The study of productivity-diversity relationships provides a window onto the variety that strictly observational studies can take. We have examples of four different kinds of observational study: (1) the single, large-scale study, (2) the single study with data compiled from multiple sources, (3) the combination of multiple studies that used a single protocol, and (4) the combination of data from studies that used disparate protocols. All are valid. All have their uses in different circumstances. Ultimately, when asking global questions, all add their piece to the puzzle.

After manipulative experiments, most people would consider a single, large-scale observational study to be the most reliable source of data. In this case, all of the information is gathered by one individual or team of individuals based on a predefined protocol. Such studies are common in productivity-diversity studies. They range from studies at very small scales to ones at regional scales. For example, Danin (1976) examined productivity-diversity relationships in herbaceous plant communities in the deserts of Israel. Using 0.1 m^2 quadrants, spread over a linear distance of less than 5 kilometers, he counted the number of species and then harvested everything to measure total standing biomass. Essentially the same sort of protocol was followed by Huston (1980) in a study of tree diversity in 0.1 ha plots, separated by up to 350 kilometers in Costa Rica, although in this case, standing biomass was estimated rather than measured directly.

Such studies, while extremely valuable, have limitations when dealing with issues that extend to scales of the entire globe. It would be impossible to do such a study on a global scale. Besides the person-hours necessary, the taxonomic expertise would be impossible to assemble. Several other approaches can be used at larger scales.

Several studies have been done as a single, large-scale study based on compiling data from disparate sources. For example, Currie and Paquin (1987) looked at patterns of tree diversity for $\frac{1}{4}° \times \frac{1}{4}°$ quadrants across North America. The diversity data came from published range maps. The productivity data came from published climate data. In this case, productivity itself was not measured or estimated. Instead, they used climate variables, such as actual evapotranspiration (an index that combines temperature and precipitation) as surrogates of productivity. In my own study with Jones of diversity in Wisconsin (Scheiner and Jones, 2002), we went a step further. The diversity data came from John T. Curtis's surveys of stands done during the 1950s and eventually published as *The Vegetation of Wisconsin* (Curtis, 1959). The productivity data came from a global climate model (BIOME-BGC; Kittel et al., 1995). That model estimates NPP on a $\frac{1}{2}° \times \frac{1}{2}°$ grid across the lower 48 states of the United States. To estimate productivity for the sampled locations, we interpolated model-output values from the surrounding four grid points.

Note the shift in these last two studies from the other examples of observational studies. In both cases, the studies depend on primary data collected by others. In both cases, at least some of the data are based on interpolations of actual observations, the range data for Currie and Paquin and the productivity estimates for Jones's and my study. In addition, our study used model estimates of productivity, while Currie and Paquin used indirect estimates of productivity. Therefore, these observations are all several steps removed from the scientists using them. Does that make them any less valid? No, it does not, because at all steps errors associated with the process can be tested. The results can be compared with other types of studies. For our Wisconsin study, we had a choice of three different global climate models. Before we made our choice, we compared the NPP estimates, estimates of productivity based on remote sensing data, and published estimates for similar community types.

Such studies can be done only by such combinations of data. In the case of our study, many of the sites surveyed by Curtis no longer exist as natural communities. In southern Wisconsin, especially, many are now houses and malls. Because he did not measure productivity when he did his survey, the only way to estimate such data is by the use of global climate models. These are examples of how background information, logic, and scientific expertise play a role in creating the totality of the evidence.

The next step up in this catalog of observational studies is the combining of data from multiple studies. This category is distinguished from the previous in that both diversity data and productivity data are collected together in each study, rather than as separate components. The gain is that both types of data are gathered at the same scale and the same time (approximately). The loss is that the data are gathered by multiple investigators, thus adding potential error and bias. Again, several variants exist.

In the ideal case, all investigators used exactly the same protocol for data collection. I would provide an example, but I know of no such case. If enough of the literature were combed, multiple studies using exactly the same protocol might be found. However, they would surely be disparate enough in other regards (e.g., biome studied, geographic extent surveyed, year of survey). Instead, we are faced with the issue of trying to combine data from disparate studies. This can be done using various sorts of standardization. One analysis (Gross et al., 2000) examined productivity-diversity relationships among grasslands and shrublands of North America. These data came from the Long-Term Ecological Research (LTER) network of sites. Because of this, the productivity data had been collected in the same fashion, for the most part. The biggest standardization issue faced was for the diversity data where different sized quadrants were used. Standardization was achieved by extrapolating up or down a species-area curve (Scheiner et al., 2000).

Another form of standardization is used in meta-analysis (Gurevitch and Hedges, 2001). The results of each study are converted into an effect size. For studies that compare two means, the effect size is the difference in the means divided by the pooled standard deviation. For continuous relationships, standardized regression coefficients can be used. The combination of studies is achieved by increasing the level of abstraction. In effect, one is now asking about the overall qualitative pattern, rather than about the details of the quantitative pattern.

For some questions, meta-analysis is not possible. Meta-analysis requires enough information to compute a standardized metric. We (Mittelbach et al., 2001) were not able to do that. Our study examined productivity-diversity relationships as a function of scale. We compiled 172 studies from the literature. These studies used a large variety of protocols, measurements of productivity and diversity, kinds of organisms, and scales. A meta-analysis does not make sense because it is not clear that even a standardized regression coefficient is measuring the same phenomena on the same scale. In seventeen cases, we could not compute a standardized coefficient because we did not have sufficient information. Yet we did not want to throw out these studies. So we resorted to an older method, vote counting. We counted the number of studies that had statistically significant hump-shaped, U-shaped, positive-linear, and negative-linear relationships. The questions we addressed were simple: are there patterns in the relationship, and does it change as a function of scale? For such broad, qualitative questions, a vote counting method is adequate. Note that we also did some analyses on standardized coefficients for subsets of the data in order to test the robustness of our more general conclusions. Again, we examined the error structure of the data using a variety of criteria and methods.

Our literature survey combines aspects of theory-test experiments and poke-at-it experiments. The theory that we tested was the claim that hump-shaped relationships were ubiquitous at all scales. We clearly showed that not to be the case. We did find that humps tend to be most common at intermediate scales where community boundaries are crossed. Beyond this question of hump ubiquity, we had no particular theory in mind when compiling our data. However, from this survey, two different activities may follow. We can use these observations to revisit the two dozen theories addressing productivity-diversity relationships. We may be able to winnow out of that list those whose predictions are consistent with the observed patterns. What was completely unexpected in our results was the observation of U-shaped patterns. They are few enough (10) that no one previously had made significant note of them. U-shaped relationships are not predicted by any current theory. Therefore, a generation of new theory is likely to result.

Observation and Causation

How do observational studies deal with the issue of direction of causation? Weren't we taught in introductory statistics that correlation does not prove causation? Sometimes it does.

Observational studies can be used to address the issue of direction of causation because mechanisms work at different scales. All theories about how greater diversity would increase productivity are based on mechanisms that operate at strictly local scales. They involve processes of competition and niche partitioning among plants in a small plot and make two predictions. First, at local scales, the relationship between productivity and diversity should be positive. Second, if this process were the only one responsible for the relationship, then patterns at greater scales should be a simple sum of patterns at lesser scales. We (Scheiner et al., 2000) have dubbed this notion of relationships of process, pattern, and scale "upward causation." So observational experiments can address causation if the alternative theories predict different patterns.

We can use observational experiments in another way that can be dubbed "the exception that proves the rule." Recall that the hypothesized relationship between productivity and diversity was prompted on the observation of a latitudinal gradient in species richness. But for vascular plants it is not a simple latitudinal gradient. While species richness is highest at the equator, it dips at about 30° N and S latitude, rises again at middle latitudes, and then falls toward the poles (Scheiner and Rey-Benayas, 1994). The dip in plant diversity at 30° is associated with the location of the major deserts of the world. Thus diversity patterns, rather than simply following a latitudinal gradient, follow a productivity gradient. Moreover, the Sonoran Desert, which is the most productive desert in the world because it has both winter and summer rainfall, is the most species-rich desert in the world. Different causative mechanisms make different predictions about the details of a pattern. Because logic tells us that rainfall in the Sonoran Desert cannot be caused by its vegetation, we can conclude that productivity is causing diversity, and not the other way around—at least at this broad, global scale.

A PRIORI VS. POST HOC EXPLANATIONS

Does it matter which came first, the data or the theory, to the extent that we "believe" either? While some would argue that it does (see Maurer, 2004 [chapter 2, this volume]), I say no (see also Mayo, 1996). That a theory is erected after the data are in hand does not weaken the strength of the agreement between data and theory. The more agreements we can assemble, especially of different kinds of data, the greater our confidence that the match is correct. Such multiple agreements are the heart of consilience (Whewell, 1858). And I hope that I have made it clear that the nature of those observations does not matter in the broad sense. Certainly, it matters in the sense that the type of observation needs to match the scale and nature of the theory. But any theory of global import is built with evidence from a variety of types and scales. Sometimes, the pieces are not recognized as such. For example, any theory about productivity-diversity relationships among vascular plants would have to take into account the well-supported observation that larger individuals almost always outcompete smaller individuals. This piece of information is embedded in many theories, mostly in unspoken ways. If any new theory were proposed that violated this precept, it would be rejected as soon as the violation was recognized. Any ecological theory presented today is, of necessity, built upon over one hundred years of ecological evidence of all types. Sorting through all of this evidence and deciding which is relevant to a particular theory, and what weight to give to each piece of evidence, is no easy task. All types of evidence deserve equal scrutiny.

It is not the type of evidence that needs to be ranked, but the type of theory. To paraphrase Carl Sagan, extraordinary claims require extraordinary evidence. For example, a claim that Earth has been visited by intelligent aliens—an extremely extraordinary claim—would require extreme evidence. Isaac Asimov stated that the only sufficient evidence in this instance would be a piece of unknown technology, something completely beyond or outside the realm of current earthly technologies. If one came upon such technology (e.g., the crashed spaceship is finally dug up in Roswell), it does not matter that the explanation (theory) was built afterward. No other theory (that I can conceive of) would explain such an observation. This example is simplified in several regards, not the least in that a single observation (the crashed spaceship) is sufficient to support the entire theory (alien visitors). In ecology, very few theories, especially those of global import, are supportable on just one or a few pieces of evidence. On the other hand, many observations are not necessarily sufficient. Despite mountains of testimony, lots of best-selling books, and a host of Web sites, acceptance of the theory of alien visitation is limited, for the most part, to nonscientists. The kinds of evidences that are available in this case are not sufficient, no matter how much there is. If evidence results from scientifically guided empirical observation, combined with background information and logic, we know what kind of observations we need to make, and we have not made them.

CONCLUSION

In this chapter, I have presented the thesis that scientific theories are built upon the consilience of the evidence. Evidence comes from scientifically-guided empirical observation, combined with background information, logic, and scientific expertise. Observations may come from manipulative experiments and observational experiments. An experiment may be designed to test a particular theory, or just to look at something potentially interesting. Because all theories are constructs consisting of many kinds of observations and logic, and because ecological theories of global import deal with phenomena, mechanisms, and processes at many different scales, all kinds of evidence must be used in building and testing these theories. What is important is not the type of observation, but whether it is appropriate for the question at hand. That is what decides whether it counts as evidence in a given instance.

3.1   Commentary

Marie-Josée Fortin

One of the most fundamental properties of nature is its complexity. This complexity expresses itself in various ways, one of which is species interaction. Nature is therefore an inexhaustible source of dynamic experiments yielding evidence. By identifying and analyzing such evidence, we restlessly quest for knowledge on how "nature" works. Our ability to identify and understand the evidence, however, is directly related to the state of our body of knowledge, i.e., theories, concepts, and paradigms. A classic example of this is Newton's apple: apples had long been falling from trees but this was not perceived as evidence for gravity until the body of global knowledge ripened. Similarly, in ecology our current ability to deal with the effects of landscape spatial heterogeneity had to wait for the development of new statistics (e.g., spatial statistics, randomization tests) and new ecological paradigms (e.g., landscape ecology, hierarchical theory).

In fact, the design of any ecological study dealing with hypothesis testing requires closed links among: (1) the questions asked, i.e., the theoretical framework; (2) the sampling or experimental design; and (3) the statistical analyses of the data. It is important to realize that these three steps interact and must be considered as a whole. Indeed, one cannot obtain more information from a data set than it contains. Even new refined and sophisticated statistical methods cannot extract information that was not gathered in the first place. Furthermore, ecologists often use statistics to blindly validate or invalidate hypotheses, thus allowing the statistics to make the decisions. This can be quite misleading, such as in cases where a significant statistical difference of means may not be ecologically important. Ecologists must therefore be aware that not all statistical evidence is ecological evidence. Hence, sampling and experimental designs should be thought of not only according to the hypotheses or goals of the study, but also with respect to the subsequent statistical analyses. By doing so, it requires the workers to comprehensively overview all the components and steps involved in the process of gaining evidence and knowledge from the system under study.

Scheiner's chapter is refreshing. It brings us back to the basics by reminding us first that evidence can be either observed or generated by field or model manipulation, and second that evidence has many facets and that only our minds construct limits for us to perceive them all and their full implications. Ecologists should therefore take the time to let their creativity grab the evidence that nature offers us.

3.2	Commentary

Manuel C. Molles, Jr.

> *When you follow two separate chains of thought, Watson, you will find some point of intersection which should approximate the truth.* —Sherlock Holmes, in *The Disappearance of Lady Frances Carfax*

How do scientists arrive at conclusions regarding structure and process in the natural world? In his provocative chapter, Scheiner says we need to pay attention to all forms of evidence. Throughout his discussion Scheiner builds on the analogy that scientists are like detectives. As detectives, scientists should not ignore any form of evidence and their conclusions are most robust when based on the totality of the evidence. Scheiner suggests further that in the scientific process, each datum is a clue. Some clues may be insignificant; some may be discarded in the end because they are inappropriate. However, the most effective detectives, those that solve the greatest number of mysteries, ignore no clues. Scheiner, echoing Holmes comment to Dr. Watson on how one approximates the truth, says, "scientific decisions are based on the consilience of the evidence."

However, Scheiner goes well beyond simply calling for a greater plurality of approaches, which others have argued for. He also lays out a basis for deciding which forms of evidence are most appropriate for which problems. Scheiner reminds the reader that all data, whether they come from designed and controlled experiments or observations of the natural world, are ultimately empirical observations. He suggests further that no particular type of observation is inherently more important or valid than any other type. However, in a particular study, different classes of data may be given different weight. Scheiner suggests that the weight given should not depend on the type of observation but on the match between the type of evidence and the question asked by the scientist.

In analyses such as this one, readers often find themselves lost in a flow of generalization. In contrast, Scheiner's presentation gathers coherence from its organization around a current issue in ecology: how does diversity vary relative to productivity? By tracing the history of this issue, Scheiner provides clear insights into the role played by different forms of evidence, including observations of geographical variation in diversity, one of the variables of interest, and productivity, the other variable of interest. As ecologists have explored the relationship between productivity and diversity, they have produced forms of evidence, ranging from the results of greenhouse and field experiments to observations of nature conducted at large and small scales. Again, Scheiner emphasizes that it is not the type of observation that determines whether an observation counts as evidence in a particular instance. The important factor to consider is whether the type of observation is appropriate to the question being addressed.

Scheiner's call for pluralism is important and his choice of questions around which to organize his presentation is timely. As I read Scheiner's chapter, I received my latest copy of the *Bulletin of the Ecological Society of America,* which included a heated debate over the relationship between productivity and diversity. Debates and disagreements among scientists show that ecology is a vital discipline and that its practitioners are actively engaged in exploration of the natural world. However, ironically, some of this published debate concerned the value of different methods of study and the validity of different forms of evidence. In addition, the discussions were tinged by accusations that one or both sides of the debate failed to appreciate alternative forms of evidence. In this social context, Scheiner's clear voice should be heard and heeded by all sides. Even today after decades of debate and controversy over methodology in ecology, the need for a diversity of approaches is too often unappreciated. While all approaches to doing science have their place, all have had their chauvinists who have asserted the primacy of their favorite method.

3.3   Rejoinder

Samuel M. Scheiner

I am flattered and heartened by the reaction to my chapter by Drs. Fortin and Molles. Nearly twenty years ago, George Salt assembled a special issue of the *American Naturalist* (vol. 122, no. 5) on research in ecology and evolutionary biology. The authors espoused various extreme positions about how to approach such research, with a divide between pro– and anti– hypothesis testing camps. In contrast, George Salt's summary was a call for moderation and a plurality of approaches. Today, such a call for plurality is accepted as the sensible approach to scientific research. I take this as a sign that our field has matured and become confident in its theoretical underpinnings. We need not be bound by a narrow philosophy of science dictating to us how we should evaluate our own data. With this self-confidence, science as a whole, and ecology and evolution in particular, we will be able to make maximal use of all of the available evidence.

REFERENCES

Aarssen, L. W. 1997. High Productivity in Grassland Ecosystems: Effected by Species Diversity or Productive Species? *Oikos* 80 : 183–184.

Adams, J. M., and F. I. Woodward. 1989. Patterns in Tree Species Richness as a Test of the Glacial Extinction Hypothesis. *Nature* 339 : 699–701.

Caley, M. J., and D. Schluter. 1997. The Relationship between Local and Regional Diversity. *Ecology* 78 : 70–80.

Connell, J. H., and E. Orias. 1964. The Ecological Regulation of Species Diversity. *Am. Nat.* 98 : 399–414.

Currie, D. J. 1991. Energy and Large-Scale Patterns of Animal- and Plant-Species Richness. *Am. Nat.* 137 : 27–49.

Currie, D. J., and V. Paquin. 1987. Large-Scale Biogeographical Patterns of Species Richness of Trees. *Nature* 329 : 326–327.

Curtis, J. T. 1959. *The Vegetation of Wisconsin.* Madison: University of Wisconsin Press.

Danin, A. 1976. Plant Species Diversity under Desert Conditions. I. Annual Species Diversity in the Dead Sea Valley. *Oecologia* 22 : 251–259.

Diamond, J. 1986. Overview: Laboratory Experiments, Field Experiments, and Natural Experiments. In Diamond, J., and T. J. Case, eds., *Community Ecology.* New York: Harper and Row.

Gough, L., C. W. Osenberg, K. L. Gross, and S. L. Collins. 2000. Fertilization Effects on Species Density and Primary Productivity in Herbaceous Plant Communities. *Oikos* 89:428–439.

Gross, K. L., M. R. Willig, L. Gough, R. Inouye, and S. B. Cox. 2000. Patterns of Species Diversity and Productivity at Different Spatial Scales in Herbaceous Plant Communities. *Oikos* 89:417–427.

Gurevitch, J., and L. V. Hedges. 2001. Meta-analysis: Combining the Results of Independent Experiments. In Scheiner, S. M., and J. Gurevitch, eds., *Design and Analysis of Ecological Experiments*. New York: Oxford University Press.

Hairston, N. G., Sr. 1989. *Ecological Experiments: Purpose, Design, and Execution*. Cambridge: Cambridge University Press

Hector, A., B. Schmid, C. Beierkuhnlein, M. C. Caldeira, M. Diemer, P. G. Dimitrakopoulos, J. A. Finn, H. Freitas, P. S. Giller, J. Good, R. Harris, P. Högberg, K. Huss-Danell, J. Joshi, A. Jumpponen, C. Körner, P. W. Leadley, M. Loreau, A. Minns, C. P. H. Mulder, G. O'Donovan, S. J. Otway, J. S. Pereira, A. Prinz, D. J. Read, M. Scherer-Lorenzen, E.-D. Schulze, A.-S. D. Siamantziouras, E. M. Spehn, A. C. Terry, A. Y. Troumbis, F. I. Woodward, S. Yachi, and J. H. Lawton. 1999. Plant Diversity and Productivity Experiments in European Grasslands. *Science* 286:1123–1127.

Hilborn, R., and M. Mangel. 1997. *The Ecological Detective: Confronting Models with Data*. Princeton: Princeton University Press.

Hull, D. L. 1988. *Science as a Process*. Chicago: University of Chicago Press.

Huston, M. A. 1980. Soil Nutrients and Tree Species Richness in Costa Rican Forests. *J. Biogeogr.* 7:147–157.

Huston, M. A. 1994. *Biological Diversity: The Coexistence of Species in Changing Landscapes*. Cambridge: Cambridge University Press.

Kittel, T. G. F., N. A. Rosenbloom, T. H. Painter, D. S. Schimel, and VEMAP Modeling Participants. 1995. The VEMAP Integrated Database for Modeling United States Ecosystem/Vegetation Sensitivity to Climate Change. *J. Biogeogr.* 22:857–862.

Kuhn, T. S. 1962. *The Structure of Scientific Revolutions*. Chicago: University of Chicago Press.

Laudan, L. 1990. *Science and Relativism*. Chicago: University of Chicago Press.

Maurer, B. A. 2004. Models of Scientific Inquiry and Statistical Practice: Implications for the Structure of Scientific Knowledge. Chapter 2 in Taper, M. L., and S. R. Lele, eds., *The Nature of Scientific Evidence: Statistical, Philosophical, and Empirical Considerations*. Chicago: University of Chicago Press.

Mayo, D. G. 1996. *Error and the Growth of Experimental Knowledge*. Chicago: University of Chicago Press.

Mayo, D. G. 2004. An Error-Statistical Philosophy of Evidence. Chapter 4 in Taper, M. L., and S. R. Lele, eds., *The Nature of Scientific Evidence: Statistical, Philosophical, and Empirical Considerations*. Chicago: University of Chicago Press.

Mittelbach, G. G., C. F. Steiner, S. M. Scheiner, K. L. Gross, H. L. Reynolds, R. B. Waide, M. R. Willig, S. I. Dodson, and L. Gough. 2001. What Is the Observed Relationship between Species and Productivity? *Ecology* 82:2381–2396.

Naeem, S., L. J. Thompson, S. P. Lawlor, and R. M. Woodfin. 1994. Declining Biodiversity Can Alter the Performance of Ecosystems. *Nature* 368:734–736.

Pärtel, M., M. Zobel, K. Zobel, and E. van der Maarel. 1996. The Species Pool and Its Relation to Species Richness: Evidence from Estonian Plant Communities. *Oikos* 75:111–117.

Pickett, S. T. A., J. Kolasa, and C. G. Jones. 1994. *Ecological Understanding.* San Diego: Academic Press.

Popper, K. R. 1959. *The Logic of Scientific Discovery.* London: Hutchinson.

Preston, F. W. 1962a. The Canonical Distribution of Commonness and Rarity: part 1. *Ecology* 43:185–215.

Preston, F. W. 1962b. The Canonical Distribution of Commonness and Rarity: part 2. *Ecology* 43:410–432.

Quinn, J. F., and A. E. Dunham. 1983. On Hypothesis Testing in Ecology and Evolution. *Am. Nat.* 122:602–617.

Rosenzweig, M. L. 1995. *Species Diversity in Space and Time.* Cambridge: Cambridge University Press.

Ruse, M. 1999. *The Darwinian Revolution.* 2nd ed. Chicago: University of Chicago Press.

Scheiner, S. M., S. B. Cox, M. R. Willig, G. G. Mittelbach, C. Osenberg, and M. Kaspari. 2000. Species Richness, Species-Area Curves, and Simpson's Paradox. *Evol. Ecol. Res.* 2:791–802.

Scheiner, S. M., and S. Jones. 2002. Diversity, Productivity, and Scale in Wisconsin Vegetation. *Evol. Ecol. Res.* 4:1097–1117.

Scheiner, S. M., and J. M. Rey-Benayas. 1994. Global Patterns of Plant Diversity. *Evol. Ecol.* 8:331–347.

Tilman, D. 1982. *Resource Competition and Community Structure.* Princeton: Princeton University Press

Tilman, D., and S. Pacala. 1993. The Maintenance of Species Richness in Plant Communities. In Ricklefs, R. E., and D. Schluter, eds., *Species Diversity in Ecological Communities: Historical and Geographical Perspectives.* Chicago: University of Chicago Press.

VanderMeulen, M. A., A. J. Hudson, and S. M. Scheiner. 2001. Three Evolutionary Hypotheses for the Hump-Shaped Species-Productivity Curve. *Evol. Ecol. Res.* 3:379–392.

Wallace, A. R. 1878. *Tropical Nature and Other Essays.* New York: Macmillan.

Whewell, W. 1858. *Novum Organon Renovatum.* London.

Wright, D. H. 1983. Species-Energy Theory: An Extension of Species-Area Theory. *Oikos* 41:496–506.

PART 2   LOGICS OF EVIDENCE

V. P. Godambe

OVERVIEW

All four papers in this part are thought provoking and well written. They discuss the core concepts of statistical evidence and inference all the way from their definitions to their applications. Indeed, because of its bewildering variety of mutually interacting factors, ecology provides a challenging subject for any kind of statistical applications. "Statistics" has many different schools of thought, although they are not necessarily sharply defined. Further, even in any given "school," there are statisticians with very different "persuasions." Such variety has caused considerable confusion; however, had there been a unified methodology, this confusion could have been avoided. At the same time, this variety has also been helpful when faced with the complex and rich problems as faced by ecologists. Practitioners can use any approach that suits the problem at hand. The papers under review here bear witness to this.

At different levels of recording information or data collection, perhaps for administrative, official, or social purposes, different kinds of statistics have been used. For instance, when recorded information was restricted to direct measures of rainfall, production, population, and the like, simple statistics such as the mean, median, and mode were used. As data collection became more sophisticated as in agriculture, biostatistics, and ecology, more advanced statistical tools such as least-squares and maximum likelihood got under way. The sophisticated data collection just mentioned was often based on designed experiments and techniques of survey sampling, both possibly involving randomization. These data were more structured than the raw observational data, and hence admitted probabilistic modeling.

Now, as I said earlier, there are many schools of thought in statistics. Yet almost all, with a few exceptions, admit some kind of probabilistic modeling. Often appropriately planned experimentation yields data that follows some "simple model." By a simple model we mean a model that has a single (often real) parameter, namely the parameter of interest, such as for instance the "pollution" of an ecosystem. This model generally would contain very few (ideally none) nuisance parameters. The basic statistical theories of evidence/inference, namely error probabilities theory (Neyman-Pearson), likelihood theories (Barnard, Birnbaum, Fisher), and Bayesian theories, were initially formulated in relation to such idealized models with no nuisance parameters. Subsequently these theories were extended to models with nuisance parameters in which these parameters could be eliminated through conditioning or marginalizing. All this certainly advanced our conceptual understanding of statistics, yet fell rather short of what practitioners actually needed. These theories were too much tied to a "model," more specifically to the "parametric likelihood function" defined by the model, while the practitioner's central interest was the parameter itself, e.g., pollution, productivity, or diversity.

Even for data obtained through a planned experiment, assumption of a model specifying a detailed likelihood function for the parameter was generally too demanding or unnatural. Yet such a detailed model was postulated, at least in the initial stages of the three fundamental theories mentioned earlier. These theories were primarily geared toward testing hypotheses; parameter estimation was somewhat secondary to them. On the other hand, for the practitioners, parameter estimation was the first requirement; testing came much later. Hence a "realistic theory" of evidence/inference should primarily aim at estimation. It should replace the assumption of a detailed likelihood model underlying the previous theories by the assumption of a semiparametric model that is characterized by just those distributional constraints required for the definition of the parameter. Such a realistic theory is obtained by generalization of the concept of the score function (and not a likelihood function). It is called the "theory of estimating functions"; here the generalized score function is characterized by the "optimal" estimating functions (Godambe, 1960, 1997). This theory is already widely used in many areas of statistics, including biostatistics. I hope the ecologists will also find the theory useful.

Using the framework of statistical methodology I've outlined above, I shall now make a few remarks about the four papers under review.

Deborah Mayo (chapter 4) discusses her reinterpretation of the Neyman-Pearson approach. She suggests that, although maligned by Bayesian thinkers in recent years, the Neyman-Pearson approach does address the issues that scientists are truly interested in. Scientists accept new ideas only after they have passed a severe test. She discusses her notion of the idea of severe tests, promoting the idea of reporting the severity at which the hypothesis passed the test. This idea of severity is tantalizingly similar to the concept of the power of the test but is in fact different because it applies post hoc.

Mayo's interpretation of Neyman-Pearson theory as a methodology of inference that takes into account the data-gathering procedure is deeply intuitive. However, to my disappointment, Mayo's emphasis seems to be exclusively testing; estimation is ignored. It is true that in Neyman-Pearson interval estimation is derived from hypothesis tests. It is also true in some sophisticated areas where cause-effect relationships are investigated, the problem is usually formulated as one of testing hypotheses. Even here it can be argued (Godambe and Thompson, 1997) that the problem is more realistically and manageably formulated as one of estimation rather than one of testing. As I said earlier, for practitioners, the problem of estimation generally has much greater priority over that of testing.

Richard Royall (chapter 5) claims that the current state of statistical thinking is muddled. The main reason for this befuddlement is because statistical thinking tries to address three distinct issues with a single technique. These critical issues are what do we do? How do we change our beliefs? And what is the evidence for one hypothesis over a competing one? He suggests that we need first to clarify the concept of evidence and then to use the correct measure of evidence to make decisions or to change beliefs. It is important to realize that evidence is a comparative measure, and hence we need two hypotheses to compare. Following Ian Hacking, he suggests the use of the likelihood ratio as the measure of the strength of evidence. His paper introduces important concepts of misleading evidence and weak evidence. He argues that evidence, even when correctly interpreted, can at times be misleading. Observations could support the wrong model. It is also quite possible that a given set of observations may not discriminate two hypotheses very well, thus leading to the possibility of weak evidence. He provides upper bounds for the probabilities of weak and misleading evidence and shows that these probabilities converge to zero as the sample size increases. This is contrasted with the fact that the probability of a type I error in the Neyman-Pearson approach remains constant no matter how much data one collects. Ramifications of the consideration of the probabilities of weak and misleading evidence are discussed in the context of the age-old problem of sample size determination in designing new studies. It is argued that evidential ideas lead to better answers than the classical frequentist approaches.

Royall's paper is characteristically clear and persuasive. Surely if we are considering only "simple" hypotheses, the likelihood ratio criterion is in most situations very intuitive. An exceptional situation is discussed later. Apart from this exception, Royall's demonstration that "the chances of getting misleading evidence are small" makes the likelihood ratio criterion very attractive. But in practice we often have nuisance parameters, and this makes the role of the likelihood ratio criterion very obscure. Even the likelihood principle, in case of nuisance parameters, is not valid. This is well illustrated by Birnbaum in his fundamental paper (Birnbaum, 1962).

What makes me doubt the universality of the likelihood ratio criterion, even for testing a simple hypothesis versus a simple alternative, is the exceptional situation, mentioned in the above paragraph. I discovered this situation while investigating the role of randomization in the context of survey sampling. It is as follows. There is a sample space S with generic points (samples) $S = \{s\}$; similarly in the parameter space $\Theta = \{\theta\}$, θ being generic points. A sample s is drawn from S with probability $p(s)$, $s \in S$. Here the probability distribution p is completely specified (known). Particularly, p is mathematically independent of θ. Yet on the basis of an observation $s = s_0$, say about two parametric points θ_1 and θ_2 in Θ, one can say that θ_1 is more plausible than θ_2! For details see Godambe (1982). I agree this is an exceptional example, but for me it is enough to raise doubts about claims of "universality" of the criterion of likelihood ratio, even while testing a simple hypothesis versus a simple alternative. Royall calls it "the law of likelihood." I wonder what sets it as a law? Science has laws. But is statistics a science or just a technology or a discipline? It is debatable.

Malcolm Forster and Elliott Sober (chapter 6) agree with Royall about the utility of the likelihood function for quantifying the strength of evidence. Royall justifies the use of the likelihood ratio based on intuition and probability bounds for weak and misleading evidence. Forster and Sober claim that prediction is at least as important a goal as is the explanation. Royall emphasizes explanation whereas Forster and Sober emphasize prediction and hence the use of the Akaike information criterion (AIC) as the measure of strength of evidence. They justify the law of likelihood as a special case of AIC comparisons when the number of parameters is the same in both models.

As does Royall, Subhash Lele (chapter 7) begins with Hacking's tenet of the law of likelihood. From the practical point of view, however, use of likelihood can be problematic on various grounds. For example, it is highly sensitive to outliers and model misspecification. Furthermore, use of likelihood requires specification of the full statistical model. In recent years, statisticians have been promoting the use of estimating functions that require specification of only a few features such as the mean and the variance function rather than the full probabilistic model. Likelihood-based approaches need to be modified to accommodate nuisance parameters. Statisticians have been using conditional, marginal, or integrated likelihood for this purpose. A natural question that arises is, can we quantify the strength of evidence in these complicated but practically important situations?

In his chapter, Lele introduces a new class of functions, named "evidence functions," that might be utilized in quantifying the strength of evidence. Using the concept and defining properties of evidence functions, Lele shows that one can construct evidence measures that are robust against outliers, that need minimal model specification, and that can handle nuisance parameters and other such practical considerations. Under this setup it is also possible to justify the use of the likelihood ratio as the optimal measure of strength of evidence in the sense of having the highest probability of strong evidence in favor of the correct hypothesis. One of the startling consequences of the use of evidence functions is that, in contrast to the likelihood ratio, the "stopping rule" appears to be relevant for quantifying the strength of evidence.

In general, in published works on hypothesis tests parameter is separated from observation by an unnatural barrier, while logically they should be together. This basic logical requirement is fulfilled in Lele's paper. Lele's evidence functions are, like estimating functions, functions of observations and parameter together. Under certain restrictions the optimality of log-likelihood ratio evidence functions is established. Lele is fully aware that further development of his work is required to reduce the dependence of his results on the assumption of a full parametric model and nuisance parameters. Hence, I just applaud the author for his bold and original paper.

As is the case with other sections, the commentators introduce balance in the discussion of the above topics. In their discussion of Mayo's paper, George Casella and Earl McCoy take up the case for the Bayesian and the classical frequentist approaches to statistical analysis. Sir D. R. Cox agrees with Royall to some degree but raises some important issues of relevance of the stopping rule, nuisance parameters, composite hypotheses, and such. Martin Curd argues for the Bayesian approach to evidence. In commenting on Forster and Sober, Robert Boik takes a strong stand in favor of Bayesian model selection, while Michael Kruse argues in favor of Mayo's approach. I agree with the point made by Boik in his commentary. The Akaike criterion is not consistent with the likelihood principle. But I would like to distinguish between "fitting a model" and "testing parametric values"; the likelihood principle is relevant only for the latter, I believe. C. C. Heyde, commenting on Lele, suggests that some of the results could be extended to dependent data situations. Paul Nelson raises some thorny problems such as comparing evidence for discrete versus continuous distributions. He concludes by challenging evidentialists to solve the important practical issue of decision making using evidence.

It is clear from reading these chapters and the commentaries that although the evidential approach has significant aesthetic appeal, has shown boldness in asking the right questions; nevertheless, it still has a long way to go before it is ready to address real-life, complex scientific issues.

REFERENCES

Birnbaum, A. 1962. On the Foundations of Statistical Inference (with discussion). *J. Am. Stat. Assn* 57:269–306.

Godambe, V. P. 1960. An Optimum Property of Regular Maximum Likelihood Estimation. *Ann. Math. Stat.* 31:1208–1211.

Godambe, V. P. 1982. Ancillarity Principle and a Statistical Paradox. *J. Am. Stat. Assn* 77:931–933.

Godambe, V. P. 1997. Estimating Functions: A Synthesis of Least Squares and Maximum Likelihood Methods. In Basawa, I., V. P. Godambe, and R. I. Taylor, eds., *Selected Proceedings of the Symposium on Estimating Functions.* Hayward, CA: Institute of Mathematical Statistics.

Godambe, V. P., and Thompson, M. E. 1997. Optimal Estimation in a Causal Framework. *J. Indian Soc. Agricultural Stat.* 49:21–46.

4 An Error-Statistical Philosophy of Evidence

Deborah G. Mayo

ABSTRACT

Despite the widespread use of error-statistical methods in science, these methods have been the subject of enormous criticism, giving rise to the popular statistical "reform" movement and bolstering subjective Bayesian philosophy of science. Given the new emphasis of philosophers of science on scientific practice, it is surprising to find they are rarely called upon to shed light on the large literature now arising from debates about these reforms—debates that are so often philosophical. I have long proposed reinterpreting standard statistical tests as tools for obtaining experimental knowledge. In my account of testing, data *x* are evidence for a hypothesis *H* to the extent that *H* passes a severe test with *x*. The familiar statistical hypotheses as I see them serve to ask questions about the presence of key errors: mistaking real effects for chance, or mistakes about parameter values, causes, and experimental assumptions. An experimental result is a good indication that an error is absent if there is a very high probability that the error would have been detected if it existed, and yet it was not detected. Test results provide a good (poor) indication of a hypothesis *H* to the extent that *H* passes a test with high (low) severity. Tests with low error probabilities are justified by the corresponding reasoning for hypotheses that pass severe tests.

Is it possible to have a general account of scientific evidence and inference that shows how we learn from experiment despite uncertainty and error? One way that philosophers have attempted to affirmatively answer this question is to erect accounts of scientific inference or testing where appealing to probabilistic or statistical ideas would accommodate the uncertainties and error. Leading attempts take the form of rules or logics relating evidence (or evidence statements) and hypotheses by measures of confirmation, support, or probability. We can call such accounts *logics of evidential relation-*

79

ship (or *E-R logics*). In this view, philosophers could help settle scientific disputes about evidence by developing and applying logics of evidential relationship. I will begin my discussion by reflecting on these logics of evidence and how they differ from what I think an adequate account of evidence requires.

LOGICS OF EVIDENTIAL RELATIONSHIP VS. ERROR STATISTICS

The leading example of a logic of evidential relationship is based on one or another *Bayesian* account or model. In the most well-known of Bayesian accounts, our inference about a hypothesis *H* from evidence *e* is measured by the probability of *H* given *e* using Bayes' theorem from probability theory. Beginning with an accepted statement of evidence *e*, scientists are to assign probabilities to an exhaustive set of hypotheses. Evidence *e confirms* or supports hypothesis *H* to the extent that the probability of *H* given the evidence exceeds the initial assignment of probability in *H*. But how can we make these initial assignments of probability (i.e., the prior probabilities)? In early attempts at building an "inductive logic" (by Carnap and other logical empiricists), assignments based on intuitive, logical principles were sought, but paradoxes and incompatible measures of evidential relationship resulted. Moreover, the inductive logicians were never able to satisfactorily answer the question of how purely logical measures of probability are to be relevant for predictions about actual experiments. Contemporary subjective Bayesian philosophers instead interpret probabilities as an agent's subjective degrees of belief in the various hypotheses. The resulting *subjective Bayesian way* furnishes the kind of logic of evidential relationship that many philosophers have sought.

Although the subjective Bayesian approach is highly popular among philosophers, its dependence upon subjective degrees of belief, many feel, makes it ill-suited for building an objective methodology for science. In science, it seems, we want to know what the data are saying, quite apart from the opinions we start out with. In trading logical probabilities for measures of belief, the problem of relevance to real world predictions remains. By and large, subjective Bayesians admit as much. Leonard (L. J.) Savage, a founder of modern "personalistic" Bayesianism, makes it very clear throughout his work that the theory of personal probability "is a code of consistency for the person applying it, not a system of predictions about the world around him" (Savage, 1972, 59). But a code for consistently adjusting subjective beliefs is unable to serve as the basis for a philosophy of evidence that can help us to understand, let alone adjudicate objectively, disputes about evidence in science.

Under E-R logics, one might wish to include the less familiar *likelihood* approach (or approaches). The likelihoodist uses the likelihood ratio as the comparative evidential-relation measure.[1] I shall have more to say about this approach as we proceed.

A second statistical methodology on which one might erect a philosophy of evidence is alternatively referred to as classical, orthodox, frequentist, or Neyman-Pearson (NP) statistics. The methods and models of NP statistics (e.g., statistical significance tests, confidence interval estimation methods) were deliberately designed so that their validity does not depend upon prior probabilities to hypotheses—probabilities that are eschewed altogether except where they may be based upon objective frequencies. Probability arises not to assign degrees of belief or confirmation to hypotheses but rather to characterize the experimental testing process itself: to express how *frequently* it is capable of discriminating between alternative hypotheses and how *reliably* it facilitates the detection of error. These probabilistic properties of experimental procedures are called *error frequencies* or *error probabilities* (e.g., significance levels, confidence levels). Because of the centrality of the role of error probabilities, NP statistics is an example of a broader category that I call *error probability statistics* or just *error statistics*.

Measures of Fit vs. Fit + Error Probabilities

A key feature distinguishing error-statistical accounts from evidential-relation logics (whether Bayesian, likelihoodist, hypothetico-deductive, or other) is that the former requires not just a measure of how well data "fit" hypotheses but also a calculation of the error probabilities associated with the fit. If you report your E-R measure of fit, the error statistician still needs to know how often such a good fit (or one even better) would arise even if the hypothesis in question were false—an error probability. For a NP error statistician, two pieces of evidence that fit a given hypothesis equally well may have very different evidential imports because of a difference in the er-

---

1. The so-called law of likelihood asserts that evidence *x* supports hypothesis *H* more than hypothesis *J* if the likelihood of *H* exceeds that of *J*. Leaders in developing a likelihood account (Hacking, Barnard, Birnbaum) have abandoned or greatly qualified their acceptance of this approach. Some likelihoodists also include prior probabilities (e.g., Edwards). Current-day likelihoodist accounts are discussed in Berger and Wolpert (1988) and Royall (1997, 2004).

ror probabilities of the procedures from which each data set was generated. For Bayesians, by contrast, because such a distinction is based on considering outcomes other than the one actually observed, it leads to "incoherence." That is because Bayesians, and also likelihoodists, accept what is called the *likelihood principle*.[2] As Edwards, Lindman, and Savage (1963, 238) put it, "Those who do not accept the likelihood principle believe that the probabilities of sequences that might have occurred, but did not, somehow affect the import of the sequence that did occur." We error statisticians do indeed: error probability considerations always refer to outcomes other than the ones observed, and such considerations are at the foundation for scrutinizing evidence. For a full discussion, see Mayo and Kruse (2001).

This distinction about the role of error probabilities is closely linked to a difference in attitude about the importance, in reasoning from data, of considering how the data were generated. E. S. Pearson puts it plainly,

> *We were regarding the ideal statistical procedure as one in which preliminary planning and subsequent interpretation were closely linked together—formed part of a single whole.* It was in this connexion that integrals over regions of the sample space [error probabilities] were required. *Certainly, we were much less interested in dealing with situations where the data are thrown at the statistician and he is asked to draw a conclusion. I have the impression that there is here a point which is often overlooked.* (Pearson, 1966, 277–278; emphasis added)

I have the impression that Pearson is correct. The main focus of philosophical discussions is on what rival approaches tell one to do once "data are thrown at the statistician and he is asked to draw a conclusion"; for example, accept or reject for a NP test or compute a posterior probability for a Bayesian. Howson and Urbach have said,

> The Bayesian theory of support is a theory of how the *acceptance as true of some evidential statement* affects your belief in some hypothesis. How you came to accept the truth of the evidence, and whether you are correct in accepting it as true, are matters that, from the point of view of the theory, are simply irrelevant, (Howson and Urbach, 1993, 419; emphasis added)

---

2. Following Edwards, Lindman, and Savage (1963), the likelihood principle may be stated by considering two experiments with the same set of hypotheses H_1 up to H_n. If D is an outcome from the first experiment and D′ from the second, then D and D′ are evidentially equivalent when $P(D'; H_i) = P(D; H_i)$ for each i.

The presumption that philosophical accounts of hypothesis appraisal begin their work with some statement of evidence *e* as given or as accepted is a key feature of logics of evidential relationship. This holdover from logical empiricist philosophies—where statements of evidence were regarded as relatively unproblematic "givens"—is a central reason that philosophers have failed to provide accounts of evidence that are relevant for the understanding and elucidating of actual scientific practice (Mayo, 2003a).

What We Really Need in a Philosophy of Evidence

In practice, where data are inexact, noisy, and incomplete, we need an account to grapple with the problems in determining if we even have evidence for a hypothesis to begin with. Thus, an adequate account must not begin with given statements of evidence but should include questions about how to generate, model, use, or discard data, and criteria for evaluating data. Even where there are no problems with the data, we need an account that makes sense of the fact that there is often disagreement and controversy as to whether data provide evidence for (or against) a claim or hypothesis of interest. An account that spews out an assessment of evidential support whenever we input statements of evidence and hypotheses does not do that. Settling or at least making progress with disputes over evidence requires appealing to considerations that are disregarded in Bayesian and other E-R accounts: considerations of reliability understood as error probabilities. Two pieces of evidence that would equally well warrant a given hypothesis according to logical measures of evidential relationship may in practice be regarded as differing greatly in their evidential value because of differences in how reliably each was produced. More specifically, as I see it, scientists seek to scrutinize whether the overall experiment from which the data arose was a reliable probe of the ways we could be mistaken in taking *x* as evidence for (or against) *H*. Scientists seem willing to forgo grand and unified schemes for relating their beliefs in exchange for a hodgepodge of methods that offer some protection against being misled by their beliefs.

This does not mean we have to give up saying anything systematic and general—as many philosophers seem to think. The hodgepodge of methods give way to rather neat statistical strategies of generating and interpreting data, and the place to look for organizing and structuring such activities, I maintain, is the host of methods and models from standard error statistics, error analysis, experimental design, and cognate methods.

Despite the increasingly widespread use of error-statistical methods in science, however, they have been the subjects of continuing philosophical controversy. Given the new emphasis philosophers of science have placed on taking their cues from actual scientific practice, it is disappointing to find them either ignoring or taking sides against error-statistical methods rather than trying to resolve these controversies, or at least helping to clarify the reasoning underlying probabilistic procedures that are so widely used to assess evidence in science. Increased availability of sophisticated statistical software has made the disputes in philosophy of statistics more pressing than ever; they are at the heart of several scientific and policy controversies—and philosophers can and should play a role in clarifying if not resolving them.

CRITICISMS OF NP STATISTICS AND THEIR SOURCES

While the confusion and controversy about NP tests have generated an immense literature, the traditional criticisms run to type. (One may identify three generations of philosophy of statistics disputes. See Mayo and Spanos, 2004.) They may be seen to originate from two premises, both false. The first deals with assumptions about what NP methods can provide, the second with assumptions about "what we really want" from a theory of statistical evidence and inference. The first set of assumptions, about what NP methods can give us, is based on what may be called the *behavioral-decision* interpretation or model of NP tests. This model regards NP tests as providing mechanical rules or "recipes" for deciding how to "act" so as to ensure that one will not behave "erroneously" too often in the long run of experience. The second assumption, about "what we really want," reflects the image of inductive inference as providing an E-R logic. It is assumed, in other words, that inductive inference must take the form of providing a final quantitative measure of the absolute or relative support or probability of hypotheses given data. But the only probabilistic quantities the NP methods provide are error probabilities of procedures. Unsurprisingly, as critics show, if error probabilities (e.g., significance levels) are interpreted as probabilities of hypotheses, misleading and contradictory conclusions are easy to generate. Such demonstrations are not really criticisms but rather are flagrant misinterpretations of NP methods.

Not all criticisms can be dismissed so readily. Other criticisms proceed by articulating criteria for testing or for appraising evidence thought to be intuitively plausible and then showing that tests that are good according to NP error criteria may not be good according to these newly defined criteria or principles. Such criticisms, although they can be subtle and persuasive, are guilty of begging the question against NP principles[3] (see Mayo, 1985a).

This type of criticism takes the following form: NP procedures cannot be used to interpret data as evidence because it leads to different results even when the evidence is the same.[4] This sounds like a devastating criticism until one looks more closely at the concept of "same evidence" that is being assumed. What the charge really amounts to is this: NP methods can lead to distinct interpretations of evidence that would not be distinguished on the basis of measures of evidential relationship that regard error probabilities as irrelevant. But NP procedures do regard differences in the error probabilities from which data arose as relevant for interpreting the evidential import of data. Hence, what at first blush appears to be a damning criticism of NP theory turns out to be guilty of begging the question against a fundamental NP principle.

But several critics seem to think that one cannot live within the strictures of NP principles, and several have argued that researchers invariably use NP tests in an inconsistent fashion that reflects "a critical defect in current theories of statistics" (Royall, 1997, xi). This common criticism has even been given a colorful Freudian twist as discussed by Gerd Gigerenzer (1993). According to Gigerenzer, researchers who use NP tests actually use a hybrid logic mixing ideas from NP tests, Fisherian tests, and Bayesian analysis—leading to a kind of Freudian "bad faith." In Gigerenzer's "Freudian metaphor," "the NP logic of hypothesis testing functions as the Superego of the hybrid logic" (Gigerenzer, 1993, 324). It insists upon prespecifying significance levels, power, and precise alternative hypotheses and then accepting or rejecting the null hypotheses according to whether outcomes fall in rejection regions of suitable NP tests. The Ego—the fellow who gets things done—is more of a Fisherian tester, we are told. The Fisherian Ego does not use prespecified significance levels but reports attained P-values after the experiment. He likewise ignores power and the alternative hypotheses altogether—despite their being an integral part of the NP methodology. The Ego violates the NP canon further by allowing himself epistemic statements

3. This is so, at any rate, for all criticisms of which I am aware. There are also criticisms (generally raised by Bayesians) that argue that NP methods fail to promote their own goal of low error probabilities. I discuss this in my rejoinder.

4. Richard Royall states it plainly: "Neyman-Pearson theory leads to different results in two situations where the evidence is the same, and in applications where the purpose of the statistical analysis is to represent and interpret the data as evidence, this is unacceptable" (Royall, 1997, 47).

that go beyond those for which the NP test (regarded as a mechanical decision rule) strictly permits; and in so doing, the Ego "is left with feelings of guilt and shame having violated the rules" (325). Finally, we get to the Bayesian Id. "Censored by both the frequentist Superego and the pragmatic Ego are statements about probabilities of hypotheses given data. These form the Bayesian Id of the hybrid logic," says Gigerenzer (325). The metaphor, he thinks, explains "the anxiety and guilt, the compulsive and ritualistic behavior, and the dogmatic blindness" associated with the use of NP tests in science. (Perhaps some might replace the Bayesian Id with the likelihoodist Id.)

Doubtless Gigerenzer is correct in claiming that statistics texts erroneously omit these philosophical and historical differences between NP tests, Fisherian tests, and Bayesian methods, and doubtless many have been taught statistics badly, but we need not turn this into an inevitability. What we need (and what I have elsewhere tried to provide) is an interpretation of statistical tests that shows how they can be seen to produce a genuine account of evidence without misinterpreting error probabilities and without being used as mechanical "cookbook" methods for outputting "acts" associated with "accept H" or "reject H."

BEYOND THE BEHAVIORAL-DECISION MODEL OF NP TESTS

The place to begin is with the first premise upon which these assaults on NP tests are based—the assumed behavioral decision model of tests. Granted, the proof by Neyman and Pearson of the existence of "best" tests, and Wald's later extension of NP theory into a more generalized decision model, encouraged the view that tests (particularly best tests) provide the scientist with a kind of *automatic rule* for deciding to accept or reject hypotheses. Nevertheless, the whole concept of "inductive behavior," a term coined by Neyman, was just Neyman's way of distinguishing these tests from the concept of "inductive inference" as understood by Bayesians as well as by Fisherians. Even that arch opponent of Neyman, subjectivist Bruno de Finetti, remarked that the expression "inductive behavior . . . that was for Neyman simply a slogan underlining and explaining the difference between his, the Bayesian and the Fisherian formulations" became, with Abraham Wald's work, "something much more substantial" (de Finetti, 1972, 176). De Finetti called this "the involuntarily destructive aspect of Wald's work" (ibid.). Pearson made it clear that he, at least, had always intended an "evidential" and not a behavioral interpretation of tests. To the statistician Allan Birnbaum, for instance, Pearson wrote in 1974:

> I think you will pick up here and there in my own papers signs of evidentiality, and you can say now that we or I should have stated clearly the difference between the *behavioral and evidential* interpretations. Certainly we have suffered since in the way the people have concentrated (*to an absurd extent often*) on behavioral interpretations. (Birnbaum, 1977, 33; emphasis added)

Pearson explicitly distanced himself from the inductive behavior concept, telling R. A. Fisher that inductive behavior "is Professor Neyman's field rather than mine" (Pearson, 1955, 207), and declaring that "no responsible statistician . . . would follow an automatic probability rule" (Pearson 1947, 192).

But even Neyman used tests to appraise evidence and reach conclusions; he never intended the result of a statistical test to be taken as a decision about a substantive scientific hypothesis. Addressing this point, Neyman declared, "I do not hesitate to use the words 'decision' or 'conclusion' every time they come handy" (Neyman, 1976, 750). He illustrates by discussing a test on a null hypotheses of no difference in means: "As a result of the tests we applied, we decided to act on the assumption (or concluded) that the two groups are not random samples from the same population" (ibid.). What needs to be understood is that Neyman regards such a decision as one of how to appraise the data for the purpose of subsequent inferences in an ongoing inquiry.[5] See also Neyman (1955). The tendency of philosophers to seek final assessments of hypotheses and theories leads them to overlook the fact that in practice evidential appraisal is part of an ongoing inquiry, and that this calls for tools that help guide us in what to try next and how to communicate data in forms that will allow others to critique and extend results. Viewed as part of a piecemeal tool for ongoing inquiry, a very different picture of the value of NP tools emerges. (See Mayo, 1990, 1996.)

Nevertheless, my sympathies are much closer to Pearson's, who would have preferred that the tests be articulated in the evidential manner in which he claims they were first intended—namely as "a means of learning" (Pearson, 1955, 206). Still, the behavioral-decision concepts, it seems to me, can be regarded as simply a way to characterize the key formal features of

---

5. Here Neyman concludes the data do not warrant the assumption that the two groups are random samples from the same population—a conclusion that then serves as input for the primary inquiry at hand.

NP tools, and understanding these features enables tests to perform the non-behavioral tasks to which tests may be put to learn from evidence in science. The task remains to explicate the nonbehavioral or evidential construal of these methods. While the reinterpretation I propose differs sufficiently from the standard NP model to warrant some new name, it retains the centerpiece of these methods: the fundamental use of error probabilities—hence the term *error statistics*.

A FRAMEWORK OF INQUIRY

To get at the use of these methods in learning from evidence in science, I propose that experimental inference be understood within a framework of inquiry. You cannot just throw some "evidence" at the error statistician and expect an informative answer to the question of what hypothesis it warrants. A framework of inquiry incorporates methods of experimental design, data generation, modeling, and testing. For each experimental inquiry we can delineate three types of models: *models of primary scientific hypotheses, models of data,* and *models of experiment.* Figure 4.1 gives a schematic representation.

A substantive scientific inquiry is to be broken down into one or more local hypotheses that make up the *primary questions* or *primary problems* of distinct inquiries. Typically, primary problems take the form of estimating quantities of a model or theory, or of testing hypothesized values of these quantities. The *experimental models* serve as the key linkage models connecting the primary model to the data—links that require not the raw data itself but appropriately *modeled data.*

In the error-statistical account, formal statistical methods relate to experimental hypotheses, hypotheses framed in the experimental model of a given inquiry. Relating inferences about experimental hypotheses to primary scientific claims is a distinct step except in special cases. Yet a third step is called for to link raw data to data models—the real material of experimental inference. The indirect and piecemeal nature of our use of sta-

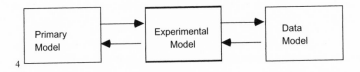

FIGURE 4.1 Models of Experimental Inquiry.

tistical methods, far from introducing an undesirable complexity into our approach, is what enables it to serve as an account of inference that is relevant to scientific practice.[6]

SEVERE TESTS AND ARGUING FROM ERROR

Despite the complexity, there is an overall logic of experimental inference that emerges: Data x *indicate* the correctness of hypothesis H, to the extent that H passes a *severe test* with x. Hypothesis H passes a severe test with x if (a) x fits H (for a suitable notion of fit)[7] and (b) the test procedure had a high probability of producing a result that accords less well with H than x does if H were false or incorrect. To infer that H is indicated by the data does not mean a high degree of probability is assigned to H—no such probabilities are wanted or needed in the error-statistical account. That data indicate that hypothesis H means that the data indicate or provide reliable evidence that hypothesis H is correct—good grounds that H correctly describes some aspect of an experimental process. What aspect, of course, depends on the particular hypothesis H in question. Generally, several checks of a given indication of H (e.g., checks of the experimental assumptions) are required. One can, if one likes, construe the correctness of H in terms of H being reliable, provided one is careful in the latter's interpretation. Learning that H is reliable would mean learning that what H says about certain experimental results will or would often be close to the results that would actually be produced—that H will or would often succeed in specified experimental applications.[8] This is *experimental knowledge.*

The reasoning used in arriving at this knowledge follows an informal pattern of argument that I call *an argument from error* or *learning from error.* The overarching structure of the argument is guided by the following thesis:

6. For much further discussion and development of the error-statistical account, see Mayo (1996).

7. I leave this notion open in order to accommodate different inference accounts. In some accounts, x fits H to the extent that $P(x; H)$ is high. Because $P(x; H)$ is often small, even if H is true, x might be said to fit H if x is more probable under H than under all (or certain specified) alternatives to H.

8. What is learned may be formally construed in terms of experimental distributions—assertions about what outcomes would be expected, and how often, if certain experiments were to be carried out. Informally, one can see this as corresponding to learning that data do or do not license ruling out certain errors and mistakes.

It is learned that an error is absent [present] when (and only to the extent that) a procedure of inquiry (which may include several tests) with a high probability of [not] detecting the error if and only if it is present [absent], nevertheless does not [does] detect the error.[9]

Not detecting the error means it produces a result (or set of results) that is in accordance with the absence of the error. Such a procedure of inquiry may be called a reliable (or highly severe) error probe. According to the above thesis, we can argue that an error is absent if it fails to be detected by a highly reliable error probe. (A corresponding assertion as to the error's absence may be said to have passed a severe test.)

It is important to stress that my notion of severity always attaches to a particular hypothesis passed or a particular inference reached. How severe is this test? is not a fully specified question until it becomes, how severe would a test procedure be, if it passed such-and-such hypothesis on the basis of such-and-such data? A procedure may be highly severe for arriving at one type of hypotheses and not another.[10] Also, severity is not equal to the formal notion of *power:* the two would give identical measures only if one was testing a (continuous) statistical hypothesis H and obtained an outcome that just misses the cutoff for rejecting H.[11]

THE ROLES OF STATISTICAL MODELS AND METHODS

Experimental inquiry is a matter of building up, correcting, and filling out the models needed for substantiating severe tests in a step-by-step manner.

9. In terms of a hypothesis H, the argument from error may be construed as follows: Evidence in accordance with hypothesis H indicates the correctness of H when (and only to the extent that) the evidence results from a procedure that with high probability would have produced a result more discordant from H, were H incorrect.

10. To illustrate, consider a diagnostic tool with an extremely high chance of detecting a disease. Finding no disease (a clean bill of health) may be seen as passing hypothesis H_1: no disease is present. If H_1 passes with so sensitive a probe, then H_1 passes a severe test. However, the probe may be so sensitive as to have a high probability of declaring the presence of the disease even if no disease exists. Declaring the presence of the disease may be seen as passing hypothesis H_2: the disease is present. If H_2 passes a test with such a highly sensitive probe, then H_2 has not passed a severe test. That is because there is a very low probability of not passing H_2 (not declaring the presence of the disease) even when H_2 is false (and the disease is absent). The severity of the test that hypothesis H_2 passes is very low. This must not be confused with a posterior probability assignment to H_2. See Mayo, 2004, and chapter 4.3 in this volume.

11. The severity with which "accept H" passes would, in this case, be equal to (or nearly equal to) the power of the test.

Error-statistical ideas and tools enter into this picture of experimental inference in a number of ways, all of which are organized around the three chief models of inquiry. They fulfill three main tasks by providing:

1. canonical models of low-level questions with associated tests and data-modeling techniques;
2. tests and estimation methods that allow control of error probabilities;
3. techniques of data generation and modeling along with tests for checking whether the assumptions of data models are met.

The three tasks just listed relate to the primary models, the models of experiment, and of data, respectively. In each case, I readily admit that the functions served by the statistical tools do not fall out directly from the mathematical framework found in statistical textbooks. There are important gaps that need to be filled in by the methodologist and philosopher of experiment.

Task (i). This is especially so for the first task, providing models of low-level questions. Explicating this task, as I see it, enjoins us to ask how scientists break down a substantive inquiry into piecemeal questions such that they can be reliably probed by statistical tests or analogs to those tests. The questions, I propose, may be seen to refer to standard types of errors. Four such standard or canonical types of errors are:

1. mistaking chance effects or spurious correlations for genuine correlations or regularities;
2. mistakes about a quantity or value of a parameter;
3. mistakes about a causal factor;
4. mistakes about the assumptions of experimental data.

Statistical models are relevant because they model patterns of irregularity that are useful for studying these errors.

Task (ii). The second task centers on what is typically regarded as statistical inference proper; namely, specifying and carrying out statistical tests (and the associated estimation procedures) or informal analogs to these tests. It is important to emphasize, however, that the error-statistical program brings with it reinterpretations of NP tests as well as extensions of their logic into informal arguments from error. The criteria for selecting tests depart from those found in classic behavioristic models of testing. One seeks not the "best" test according to the low error probability criteria alone, but rather sufficiently informative tests.

What directs the choice of a test reflects not merely a concern to control a test's error probabilities but also a desire to ensure that a test is based on a plausible "fit" or *distance measure* (between data and hypotheses). The recognition of these twin concerns allows answering a number of criticisms of NP tests. While I admit this takes us beyond the usual formulations of NP tests, there is plenty of evidence that Pearson, at least, intended such features in the original construction of NP tests:

After setting up the *test* (or null) *hypothesis,* and the *alternative hypotheses* against which "we wish the test to have maximum discriminating power" (Pearson, 1947, 173), Pearson defines three steps in specifying tests:

> Step 1. We must specify the experimental probability set [the sample space, discussed in chapter 1 of this volume)] the set of results which could follow on repeated application of the random process used in the collection of the data . . .
>
> Step 2. We then divide this set [of possible results] by a system of ordered boundaries . . . such that as we pass across one boundary and proceed to the next, we come to a class of results which makes us more and more inclined, on the information available, to reject the hypothesis tested in favour of alternatives which differ from it by increasing amounts. (Pearson, 1947, 173)

Results make us "more and more inclined" to reject *H* as they get further away from the results expected under *H;* that is, as the results become more probable under the assumption that some alternative *J* is true. This suggests that one plausible measure of inclination is the likelihood of *H*—the probability of a result *e* given *H.* We are "more inclined" toward *J* as against *H* to the extent that *J* is more likely than *H* given *e.*

NP theory requires a third step—ascertaining the error probability associated with each measure of inclination (each "contour level"):

> Step 3. We then, if possible, associate with each contour level the chance that, if [*H*] is true, a result will occur in random sampling lying beyond that level (Pearson, 1947, 173).[12]

For example, step 2 might give us the likelihood or the ratio of likelihoods of hypotheses given evidence, and at step 3 we may calculate the probabil-

---

12. Where this is not achievable (e.g., certain tests with discrete probability distributions) the test can associate with each contour an upper limit to this error probability.

ity of getting so high a likelihood ratio, say in favor of *J* against hypothesis *H*, when in fact *H* is true.

Pearson explains that in the original test model, step 2 (using likelihood ratios) did precede step 3, and that is why he numbers them this way. Only later, he explains, did formulations of the NP model begin first by fixing the error value for step 3 and then determining the associated critical bounds for the rejection region.[13]

If the rationale were solely error probabilities in the long run, the need to first deliberate over an appropriate choice of measuring distance at step 2 would indeed drop out, as in the behavioral-decision formulation. In the behavioral model, having set up the hypotheses and sample space (step 1), there is a jump to step 3, fixing the error probabilities. From step 3 we can calculate how the test, selected for its error probabilities, is dividing the possible outcomes. But this is different from having first deliberated at step 2 as to which outcomes are "further from" or "closer to" *H* in some sense, and thereby should incline us more or less to reject *H*. The resulting test, while having low error probabilities, may fail to ensure that the test has an increasing chance of rejecting *H* the more the actual situation deviates from the one that *H* hypothesizes. The resulting test may even be irrelevant to the hypothesis of interest. The reason that many counterintuitive tests appear to be licensed by NP methods, e.g., certain mixed tests,[14] is that tests are couched in the behavioral framework in which the task Pearson intended for step 2 is absent.[15]

Task (iii). The third task involves pretrial planning to generate data likely to justify needed assumptions, and after-trial checking to test if the assumptions are satisfactorily met. Often this is accomplished by deliberately

13. Pearson warns that "although the mathematical procedure may put Step 3 before 2, we cannot put this into operation before we have decided, under Step 2, on the guiding principle to be used in choosing the contour system. That is why I have numbered the steps in this order" (Pearson, 1947, 173).

14. In a mixed test certain outcomes instruct one to apply a given chance mechanism and accept or reject *H* according to the result. Because long-run error rates may be improved using some mixed tests, it is hard to see how a strict follower of NP theory (where the lower the error probabilities the better the test) can inveigh against them. This is not the case for one who rejects the behavioral model of NP tests as do Pearson and "error statisticians." An error statistician could rule out the problematic mixed tests as being at odds with the aim of using the data to learn about the causal mechanism operating in a given experiment.

15. Exceptions exist, e.g., the exposition of tests in Cox and Hinkley (1974) and in Kempthorne and Folks (1971), in which test statistics are explicitly framed in terms of distance measures.

introducing statistical considerations (e.g., via randomization). We noted earlier how Pearson linked the importance of error probabilities to the fact that the NP account regards the "preliminary planning and subsequent interpretation" as closely linked. By considering ahead of time a test's probabilities of detecting discrepancies of interest, one can avoid carrying out a study with little or no chance of teaching one what one wants to learn; for example, one can determine ahead of time how large a sample would need to be in a certain test to have a reasonably high chance (power) of rejecting H when in fact some alternative J is true. Few dispute this (before-trial) function of error probabilities, it seems.

But there is a second connection between error probabilities and preliminary planning, and this explains their relevance even after the data are in hand. It is based on the supposition that in order to interpret correctly the bearing of data on hypotheses one must know the procedure by which the data got there; and it is based on the idea that a procedure's error probabilities provide this information. It is on this after-trial function that I want to focus; for this is what non-error-statistical approaches (those accepting the likelihood principle) deny.[16]

Although there are several (after-trial) uses of error probabilities, they may all be traced to the fact that error probabilities are properties of the procedure that generated the experimental result. This permits error probability information to be used as a key by which available data open up answers to questions about the process that produced them. Error probability information informs whether given claims are or are not mistaken descriptions of the data-generating procedure. It teaches us how typical (or how rare) given results would be under varying hypotheses about the experimental process. A central use of this information is to determine what it would be like if various hypotheses about the underlying experimental process misdescribed a specific experimental process. (It teaches us what it would be like were it a mistake to suppose a given effect were nonsystematic or due to chance, what it would be like were it a mistake to attribute the effect to a given factor, what it would be like were it a mistake to hold that a given quantity or parameter had a certain value, and what it would be like were it

16. Some (e.g., Hacking, 1965) have suggested that error probabilities, while acceptable for before-trial planning, should be replaced with other measures (e.g., likelihoods) after the trial. Pearson (1966) takes up and rejects this proposal, raised by Barnard in 1950, reasoning that "if the planning . . . is based on a study of the power function of a test and then, having obtained our results, we do not follow the first rule but another, based on likelihoods, what is the meaning of the planning?" (Pearson, 1966, 228).

a mistake to suppose experimental assumptions are satisfactorily met.) Statistical tests can then be designed so as to magnify the differences between what it would be like under various hypotheses.

The need for such information develops out of an awareness that we can easily be misled if we look only at how well data fit hypotheses, ignoring sampling rules (e.g., rules for when to stop collecting data or *stopping rules*) and failing to take into account whether hypotheses were deliberately selected for testing because they fit the data (hypothesis specification). Although these features of sampling and specifying hypotheses do not alter measures of fit such as likelihood ratios, they do alter error probabilities. So requiring a report of error probabilities alerts us when an altered appraisal of the evidence is required. That is why fitting, even being the best-fitting hypothesis, is not enough for the error statistician. Pearson and Neyman (1930) explain that

> if we accept the criterion suggested by the method of likelihood it is still necessary to determine its sampling distribution in order to control the error involved in rejecting a true hypothesis, because a knowledge of l [the likelihood ratio] alone is not adequate to insure control of this error. (Pearson and Neyman, 1930, 106) [17]

Examples of how likelihood ratios alone fail to control error probabilities are well known. One type of case is where hypothesis *H* is the common null hypothesis that asserts an observed correlation *e* results from mere "chance." (*H* asserts, in effect, that it would be an error to regard *e* as the result of a genuine rather than a chance correlation.) Suppose that *P(e; H)* is small. Since we can always find an alternative hypothesis *J* that perfectly fits the data (so the likelihood of *J* is maximal), we can always get a high likelihood ratio in favor of *J* against *H*. Because one could always do this even when *H* is true, the probability of finding support for *J* against *H erroneously* is 1. (One always denies the error asserted in *H*, even though, in fact, the error is occurring.) It is for this reason that many who were sympathetic to likelihood testing, including leaders of that approach, e.g., Hacking (1972), Barnard (1972), and Birnbaum (1969), gave it up.

---

17. Let *l* be the ratio of the likelihood of *H* and an alternative hypothesis *J* on given data x. We cannot say that because *l* is a small value that "we should be justified in rejecting the hypothesis" *H*, because "in order to fix a limit between 'small' and 'large' values of l we must know how often such values appear when we deal with a true hypothesis. That is to say we must have knowledge of the chance of obtaining [so small a likelihood ratio] in the case where the hypothesis tested [*H*] is true" (Pearson and Neyman, 1930, 106).

Thus, it seems that the likelihood concept cannot be construed so as to allow useful appraisal, and thereby possible control, of probabilities of erroneous interpretation (Birnbaum, 1969, 128). Bayesians face the analogous problem, even with the addition of prior probabilities, as shown by Armitage (1962). (For a discussion, see Mayo, 1996, chap. 10; see also Mayo and Kruse, 2001; Mayo, 2003b.)

A test's error probabilities do matter, not because of an interest in having a good track record, but because of their role in informing us of the process that produced the data in front of us. We use them as standards to weed out erroneous interpretations of data. Those who insist that a genuine evidential appraisal is one of the E-R measures alone are demanding we renounce this use of error probabilities. This is something that scientists interested in interpreting data objectively are or should be unwilling to do.

CONCLUDING REMARKS

To summarize, the key difference between standard NP methods and E-R logics is that the former desires and is able to control error probabilities, whereas the latter is not. Criticisms of NP tests that are not merely misinterpretations stem from supposing that (1) the reason error probabilities matter in NP tests is solely from an interest in a low probability of erroneous "acts" in the long run and (2) an appropriate theory of evidence must provide an E-R measure. I have argued against both of these assumptions.

Error probabilities are used to assess the severity with which primary hypotheses of interest are tested by means of statistical evidence. By reporting on error probabilities (together with outcomes and sample sizes), the error statistician is providing information by which we can assess (at least approximately) whether data provide reliable evidence for experimental claims. Knowledge of the error-probabilistic properties of a test procedure—however that test was chosen—is the basis for criticizing a given test (e.g., as too sensitive or not sensitive enough) and for finding certain interpretations of evidence unwarranted. Nothing analogous can be said for criticizing subjective degrees of belief. Systematic rules may be specified to guide the appropriate interpretation of particular statistical results. Two such "meta-statistical" rules that I propose are directed at interpreting rejections (rule of rejection) and acceptances (rule of acceptance). (See in Mayo, 1996, chap. 11. The reader will find considerable elaboration and updates of these ideas in chapter 4.3, my rejoinder to the commentaries on this chapter.)

The error-statistical account licenses claims about hypotheses that are and are not indicated by tests without assigning quantitative measures of support or probability to those hypotheses. To those E-R theorists who insist that every uncertain inference have a quantity attached, our position is that this insistence is seriously at odds with the kind of inferences made every day, in science and in our daily lives. There is no assignment of probabilities (or other E-R measures) to the claims themselves when we say such things as: the evidence is a good (or a poor) indication that light passing near the sun is deflected, that treatment X prolongs the lives of AIDS patients, that certain dinosaurs were warm-blooded, that my four-year-old can read, that metabolism slows down when one ingests fewer calories, or any of the other claims that we daily substantiate from evidence. What there is, instead, are arguments that a set of errors have (or have not) been well ruled out by appropriately severe tests.

4.1	Commentary

Earl D. McCoy

Scientists need philosophers to show them how science should operate. A deeper understanding of how evidence can be used to generate knowledge ultimately can provide scientists with more effective techniques for doing so. I would guess, however, that most scientists would be hard pressed to suggest how philosophical analysis actually has contributed to the incorporation of more effective techniques into their particular field of study. I also would guess that most scientists believe that philosophical analysis is the philosophers' game, with no real relevance to scientists' day-to-day practices. Of course, philosophers would retort that this position is borne out of a naïve viewpoint, and they are right. But if my guesses are correct, then what is to blame for these attitudes among most scientists? To be provocative, I will suggest that at least part of the blame must rest with philosophers. They sometimes seem to take positions on important issues that, even if philosophically consistent and well reasoned, appear to scientists to be out of touch with reality. I will suggest further, therefore, that just as scientists need philosophers, philosophers need scientists, to remind them of how science actually does operate. Deborah Mayo's chapter reinforces these two suggestions admirably.

Mayo first sets up the Bayesian model as the leading example of a logic of evidential relationship. She notes, however, the increasingly widespread use of error-statistical methods in science. She then proceeds to set out a series of arguments to convince her readers to adopt the error-statistical (Neyman-Pearson, frequentist) model instead of the Bayesian (including the special-case maximum-likelihood) model. The principal advantage of the error-statistical model, she thinks, is that calculation of error probabilities is at its core. Indeed, whether intrinsic and supposedly unbiased error probabilities or extrinsic and often subjective probabilities based on experiences and beliefs are more relevant to the practice of science is the root of the debate between advocates of the two kinds of models. Mayo's arguments in favor of the error-statistical model are compelling. But two points strike me as strange. First, most scientists I know would not consider the Bayesian model to be the leading example of a logic of evidential relationship at all, but rather the error-statistical model. Second, while Mayo is encouraging a move away from the Bayesian model and toward the error-statistical model, increasing numbers of scientists are paying much closer attention to the Bayesian perspective on statistical analysis (see Malakoff, 1999). In my own discipline, ecology, several recent major examples of this trend can be noted (see, e.g., Hilborn and Mangel, 1997). What could lead to two such distinct, even contradictory, impressions of current scientific practice?

The answer to this question is not simple and may hinge on even deeper aspects of scientific inquiry. Mayo's arguments in favor of the error-statistical model appear to assume the central importance of the classic reductionist approach to scientific understanding. Many practicing scientists, on the other hand, are questioning the relevance of this approach to their research. Scientists who adopt this viewpoint often are those who need to incorporate social and environmental complexities into their analyses, rather than eliminate them; or who need to provide answers to questions that do not lend themselves readily to hypothesis testing; or who need to provide policy makers with realistic probabilities of uncommon but high-consequence events. In ecology, for example, practicing scientists often have suggested that the long time scales, lack of replication, and lack of controls inherent in many ecological studies prevent effective use of the classic reductionist approach. Many scientists, therefore, may view Mayo's arguments as irrelevant, even while agreeing with them in principle. In the best of circumstances, these scientists might advocate the error-statistical model. In the increasingly pervasive climate of uncertainty and urgency present in a variety of disciplines, however, the same scientists might see the Bayesian method (and other "unusual" methods, such as risk analysis and exploratory data analysis) as potentially useful.

Given this philosophical divide, the rational course may be to keep both the baby and the bath water. For one thing, the Bayesian model is not going to challenge the preeminence of the error-statistical model among scientists in the foreseeable future, because frequentist statistics have an established track record and Bayesian statistics do not. Furthermore, the two models may have different purposes: the error-statistical model may be the better approach for "basic" science (if such a thing really exists), but it may not be for "applied" science. The error-statistical model may not be the better approach for applied science because time may not permit us to ignore prior knowledge in solving applied problems and because the successful solution of many applied problems may not even require the development of full knowledge from the evidence at hand. I suggest that we simply admit that no way of knowing is perfect, keep both models, and use each of them appropriately and cautiously.

4.2	Commentary

George Casella

The thesis of Professor Mayo is that "error statistics," an approach that both assesses fit and attaches frequentist error probabilities to inferences, is the proper way for a statistician to produce an inference. Although I basically agree with this view, the framework that Mayo outlines may be too rigid to be applied in problems that the practicing statistician (data analyst, data miner!) faces today. However, we do need to apply these principles, but in a way that can actually be used in today's practice.

Mayo makes an eloquent argument in favor of the Neyman-Pearson (error statistics) school over the Bayesian (evidential relationship) school. But both approaches have their problems, and neither one is appropriate for all experiments. The Neyman-Pearson approach results in inferences that are valid against a long-run, objective assessment. In theory this is quite desirable, but in practice embedding an experiment in long-run trials doesn't always make sense. (For example, I recall an animal scientist whose data consisted of records of all dairy cattle in a particular country. As he noted, the experiment is not likely to be repeated.) Moreover, the classical Neyman-

This research was supported by NSF Grant No. DMS 9971586.

Pearson conclusion (accept or reject) is almost universally ignored by subject matter researchers in favor reporting the P-value.

The Bayesian (hierarchical) approach to modeling is well suited for complex problems, but the associated inference is consistently mired in its lack of objectivity. Although there are (seemingly countless) attempts at "objective" Bayes inference, even Bayesians cannot seem to agree on this, not to mention subject matter scientists.

But what do practicing statisticians need? How to proceed? Almost any reasonably complex model is a mixture of empirical and subjective components. When evaluating such models, the statistician must consider both frequency and Bayesian measures. Moreover, it may be the case that the empirical and subjective pieces are so intertwined as not to allow separate evaluations (for example, in empirical Bayes models the prior parameters are estimated from the data). Is such a model best evaluated by one or the other camp, or by a mixture? Mayo outlines a "framework of inquiry" that is compelling in its ideas, frustrating in its vagueness, and unworkable in its rigidity. However, the advice here is too good to disregard. Rather, it needs to be relaxed and expanded to better reflect what is happening in practice. Let us consider the three tasks of the framework of inquiry.

Task (i) instructs us to break large questions into small ones that can better be answered by data models. However, keep in mind that such low-level models are usually constructed after the data are seen, when helping an experimenter to make sense of the data. Thus, inferences will not be made using data collected from a simple model, but will be about a low-level model in the context of the entire experiment.

Task (ii) concerns the types of inferences to be made for the models of task (i). There is some good advice here, seemingly aimed at combating the accept/reject shortcomings of the Neyman-Pearson inference, but it is at a vague level and the possibility of its implementation is not clear. In particular, Mayo is describing an approach similar to Kiefer's conditional confidence (Kiefer, 1977). Despite a large amount of work on this topic (see, for example, Casella, 1988, or Goutis and Casella, 1995), it is not yet ready for prime time. Work by Berger, Brown, and Wolpert (1994) and Berger, Boukai, and Wang (1997) has promise, but is still confined to more stylized problems than one meets in practice.

Task (iii) seems to say, "do a power calculation first" and, after the data have been gathered, "check residuals." In general, this is very sound advice.

What puzzles me is that I cannot see why the three tasks cannot be accomplished within a Bayesian framework too (not instead). The low-level modeling of task (i) can be accomplished equally well within the Bayesian paradigm, and the power calculations of task (iii) are also straightforward for a Bayesian, who will usually agree to average over the sample space before the data are seen. This leaves us with how to assess error probabilities and fit. But this can be accomplished by looking at posterior probabilities and credible regions, along with a sensitivity analysis. If one accepts that the Bayes inference is based on a subjective premise, and the effect of the premise can be assessed through a sensitivity analysis, then the error assessment of task (ii) can be done equally well within the Bayesian framework. Then, using both Neyman-Pearson and Bayesian tools, we can apply these very sound principles to the type of complex models that are being used in practice.

4.3	Rejoinder

Deborah G. Mayo

I am grateful to professors McCoy and Casella for their interesting and enlightening comments on my paper, and grateful too for this chance to clear up some misunderstandings and lack of clarity as to what I am advocating. My response will discuss: (a) flawed philosophical assumptions about objectivity in science, (b) problems with what might be called the "conditional error probability (CEP)" movement, and (c) an illustration of the severity reinterpretation of NP tests.

THE "OLD" (LOGICAL POSITIVIST) IMAGE OF SCIENCE:
IT IS TIME TO REJECT THE FALSE DILEMMA

To begin with, I find the remarks of both these scientists to be premised on a false dilemma based on a discredited philosophy of science and of scientific objectivity, and in this first part of my comments, I will attempt to explain (1) how to escape the horns of their dilemma, and (2) why doing so is so important precisely for the practicing scientists with whom both claim to be concerned. The false dilemma stems from the supposition that either scientific inference is "reducible" to purely objective, value-free methods or else they are ineluctably subjective, relative, and more a matter of belief and opinion than of strictly empirical considerations. When this false dilemma is coupled with the assumption that the "error-statistical" philosophy of statistics goes hand in hand with the former ("reductionist") position, it is easy to see why McCoy and Casella argue from the impossibility of a value-free science to conclude that scientists must embrace a subjective-relativist position, as reflected, for instance, in subjective Bayesian statistics. For example, McCoy alleges that my "arguments in favor of the error-statistical model appear to assume . . . the classic reductionist approach" to science, in contrast to practicing scientists who, recognizing the "need to incorporate social and environmental complexities into their analyses" are led to use Bayesian methods, wherein such complexities may find cover under subjective degrees of belief. Casella too argues that since in scientific practice "it may be the case that the empirical and subjective pieces are so intertwined as not to allow separate evaluations," it follows that scientists need to use both "frequency and Bayesian measures." Both arguments are fallacious.

Granted, logical positivist philosophers had long encouraged the traditional or "old image" of a purely objective, value-free science that underwrites McCoy and Casella's remarks, but the philosophies of science and statistics I favor are based on denying both disjuncts that they appear to regard as exhaustive of our choices. As such, these charges have no relevance to my position. As I make clear in my paper, I deny the classic logical positivist position that a genuinely objective scientific inference must be reducible to a logic or algorithm applied to (unproblematically) given statements of data and hypotheses. At the same time I deny the assumption of many a postpositivist (led by Kuhn and others) that the untenability of any such algorithm or "logic of evidential-relationship" (E-R logic) leaves us with no choice but to embrace one of the various stripes of relativistic or subjectivistic positions, as found both in the context of general philosophy of knowledge (social-constructivism, psychologism, postmodernism, anarchism) and in philosophy of statistics (e.g., subjective Bayesianism).

We can readily agree with Casella that "any reasonably complex model is a mixture of empirical and subjective components" while adamantly denying his inference that therefore "the statistician must consider both frequency and Bayesian measures." It is a mistake, a serious one, to suppose that the empirical judgments required to arrive at and evaluate complex models are no more open to objective scrutiny than are subjective Bayesian degrees of belief. The former judgments, in my view, are not even captured by prior probabilities in hypotheses, and they are open to criticism in ways that subjective degrees of belief are not. For McCoy to imply, as he does, that the error-statistical model requires us "to ignore prior knowledge" is to misunderstand the crucial ways in which background knowledge must be used in each of the tasks of collecting, modeling, and interpreting data. Summaries of subjective degrees of belief may seem to offer a more formal way of importing information into inference, but this way fails to serve the error statistician's goals.

Moreover, from the recognition that background beliefs and values may enter at every stage of scientific inference, as I have long argued (e.g., Mayo, 1989, 1991, 1996), we can develop explicit methods for critically evaluating results of standard statistical tests. Such a metastatistical scrutiny enables us to discern what the data do and do not say about a phenomenon of interest, often in terms of the existence and direction of a discrepancy from a hypothesized value (I take this up below). I am unaware of an existing subjective Bayesian approach that achieves these ends or one that promotes the critical scrutiny needed to understand and adjudicate conflicting interpretations of evidence.

Once again, I could not agree more with Casella and McCoy in their demand that any adequate account be able to deal with what Casella calls the "problems that the practicing statistician (data analyst, data miner) faces today," and to grapple with the "lack of replication, and lack of controls inherent in many ecological studies" that McCoy refers to (on data mining, see Spanos, 2000). However, it is precisely because of the complexities and limitations in actual experimental practice that scientists are in need not of a way to quantify their subjective beliefs as Casella and McCoy recommend (much less to allow their opinions to color the inference as they do in a posterior degree-of-belief calculation), but rather of a way to check whether they are being *misled* by beliefs and biases, in scrutinizing both their own data and that of other researchers. An adequate account of scientific inference must provide a methodology that permits the consequences of these uncertainties and knowledge gaps to be criticized by others, not permit them to be kept buried within a mixture of subjective beliefs and prior opinions.

In this connection, I must protest McCoy's allegation that scientists faced with "the increasingly pervasive climate of uncertainty and urgency" of practice may view my arguments as "irrelevant." The truth is that every step of my error-statistical (re)interpretation of standard (Neyman-Pearson) statistical methods grew directly out of the real and pressing concerns that arise in applying and interpreting these methods, especially in the face of the "uncertainty and urgency" of science-based policy regarding risky technologies and practices. (See, for instance, my sketch of the "rule of acceptance" below). Where McCoy touts the subjective Bayesian model because it licenses scientists in providing policy makers with probabilities of "uncommon but high-consequence events," for many of us, the increasing dependence on such "expert" assessments that cannot be readily critiqued, and which too often have been seriously wrong, is more a cause for alarm than it is a point in favor of the Bayesian model. (To my mind, the very reason that a philosopher such as myself was invited to the conference upon which this volume is based is precisely the concern engendered by the increased tendency of policy makers to rely on Bayesians who are only too happy to give us "expert" assessments of such things as how improbable it is for a given species to go extinct, or for a given transgenic food to adversely affect untargeted species, and so on, never mind the uncertainties, the effects not yet even tested, etc.)

As I see it, the appropriateness of tools for responsible scientific practice should be measured not according to whether they can cough up the kinds of numbers that satisfy policy makers—who may not be fully aware of the uncertainties and complexities involved—but according to how good a job they do in interpreting and communicating the nature of the uncertainties and errors that have and have not been well ruled out by existing evidence (e.g., "there is evidence from the lab that a genetically modified crop does not harm untargeted species, but we have yet to test it in the field"). While this may result in far less tidy reports than some might wish, this may be the only responsible scientific report that can be given, and it should be left to the public and to policy makers to decide what management decisions are appropriate in the face of existing uncertainties (e.g., taking precautions).

In practice, adopting the premise that McCoy and Casella blithely assume—that unless statistical inference is value-free we are forced to embrace subjectivism—exacerbates the problem of holding risk managers accountable to the public who are affected most by science-based policies that are made in the "climate of uncertainty and urgency." If statistical inferences on which risk judgments are based are really as inextricably entwined with social policy opinions as my commentators appear to assume, there would be little grounds for criticizing a risk assessment as a misinterpretation of data, in the sense of incorrectly asserting what the data indicate about the actual extent of a given risk. Statistical methods become viewed as largely tools for manipulation rather than as instruments to help achieve an unbiased adjudication of conflicting assessments. In the current climate of widespread public suspicion that "scientific expertise" can be bought in support of one's preferred policy, what is needed most is neither tidy degree-of-belief judgments nor despairing evidence-based risk assessments. What are needed are principled grounds to solicit sufficient information to enable statistical reports to be criticized and extended by others. This requires, in the realm of statistical reports, information about the *error probabilities* of methods—the very thing that is absent from subjective Bayesian accounts. This leads to my second set of remarks.

THE THREE TASKS: WHY BAYESIANS (AND OTHER CONDITIONALISTS) ARE WRONG FOR THE JOB

I am struck by Casella's suggestion that he "cannot see why the three tasks" that I list for applying error-statistical methods to scientific inquiry "cannot be accomplished within a Bayesian framework" as well. Perhaps they can, but it is important to see that they will no longer be the same tasks as those understood by the error statistician. I will focus on task (ii) (carrying out error-statistical tests), but a few words about tasks (i) and (iii) are in order.

A central obstacle to carrying out the kind of piecemeal analysis of task (i) within a Bayesian account is that Bayes' theorem requires one to work with "a single probability pie" (Mayo, 1997a, 231), as it were. One cannot escape the assessment of the Bayesian "catchall factor," $P(x \mid \text{not-}H)$, in making inferences about H, where not-H will include all hypotheses other than H, including those not yet even dreamt of. The error-statistical assessment, by contrast, permits probing local hypotheses without having to consider, much less assign probabilities to, an exhaustive space of alternative hypotheses.

When it comes to testing the assumptions of statistical test data [task (iii)] the error statistician has a principled justification for appealing to pretrial methods, e.g., randomization, to generate data satisfying error-statistical assumptions, as well as a methodology for running posttrial tests of assumptions (e.g., independence and identical distribution). Once the data X are in hand, the likelihood principle entails the irrelevance of the procedures generating X (e.g., stopping rules, randomization) that do not alter likelihoods even though they can dramatically alter error probabilities (Mayo, 1996, 2003a; Mayo and Kruse, 2001). Moreover, if probability models reflect subjective degrees of belief (or ignorance), how are they to be put to a test? What can it even mean for them to be "wrong" (Lindley, 1976)? Even nonsubjective Bayesians are limited to appeals to "robustness" and asymptotics that come up short.

I now turn to focus in some detail on task (ii), carrying out statistical tests in a manner that avoids the shortcomings of the rigid NP model.

THE NEW CONDITIONAL ERROR PROBABILITY (CEP) MOVEMENT

Let me readily grant Cassella's charge that my proposals in chapter 4 remained at "a vague level"—something I rectify somewhat in part (c) of these comments. As for Casella's statement that I am "describing an approach similar to Kiefer's conditional confidence (Kiefer, 1977)," I cannot say (though I hope to probe this further). However, I am less sanguine that the goals of my reinterpretation could be met by the contemporary "conditional error probability" (CEP) movement to which Cassella also alludes, e.g., as promoted by Jim Berger, even if extended beyond it's current "stylized" examples.

As intriguing as I find these Bayesian-error-statistical links, the notion of "conditional error probabilities" that arises in this work refers not to error probabilities in the error statistician's sense but rather to posterior probabilities in hypotheses. These differ markedly from NP error probabilities, which refer only to the distribution of test statistic $d(X)$, i.e., the *sampling distribution.* True, my use of the sampling distribution (for a post-data evaluation of severity) goes beyond what is officially prescribed in the NP model; nevertheless, all of my probabilistic assessments are in terms of the distribution of $d(X)$, and do not involve any posterior probability assignment. Hence, it is misleading to refer to the conditionalist's posterior probability assignments as "conditional error probabilities"; to do so is sure to encourage, rather than discourage, the confusion that already abounds in interpreting error probabilities such as observed significance levels or P-values.

Calling these posterior probabilities in H "error probabilities" or even "conditional error probabilities" is misleading. The posterior probabilities of H, arrived at via one of these conditional approaches (e.g., by stipulating an "uninformative prior") will not provide control of error probabilities as understood in the NP testing account. Accordingly, they will not underwrite the severity assessments that are the linchpins of my approach. Ironically, however, CEPs are often billed as "correcting" frequentist error probabilities by introducing priors that are allegedly kosher even for a frequentist.

Consider a very common illustration. In a normal distribution (two-sided) test of H_0: $\mu = 0$ versus H_1: $\mu \neq 0$, "at least 22%—and typically over 50%—of the corresponding null hypotheses will be true" if we assume that "half of the null hypotheses are initially true," conditional on a .05 statistically significant result (Berger, 2003). This is taken to show the danger in interpreting P-values as error probabilities, but note the shift in meaning of "error probability." The assumption is that the correct error probability is given by the proportion of true null hypotheses (in a chosen population of nulls). But an error statistician would (or should) disagree.

Take the recent studies reporting statistically significant increases in breast cancer among women using hormone replacement therapy HRT for 5 years or more. Let us suppose P is .02. (In a two-sided test this is approximately a .05 P-value. Because the 1-sided test has much less of a discrepancy between P-values and posteriors, the critics focus on the 2-sided, and I do as well.)

The probability of observing so large an increase in disease rates when H_0 is true and HRT poses no increased risk is small (.05). Given that the assumptions of the statistical model are met, the error-statistical tester takes this to indicate a genuine increased risk (e.g., 2 additional cases of the disease per 10,000). To follow the CEP recommendations, it would seem, we must go on to consider a pool of null hypotheses from which H_0 may be seen to belong, and consider the proportion of these that have been found to be true in the past. This serves as the prior probability for H_0. We are then to imagine repeating the current significance test p over all of the hypotheses in the pool we have chosen, and look to the posterior probability of H_0 to determine whether the original assessment of a genuine increased risk is or is not misleading. But which pool of hypotheses should we use? Shall we look at all those asserting no increased risk or benefit of any sort? Or only of cancer? In men and women? Or women only? Of breast cancer or other related cancers? Moreover, it is hard to see how we could ever know the proportion of nulls that are true rather than merely those that have thus far not been rejected by statistical tests.

Finally, even if we agreed that there was a 50% chance of randomly selecting a true null hypothesis from a given pool of nulls, that would still not give the error statistician's frequentist prior probability of the truth of the given null hypothesis, e.g., that HRT has no effect on breast cancer risks. Either HRT increases cancer risk or not. Conceivably, the relevant parameter, say the increased risk of breast cancer, could be modeled as a random variable (perhaps reflecting the different effects of HRT in different women), but its distribution would not be given by computing the rates of other benign or useless drugs!

These allegedly frequentist priors commit what might be called the fallacy of instantiating probabilities:

$p\%$ of the null hypotheses in a given pool of nulls are true.
This particular null hypothesis H_0 was randomly selected from this pool.
Therefore $P(H_0$ is true$) = p$.

Admittedly, most advocates of Bayesian CEP, at least if pressed, do not advocate looking for such a frequentist prior (though examples abound in the literature), but instead advocate accepting from the start the "objective" Bayesian prior of .5 to the null, the remaining .5 probability being spread out over the alternative parameter space. But seeing how much this would influence the Bayesian CEP and how this in turn would license discounting the evidence of increased risk should make us that much more leery of assuming them from the start. Now, the Bayesian significance tester wishes to start with a fairly high prior to the null—the worry being that otherwise a rejection of the null would amount to reporting that a fairly improbable hypothesis has become even more improbable. However, it greatly biases the final assessment toward finding no discrepancy. In fact, with reasonably large sample size n, a statistically significant result leads to a posterior probability in the null that is higher than the initial "ignorance" prior of .5 (see, e.g., Berger, 2003).

Hence, what the error statistician would regard as a good indication of a positive discrepancy from 0 would, on this conditional approach, not be taken as any evidence against the null at all. While conditionalists view this as rectifying the alleged unfairness toward the null in a traditional significance test, for the error statistician this would conflict with the use of error probabilities as postdata measures of the *severity* of a test. What the severity tester would regard as good evidence for a positive discrepancy from H_0 is no longer so regarded because the evidence of a discrepancy has been washed away by the (high) prior probability assignment to H_0 (Mayo, 2003b).

Note the consequences of adopting such an approach in the "pervasive climate of uncertainty" about which McCoy is concerned. Uncertainty about a positive risk increase due to an unfamiliar technology would seem to license the "ignorance prior" probability of .5 to hypothesis H_0: 0 risk increase. Accepting the conditionalist strategy exacerbates the problem that already exists in finding evidence of risk increases with low-probability events, except that there would be no grounds for criticizing the negative report of low confirmation of risk.

The severity assessment *is* applicable to those cases where a legitimate frequentist prior probability distribution exists,[18] and the aim is inferring claims about posterior probabilities. In such cases, the severity assessment would be directed at the reliability of these probability assignments and at

18. Editors' note: see Goodman, 2004 (chapter 12 of this volume), for another discussion of the use of empirical priors.

checking for mistakes in the various ingredients of the Bayesian calculation involved.

Given that I find myself so often seconding Casella's own criticisms of (CEP) attempts (e.g., Casella and Berger, 1987), I remain perplexed that he seems to regard at least the broad line of the CEP movement as a promising way to resolve problems in the philosophy of statistics. Why not prefer, instead, a resolution that does not entail renouncing frequentist principles and philosophy?

To be fair, Casella appears to harbor an assumption that is nearly ubiquitous among contemporary critics of frequentist methods: namely that the way to supply NP tests with a viable postdata inferential interpretation is by developing strategies wherein NP error probabilities coincide with postdata probabilities on hypotheses.

The underlying assumption is that for error probabilities to be useful for inference they need to provide postdata measures of confirmation to hypotheses—i.e., E-R measures—and that such measures are afforded by posterior probability assignments to hypotheses. The central thesis of my chapter, however, is to reject this assumption. My postdata interpretation, in contrast to that of the conditionalist, uses NP error probabilities to assess the severity with which a particular inference passes a test with data x; and it is a mistake to construe a postdata severity assessment as a measure of support, probability, or any other E-R measure to the statistical hypotheses involved!

THE SEVERITY REINTERPRETATION OF
STATISTICAL HYPOTHESIS TESTS

Now to go at least part way towards addressing Casella's charge that my reinterpretation of NP methods remains "at a vague level." The quickest way to put some flesh on the ideas of my paper, I think, will be by means of a very familiar statistical test—the one-sided version of the test we just considered.

Example: Test T_α. Suppose that a random sample of size n, $X = (X_1, \ldots, X_n)$, where each X_i is normally distributed with unknown mean μ and known standard deviation $\sigma = 1$, is used to conduct a *one-sided test* of the hypotheses

$H_0: \mu = \mu_0$ against $H_1: \mu > \mu_0$.

The NP test is a rule that tells us for each possible outcome $x = (x_1, \ldots, x_n)$ whether to "accept H_0" or "reject H_0 and accept H_1." The rule is defined in terms of a test statistic or distance measure $d(X)$

$$d(X) = \frac{\overline{X} - \mu}{\sigma/\sqrt{n}},$$

where $\overline{X}$ is the sample mean. In particular, the uniformly most powerful (UMP) test with significance level α, denoted by T_α, is

T_α: reject H_0, if $d(x) \geq c_\alpha$.

Equivalently,

T_α: reject H_0, if $\overline{X} \geq \mu_0 + c_\alpha \sigma_x$,

where $\sigma_X = \dfrac{\sigma}{\sqrt{n}}$. Setting $\alpha = .02$ corresponds to a c_α of approximately 2, and so $d(X)$ would be *statistically significant* (at the .02 level) whenever $d(X) > 2$.

To have a specific numerical example, let $n = 100$, so $\sigma_x = 0.1$ (since we are given that $\sigma = 1$), and we have

$T_{.02}$: reject H_0 whenever $d(X) \geq 2$.

THE FORM OF THE REINTERPRETATION

Although NP tests are framed in terms of a hypothesis being "rejected" or "accepted," both results will correspond to "passing" some hypothesis or, rather, some hypothesized discrepancy, enabling a single notion of severity to cover both. That is, we use the observed x that leads to "reject" or "accept" H_0 in order to assess the *actual severity* with which specific discrepancies from H_0 pass the given test. "Reject H_0" in this one-sided test will license inferences about the extent of the positive discrepancy that is *indicated* by data x; whereas "accept H_0" will correspond to inferences about the extent of the discrepancy from H_0 that is *ruled out:*

Reject H_0 (with x) licenses inferences of the form $\mu > \mu'$.
Accept H_0 (with x) licenses inferences of the form $\mu \leq \mu'$.

It is very important to emphasize that this is *not* to change the null and alternative hypotheses associated with test T_α; they are the same ones given at the outset. Rather, these are postdata inferences that might be considered in interpreting the results of running T_α. The particular inference that is *warranted* will depend on the corresponding *severity* assessment. This brings out a crucial contrast between a severity assessment and the probabilities of type I and type II errors (or power), namely that a severity assessment is always relative to a particular hypothesis or inference one is entertaining. The question "Is T_α a severe test?" is not well posed until one specifies the particular inference being entertained and with what test result. This will become clearer as we proceed.

Severity
To implement the reinterpretation, we need to remind ourselves of the definition of severity. The basic idea of a hypothesis H passing a severe test with data x is the requirement that it be highly improbable that H would have passed so well if in fact H is false (e.g., if a specified discrepancy from H exists). More precisely, a hypothesis H passes a severe test with data x if (and only if),

(i) x agrees with H (for a suitable measure of agreement or fit), and
(ii)   with very high probability, the test would have produced an outcome that fits H *less well* than x does if H were false or incorrect.

Alternatively, we can write clause (ii) as

(ii′) there is a very *low* probability that the test would have produced an outcome that fits H *as well* as x does, if H were false or incorrect (Mayo, 1996).

The basic tenet underlying the interpretation of test results is that inferences are warranted just to the extent that they have passed severe tests. We can call this the severity principle:
Severity principle: Data x provide a good indication of, or good evidence for, hypothesis H to the extent that H passes a severe test with x.
Rules of Rejection and Acceptance: The postdata interpretation for the results of test T_α may be expressed in terms of two rules associated with "reject H_0" and "accept H_0" respectively.
Rules of Rejection (RR) for T_α (i.e., $d(x) > c_\alpha$):

(a) If there is a very *low* probability of obtaining so large a $d(X)$ even if $\mu \leq \mu'$ then $d(x)$ passes hypothesis $\mu > \mu'$ with *high* severity; hence, by the *severity principle, $d(x)$ provides *good evidence* for $\mu > \mu'$.

The severity principle also tells us which discrepancies we would *not* be warranted to infer, and we can state this in a companion rule:

(b) If there is a very *high* probability of obtaining so large a $d(x)$ even if $\mu \leq \mu'$ then $d(x)$ passes hypothesis $\mu > \mu'$ with *low* severity; hence, by the *severity principle, $d(x)$ provides *poor evidence* for $\mu > \mu'$.

Analogously, we can state rules for interpreting "accept H_0", this time combining both parts (a) and (b):
Rules of Acceptance (RA) for test T_α (i.e., $d(x) \leq c_\alpha$:—a "negative" result):

(a) and (b): If there is a very *high (low)* probability that $d(x)$ would have been larger than it is, (even) if $\mu > \mu'$, then $d(x)$ passes hypothesis $\mu \leq \mu'$ with *high (low)* severity; hence, by the severity principle, $d(x)$ provides *good (poor) evidence* for $\mu \leq \mu'$.

Defining Severity for T_α: The severity with which hypothesis H passes T_α with an observed $d(x_0)$, abbreviated as Sev$(T_\alpha; H, d(x_0))$, where x_0 is the actual outcome:

The Case of Rejecting H_0 (i.e., $d(x_0) > c_\alpha$):
Sev$(T_\alpha; \mu > \mu', d(x_0)) = P(d(X) \leq d(x_0); \mu > \mu'$ is false).

The Case of Accepting H_0 (i.e., $d(x_0) < c_\alpha$):
Sev$(T_\alpha; \mu \leq \mu', d(x_0)) = P(d(X) > d(x_0); \mu \leq \mu'$ is false).

In both cases, we calculate severity under the point value μ' giving the lower bounds for severity.[19]

PARTICULAR APPLICATIONS AND REMARKS

To get a feel for applying these rules using the definition above, consider some particular outcomes from test $T_{.02}$ with $n = 100$. Recall that

19. Editors' note: similar to the power curve, this statement actually defines a curve, not a single probability number, as μ' varies.

$T_{.02}$: reject H_0 whenever $d(x) \geq 2$ (rounding for simplicity).

Suppose we are testing

H_0: $\mu = 0$ against H_1: $\mu > 0$.

Consider first some examples for applying RR: Let $\bar{x} = 0.2$—an outcome just at the cutoff for rejecting H_0. We have $P(\bar{X} \leq 0.2; \mu = 0) = P(Z \leq 2) = .977$, and thus from rule RR we can infer with severity .98 that $\mu > 0$. Further, because $P(\bar{X} \leq 0.2; \mu = .1) = P(Z \leq 1) = .84$ we can infer with severity .84 that $\mu > 0.1$.

Note in general for tests of type T_α, that by stipulating that H_0 be rejected only if $d(X)$ is statistically significant at some small level α, it is assured that such a rejection—*at minimum*—warrants hypothesis H_1, that $\mu > \mu_0$. Rule RR(a) makes this plain.

Severity vs. Power
At first glance, severity requirement (ii) may seem to be captured by the notion of a test's power. However, the severity with which alternative $\mu > \mu'$ passes a test is not given by, and is in fact inversely related to, the test's power to reject H_0 in favor of μ'. For instance, the severity with which $\bar{x} = 0.2$ passes hypothesis H: $\mu > 0.4$ is .03, so it is a very poor indication that the discrepancy from 0 is this large. However, the power of the test against 0.4 is high, .97 (it is in the case of "accept H_0" that power correlates with severity, as seen below).

Beyond the Coarse Interpretation of the NP Test
If $d(X)$ exceeds the cut-off 2, this is reflected in the severity assessments. For example, if $\bar{x} = 0.3$, we have from rule RR:

infer with severity .999 that $\mu > 0$, and also
infer with severity .98 that $\mu > 0.1$.

Consider now the case where test T_α outputs "accept H_0." We know that failing to reject H_0 does not license the inference to H_0—that μ is *exactly* 0—because the probability of such a negative result is high even if there is *some* positive discrepancy from 0. Hence, the well-known admonition that "no evidence against (the null) is not the same as evidence for (the null)!" Rule RA explains this in terms of severity. However, RA may also be used to find a μ' value that *can* be well ruled out.

Suppose $\bar{x} = 0.1$. $P(\bar{X} > 0.1; \mu = 0.05) = P(Z > 0.5) = 0.31$, and from rule RA we can infer with severity (only) .31 that $\mu < 0.05$. But also, because $P(\bar{X} > 0.1; \mu = 0.2) = P(Z > -1) = .84$, we can infer with severity .84 that $\mu < 0.2$.

Thus, imagine on the basis of $\bar{x} = 0.1$, someone proposes to infer "This result is evidence that if there are any discrepancies from 0, it is less than 0.05." Since the severity with which $\mu < 0.05$ passes this test is only 0.3, we would criticize such an interpretation as unwarranted—and we would do so on objective grounds. One might go on to note that the evidence can only be regarded as passing a much less informative claim, i.e., $\mu < 0.2$, with reasonable severity (e.g., .84). By making this type of critique systematic, the severity assessment supplies a powerful tool for critically evaluating negative reports, especially in cases of statistical risk assessment (e.g., Mayo, 1985b, 1989, 1991, 1996). Using severity to interpret nonsignificant results thereby supplies a data dependency that is missing from recent attempts at "postdata power analyses" (Mayo and Spanos, 2000, 2004).

Answering Central Criticisms of NP tests

This postdata reinterpretation gives the following response to the alleged arbitrariness in choosing the probabilities for type I and II errors. Predata, the choices should reflect the goal of ensuring the test is capable of licensing given inferences severely. We set the "worst case" values accordingly: small α ensures, minimally, that any rejection of H_0 licenses inferring *some* discrepancy from H_0; and high power against μ' ensures that *failing* to reject H_0 warrants inferring $\mu \leq \mu'$ with high severity. Postdata, the severity assessment allows for an objective interpretation of the results of whatever test happened to be chosen. While statistical tests cannot themselves tell us which discrepancies are of substantive importance, the concept of severity lets us use the test's properties to assess both the discrepancies that have, and that have not, been passed severely. This is a powerful source for critically evaluating statistical results—something sorely needed in a great many applications of tests.

Summary of the Rules

We can summarize rules RR and RA for the case of test T_α with result $d(x_0)$ as follows:

RR: reject H_0 (i.e., $d(x_0) > c_\alpha$)

infer with severity $1 - \varepsilon$: $\mu > \bar{X} - k_\varepsilon \sigma / \sqrt{n}$, where $P(Z > k_\varepsilon) = \varepsilon$

or

infer with severity $1 - \varepsilon: \mu > \mu_0 + \gamma$, where $\gamma = (d(x_0) - k_\varepsilon)\sigma/\sqrt{n}$.[20]

RA: accept H_0 (i.e., $d(x_0)) \leq c_\alpha$):

infer with severity $1 - \varepsilon: \mu \leq \overline{X} - k_\varepsilon\sigma/\sqrt{n}$, where $P(Z > k_\varepsilon) = \varepsilon$

or

infer with severity $1 - \varepsilon: \mu \leq \mu_0 + \gamma$, where $\gamma = (d(x_0) + k_\varepsilon)\sigma/\sqrt{n}$.

In this form, there is a clear connection with NP confidence intervals, often recommended for a postdata inference. There are important differences, however, most notably that confidence intervals (in addition to having a problematic postdata interpretation) treat all values in the interval on par, whereas each value in the interval corresponds to inferences (about discrepancies from that value) with different degrees of severity. A full comparison between a severity assessment and confidence intervals calls for a separate discussion (Mayo and Spanos, 2000).

Also beyond the scope of these comments is a full discussion of how to use (and avoid misusing) the admittedly unfamiliar expression *infer with severity* $1 - \varepsilon$ *that* $\mu > \mu'$ $[\mu_0 + \gamma]$. One warning must suffice: $1 - \varepsilon$ should *not* be construed as a degree of support or other E-R measure being accorded to $\mu > \mu'$. It is just a shorthand for the longer claim: hypothesis $\mu > \mu'$ passes test T with severity $1 - \varepsilon$.

Examples of the kinds of questions that this reinterpretation allows one to pose are: *How severely does a given hypothesis pass T with a given result?* and *Which alternatives to H_0 pass T at a given level of severity?* Other types of questions can also be underwritten with this postdata interpretation, and it would be a matter for practitioners to consider which are most useful for given applications.

REFERENCES

Armitage, P. 1962. Contribution to Discussion. In Savage, L. J., ed., *The Foundations of Statistical Inference: A Discussion.* London: Methuen.

Barnard, G. A. 1962. Contribution to discussion. In Savage, L. J., ed., *The Foundations of Statistical Inference: A Discussion.* London: Methuen.

20. Here Z is the standard normal variate.

Barnard, G. A. 1972. Review of *The Logic of Statistical Inference,* by I. Hacking). *B. J. Phil. Sci.* 23:123–132.

Berger, J. O. 2003. Could Fisher, Jeffreys, and Neyman Have Agreed on Testing? *Stat. Sci.* 18:1–12.

Berger, J. O., B. Boukai, and Y. Wang. 1997. Unified Frequentist and Bayesian Testing of a Precise Hypothesis (with discussion). *Stat. Sci.* 12:133–160.

Berger, J. O., L. D. Brown, and R. L. Wolpert. 1994. A Unified Conditional Frequentist and Bayesian Test for Fixed and Sequential Simple Hypothesis Testing. *Ann. Stat.* 22:1787–1807.

Berger, J. O., and Wolpert, R. L. 1988. *The Likelihood Principle.* 2nd ed. Hayward, CA: Institute of Mathematical Statistics.

Birnbaum, A. 1969. Concepts of Statistical Evidence. In Morgenbesser, S., P. Suppes, and M. White, eds., *Philosophy, Science, and Method: Essays in Honor of Ernest Nagel.* New York: St. Martin's Press.

Birnbaum, A. 1977. The Neyman-Pearson Theory as Decision Theory, and as Inference Theory; with a Criticism of the Lindley-Savage Argument for Bayesian Theory. *Synthese* 36:19–49.

Carnap, R. 1962. *Logical Foundations of Probability.* Chicago: University of Chicago Press.

Casella, G. 1988. Conditionally Acceptable Frequentist Solutions (with discussion). In Gupta, S. S., and J. O. Berger, eds. *Statistical Decision Theory IV.* New York: Springer.

Casella, G., and R. L. Berger. 1987. Commentary of J. O. Berger and M. Delampady's "Testing Precise Hypotheses." *Stat. Sci.* 2:344–347.

Cox, D. R., and D. V. Hinkley. 1974. *Theoretical Statistics.* London: Chapman and Hall.

de Finetti, B. 1972. *Probability, Induction, and Statistics: The Art of Guessing.* New York: Wiley.

Edwards, W., H. Lindman, and L. Savage. 1963. Bayesian Statistical Inference for Psychological Research. *Psych. Rev.* 70:193–242.

Gigerenzer, G. 1993. The Superego, the Ego, and the Id in Statistical Reasoning. In Keren, G. and C. Lewis, eds., *A Handbook for Data Analysis in the Behavioral Sciences: Methodological Issues.* Hillsdale, NJ: Erlbaum.

Goutis, C. and Casella, G. 1995. Frequentist Post-data Inference. *Internat. Stat. Rev.* 63:325–344.

Hacking, I. 1965. *Logic of Statistical Inference.* Cambridge: Cambridge University Press.

Hacking, I. 1972. Likelihood. *Br. J. Phil. Sci.* 23:132–137.

Hilborn, R., and M. Mangel. 1997. *The Ecological Detective: Confronting Models with Data.* Princeton: Princeton University Press.

Howson, C., and P. Urbach. 1993. *Scientific Reasoning: The Bayesian Approach.* 2nd ed. La Salle, IL: Open Court.

Kempthorne, O., and L. Folks. 1971. *Probability, Statistics, and Data Analysis.* Ames: Iowa State University Press.

Kiefer, J. 1977. Conditional Confidence Statements and Confidence Estimators (with discussion). *J. Am. Stat. Assn* 72:789–827.

Lindley, D. V. 1976. Bayesian Statistics. In Harper, W. L., and C. A. Hooker, eds., *Foundations and Philosophy of Statistical Inference.* Dordrecht: Reidel.

Malakoff, D. 1999. Bayes Offers a "New" Way to Make Sense of Numbers. *Science* 286:1460–1464.

Mayo, D. G. 1983. An Objective Theory of Statistical Testing. *Synthese* 57 (pt. 2): 297–340.

Mayo, D. G. 1985a. Behavioristic, Evidentialist, and Learning Models of Statistical Testing. *Phil. Sci.* 52:493–516.

Mayo, D. G. 1985b. Increasing Public Participation in Controversies Involving Hazards: The Values of Metastatistical Rules. *Sci. Tech. Human Values* 10:55–68.

Mayo, D. G. 1989. Toward a More Objective Understanding of the Evidence of Carcinogenic Risk. *PSA: Proceedings of the BiennialMeeting of the Philosophy of Science Association,* 1988, vol. 2: 489–503.

Mayo, D. G. 1990. Did Pearson Reject the Neyman-Pearson Philosophy of Statistics? *Synthese* 90:233–262.

Mayo, D. G. 1991. Sociological versus Metascientific Views of Risk Assessment. In Mayo, D., and R. Hollander, eds., *Acceptable Evidence: Science and Values in Risk Management.* New York: Oxford University Press.

Mayo, D. G. 1996. *Error and the Growth of Experimental Knowledge.* Chicago: University of Chicago Press.

Mayo, D. G. 1997a. Duhem's Problem, The Bayesian Way, and Error Statistics, or "What's Belief Got to Do with It?"; and response to Howson and Laudan. *Phil. Sci.* 64:222–244, 323–333.

Mayo, D. G. 1997b. Error Statistics and Learning from Error: Making a Virtue of Necessity. *Phil. Sci.* 64 (supp.): S195–S212.

Mayo, D. G. 2003a. Severe Testing as a Guide for Inductive Learning. In Kyburg, H., and M. Thalos, eds., *Probability Is the Very Guide in Life.* Chicago: Open Court.

Mayo, D. G. 2003b. Commentary on J. O. Berger's Fisher Address ("Could Fisher, Jeffreys, and Neyman Have Agreed?"). *Stat. Sci.* 18:19–24.

Mayo, D. G. 2004. Evidence as Passing Severe Tests: Highly Probable vs. Highly Probed Hypotheses. In Achinstein, P., ed., *Scientific Evidence: Philosophical Theories and Applications.* Baltimore: Johns Hopkins University Press.

Mayo, D. G., and M. Kruse. 2001. Principles of Inference and Their Consequences. In Cornfield, D., and J. Williams, eds., *Foundations of Bayesianism.* Dordrecht: Kluwer.

Mayo, D. G., and A. Spanos. 2000. *A Post-data Interpretation of Neyman-Pearson Methods Based on a Conception of Severe Testing.* London: Tymes Court.

Mayo, D. G., and A. Spanos. 2004. Methodology in Practice: Statistical Misspecification Testing. *Phil. Sci.* (vol. 2 of PSA 2002 proceedings).

Neyman, J. 1955. The Problem of Inductive Inference. Part 1. *Comm. Pure Appl. Math.* 8:13–46.

Neyman, J. 1976. Tests of Statistical Hypotheses and Their Use in Studies of Natural Phenomena. *Comm. Stat. Theor. Methods* 8:737–751.

Pearson, E. S. 1947. The Choice of Statistical Tests Illustrated on the Interpretation of Data Classed in a 2 × 2 Table. *Biometrika* 34:139–167. Reprinted in Pearson (1966).

Pearson, E. S. 1950. On Questions Raised by the Combination of Tests Based on Discontinuous Distributions. *Biometrika* 37:383–398. Reprinted in Pearson (1966).

Pearson, E. S. 1955. Statistical Concepts in Their Relation to Reality. *J. Roy. Stat. Soc.,* ser. B, 17:204–207.

Pearson, E. S. 1966. *The Selected Papers of E. S. Pearson.* Berkeley: University of California Press.

Pearson, E. S., and Neyman, J. 1930. On the Problem of Two Samples. *Bull. Acad. Pol. Sci.,* 73–96. Reprinted in Neyman and Pearson (1967).

Royall, R. M. 1997. *Statistical Evidence: A Likelihood Paradigm.* London: Chapman and Hall.

Royall, R. M. 2004. The Likelihood Paradigm for Statistical Evidence. Chapter 5 in Taper, M. L., and S. R. Lele, eds., *The Nature of Scientific Evidence: Empirical, Statistical, and Philosophical Considerations.* Chicago: University of Chicago Press.

Savage, L. J. 1972. *The Foundations of Statistics.* New York: Dover.

Spanos, A. 2000. Revisiting Data Dining: "Hunting" with or without a License. *J. Econ. Methodology* 7:231–264.

5 The Likelihood Paradigm for Statistical Evidence

Richard Royall

ABSTRACT

Statistical methods aim to answer a variety of questions about observations. A simple example occurs when a fairly reliable test for a condition or substance, C, has given a positive result. Three important types of questions are: (1) Should this observation lead me to believe that C is present? (2) Does this observation justify my acting as if C were present? (3) Is this observation evidence that C is present? We distinguish among these three questions in terms of the variables and principles that determine their answers. Then we use this framework to understand the scope and limitations of current methods for interpreting statistical data as evidence. Questions of the third type, concerning the evidential interpretation of statistical data, are central to many applications of statistics in science. We see that for answering them current statistical methods are seriously flawed. We find the source of the problems, and propose a solution based on the law of likelihood. This law suggests how the dominant statistical paradigm can be altered so as to generate appropriate methods for (i) objective representation and measurement of the evidence embodied in a specific set of observations, as well as (ii) measurement and control of the probabilities that a study will produce weak or misleading evidence.

INTRODUCTION

An important role of statistical analysis in science is interpreting observed data as evidence—showing "what the data say." Although the standard statistical methods (hypothesis testing, estimation, confidence intervals) are routinely used for this purpose, the theory behind those methods contains

119

no defined concept of evidence and no answer to the basic question "When is it correct to say that a given body of data represents evidence supporting one statistical hypothesis over another?" or to its sequel, "Can we give an objective measure of the *strength* of statistical evidence?" Because of this theoretical inadequacy, the use of statistical methods in science is guided largely by convention and intuition and is marked by unresolvable controversies (such as those over the proper use and interpretation of *P*-values and adjustments for multiple testing).

We argue that the law of likelihood represents the missing concept and that its adoption in statistical theory can lead to a frequentist methodology that avoids the logical inconsistencies pervading current methods while maintaining the essential properties that have made those methods into important scientific tools.

STATISTICAL EVIDENCE

By "statistical evidence," we mean observations that are interpreted under a probability model. The model consists of a collection of probability distributions, and the observations are conceptualized as having been generated from one of the distributions.

For example, a subject is given a diagnostic test for a disease and the result is positive. This observation might be interpreted as a realization of a random variable X whose possible values are 1 (positive) and 0 (negative). The distribution of X is determined by whether the subject does or does not have the disease, as shown in the following table of probabilities.

Test Result

	$X = 1$	$X = 0$
disease present	.94	.06
disease absent	.02	.98

This simple model has only two probability distributions (given in the two rows of the table). In this context, the observed test result is an example of statistical evidence.

Three Questions

Statistics is the discipline concerned with statistical evidence—producing, modeling, interpreting, communicating, and using it. We will focus on an area of statistics that is central to its role in science, interpreting and communicating statistical evidence per se. To distinguish this problem area from some other branches of statistics, and to introduce its essential principle, we consider the above diagnostic test, and three conclusions about disease status that might be appropriate after a positive result has been observed:

This person probably has the disease.
This person should be treated for the disease.
This test result is evidence that this person has the disease.

How can we determine which, if any, of these conclusions are correct?

The first is a statement about the present state of uncertainty concerning the subject's disease status, i.e., the conditional probability of disease, given the positive test result. It states that this probability is greater than .5. The above model does not determine whether this conclusion is true. This is because the conditional probability of disease (given the positive test result) depends not only on the probabilities that comprise the model, but also on a quantity that is not represented in this model. That missing quantity is the probability of disease before the test (the prior probability). Precisely *how* the present uncertainty depends on the prior is detailed in elementary probability theory by Bayes' theorem. For example, if the test was used in a mass screening program for a rare disease, and if the subject is simply one of those whose results were positive, then the prior probability is the prevalence of the disease in the screened population. And if that probability is less than .021, then Bayes' theorem shows that the present probability of disease is still less than .5, so that, although the test is positive, the first conclusion is wrong—this person probably does *not* have the disease. But if, instead of an anonymous participant in the screening program, she is a patient whose symptoms implied a disease probability of .10 before the test, then the probability of disease is now .84, and the first conclusion is quite correct.

The correctness of the second statement ("This person should be treated for the disease") depends on the present probability of disease, so that it too depends on the prior probability. But it depends on other factors as well, such as the costs of treating and of not treating, both when the disease is present and when it is not. Even when the first conclusion is wrong, the second might be correct. This would be true if the treatment is a highly effective one, with little cost or risk, while not treating a patient with the disease is disastrous. But under different conditions of prior uncertainty and costs, the opposite conclusions might be appropriate—it might be best that the subject *not* be treated, even though she probably has the disease.

The third conclusion, unlike the first two, requires for its appraisal nothing more than the probability model represented by the table. Under that model, the third conclusion is correct, regardless of the disease probability before the test and regardless of the costs associated with whatever treatment decisions might be made—the positive test result *is* evidence that this person has the disease. This strongly intuitive conclusion is certified by the basic rule for interpreting statistical evidence, which will be stated in the next section.

Each of the three conclusions represents an answer to a different question:

What should I believe?
What should I do?
How should I interpret this body of observations as evidence?

These three questions define three distinct problem areas of statistics.

It is the third problem area, proper interpretation of statistical data as evidence, that we are concerned with in this paper. It is a critical question in scientific research, and, as the simple diagnostic test example shows, it is the only one of the three questions that can be answered independently of prior beliefs.

The Law of Likelihood

We have seen that, after the positive test result has been observed, the conclusion that the subject probably does not have the disease is appropriate when the prior probability is small enough. Similarly, the conclusion that the best course of action is not to treat for the disease is appropriate under certain conditions on the prior probability and the potential costs. But to interpret the positive test result as *evidence that the subject does not have the disease* is never appropriate—it is simply and unequivocally wrong. Why is it wrong?

The interpretation is wrong because it violates the fundamental principle of statistical reasoning. That principle, the basic rule for interpreting statistical evidence, is what Hacking (1965, 70) named the law of likelihood. It states:

If hypothesis A implies that the probability that a random variable X takes the value x is $p_A(x)$, while hypothesis B implies that the probability is $p_B(x)$, then the observation $X = x$ is evidence supporting A over B if and only if $p_A(x) > p_B(x)$, and the likelihood ratio, $p_A(x)/p_B(x)$, measures the strength of that evidence.

[handwritten marginalia: What should I believe? What should I do? Is this evidence (cumulative)?]

[handwritten note at bottom: Depends on what is the a priori probability?]

This says simply that if an event is more probable under hypothesis A than hypothesis B, then the occurrence of that event is evidence supporting A over B—the hypothesis that did the better job of predicting the event is better supported by its occurrence. It further states that the *degree* to which occurrence of the event supports A over B (the strength of the evidence) is quantified by the ratio of the two probabilities.

When uncertainty about the hypotheses, before $X = x$ is observed, is measured by prior probabilities, $P(A)$ and $P(B)$, the law of likelihood can be derived from elementary probability theory. In that case, the quantity $p_A(x)$ is the conditional probability that $X = x$, given that A is true, $P(X = x|A)$, and $p_B(x)$ is $P(X = x|B)$. The definition of conditional probability implies that

$$\frac{P(A|X = x)}{P(B|X = x)} = \frac{p_A(x)P(A)}{p_B(x)P(B)}.$$

This formula shows that the effect of the statistical evidence (the observation $X = x$) is to change the probability ratio from $P(A)/P(B)$ to $P(A|X = x)/P(B|X = x)$. The likelihood ratio, $p_A(x)/p_B(x)$, is the exact factor by which the probability ratio is changed. If the likelihood ratio equals 5, then the observation $X = x$ constitutes evidence just strong enough to cause a fivefold increase in the probability ratio. Note that the strength of the evidence is independent of the prior probabilities. (The same argument and conclusion apply when $p_A(x)$ and $p_B(x)$ are not probabilities but probability densities at x.)

The likelihood ratio is a precise and objective numerical measure of the strength of statistical evidence. Practical use of this measure requires that we learn to relate it to intuitive verbal descriptions such as "weak," "fairly strong," "very strong," etc. The values 8 and 32 have been suggested as benchmarks for likelihood ratios—observations with a likelihood ratio of 8 (or 1/8) constitute moderately strong evidence, and observations with a likelihood ratio of 32 (or 1/32) are strong evidence. These benchmark values come from considering the various possible results of one of the simplest of experiments (Royall, 1997) and are similar to others that have been suggested (Jeffreys, 1961; Edwards, 1972; Kass and Raftery, 1995).

Misleading Evidence
The positive result on our diagnostic test, with a likelihood ratio (LR) of .94/.02 = 47, constitutes strong evidence that the subject has the disease. This interpretation of the test result is correct, regardless of that subject's actual disease status. If she does not have the disease, then the evidence is mis-

leading. We have not made an error—we have interpreted the evidence correctly. *It is the evidence itself that is misleading.*

Statistical evidence, properly interpreted, can be misleading. But we cannot observe strong misleading evidence very often. In our example, if the disease is not present table 1 shows that the probability of a (misleading) positive test is only .02. It is easy to prove that in other situations the probability of observing misleading evidence this strong or stronger (LR $\geq$ 47) can be slightly greater, but it can never exceed $1/47 = .0213$. We can state a *universal bound on the probability of misleading evidence:* If hypothesis A implies that the probability that a random variable X has one probability density (or mass) function, $f_A(\cdot)$, while hypothesis B implies another, $f_B(\cdot)$, then if A is true the probability of observing evidence supporting B over A by a factor of k or more cannot exceed $1/k$. That is, $P_A(f_B(X)/f_A(X) \geq k) \leq 1/k$ (Royall, 1997).

This bound has been noted by various authors (e.g., Smith, 1953; Birnbaum, 1962). Much tighter bounds apply in important special cases. For example, when the two distributions are normal with different means and a common variance, the universal bound $1/k$ can be replaced by the much smaller value, $\Phi(-\sqrt{2\ln(k)})$, where Φ is the standard normal distribution function. In that case the probabilities of misleading evidence, for the proposed benchmarks $k = 8$ and $k = 32$ (representing "pretty strong" and "strong" evidence respectively), cannot exceed .021 and .004 (Royall, 1997).

The universal bound $1/k$ applies even if we deliberately seek evidence supporting B over A by continuing to make observations on X until our sample gives a likelihood ratio of at least k in favor of B. When A is true, the probability is at most $1/k$ that we will succeed, sooner or later finding that the accumulated observations represent strong evidence in favor of B. That is, the probability is at least $1 - 1/k$ that we will *never* succeed, sampling forever without once finding that our data represent strong evidence in favor of B (Robbins, 1970).

The Likelihood Function

As a second example of statistical evidence, consider observing a sequence of tosses of a 40¢ coin. The coin is asymmetric, consisting of an ordinary quarter, nickel, and dime that have been glued together so that the heads of the dime is on one side and the tails of the quarter is on the other. If we model the tosses as independent trials with a common probability of heads, θ, then every value of θ between 0 and 1 determines a different probability distribution. Under this model a sequence of observations, such as 1, 1, 0, 1, 1, 0, . . . (heads = 1, tails = 0) represents statistical evidence. The probabil-

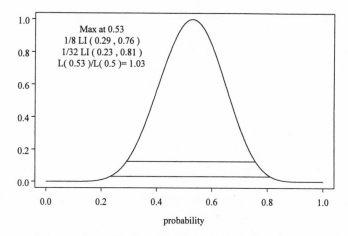

Max at 0.53
1/8 LI (0.29 , 0.76)
1/32 LI (0.23 , 0.81)
L(0.53)/L(0.5)= 1.03

probability

FIGURE 5.1 Likelihood for probability of heads with 9 heads observed in 17 tosses.

ity that a sequence of n tosses will produce observations $x_1, x_2, \ldots, x_n$, is $\prod_{i=1}^{n} \theta^{x_i}(1 - \theta)^{1-x_i} = \theta^k(1 - \theta)^{n-k}$ where $k = \sum_{1}^{n} x_i$, the number of heads. For a given set of observations this probability is a function of the variable θ and is called the *likelihood function:*

$$L(\theta) = \theta^k(1 - \theta)^{n-k} \qquad 0 \le \theta \le 1.$$

The law of likelihood gives this function its meaning: $L(\theta)$ determines the relative support for every pair of values of θ, the probability of heads. Observation of a sequence of n tosses that contains k heads represents evidence supporting the hypothesis that $\theta = \theta_1$ over the hypothesis that $\theta = \theta_2$ by the factor $L(\theta_1)/L(\theta_2)$.

Two likelihood functions that differ only by a constant multiple are equivalent, because they give identical likelihood ratios for all pairs of values of the parameter. That is, the likelihood function is defined only up to an arbitrary constant multiplier.

To produce some evidence about the probability of heads for my 40¢ piece, I performed a simple experiment. I tossed the coin 17 times and noted the results. Heads appeared 9 times. To see what these observations say about the probability of heads, we look at the likelihood function, $\theta^9(1 - \theta)^8$, which is shown in figure 5.1. It is a graphical representation of this statistical evidence. We have scaled this function so that its maximum value is 1. We have noted the value of θ that is best supported by the observations ($9/17 = .53$), as well as the $1/8$ and $1/32$ likelihood intervals (LIs).

These intervals, where the scaled likelihood is at least 1/8 (1/32), represent the values of θ that are consistent with the observations at the levels of the two benchmarks, 8 and 32—if a value is in the 1/8 interval, (0.29, 0.76), then there is no alternative that is better supported by a factor of 8 or more (fairly strong evidence).

The Likelihood Principle

Suppose two instances of statistical evidence generate the same likelihood function. According to the law of likelihood, this means that for every pair of parameter values, θ_1 and θ_2, the strength of the evidence in support of θ_1 vis-à-vis θ_2, $L(\theta_1)/L(\theta_2)$, is the same in both instances. Both will have the same impact on any prior probability distribution for θ. In this powerful sense, they are equivalent. On the other hand, if two instances of statistical evidence are equivalent, then all of the likelihood ratios, $L(\theta_1)/L(\theta_2)$, must be equal, which means that the two likelihood functions are the same. Therefore, the law of likelihood implies that two instances of statistical evidence are equivalent if and only if they generate the same likelihood function. This proposition is called the likelihood principle (Birnbaum, 1962; Edwards, Lindman, and Savage, 1963). It means that *the likelihood function is a mathematical representation of the statistical evidence per se.*

Our experiment with the 40¢ piece produced some observations. The likelihood function shows what those observations mean, as evidence about the probability of heads, in the context of the model that we are using. It also shows that, within this context, certain aspects of the observations have no effect on their evidential meaning and are, in that sense, irrelevant. For example, I actually observed a particular sequence of heads and tails: 1, 1, 0, 1, 1, 0, 0, 0, 1, 0, 1, 1, 1, 0, 0, 0, 1. Other observations, consisting of 9 heads and 8 tails in some different order, would give the same likelihood function, $\theta^9(1 - \theta)^8$. Those observations would constitute evidence of exactly the same strength as ours in support of any value, θ_1, versus any other, θ_2, and would therefore represent evidence that is equivalent to ours. The meaning of the observations, as evidence about the probability of heads, depends only on the number of heads and the number of tails. The order of the observations does not affect any of the likelihood ratios (i.e., it does not affect the likelihood function)—the order is irrelevant.

Besides showing that certain characteristics of the observations are irrelevant to their proper interpretation as evidence, the likelihood principle reveals that certain aspects of the experimental procedure that produced those observations are also irrelevant. My observation of 9 heads in 17 trials might have been generated by fixing the number of trials at 17, observing the (random) number of heads, say H, and finding $H = 9$. But it might have been produced in another way; instead of stopping after a fixed number of trials, I could have generated these observations by fixing the number of heads to be observed at 9, observing the (random) number of trials, say N, required to produce 9 heads, and finding $N = 17$. The probability of observing 9 heads in seventeen trials is $P(H = 9) = \binom{17}{9}\theta^9(1 - \theta)^8$ in the first experiment, and $P(N = 17) = \binom{16}{8}\theta^9(1 - \theta)^8$ in the second. Although these probabilities differ, the likelihood functions are the same (proportional to $\theta^9(1 - \theta)^8$). Thus, it does not matter which procedure I actually used (stopping after 17 trials, or after 9 heads). What these observations mean, as evidence about the probability of heads, is shown in figure 5.1: it is the same in both cases. The stopping rule does not affect the likelihood function—for interpreting the observed data as evidence, the stopping rule is irrelevant.

This conclusion, irrelevance of the stopping rule, implies that conventional frequentist statistical methods are not appropriate *for interpreting the observed data as evidence.* This is because those methods, when applied to our observations (9 heads in 17 trials) give different results under the two stopping rules. For example, the observed proportion of successes, 9/17, is an unbiased estimator of θ under the first stopping rule, but not under the second. Furthermore, this estimator has different standard errors under the two stopping rules. Confidence coefficients and P-values also differ, depending on which stopping rule was used. P-values do not measure the strength of the evidence, because the evidence is the same in both cases, but the P-values (for testing the hypothesis that θ is .8 vs. some smaller value, for example) are different. Confidence intervals do not show "what the data say" about θ, because the data say the same thing in both cases, while the confidence intervals are different.

We will consider this example and the radical conclusion that it illustrates in more detail in the next section. The irrelevance of the stopping rule, in the evidential interpretation of observed data, will be illustrated again below.

STATISTICS IN SCIENCE: THE MISSING LINK

Statistics today is in a conceptual and theoretical mess. The discipline is divided into two rival camps, the frequentists and the Bayesians, and neither camp offers the tools that science needs for objectively representing and inerpreting statistical data as evidence. In this section, we will identify the theoretical deficiency that has produced this methodological one, and see how to correct both.

Bayesian statistics is primarily concerned with the question of how one's beliefs should change in response to new statistical evidence; that is, its focus is on the first of the three questions listed above ("What should I believe?"). As a leading Bayesian put it, "The main subject matter of statistics is the study of how data sets change degrees of belief; from prior, by observation of A, to posterior. They change by Bayes' theorem" (Lindley, 1965, 30). The result of a Bayesian statistical analysis is a (posterior) probability distribution for the parameter. Now, the posterior distribution is determined by the prior and the likelihood function *together* (as explained by Bayes' theorem). Therefore, as shown in the diagnostic test example above, a Bayesian analysis requires, in addition to the probability model for the observed data (which determines the likelihood function), a (prior) probability distribution for the parameter. A frequentist model for my observations on the 40¢ coin represents the tosses as independent trials with some unknown probability, θ, of heads. A Bayesian model must supplement this with a prior probability distribution for θ. The two camps use the same model for the probability distribution of the observable random variables (the results of the tosses); but the Bayesian requires, in addition, a prior probability distribution for the parameter. This prior distribution represents the experimenter's state of uncertainty about the parameter before the observations are made (Edwards, Lindman, and Savage, 1963).

The observations affect the posterior probability distribution only through the likelihood function: for a given prior distribution, if different observations (perhaps from different experiments) produce the same likelihood function, then they produce the same posterior probability distribution. Thus, the Bayesian approach leads inevitably to the likelihood principle. But while they have embraced the principle, Bayesians have shown little interest in likelihoods per se, i.e., likelihoods without prior probabilities. Savage himself wrote (1962, 307), "I, myself, came to take personalistic statistics, or Bayesian statistics as it is sometimes called, seriously only through recognition of the likelihood principle. I suspect that once the likelihood principle is widely recognized, people will not long stop at that halfway house." More recently, Berger and Wolpert (1988, 124) argued that "sensible use of the likelihood function seems possible only through Bayesian analysis," and Lindley (1992, 415) went so far as to proclaim that "the only satisfactory measures of support are probability-based. Likelihood will not do."

The style of uncertainty!

Bayesian efforts to show what the data say have concentrated on (i) searching for prior probability distributions to represent total ignorance (the state of knowledge of an ideal ignoramus) or, failing that, (ii) arguing that certain distributions should be adopted as standard "reference priors" (Bernardo, 1979; Berger and Bernardo, 1992). Posterior probability distributions corresponding to such priors would then supposedly show what the data say. That is, Bayesians have tried to use posterior probability distributions, appropriate to question 1 ("What should I believe?") to answer question 3 ("How should I interpret this body of observations as evidence?"). Non-Bayesians generally judge these attempts to have failed (e.g., Edwards, 1969; 1992, sec. 4.5). The reason for this failure is found in the law of likelihood— they have failed because probabilities represent and measure degrees of belief or uncertainty, not evidence. The law of likelihood reveals that *evidence has a different mathematical form than uncertainty.* It is likelihood ratios, not probabilities, that represent and measure statistical evidence (Royall, 1997, secs. 1.13, 8.6). It is the likelihood function, and not any probability distribution, that shows what the data say.

Science, looking to statistics for objective ways to represent and quantify evidence, has not embraced Bayesian methods. One obvious reason is the need for prior probabilities in Bayesian analyses. Since these probabilities are usually personal, or subjective (the quest for objective "ignorance" priors or widely acceptable "reference" priors having failed), they are widely seen as incompatible with the scientific need for objectivity. As Efron (1986, 4) put it, "The high ground of scientific objectivity has been seized by the frequentists."

But all is not well with statistics in science. As we saw in the previous section, there is something fundamentally wrong with today's standard frequentist methodology for evidential interpretation of scientific data. Allan Birnbaum (1970), while advocating the use of this methodology, acknowledged that it consists of "an incompletely formalized synthesis of ingredients borrowed from mutually incompatible theoretical sources."

The theoretical problems have created practical ones, as evidenced by the endless controversy over the proper use and interpretation of statistical hypothesis testing in science (Morrison and Henkel, 1970; Cohen, 1994; Bower, 1997; Thompson, 1998; Goodman, 1998; Sterne and Smith, 2001). This controversy flows from the discord between theory and practice—frequentist statistical *theory* is based on Neyman's (1950) behavioristic view that a hypothesis test is a procedure for choosing between the two hypotheses (with evidential interpretation explicitly disallowed), while science's main use of hypothesis tests is for showing the direction and strength of statistical evidence. The theory is aimed at our question 2 ("What should I do?"), but the methods are used to answer question 3. (For further discussion and details see Royall, 1997, chaps. 2, 3.)

The frequentists' claim to occupy the high ground of objectivity has even been challenged. In fact, as Edwards, Lindman, and Savage (1984, 59) pointed out, dependence on stopping rules makes conventional frequentist statistical procedures (which they referred to as "classical procedures") subjective: "The irrelevance of stopping rules is one respect in which Bayesian procedures are more objective than classical ones. Classical procedures . . . insist that the intentions of the experimenter are crucial to the interpretation of data." These authors illustrated their point with an example like our observation of 9 heads in 17 tosses of a 40¢ piece. Suppose you and I were collaborators on that experiment, but we disagreed about which stopping rule should be used. I wanted to fix the number of tosses (at 17), but you wanted to fix the number of heads (at 9). Instead of postponing the experiment until we could agree on a stopping rule, we decided to begin sampling and to continue so long as both stopping rules said that we should (i.e., until we had either 9 heads or 17 observations, whichever came first). At that point we would decide which rule to apply. Because the ninth head occurred on the seventeenth toss (so *both* rules told us to stop), the decision was never made. But our data cannot be analyzed using conventional frequentist methods until it is made, i.e., until someone determines *whose will would have prevailed if different results had been observed*—if the ninth head had occurred earlier, would you have persuaded me to stop the study, or would I have persuaded you to continue? If, after 17 tosses, we had not yet observed 9 heads, would I have persuaded you to stop? Although both stopping rules assign exactly the same probability, $\theta^9(1 - \theta)^8$, to the actual observations, (1, 1, 0, 1, 1, 0, 0, 0, 1, 0, 1, 1, 1, 0, 0, 0, 1), conventional frequentist methods imply, quite wrongly, that what these observations *mean,* as evidence about the tendency of the coin to fall heads, depends on whose will would have prevailed, yours or mine, if different observations had been made.

Why does science continue to use frequentist methods, despite their well-known logical defects? We have seen one important reason—the available alternative, Bayesian statistics, has a more conspicuous subjective component. But there is another, more practical reason: Science has embraced these methods because they provide something that science needs. Specifically, *frequentist statistical methods provide explicit objective measure and control of the frequency of errors (and Bayesian methods do not)*. This is illustrated most clearly in the formulas that are routinely used to determine sample size when a scientific study is planned. The study is conceptualized as a procedure for choosing between two hypotheses, with potential errors of two types. Probabilities of the two types of error are set at specified target levels. These are plugged into a formula that gives the required sample size. The researcher cannot eliminate the *possibility* of errors. But by using frequentist statistical methods, he can calculate and control their *probability*.

Can statistics avoid the logical inconsistencies that pervade current frequentist methods for interpreting data as evidence, while still providing what science requires—objective measure and control of the risk of unsatisfactory results? Yes. The key to accomplishing these two goals is to recognize that probabilities, which properly measure the uncertainties associated with a given procedure for generating statistical evidence, are not appropriate for measuring the strength of evidence produced. As Fisher (1959, 93) put it, "As a matter of principle, the infrequency with which, in particular circumstances, decisive evidence is obtained, should not be confused with the force, or cogency, of such evidence."

We must distinguish between the strength of evidence and the probability that a procedure will produce evidence of a given strength. The problem with current frequentist theory is that, lacking an explicit concept of evidence, it attempts to use the same quantities (probabilities) to measure both the chance of errors and the strength of observed evidence.[1] The solution, presented in the next section, is found in the law of likelihood, which embodies the explicit, objective, quantitative concept of evidence that is missing from current frequentist theory, and which explains that it is likelihood ratios, not probabilities, that measure evidence. Probabilities measure uncertainty; likelihood ratios measure evidence.

TWO APPROACHES TO PLANNING A SIMPLE EXPERIMENT

In this section, we consider the problem of deciding how many observations will be made in a scientific study or experiment. The Neyman-Pearson formulation and solution to this problem represents the paradigm that guides modern frequentist statistical theory (Royall, 1997, chap. 2). We first describe that paradigm in its simplest, clearest form. Then we see how it is changed when the problem is reformulated in terms of statistical evidence.

1. For further discussion, see Royall (1997, chap. 5).

Neyman-Pearson Paradigm

Purpose: The experiment is a procedure for choosing between two hypotheses. We will make some observations, and they will determine which hypothesis is chosen.

Probability model: Independent random variables $X_1, X_2, \ldots, X_n$, identically distributed as X, will be observed. The model consists of two probability distributions for X, corresponding to simple hypotheses H_1 and H_2.

Objective of statistical analysis: We wish to use the observations to choose between H_1 and H_2.

Desiderata: Noting that we can make errors of two types, choosing H_1 when H_2 is true, and vice-versa, we want:

(a) To measure and control the error probabilities. (We want to be pretty sure (probability $\geq 1 - \alpha$) that we won't choose H_2 when H_1 is true and pretty sure (probability $\geq 1 - \beta$) that we won't choose H_1 when H_2 is true.)

(b) To minimize sample size subject to (a).

The immediate goal of many scientific studies is not to choose among hypotheses, but to generate empirical evidence about them. This evidence will be communicated in reports and journal articles. Then various parties will use it, in combination with other information, as well as judgments about the consequences of alternative actions, in making a variety of choices and decisions, many that were not even imagined by the authors of the study. An epidemiological study, for example, might produce strong statistical evidence that cigarette smokers have a much greater risk of dying from a certain type of cancer than nonsmokers. The published evidence will be used by many parties—legislators, lawyers, tobacco farmers, smokers, nonsmokers, researchers in other fields (oncology, genetics, etc.), insurance companies—in making innumerable choices and decisions. From this perspective, a more realistic formulation of the problem of planning a scientific study uses the evidence-generating paradigm.

Evidence-Generating Paradigm

Purpose: The experiment is a procedure for generating empirical evidence about the two hypotheses. We will make some observations and interpret them as evidence.

Probability model: Independent random variables $X_1, X_2, \ldots, X_n$, identically distributed as X, will be observed. The model consists of two probability distributions for X, corresponding to simple hypotheses H_1 and H_2.

Objective of statistical analysis: We wish to interpret the observations as evidence regarding H_1 vis-à-vis H_2.

Desiderata: We will make no errors—we will interpret the evidence correctly. But the evidence itself can be unsatisfactory in two ways: It can be weak (evidence that does not strongly support either hypothesis), or it can be misleading (strong evidence for H_1 when H_2 is true or vice-versa). We want:

(a) To measure and control the probabilities of observing weak or misleading evidence. (If either hypothesis is true we want to be pretty sure (probability $\geq 1 - W$) that we won't find weak evidence, and pretty sure (probability $\geq 1 - M$) that we won't find strong evidence in favor of the other one, or misleading evidence.)

(b) To minimize sample size subject to (a).

To compare this evidence-generating paradigm to the preceding Neyman-Pearson decision-making paradigm, we consider the familiar example of observing normally distributed random variables with known standard deviation. Hypothesis H_1 specifies one value for the mean, and H_2 specifies another value, larger by a fixed multiple, Δ, of the standard deviation (i.e., $\mu_2 = \mu_1 + \Delta\sigma$).

First, we apply the Neyman-Pearson approach. For a given probability of the first type of error, α, the probability of the second type, β, is determined by the sample size n: $\beta(n) = \Phi(z_{1-\alpha} - n^{1/2}\Delta)$ where $z_{1-\alpha}$ is the $100(1 - \alpha)$th percentile of the standard normal distribution. The standard formula (e.g., Pagano and Gauvreau, 1993, 225) shows that the minimum sample size that will control the error rates at the target values (α, β) is $n = (z_{1-\alpha} + z_{1-\beta})^2/\Delta^2$. For example, if we take $\alpha = \beta = .05$, then for $\Delta = 1$, $n = (1.645 + 1.645)^2 = 10.8$; 11 observations are sufficient.

If we take the evidential approach, with the suggested benchmark value $k = 8$, representing pretty strong evidence, then the probability of strong misleading evidence (likelihood ratio greater than 8 in favor of the false hypothesis) is the same under both hypotheses. For a sample of n observations, this probability is: $M(n) = \Phi(-\sqrt{n}\Delta/2 - \ln(8)/(\sqrt{n}\Delta))$.

The probability of weak evidence (likelihood ratio between 1/8 and 8) is also the same under both hypotheses:

$$W(n) = \Phi(\sqrt{n}\Delta/2 + \ln(8)/(\sqrt{n}\Delta)) - \Phi(\sqrt{n}\Delta/2 - \ln(8)/(\sqrt{n}\Delta)).$$

These two functions, $M(n)$ and $W(n)$, are shown in figure 5.2 for $\Delta = 1$. Dashed lines indicate the Neyman-Pearson error probabilities, $\alpha = .05$ (constant) and β_n, for comparison. Note that $\beta_n = .05$ when $n = 10.8$, as found in the previous paragraph.

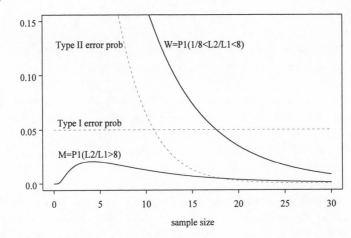

FIGURE 5.2 Probabilities of weak and misleading evidence with normal means (Differences = one standard deviation).

Several things are clear from figure 5.2.

(a) Both probabilities, M and W, can be held below any desired positive bound, no matter how small, by making enough observations.

(b) The probability of misleading evidence, M, is very small for small n, *increasing* with n until reaching a maximum (at $n = 2 \ln(8)/\Delta^2$), and decreasing thereafter.

(c) The *maximum* probability of misleading evidence, over all values of n, is small. (It is actually $\Phi(-\sqrt{2 \ln(8)})$, or .021.) It is attained when the sample size is so small that the probability of weak evidence is large $(W(n) = \Phi(\sqrt{2 \ln(8)})) = .979$).

(d) Because of (c), the sample size calculation is driven by the constraint on the probability of weak evidence. We need large samples in order to have a good chance of getting strong evidence. *The nature of statistical evidence is such that the chances of getting misleading evidence are small at all sample sizes.* This remains true in general: for any $k > 1$, the maximum probability of misleading evidence (over all n) is $\Phi(-\sqrt{2 \ln(k)})$. Even for the unreasonably small benchmark value of $k = 4$, this probability cannot exceed .05.

(e) There are essential differences between the "analogous" quantities α and M. We can fix α, the probability that when H_1 is true we will choose H_2, at any level we like. But as noted in (d) there are natural limits on M, the probability that when H_1 is true we will find strong evidence in favor of H_2.

(f) The standard calculation gives a sample size that is too small to ensure that the experiment will, with high probability, produce strong evidence about these two hypotheses. At $n = 11$, where β, the probability of an error of the second type, falls just below .05, the probability of finding only weak evidence is about three times as great: $W(11) = .14$.

PLANNING A STUDY VS. INTERPRETING OBSERVATIONS AS EVIDENCE

The probabilities of weak and misleading evidence are important in planning a study, but they play no role in the proper interpretation of the study's results. In Hacking's (1965) words, probabilities are for "before trial betting," while likelihoods are for "after trial evaluation." An example can make the distinction more tangible: Suppose we are going to observe normal random variables with unit standard deviation, and we are interested in the mean, θ. We are particularly interested in the two hypotheses H_1: $\theta = 0$ and H_2: $\theta = 1$. Since the difference between these means is one standard deviation ($\Delta = 1$), figure 5.2 applies. Figure 5.2 shows that if we want to ensure that the probability of observing weak evidence with respect to these two hypotheses is less than .05, we must make at least $n = 18$ observations, and that with this sample size the probability of our observing misleading evidence (strong evidence in favor of one hypothesis when the other is true) is only $M(18) = .0045$.

I carried out such a study, generating 18 normal random variables with a common mean θ and unit standard deviation. The sample mean was 0.233, and the likelihood function, $L(\theta) = \exp[-18(0.233 - \theta)^2/2]$ (shown in figure 5.3), represents the evidence about the value of θ. These observations represent strong evidence supporting H_1 over H_2 ($L(0)/L(1) = 120$).

The fact that we considered only two specified values of θ when we planned the study does not preclude our examining the evidence as it relates to others, and figure 5.3 shows that some intermediate values are better supported than either $\theta = 0$ or $\theta = 1$, but the evidence supporting those values over $\theta = 0$ is weak. Nor does the fact that we planned to make only $n = 18$ observations preclude our deciding to make some more.

I made 4 more observations, looking at the evidence (the likelihood function) after each one. At that point I decided to double the sample size originally planned, and made 14 more observations. Figure 5.4 shows the likelihood function for all 36 observations. The evidence in favor of H_1 vs. H_2 is still strong ($L(0)/L(1) = 120$).

Finally, just to see what would happen, I decided to increase the sample

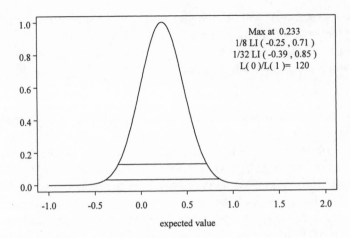

FIGURE 5.3 Likelihood for normal mean, $n = 18$ observations.

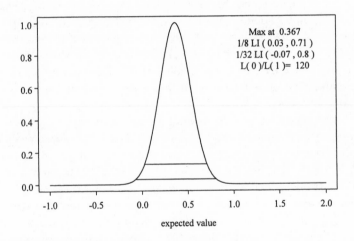

FIGURE 5.4 Likelihood for normal mean, $n = 36$ observations.

to $n = 100$. The evidence supporting $\theta = 0$ over $\theta = 1$ is now overwhelming, but figure 5.5 shows that we have strong evidence supporting values near $1/4$ over $\theta = 0$.

What these 100 observations say about the value of θ is shown in figure 5.5. The probabilities in figure 5.2, which I used in choosing the initial sample size, are not relevant to the interpretation of this evidence. (For more on this point see Royall, 1997, sec. 4.5; 2000, rejoinder.) Nor is the fact that I considered only the two values $\theta = 0$ and $\theta = 1$ at the planning stage. The

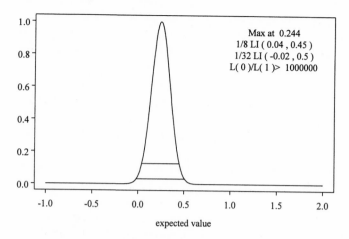

FIGURE 5.5 Likelihood for normal mean, $n = 100$ observations.

same is true of the stopping rule—whatever these observations mean, as evidence about their expected value θ, it is independent of the stopping rule. This evidence is not weakened (or otherwise affected) by the fact that I "peeked" at the data along the way, or that my original target likelihood ratio was only 8. Such facts are critical for determining confidence coefficients, P-values, Neyman-Pearson error probabilities (α, β), etc., but they have no effect on the likelihood function and no valid role to play in interpreting these 100 observations as evidence about θ.[2]

CONCLUSION

In a critique of Neyman-Pearson theory John Pratt (1961) rightly observed, "of course, skillful people can do useful statistics using Neyman and Pearson's formulation. But so can they using Fisher's or Jeffreys', or minimax decision theory, or subjective probability and Bayes' Theorem." Nevertheless, when the purpose of a statistical analysis is to represent and interpret data as evidence, all of these methods have serious shortcomings. These can be overcome within the frequentist approach by making explicit what has heretofore been treated only implicitly and intuitively—the concept of statistical

2. These observations are actually 100 random normal deviates with mean $\theta = 0.3$, and standard deviation 1.

evidence. The concept embodied in the law of likelihood, and represented in the terms "weak evidence" and "misleading evidence" that are central to the likelihood "evidence-generating" paradigm, can lead to a body of statistical theory and methods that:

1. requires probability models for the observable random variables only (and is in that sense frequentist, not Bayesian);
2. contains a valid, explicit, objective measure of the strength of statistical evidence;
3. provides for explicit, objective measure (and control) of the probabilities of observing weak or misleading evidence.

Under the likelihood paradigm, changing the stopping rule changes the probabilities of observing weak or misleading evidence, but these probabilities are used for planning only and do not affect the objective assessment of the observations as statistical evidence. There is no sound statistical reason to stop a study when the evidence is weak—it is quite appropriate to choose an initial sample size (perhaps on the basis of the *probability* of finding strong evidence, as in figure 5.2), evaluate the evidence in the observed sample, and then decide whether or not to make more observations. The scientist who carefully examines and interprets his observations (via likelihood functions) as they arise is behaving appropriately. He is not weakening or damaging the statistical evidence, he is not spending or using up statistical capital, he should not be chastized for peeking at the data, and there is no valid basis for levying a statistical fine on his study.

| 5.1 | Commentary |

D. R. Cox

First, it should be stressed that the law of likelihood is a very appealing working hypothesis, but in its strong sense it is no more than that. From several points of view two sets of data having equivalent likelihood functions should be regarded as providing the same evidence so long as the probability model from which they are derived is a secure base for interpretation. It is not at all so obvious that the same is true when the likelihoods come from different probability models. Indeed, it is a basic precept of the design of experiments and of survey sampling that the way the data are obtained should be taken account of in analysis and in that sense dependence on a stopping rule, although a nuisance, is to be expected. Of course in many, but not all, contexts the dependence is so slight that, however interesting its presence or absence from a conceptual point of view, its practical relevance when treated correctly is negligible.

It is entirely sensible to begin the discussion of fundamental issues by looking at a very idealized situation, one with no nuisance parameters and just two possible hypotheses. Yet to some extent an approach has to be judged by the range of issues to which it will contribute. In some special cases the presence of nuisance parameters can be evaded by suitable modifications of the likelihood function, obtained, for example, by conditioning or marginalizing, but to be valuable an approach needs to go far beyond that. In some ways more serious is the matter of questions made precise only in the light of the data. For example, a particular value of a parameter may be compared with the maximum likelihood estimate. It is not at all clear that a given numerical value of the likelihood ratio has the same interpretation as in the simpler case. In both cases further calibration of likelihood-based statistics seems needed, and it is of course at this point that Richard Royall and I part company.

The discussion of the choice of sample size is illuminating, although for many purposes a formulation in terms of estimation is more appealing. That is, the objective is regarded as estimation of the parameter with a concentration of the likelihood function that depends in an approximately preassigned way on the location of the likelihood. Then some difficulties disappear. The final comment in the paper justifying continuous inspection of the data when data accrue sequentially also has much appeal. Indeed, I believe that many statisticians approaching statistics from a broadly frequentist perspective are uneasy at notions such as "spending error rates," perhaps because these treat notions of error rate as more than just hypothetical concepts used for calibrating measures of uncertainty against performance in idealized situations. While in some situations there may be compelling quasi-political arguments, as well as cost considerations, pointing against too frequent an analysis, in principle it is hard to see an argument at a completely fundamental level.

A full discussion of the evaluation of evidence needs to include procedures for checking the adequacy of any proposed model. These can be forced into a likelihood formulation only in a very restrictive sense.

Two further references that readers may find helpful are Barnard (1947) and Birnbaum (1969). Barnard was the first, I believe, to give the interesting inequality relating error rates to likelihood ratio. Birnbaum describes in

An approach has to be judged by the range of issues to which it will contribute.

more detail why after much careful thought he had come to the conclusion that, the likelihood principle not withstanding, "sensible" confidence intervals were the most fruitful approach to the evaluation of evidence.

5.2	Commentary

Martin Curd

Professor Royall's paper is both elegant and challenging. In my remarks, I focus on the contrast between the likelihood conception of evidence and the Bayesian account. I do this because, despite the rarity of the Bayesian approach among professional statisticians (Moore, 1997), it has been a dominant influence on recent attempts by philosophers to understand the confirmation of scientific theories (Horwich, 1982; Salmon, 1990; Earman, 1992; Howson and Urbach, 1993). My main concern is whether the likelihood paradigm of evidence can be plausibly extended from the realm of statistics to scientific theories in general.

According to the standard Bayesian approach, a hypothesis H is confirmed by an observation E (and thus E is evidence for H) if and only if the posterior probability $P(H|E)$ is greater than the prior probability $P(H)$. The two are related by the Bayesian equation $P(H|E) = [P(H) \times P(E|H)]/P(E)$, where $P(E) > 0$. $P(E|H)$ is the likelihood of H, and $P(E)$ the expectedness of the evidence. (To avoid clutter, the reference to background knowledge in each of the factors in Bayes' equation has been omitted.) Although other measures have been proposed, Bayesians often take the strength of E as evidence for H (the degree of confirmation that E confers on H) to be some positive increasing function of $P(H|E) - P(H)$, the difference between the posterior probability of H and its prior probability (Howson and Urbach, 1993; Christensen, 1999). Of the three terms on the right-hand side of Bayes' equation, the likelihood $P(E|H)$ is the least controversial since scientific theories are often formulated such that (in conjunction with appropriate auxiliary hypotheses and background assumptions) we can deduce from them the probability of events such as E. Thus, $P(E|H)$ can be regarded as being objective. But what about $P(H)$ and $P(E)$? Some Bayesians are unashamedly subjectivist about $P(H)$ and regard the prior probability of H as simply the degree of belief that a particular scientist happens to have in H. Others have tempered this subjectivism by placing constraints on permissible priors (for example, no one is allowed to assign a prior of zero to any noncontradictory hypothesis, since this would prevent that hypothesis from ever being confirmed) and by regarding $P(H)$ as our informed judgment about the plausibility of H in light of such factors as simplicity, explanatory power, and the track record of similar hypotheses in the past. Bayesians also appeal to the "swamping" argument in mitigation of the charge of subjectivism. For, as each new piece of evidence E_i is considered, Bayesians replace the "old prior" $P(H)$ with the "new prior" $P(H|E_i)$, calculated using Bayes' equation. Although scientists may initially assign widely different (nonzero) priors to H, the values of $P(H|E_i)$ will converge as evidence accumulates (as long as the scientists agree on the likelihoods). Thus, subjective disagreements about the plausibility of hypotheses (reflected in different initial assignments of priors) have a diminishing influence on the confirming power of new evidence. In the long run, the initial priors become irrelevant.

Much ink has been spilled on the difficulty of interpreting $P(E)$ since Glymour (1980) first drew attention to the problem of old evidence; that is, the problem of assigning to $P(E)$ a value less than 1 when E becomes known and thus presumably certain for the scientist who knows it. One approach is to expand $P(E)$ as a sum of terms of the form $P(H_i) \times P(E|H_i)$, where the set $\{H_i\}$ includes H and all the logically possible alternative hypotheses that also predict E. But in practice, the number of alternatives that have been formulated is small, and so typically we have to invoke the notorious catchall hypothesis, H_c, to complete the set. Unfortunately, the catchall is not a definite hypothesis at all but says merely "none of the above"; and so we cannot use it to calculate $P(E|H_c)$ and thus complete the calculation of $P(E)$. Another approach (Howson and Urbach, 1993) is to say that $P(E)$ is the probability that would be assigned to E independently of consideration of H. We are enjoined to imagine that we delete H (and anything that entails H) from our corpus of belief and then see what probability would be assigned to E given our background knowledge. While the details of this deletion procedure are hard to make precise, there are cases in which our background information can provide a reasonable basis for estimating $P(E)$. (See the discussion of the raven paradox at the end of this commentary.)

Royall thinks that the standard Bayesian account of evidence is fundamentally flawed because of its reliance on subjective probabilities such as $P(H)$. He argues that while Bayes' equation describes how one should change one's beliefs in response to new information, it is a mistake to think that any function of prior and posterior probabilities could measure (objective) evidential strength. Rather, for rival hypotheses H and J, that role is played by the likelihood ratio, $P(E|H)/P(E|J)$. As stated by Royall, the law of likelihood says two things. First, it gives a sufficient condition for an event, E, to confer greater evidential support on hypothesis H than on hypothesis J. It says that if $P(E|H) > P(E|J)$, then E supports H more than it does J. The second part says that the likelihood ratio measures the strength of that comparative support.

A striking feature of the likelihood paradigm is its insistence that evidential support is essentially comparative. If Royall is right, there is literally no such thing as the support that a prediction E gives to a hypothesis H; there is only the degree of support that E gives to H as compared with a rival hypothesis, J. Note that there is an important difference between evidential support being by its very nature comparative and its being contextual. Bayesians agree that evidential support is contextual since the derivation of predictions from H and the estimation of $P(E)$ will depend on other hypotheses and background information. But although the Bayesian algorithm requires contextual input, its output (telling us the degree to which E supports H) is noncomparative: it is not limited, in principle, to comparing E's support of H with E's support of J. That we can make comparative judgments in contexts where absolute measures are difficult or practically impossible should occasion no surprise. For example, we can judge which of two sticks is the longer or which of two objects is the heavier without having first to ascertain the length or weight of each. But in these cases, underlying a comparative measure there is a pair of absolute values that grounds the comparison. Why, according to the likelihood paradigm, should statistical evidence be so different in this regard? How, one might naïvely ask, can we make sense of the notion that E supports H more than it does J if there is no such thing as E's support for H or E's support for J? Moreover, if we try to generalize the likelihood conception of comparative evidence into a general account of evidence for scientific theories, its scope (unlike that of the Bayesian account) would be severely limited. For it would apply only to the comparison of pairs of rival theories with respect to the same evidence, leaving unanswered such questions as "Is H supported more strongly by evidence E than J is by evidence F?"

As stated by Royall, the law of likelihood gives a sufficient condition for greater evidential support, not a necessary condition, and so it implies nothing about cases in which the likelihoods of competing hypotheses are the same. But it would appear to be in the spirit of the likelihood account that, when the likelihoods $P(E|H)$ and $P(E|J)$ are equal, neither H nor J receive greater support from E. Either each receives no support from E or they each receive the same degree of support. As far as E is concerned, H and J are evidentially on a par. Similarly, for separate pieces of evidence E and F, if $P(E|H)$ and $P(F|H)$ are the same, so is their confirming power for H. But this consequence of the likelihood paradigm—same likelihood, same confirming power—seems contrary to our intuitions about evidence. To illustrate the problem for the likelihood account, consider two cases: the problem of irrelevant conjunction and the raven paradox.

The problem of irrelevant conjunction arises when we have two hypotheses, H and $(H \& I)$, where I is a contingent statement that is logically independent of H and irrelevant to E. For simplicity's sake, I shall focus on the case in which H is deterministic, but the same problem can arise when H is statistical. H (in conjunction with background information) entails a true observational prediction, E. It follows that the augmented theory $(H \& I)$ also makes the same prediction, and the two likelihoods, $P(E|H)$ and $P(E|H \& I)$, are each equal to 1. Nonetheless, it is widely held that E confirms $(H \& I)$ less strongly than it does H. For example, the observation of a white swan provides weaker support for "All swans are white and some gazelles are brown" than it does for "All swans are white." The Bayesian account can do justice to this intuition by appealing to the role of prior probabilities in confirmation. For on the Bayesian analysis we have $P(H|E) - P(H) = P(H) \times [(1 - P(E))/P(E)]$, and $P(H \& I|E) - P(H \& I) = P(H \& I) \times [(1 - P(E))/P(E)]$. The factor in the square brackets is the same for both theories, and so their respective degrees of confirmation are proportional to the prior probabilities, $P(H \& I)$ and $P(H)$. Since $(H \& I)$ entails H, and I is a contingent statement that is independent of H, $P(H \& I)$ must be less than $P(H)$. Thus, E confirms $(H \& I)$ by a smaller amount than it confirms H; adding the irrelevant conjunct to H lowers the confirmation provided by E. In this respect, then, the Bayesian approach to confirmation appears to be superior to the likelihood account.

The key to the Bayesian solution of the raven paradox lies in the different values assigned to $P(E)$ by our background knowledge. Let H be the hypothesis "All ravens are black," which we shall write as "All Rs are B." Since H is logically equivalent to its contrapositive, "All non-Bs are non-Rs," it would seem that H should be confirmed not only by the observation of a black raven but also by the observation of a nonblack nonraven (such as a white shoe). Bayesians see no paradox here. They argue that the observation of a white shoe does confirm H, but only to a very small degree, a degree that is much smaller than the confirmation conferred on H by the observation of a black raven. In this way, Bayesians explain why many people regard the raven case as paradoxical when they first encounter it; for it is understandable that most of us are unable to distinguish a very low degree of confirmation from no confirmation at all.

Following Horwich (1982) we adopt a notation that reflects the manner in which evidence is collected: $(R * B)$ is the discovery that a randomly selected object that is already known to be a raven is black; $(\sim B * \sim R)$ is the discovery that a randomly selected object that is already known to be nonblack is a nonraven. The asterisk indicates which component of each paired observation is made first. To repeat, the observation reports include information about the method used to generate the report. That said, it must be the case that both likelihoods, $P(R * B | H)$ and $P(\sim B * \sim R | H)$, are equal to 1; for, if all ravens are black, then the probability that a raven will turn out to be black is 1; similarly, given H, the probability that a nonblack thing will turn out to a nonraven is also 1. Thus, the Bayesian comparison of the confirming power of the observation report $(R * B)$ with the confirming power of the observation report $(\sim B * \sim R)$ depends on the inverse ratio of their probabilities. In order to estimate those probabilities, we need to specify some background information. Let x be the fraction of things in the universe that are ravens, let y be the fraction of things that are black, and let α be the fraction of things that initially are believed to be black ravens. This yields $P(R * B) = \alpha/x$, and $P(\sim B * \sim R) = [(1 - y) - (x - \alpha)]/(1 - y)$. Comparison of these two expressions shows that $(R * B)$ must support H more strongly than does $(\sim B * \sim R)$ if x is greater than α and $(1 - y)$ is greater than x. Thus, if we do not already believe that all ravens are black, $(R * B)$ is stronger evidence for H than is $(\sim B * \sim R)$ as long as we also believe that nonblack things are more abundant than ravens. Both sorts of observation should increase our confidence in H, but finding black ravens provides stronger support.

It is instructive to compare this Bayesian treatment of the raven paradox with the likelihood analysis in Royall (1997). On Royall's likelihood analysis, the likelihoods of $(R * B)$ and $(\sim B * \sim R)$ are not both equal to 1. They differ because the likelihood of each observation report is assessed with respect to H as compared with the rival hypothesis J (according to which the proportion of ravens that are black is some fraction less than 1); and, in the case of $(\sim B * \sim R)$ reports, the calculation of the likelihood $P(\sim B * \sim R | J)$ depends on the same background information about the relative abundance of ravens and nonblack things as in the Bayesian analysis. Royall agrees that when the sampling procedure yields reports of the form $(\sim B * \sim R)$, the observation of a nonblack nonraven has the power to confirm H; and that when the number of nonblack things in the universe vastly exceeds the number of ravens, the confirming power of such observations is very weak. But these conclusions about evidential support, based on the likelihood paradigm, are essentially comparative: observations of the type $(\sim B * \sim R)$ provide only marginally stronger support for H than for J. Similarly, observations of the type $(R * B)$ support H more strongly than they do J. Unlike the

Bayesian account, the likelihood analysis yields no conclusion about the relative confirming power for H of the two different types of observation. Because it is limited to judging the relative support given to rival hypotheses by the same observation report, the likelihood analysis leaves unanswered the crucial question that lies at the heart of the raven paradox: why do reports of the form $(R * B)$ confirm H much more strongly than do reports of the form $(\sim B * \sim R)$?

5.3   Rejoinder

Richard Royall

Professor Cox rejects the law of likelihood out of hand, expressing the opinion that it represents nothing more profound than a "working hypothesis." He goes on to describe some aspects of my paper that he finds appealing and some that he does not.

It is disappointing that Cox chooses to reveal so little about the rationale for his judgments. For instance, he apparently embraces a "basic precept of the design of experiments . . . that the way the data are obtained should be taken account of in analysis." In my opinion it would be quite useful to the discipline of statistics if Cox would formalize and elaborate on this "basic precept"[3] in such a way that its implications could be seen (and its validity tested) in specific examples such as the one I use to illustrate the irrelevance of stopping rules in analyses *whose purpose is to interpret observed data as evidence.* Does his precept imply that the meaning of the coin-toss observations that you and I made together, (1, 1, 0, 1, 1, 0, 0, 0, 1, 0, 1, 1, 1, 0, 0, 0, 1), as evidence about the tendency of the 40¢ coin to fall heads depends on whether we stopped the experiment because we had made 17 tosses (my preferred stopping rule), or because we had seen 9 heads (your rule)? . . . or because the coin fell apart? . . . or because it was time for tea? If his precept does indeed have such implications then it would suggest to me that Cox should correct the precept, not reject the law of likelihood.

In the meantime, statistics remains in a theoretical and conceptual mess —a "synthesis of ingredients borrowed from mutually incompatible theoretical sources" (Birnbaum, 1970, 1033). With no accepted principles to

3. In Royall (1991), I tried to formalize and criticize such a precept, which I called the "randomization principle."

guide statistical reasoning, we can offer scientists who are perplexed by our controversies (such as the one about "spending error rates") nothing more than conflicting expert judgments about what is sensible. I cannot share Professor Cox's satisfaction with this state of our discipline.

I now reply to comments by Martin Curd. The law of likelihood answers a fundamental question about empirical evidence: When does an observation constitute evidence supporting one hypothesis vis-à-vis another? The law says it is when the two hypotheses imply different probabilities for the observation. It says that the hypothesis that implies the greater probability is the better supported and that the probability ratio measures the strength of the evidence.

Professor Curd contrasts this concept of evidence to the Bayesian account, where only one hypothesis is made explicit, the question being, When does an observation constitute evidence supporting a hypothesis? The Bayesian answer is, When the observation has the effect of increasing the probability of the hypothesis. Preferring the Bayesian account, Professor Curd challenges the law of likelihood by presenting two examples where it purportedly leads to conclusions that seem "contrary to our intuitions about evidence."

I want to suggest that the law of likelihood is not a threat, or even an alternative, to the Bayesian view, but that to the contrary it constitutes the essential core of the Bayesian account of evidence. My claim is that the Bayesian who rejects the law of likelihood undermines his own position. Then I will argue that this act of self-destruction is unwarranted, because Professor Curd's argument leading to rejection of the law springs from a misunderstanding. In this discussion I will assume that the probabilities obey the usual (Kolmogorov) axioms. Since Bayes' theorem is derived from those axioms, I find it hard to imagine a Bayesian who rejects them.

According to Professor Curd, the Bayesian interpretation is that an observation E as evidence for H when $P(H|E) > P(H)$. His analysis assumes the existence of the three terms, $P(E|H)$, $P(H)$, and $P(E)$, that appear on the right-hand side of what he calls Bayes' equation,

$$P(H|E) = P(E|H)P(H)/P(E).$$

Now if H has probability $P(H)$ then the axioms imply that its negation $\sim H$ has probability $1 - P(H)$, and that furthermore $P(E)$ can be expressed as[4]

---

4. This result is a special case of what is sometimes called the theorem on total probability, which is essential to Bayes' theorem (Kolmogorov, 1956, sec. 4).

$$P(E) = P(E|H)P(H) + P(E|{\sim}H)[1 - P(H)].$$

Rearranging this expression, we see from the three quantities that the Bayesian must supply, $P(E|H)$, $P(H)$, and $P(E)$, we can deduce the value for the probability of E that is implied by the hypothesis [5] $\sim H$:

$$P(E|{\sim}H) = [P(E) - P(E|H)P(H)]/[1 - P(H)].$$

If we return to the critical inequality, $P(H|E) > P(H)$, and replace $P(H|E)$ with the expression given by Bayes' equation, $P(H|E) = P(E|H)P(H)/P(E)$, we see that the Bayesian interprets E as evidence for H if and only if $P(E|H) > P(E)$. And if in this last inequality we substitute the expression for $P(E)$ displayed above, we see that the Bayesian's criterion is equivalent to $P(E|H) > P(E|H)P(H) + P(E|{\sim}H)[1 - P(H)]$, which is equivalent to

$$P(E|H) > P(E|{\sim}H).$$

That is, $P(H|E) > P(H)$ if and only if $P(E|H) > P(E|{\sim}H)$.

When is E "evidence for H" in the Bayesian scheme? When does $P(H|E)$ exceed $P(H)$? It is precisely when H implies a greater probability for E than the alternative hypothesis $\sim H$ does. It is precisely when the law of likelihood says that E is evidence supporting H over $\sim H$.

The Bayesian may prefer not to make $P(E|{\sim}H)$ explicit—he may prefer to leave this object buried within $P(E)$. But whether he chooses to make it explicit or not, the value of $P(E|{\sim}H)$ is determined by $P(E|H)$, $P(E)$, and $P(H)$. And when all the cards are laid on the table, the winner of the Bayesian's game is decided according to the law of likelihood—the hypothesis (H or $\sim H$) that is better supported by the observation E is the one that implies the greater probability for that observation. *Hypothesis H "is confirmed by E" if and only if E is more probable if H is true than if H is false.* The Bayesian's qualitative conclusion (e.g., that E is evidence *for* H rather than against it) must conform to the law of likelihood.

---

5. The same deviation applies when there is a set of disjoint hypotheses $\{H_i\}$. If $\sum P(H_i) <$ 1, then from $P(H_i)$, $P(E|H_i)$, and $P(E)$, we can deduce the probability of E under the "notorious catchall hypothesis, H_c," i.e., $P(E|H_c) = [P(E) - \sum P(E|H_i)P(H_i)]/[1 - \sum P(H_i)]$. The Bayesian cannot simultaneously claim (i) to know the terms on the right-hand side of this equation and (ii) that $P(E|H_c)$ is nonexistent or unknowable without violating the axioms of probability theory.

But are the two accounts of evidence in *quantitative* accord? One says un-equivocally that the strength of the evidence for H versus $\sim H$ is measured by the likelihood ratio $P(E|H)/P(E|\sim H)$.[6] The other, according to Professor Curd, is less definite: "Bayesians often take the strength of E as evidence for H to be some positive increasing function of $P(H|E) - P(H)$." As we have seen, the difference $P(H|E) - P(H)$ does point in the right direction. But as a measure of the *strength* of the evidence it is curious. It says that E cannot be strong evidence for H when $P(H)$ is large, i.e., when there is strong prior evidence for H. Thus E can be strong evidence for H when it flies in the face of extensive previous experience ($P(H)$ is small), but not when it is consis-tent with that experience ($P(H)$ near 1). Now, it is plausible that E is, in some sense, more valuable, more important, or more newsworthy when it elevates a previously implausible hypothesis to respectability than when it merely confirms what was already believed. But that E is *stronger evidence* in the former case is not at all clear.

This counterintuitive aspect of the way the difference, $P(H|E) - P(H)$, depends on the prior probability $P(H)$ suggests that the Bayesian should adopt some other measure. In fact the Bayesian I. J. Good (1968) proposed some desiderata for an evidence measure and proved they imply that the measure should be an increasing function (the logarithm) of the likelihood ratio $P(E|H)/P(E|\sim H)$, which is independent of $P(H)$.

Thus, it is clear to me that the Bayesian cannot escape either the qualita-tive or quantitative conclusions of the law of likelihood. Professor Curd, on the other hand, sees a problem with "the likelihood account" of evidence, and he gives two examples intended to illustrate that problem. What I think they actually illustrate is the misconception that is expressed in his state-ment of the problem: "for separate pieces of evidence E and F, if $P(E|H)$ and $P(F|H)$ are the same, so is their confirming power for H. But this con-sequence of the likelihood paradigm—same likelihood, same confirming power—seems contrary to our intuitions about evidence."

What *does* the law of likelihood say when one hypothesis attaches the same probability to two different observations? It says absolutely nothing. The law of likelihood applies when two different hypotheses attach proba-bilities to the same observation. It states that the ratio of the probabilities, $P(E|H)/P(E|J)$, measures the evidence in the observation E for hypothesis H

---

6. The answer to Professor Curd's question, "Is H supported more strongly by evidence E than J is by evidence F?" is straightforward: E supports H versus $\sim H$ more strongly than F supports J versus $\sim J$ if and only if $P(E|H)/P(E|\sim H) > P(F|J)/P(F|\sim J)$.

vis-à-vis hypothesis J. Although it says that their ratio measures the evidence, the law attaches no evidential meaning to either probability in isolation. In the absence of an alternative hypothesis, $P(E|H)$ is not the support for H, the strength of the evidence for H, or the confirming power of E for H.

Consider an example: Let H be the hypothesis that the proportion of white balls in an urn is 1/2. For the two observations

E: 1 white ball in 1 draw and
F: 10 or fewer white balls in 21 independent draws,

the probabilities $P(E|H)$ and $P(F|H)$ are the same—both equal 1/2. But the law of likelihood says nothing about the "confirming power" of E for H or of F for H. It does address the interpretation of these observations as evidence for H vis-à-vis any other hypothesis that also implies probabilities for them, such as the hypothesis J that the proportion of white balls is 1/4. Since $P(E|J) = .25$ and $P(F|J) = .994$, the likelihood ratios are $P(E|H)/P(E|J) = 2.0$ and $P(F|J)/P(F|H) = 1.987$, so the law of likelihood says that E supports H over J, that F supports J over H, and that the evidence has nearly the same (weak) strength in both cases.

Now consider a different alternative to H. Let K be the hypothesis that the proportion of white balls is 3/4 so that $P(E|K) = .75$ and $P(F|K) = .0064$. Observation E is weak evidence supporting K over H (likelihood ratio = 1.5), while F is strong evidence for H versus K (likelihood ratio = 78). Thus, although observations E and F are equally probable under hypothesis H, E is evidence *for* H vis-à-vis hypothesis J, and *against* H vis-à-vis K. And in each case the observation F has the opposite evidential meaning.

The likelihood view is that observations like E and F have no valid interpretation as evidence in relation to the single hypothesis H. I have discussed elsewhere (Royall, 1997, sec. 3.3) the futility of efforts to attach evidential meaning to an observation in relation to a single hypothesis that is not logically incompatible with the observation.

Finally, I want to compare the Bayesian and likelihood analyses of the raven paradox. Professor Curd discusses the paradox in terms of the proportions in a 2×2 table:

	black	nonblack	
ravens	α	$x - \alpha$	x
nonravens	$y - \alpha$	$1 - x - y + \alpha$	$1 - x$
	y	$1 - y$	1

Here x represents the proportion of the objects in the population under consideration that are ravens, y is the proportion of objects that are black, and α is the proportion that are black ravens.

An observation $R * B$ represents an object drawn at random from the first row that is found to come from the first column (a raven that is found to be black), and an observation $\sim B * \sim R$ represents a draw from the second *column* that is found to come from the second row (a nonblack thing that proves to be a nonraven). The likelihood analysis is as follows:

The hypothesis H (all ravens are black) asserts that α equals x, implying that the probability of $R * B$ is 1.

The alternative hypothesis, J, asserts that α has a value less than x, which implies that the probability of $R * B$ is $\alpha/x < 1$.

Therefore the law of likelihood says the observation $R * B$ is evidence supporting H over J by the factor: $P(R * B|H)/P(R * B|J) = 1/(\alpha/x) = x/\alpha$.

Similarly H implies that the probability of $\sim B * \sim R$ is 1, while J implies that the probability is $(1 - x - y + \alpha)/(1 - y) < 1$, so the law of likelihood says the observation $\sim B * \sim R$ is evidence supporting H over J by

$$\frac{P(\sim B * \sim R|H)}{P(\sim B * \sim R|J)} = \frac{1}{(1 - x - y + \alpha)/(1 - y)} = \frac{1 - y}{1 - x - y + \alpha}.$$

These are exactly the same expressions on which Professor Curd bases his Bayesian analysis. His measures of "the relative confirming power for H of the two different types of observation" are just the two likelihood ratios.

As I noted earlier, the Bayesian may choose not to speak of the alternative to H. His use of the expression "the relative confirming power for H of the two different types of observation," without reference to an alternative, obscures the fact (revealed in the appearance of the quantity α in his formulae) that the observation can properly be said to confirm H only in relation to an alternative hypothesis that assigns lower probability to that observation than H does. It obscures the fact that his statement "reports of the form $(R * B)$ confirm H much more strongly than do reports of the form $(\sim B * \sim R)$" means nothing more or less than that the first report's likelihood ratio, $P(R * B|H)/P(R * B|J) = x/\alpha$, is the greater.

The Bayesian's failure to make explicit the alternative hypothesis that is implicit in his analysis obscures the fact that the rock on which his analysis is built is the law of likelihood.

REFERENCES

Barnard, G. A. 1947. Review of *Sequential Analysis,* by A. Wald. *J. Am. Stat. Assn* 42: 658–664.

Berger, J. O., and Bernardo, J. M. 1992. On the Development of Reference Priors. In J. M. Bernardo, J. O. Berger, A. P. Dawid, and A. F. M. Smith, eds., *Bayesian Statistics 4.* New York: Oxford University Press.

Berger, J. O., and Wolpert, R. L. 1988. *The Likelihood Principle.* 2nd ed. Hayward, CA: Institute of Mathematical Statistics.

Bernardo, J. M. 1979. Reference Posterior Distributions for Bayesian Inference (with discussion). *J. Roy. Stat. Soc.,* ser. B, 41: 113–147.

Birnbaum, A. 1962. On the Foundations of Statistical Inference (with discussion). *J. Am. Stat. Assn* 57: 269–326.

Birnbaum, A. 1969. Concepts of Statistical Evidence. In. Morgenbesser, S., P. Suppes, and M. White, eds., *Philosophy, Science, and Method: Essays in Honor of Ernest Nagel.* New York: St. Martin's Press.

Birnbaum, A. 1970. Statistical Methods in Scientific Inference. *Nature* 225: 1033.

Bower, B. 1997. Null Science: Psychology's Status Quo Draws Fire. *Science News* 151: 356–357.

Christensen, D. 1999. Measuring Confirmation. *J. Phil.* 96: 437–461.

Cohen, J. 1994. The Earth Is Round ($p < .05$). *Am. Psych.* 49: 997–1003.

Earman, J. 1992. *Bayes or Bust? A Critical Examination of Bayesian Confirmation Theory.* Cambridge: MIT Press.

Edwards, A. W. F. 1969. Statistical Methods in Scientific Inference. *Nature* 222: 1233–1237.

Edwards, A. W. F. 1972. *Likelihood: An Account of the Statistical Concept of Likelihood and its Application to Scientific Inference.* Cambridge: Cambridge University Press.

Edwards, A. W. F. 1992. *Likelihood* (expanded ed.). Baltimore: Johns Hopkins University Press.

Edwards, W., H. Lindman, and L. Savage. 1984. Bayesian Statistical Inference for Psychological Research. In Kadane, J. B., ed., *Studies in Bayesian Econometrics.* (Originally published in *Psych. Rev.* 70: 193–242.)

Efron, B. 1986. Why Isn't Everyone a Bayesian? (with discussion), *Am. Stat.* 40: 1–11.

Fisher, R. A. 1959. *Statistical Methods and Scientific Inference.* 2nd ed. New York: Hafner.

Glymour, C. 1980. *Theory and Evidence.* Princeton: Princeton University Press.

Good, I. J. 1968. Corroboration, Explanation, Evolving Probability, Simplicity, and a Sharpened Razor. *Br. J. Phil. Sci.* 19: 123–143.

Goodman, S. N. 1998. Multiple Comparisons, Explained. *Am. J. Epidemiol.* 147 : 807–812.

Hacking, I. 1965. *Logic of Statistical Inference,* New York: Cambridge University Press.

Horwich, P. 1982. *Probability and Evidence.* Cambridge: University of Cambridge Press.

Howson, C., and Urbach, P. 1993. *Scientific Reasoning: The Bayesian Approach.* 2nd ed. Peru, IL: Open Court.

Jeffreys, H. 1961. *Theory of Probability.* 3rd ed. Oxford: Oxford University Press.

Kass, R. E., and A. E. Raftery. 1995. Bayes Factors. *J. Am. Stat. Assn* 90 : 773–795.

Kolmogorov, A. N. 1956. *Foundations of Probability.* Trans. N. Morrison. New York: Chelsea.

Lindley, D. V. 1965. *Introduction to Probability and Statistics: Part I,* Cambridge: Cambridge University Press.

Lindley, D. V. 1992. Discussion of Royall, R. M.: The Elusive Concept of Statistical Evidence. In Bernardo, J. M., J. Berger, A. P. Dawid, and A. F. M. Smith, eds., *Bayesian Statistics 4.* New York: Oxford University Press.

Moore, D. S. 1997. Bayes for Beginners? Some Reasons to Hesitate. *Am. Stat.* 51 : 254–261.

Morrison, D. E., and Henkel, R. E. 1970. *The Significance Test Controversy.* Chicago: Aldine.

Neyman, J. 1950. *First Course in Probability and Statistics.* New York: Henry Holt.

Pagano, M., and K. Gauvreau. 1993. *Principles of Biostatistics,* Belmont, CA: Duxbury,

Pratt, J. W. 1961. Review of *Testing Statistical Hypotheses,* by E. L. Lehmann (1959). *J. Am. Stat. Assn* 56 : 163–166.

Robbins, H. 1970. Statistical Methods Related to the Law of the Iterated Logarithm. *Ann. Math. Stat.* 41 : 1397–1409.

Royall, R. M. 1991. Ethics and Statistics in Randomized Clinical Trials (with discussion). *Stat. Sci.* 6 : 52–88.

Royall, R. M. 1997. *Statistical Evidence: A Likelihood Paradigm.* London: Chapman and Hall.

Royall, R. M. 2000. On the Probability of Observing Misleading Statistical Evidence (with discussion). *J. Am. Stat. Assn* 95 : 760–780.

Salmon, W. C. 1990. Rationality and Objectivity in Science, or Tom Kuhn Meets Tom Bayes. In Savage, C. W., ed., *Scientific Theories.* Minneapolis: University of Minnesota Press.

Savage, L. J. 1962. Discussion of A. Birnbaum, On the Foundations of Statistical Inference. *J. Am. Stat. Assn* 53 : 307–308.

Smith, C. A. B. 1953. The Detection of Linkage in Human Genetics. *J. Roy. Stat. Soc.,* ser. B,15 : 153–192.

Sterne, J. A. C., and Smith, G. D. 2001. Sifting the Evidence: What's Wrong with Significance Tests? *BMJ* 322 : 226–231.

Thompson, J. R. 1998. Invited commentary on Multiple Comparisons and Related Issues in the Interpretation of Epidemiologic Data. *Am. J. Epidemiol.* 147 : 801–806.

6 Why Likelihood?

Malcolm Forster and Elliott Sober

ABSTRACT

The likelihood principle has been defended on Bayesian grounds, on the grounds that it coincides with and systematizes intuitive judgments about example problems, and by appeal to the fact that it generalizes what is true when hypotheses have deductive consequences about observations. Here we divide the principle into two parts—one qualitative, the other quantitative —and evaluate each in the light of the Akaike information criterion (AIC). Both turn out to be correct in a special case (when the competing hypotheses have the same number of adjustable parameters), but not otherwise.

INTRODUCTION

Mark Antony said that he came to bury Caesar, not to praise him. In contrast, our goal is neither to bury the likelihood concept nor to praise it. Instead of praising it, we will present what we think is an important criticism. However, the upshot of this criticism is not that likelihood should be buried, but a justification of likelihood, properly understood.

Before we get to our criticism of likelihood, we should say that we agree with the criticisms that likelihoodists have made of Neyman-Pearson-Fisher statistics and of Bayesianism (Edwards, 1987; Royall, 1997). In our opinion, likelihood looks very good indeed when it is compared with these alternatives. However, the problem of a positive defense of likelihood remains. Royall begins his excellent book with three kinds of justification.

We would like to thank Ken Burnham, Ellery Eells, Branden Fitelson, Ilkka Kieseppä, Richard Royall, and the editors of this volume for helpful comments on an earlier draft.

He points out, first, that the likelihood principle makes intuitive sense when probabilities are all 1s and 0s. If the hypothesis H_1 says that what we observe *must* occur, and the hypothesis H_2 says that it *cannot*, then surely it is clear that the observations strongly favor H_1 over H_2. The likelihood principle seems to be an entirely natural generalization from this special case; if O strongly favors H_1 over H_2 when $P(O|H_1) = 1$ and $P(O|H_2) = 0$, then surely it is reasonable to say that O favors H_1 over H_2 if $P(O|H_1) > P(O|H_2)$.

Royall's second argument is that the likelihood ratio is precisely the factor that transforms a ratio of prior probabilities into a ratio of posteriors. This is because Bayes' theorem (and the definition of conditional probability from which it derives) entails that

$$\frac{P(H_1|O)}{P(H_2|O)} = \frac{P(O|H_1)P(H_1)}{P(O|H_2)P(H_2)}.$$

It therefore makes sense to view the likelihood ratio as a valid measure of the evidential meaning of the observations.

Royall's third line of defense of the likelihood principle is to show that it coincides with intuitive judgments about evidence when the principle is applied to specific cases. Here Royall is claiming for likelihood what philosophers typically claim in defense of the explications they recommend. For example, Rawls (1971) said of his theory of justice that his account is to be assessed by seeing whether the principles he describes have implications that coincide with our intuitive judgments about what is just and what is not in specific situations.

While we think there is value in each of these three lines of defense, none is as strong as one might wish. Even if the likelihood principle has plausible consequences in the limit case when all probabilities are 1s and 0s, the possibility exists that other measures of evidential meaning might do the same. Though these alternatives agree with likelihood in the limit case, they may disagree elsewhere. If so, the question remains as to why likelihood should be preferred over these other measures. As for Royall's second argument, it does show why likelihood makes sense if you are a Bayesian; but for anti-Bayesians such as Royall himself (and us), a very important question remains —when the hypotheses considered cannot be assigned objective probabilities, why think that likelihood describes what the evidence tells you about them? As for Royall's third argument, our hesitation here is much like the reservations we have about the first. Perhaps there are measures other than likelihood that coincide with what likelihood says in the cases Royall considers but disagree elsewhere.[1]

In any event, friends of likelihood might want a stronger justification than the three-point defense that Royall provides. Whether there *is* anything more that can be said remains to be seen. In this regard, Edwards (1987, 100) endorsed Fisher's (1938) claim that likelihood should be regarded as a "primitive postulate"; it coincides with and systematizes our intuitions about examples, but nothing more can be said in its behalf. This, we should note, is often the fate of philosophical explications. If likelihood is epistemologically fundamental, then we should not be surprised to find that it cannot be justified in terms of anything that is more fundamental. It would not be an objection to the likelihood concept if it turned out to be an item of rock-bottom epistemology.

Royall follows Hacking (1965) in construing the likelihood principle as a two-part doctrine. There is first of all the idea, noted above, which we will call the qualitative likelihood principle:

(QUAL) O favors H_1 over H_2 if and only if $P(O|H_1) > P(O|H_2)$.

Notice that this principle could be expressed equivalently by saying that the likelihood ratio must be greater than unity or that the difference in likelihoods must be greater than 0.

Hacking adds to this a second and logically stronger claim—that the likelihood ratio measures the degree to which the observations favor one hypothesis over the other:

(DEGREE) O favors H_1 over H_2 to degree x if and only if O favors H_1 over H_2 and $P(O|H_1)/P(O|H_2) = x$.

1. Here we should mention Royall's demonstration (which he further elaborates in Royall, 2001) that in a certain class of problems, a bound can be placed on the probability that the evidence will be misleading when one uses a likelihood comparison to interpret it. That is, for certain types of competing hypotheses H_1 and H_2, if H_1 is true, a bound can be placed on the probability that one's data will be such that $P(\text{Data}|H_1) < P(\text{Data}|H_2)$. We agree with Royall that this is welcome news, but we do not think that it shows that a likelihood comparison is the uniquely correct way to interpret data. The question remains whether other methods of interpreting data have the same property. Furthermore, we think it is important to recognize a point that Royall (forthcoming) makes in passing: that it is no criticism of likelihood that it says that a false hypothesis is better supported than a true one when the data are misleading. In fact, this is precisely what likelihood *should* say, if likelihood faithfully interprets what the data indicate; see Sober (1988, 172–83) for discussion.

[handwritten note:] A false hypothesis is better supported than a true one when the data are misleading!

Obviously, (DEGREE) includes (QUAL) as a special case, but not conversely. However, even if (QUAL) were correct, a further argument would be needed to accept (DEGREE). Why choose the likelihood ratio, rather than the difference, or some other function of likelihoods, as one's measure of strength of evidence? There are many alternatives to consider, and the choice makes a difference because pairs of measures frequently fail to be ordinally equivalent (see Fitelson, 1999). Consider, for example, the following four likelihoods:

$$P(O_1|H_1) = .09, P(O_1|H_2) = .02$$
$$P(O_2|H_3) = .8, P(O_2|H_4) = .3.$$

If we measure strength of evidence by the ratio measure, we have to say that O_1 favors H_1 over H_2 more strongly than O_2 favors H_3 over H_4. However, if we choose the difference measure, we get the opposite conclusion.[2] Royall does not take up the task of defending (DEGREE). Our discussion in what follows will focus mainly on the principle (QUAL), but we will have some comments on the principle (DEGREE) as well.

There is a third element in Royall's discussion that we should mention, one that differs from both (QUAL) and (DEGREE). This is his criterion for distinguishing Os *weakly* favoring H_1 over H_2 from Os *strongly* favoring H_1 over H_2:

(R) O strongly favors H_1 over H_2 if and only if O favors H_1 over H_2 and $P(O|H_1)/P(O|H_2) > 8$.

O weakly favors H_1 over H_2 if and only if O favors H_1 over H_2 and $P(O|H_1)/P(O|H_2) \leq 8$.

As was true of (DEGREE), principle (R) includes (QUAL) as a special case, but not conversely. Furthermore, (R) and (DEGREE) are logically independent. Royall recognizes that the choice of cutoff specified in (R) is conventional. In what follows, our discussion of (QUAL) will allow us to make some critical comments on (R) as well.

One of Royall's simple but extremely important points is that it is essen-

---

2. Even if we restrict (DEGREE) to comparisons of hypotheses relative to the same data, it remains true that (DEGREE) is logically stronger than (QUAL). In our example, the ratio and difference measures disagree even when $O_1 = O_2$.

tial to distinguish carefully among three questions that you might want to address when evaluating the testimony of the observations:

1. What should you do?
2. What should you believe?
3. What do the observations tell you about the hypotheses you're considering?

Royall argues that Neyman-Pearson statistics addresses question (1). However, if this *is* the question that the Neyman-Pearson approach tries to answer, then the question falls in the domain of decision theory, in which case utilities as well as probabilities need to be considered. Question (2) is the one that Bayesians address. Question (3) is different from (2) and also from (1); (3) is the proper province for the likelihood concept. Royall considers the possibility that Fisher's idea of statistical significance might be used in Neyman-Pearson statistics to address question (3). However, a good part of Royall's book is devoted to showing that the likelihoodist's answer to question (3) is better.

We will argue in what follows that (3) needs to be subdivided. There are at least two different questions you might ask about the bearing of evidence on hypotheses:

3a. What do the observations tell you about the *truth* of the hypotheses you're considering?
3b. What do the observations tell you about the *predictive accuracy* of the hypotheses you're considering?

Question (3a) is what we think Royall has in mind in his question (3). The second concept, of predictive accuracy, is something we'll discuss later. It is an important feature of this concept that a false model is sometimes more predictively accurate than a true one. The search for truth and the search for predictive accuracy are different; this is why we separate questions (3a) and (3b).

To develop these ideas, we now turn to a type of inference problem that is, we think, the Achilles' heel of the likelihood approach. This is the problem of "model selection." By a "model," we mean a statement that contains at least one adjustable parameter. Models are composite hypotheses. Consider, for example, the problem of deciding whether the dependent variable y and the independent variable x are related linearly or parabolically:

(LIN) $y = a + bx + u$
(PAR) $y = a + bx + cx^2 + u.$

In these models, a, b, and c are adjustable parameters, while u is an error term with a normal distribution with zero mean and a fixed variance. Fixing the parameter values picks out a specific straight line or a specific parabola.

How can likelihood be used to choose between these models? That is, how are we to compare $P(\text{Data}|\text{LIN})$ and $P(\text{Data}|\text{PAR})$? As Royall says, there are no general and entirely valid solutions to the problem of assessing the likelihoods of composite hypotheses. Let us consider some alternative proposals and see how they fare. The model LIN is a family composed of the infinite set of straight lines in the x-y plane. Strictly speaking, the likelihood of LIN is an average of the likelihoods of all these straight lines:

$$P(\text{Data}|\text{LIN}) = \sum_i P(\text{Data}|\text{Li})P(\text{Li}|\text{LIN}).$$

If we have no way to figure out how probable different straight lines are, conditional on LIN, then we cannot evaluate the likelihood. Suppose, however, that every straight line has the same probability as every other, conditional on LIN. In this case $P(\text{Data}|\text{LIN}) = 0$ and the same is true of $P(\text{Data}|\text{PAR})$ (Forster and Sober, 1994). This is an unsatisfactory conclusion, since scientists often believe that the data discriminate between LIN and PAR.[3]

3. Rosenkrantz (1977) proposed that average likelihoods (based on a uniform prior) would favor simpler models, while Schwarz (1978) provided a mathematical argument for that conclusion, together with an approximate formula (BIC, the Bayesian information criterion) for how the average likelihood depends on the maximum likelihood and the number of adjustable parameters. However, Schwarz's derivation is suspect in examples like the one we discuss. Uniform priors are improper in the sense that the probability density cannot integrate to 1. If the density is zero, then it integrates to zero. If it is nonzero, then it integrates to infinity, no matter how small. That is, the density is improper because it cannot be normalized. For the purpose of calculating the posterior probability, this does not matter because if the density is everywhere equal to some arbitrary nonzero constant, then the posterior density of any curve is proportional to its likelihood, and the arbitrary constant drops out when one normalizes the posterior (which one should do whenever possible). So Bayesians got used to the idea that improper priors are acceptable. However, they are not acceptable for the purpose of calculating average likelihoods because there is no such thing as the normalization of likelihoods (they are not probabilities). Of course, one could instead assume that the arbitrary constant is the same for all models. Then each average likelihood is proportional to the sum (integral) of all the likelihoods. But to compare integrals of different dimensions is like comparing the length of a line with the area of a rectangle—it makes little sense when there is no principled way of specifying the units of measurement (Forster and Sober, 1994). For ex-

This example, we should mention, also illustrates a problem for Bayesianism, one that Popper (1959) noted. Because LIN is nested inside of PAR, it is impossible that $P(\text{LIN}|\text{Data}) > P(\text{PAR}|\text{Data})$, no matter what the data say. When scientists interpret their data as favoring the simpler model, it is impossible to make sense of this judgment within the framework of Bayesianism.[4]

An alternative is to shift focus from the likelihoods of LIN and PAR to the likelihoods of L(LIN) and L(PAR). Here L(LIN) is the likeliest straight line, given the data, and L(PAR) is the likeliest parabola. The suggestion is to compare the families by comparing their likeliest special cases. The problem with this solution is that it is impossible that $P[\text{Data}|\text{L(LIN)}] > P[\text{Data}|\text{L(PAR)}]$. This illustrates a general point—when models are nested, it is almost certain that more complex models will fit the data better than models that are simpler. However, scientists don't take this as a reason to conclude that the data always favor PAR over LIN. Instead, they often observe that simplicity, and not just likelihood, matters in model selection. If L(LIN) and L(PAR) fit the data about equally well, it is widely agreed that one should prefer L(LIN). And if LIN is compared with a polynomial that has 100 terms, scientists will say that even if L(POLY-100) fits the data *much* better than L(LIN) does, that *still* one might want to interpret the data as favoring L(LIN). Likelihood is an incomplete device for interpreting what the data say. It needs to be supplemented by due attention to simplicity. But how can simplicity be represented in a likelihood framework?

---

ample, if one compares the very simple model $y = a$ with $y = a + bx$, and then with $y = a + 2bx$, the added b-dimension is scaled differently. So, when one integrates the likelihood function over parameter space, the result is different (unless one can justify the formula $y = a + bx$ uniquely as the principled choice).

An uneasiness about Schwarz's derivation has since led some Bayesians to invent other ways to compute average likelihoods (see Wasserman, 2000, for an easy technical introduction to the Bayesian literature). One proposal, which also gets around our objection, is the theory of "intrinsic Bayes factors" due to Berger and Pericchi (1996). The idea is to "preconditionalize" on each datum in the data set and average the results to obtain a well-conditioned and "approximately" uniform "prior." Then the average likelihoods are nonzero and can be compared straightforwardly. First, we note that this is not a vindication of Schwarz's argument because it yields a different criterion. Second, this solution appears to us to be unacceptably ad hoc as well.

4. Bayesians sometimes address this problem by changing the subject. If we define PAR* to be the family of parabolas that does not include straight lines (i.e., c is constrained to be nonzero), then the axioms of probability do not rule out the possibility that LIN might have a higher probability than PAR*. However, it remains unclear what reason a Bayesian could have for thinking that $P(c = 0) > P(c \neq 0)$.

We think that it cannot, at least not when simplicity is a consideration in model selection.[5] This is our criticism of likelihood. However, there is another inferential framework in which the role of simplicity in model selection makes perfect sense. This framework was proposed by statistician H. Akaike (one of his earliest articles is Akaike, 1973; one of his latest is Akaike, 1985; see Sakamoto, Ishiguro, and Kitagawa, 1986, for a thorough introduction to his method, and Burnham and Anderson, 1998, for some scientific applications of Akaike's approach). The Akaike framework assumes that inference has a specific goal; the goal is not to decide which hypothesis is most probably true, or most likely, but to decide which will be most predictively accurate.

What does it mean to talk about the predictive accuracy of a model, like LIN? Imagine that we sample a set of data points from the true underlying distribution and use that data to find the best-fitting straight line, namely L(LIN). We then use L(LIN) to predict the location of a new set of data. We draw these new data and see how close L(LIN) comes to predicting their values. Imagine repeating this process many times, using an old data set to find L(LIN) and then using that fitted model to predict new data. The average closeness to new data (as measured by the per-datum log-likelihood) is LIN's predictive accuracy. If the mean function for the true underlying distribution is in fact linear, LIN may do poorly on some of these trials, but on average it will do well. On the other hand, if the mean function for the true underlying distribution is highly nonlinear, LIN may do fairly well occasionally, but on average it will do a poor job of predicting new data. Obviously, the predictive accuracy of a model depends on what the true underlying distribution is. However, in making an inference, we of course don't know in advance what the truth is. Maximizing predictive accuracy might be a sensible goal, but so far it appears to be epistemologically inaccessible. Is it possible to figure out, given the single data set before us, how predictively accurate a model is apt to be?

Akaike proved a surprising theorem,[6] one that shows that predictive ac-

5. In some inference problems, simplicity or parsimony can be shown to be relevant because simplicity influences likelihood. Phylogenetic inference is a case in point; see Sober (1988) for discussion.

6. Akaike's theorem rests on some assumptions: a Humean "uniformity of nature assumption" (that the old and new data sets are drawn from the same underlying distribution), and a surprisingly weak "regularity" assumption that implies (among other things) that the true distribution of the parameter estimates, when the number of data n is sufficiently large, is a multivariate normal distribution with a covariance matrix whose terms are inversely proportional to n. The central limit theorems in their various forms (Cramér, 1946) entail a similar result for the distributions of the sums of random variables. The full details of the normality assumption are complex, and we refer the interested reader to Sakamoto, Ishiguro, and Kita-

curacy is epistemologically accessible. He showed that an unbiased estimate of a model's predictive accuracy can be obtained by taking the log-likelihood of its likeliest case, relative to the data at hand and correcting that best-case likelihood with a penalty for complexity: an unbiased estimate of the predictive accuracy of model M is equal to Log $P[\text{Data} \mid L(M)] - k$, where k is the number of adjustable parameters in the model (see Forster, 1999, for a more exact description of the meaning of k). LIN contains 2 adjustable parameters, while PAR contains 3, and POLY-100 contains 101.[7] Akaike's theorem says that likelihood provides information about the predictive accuracy of a model, but the information is always distorted. Likelihood is like a bathroom scale that always tells you that you are lighter than you are. Its outputs are evidentially relevant, but they need to be corrected.

We now can explain our earlier remark that a true model can be less predictively accurate than a false one. Suppose you know that the relationship of x and y is nonlinear and parabolic. It *still* can make sense to use LIN to predict new data from old, if L(LIN) fits the data about as well as L(PAR). The truth can be a misleading predictor. It is a familiar fact that idealizations are valuable in science when a fully realistic model is either unavailable or mathematically intractable. The Akaike framework reveals an additional virtue that idealizations can have—even when we possess a fully realistic (true) model, a (false) idealization can be a better predictor (Forster and Sober, 1994; Sober, 1999; Forster, 2000a).[8]

As we mentioned earlier, model selection is the Achilles' heel of likelihood. Yet Akaike's theorem describes a general circumstance in which likelihood provides an unbiased estimate of a model's predictive accuracy—*when two models have the same number of parameters, the likelihoods of their likeliest cases provide an unbiased indication of which can be expected to be more predictively accurate.* Likelihood needn't be viewed as a primitive postulate. We needn't resign ourselves to the idea that we value likelihood for its own sake. If predictive accuracy is your goal, likelihood is one relevant consideration because it helps you estimate a model's predictive accuracy.

---

gawa (1986), which provides the simplest technical introduction to these details. Akaike's result can also hold exactly for small sample sizes when additional conditions are met (see, e.g., Kieseppä, 1997).

7. The number of adjustable parameters should also include the variance of the (assumed) error distribution and any other free parameters used to define it. However, we have chosen to ignore this complication because it is not relevant to the main point of this essay.

8. Recall that we defined "model" as a statement containing at least one adjustable parameter. Our point about idealizations would not be correct for statements containing no adjustable parameter.

And when the models under consideration are *equally* complex, likelihood is the *only* thing you need to consider. Likelihood is a means to an end and is justified relative to that end.[9]

Not only does the qualitative likelihood principle receive a circumscribed justification from the Akaike framework; we can use Akaike's theorem to evaluate the (DEGREE) and (R) principles as well. The theorem provides an unbiased criterion for when one model will be more predictively accurate than another:

M_1 is estimated to be more predictively accurate than M_2 if and only if
$\log\text{-}P[\text{DATA} \,|\, L(M_1)] - k_1 > \log\text{-}P[\text{DATA} \,|\, L(M_2)] - k_2$.

This can be rewritten as

M_1 is estimated to be more predictively accurate than M_2 if and only if
$P[\text{DATA} \,|\, L(M_1)]/P[\text{DATA} \,|\, L(M_2)] > \exp(k_1 - k_2)$.

If $k_1 = k_2$, this can be stated equivalently by saying that the likelihood ratio must be greater than 1 or by saying that the difference in log-likelihoods must be greater than 0.[10] However, if $k_1 \neq k_2$, a *ratio* criterion can be formulated, but there is no equivalent criterion that can be stated purely in terms of likelihood *differences*.[11] This helps distinguish some measures of strength of evidence from others, as (DEGREE) requires.

9. Our point is in accord with Akaike's (1973) observation that AIC is an "extension of the maximum likelihood principle."

10. The most important special case here is when $k_1 = 0 = k_2$, which is just the nonmodel selection problem of comparing two specific point hypotheses. In this case, there are many independent reasons in favor of a likelihood ratio law of this form, including the classical Neyman-Pearson theorems that prove that a decision rule based on such a rule is the most powerful test of its size (see Hogg and Craig, 1978, 246). The limitation of these theorems is that they presuppose a very simple 0 or 1 measure of discrepancy between a hypothesis and the truth. Lele (2004 [chapter 7 of this volume]) develops an alternative analysis based on more interesting discrepancy measures, which also speaks in favor of the likelihood ratio criterion.

11. If it were granted that the degree of evidence depends only on the likelihoods in some way, then there would be an independent reason for not using the difference measure. For in the case of continuous variables, likelihoods are equal to the probability *density* of an observed quantity x times an arbitrary multiplicative constant (Edwards, 1987). To understand the reason for this arbitrary factor, consider a transformation of the variable x, $x' = f(x)$, for some one-to-one function f. Probabilities are invariant under such transformations, so con-

Similar remarks apply to Royall's principle (R). If $k_1 = k_2$, Royall's stipulation in (R) of the number 8 as the cutoff separating strong from weak evidence favoring H_1 over H_2 is a possibility (though other cutoffs are possible as well). However, when $\exp(k_1 - k_2) > 8$, the difference between strong and weak evidence cannot be defined by the proposed cutoff of 8. Akaike's theorem does not determine how the distinction between weak and strong evidence should be drawn, but it does restrict the search to criteria defined in terms of likelihood ratios.

Our defense of (DEGREE) is not a defense of everything it implies. Remember that (DEGREE) equates the degree to which O favors H_1 over H_2 with the likelihood ratio, and this implies two things: (a) that the likelihood ratio is the correct way of capturing how the degree of fit between O and H_1 and between O and H_2 influences the degree to which O favors H_1 over H_2, and (b) that nothing else influences the degree to which O favors H_1 over H_2. Implication (a) rules out the possibility that any other measures of fit, such as the sum of the absolute values of the residuals between the data and the mean curve of the hypothesis, affect the relative degree of support. For AIC, H_1 and H_2 are the likeliest hypotheses $L(M_1)$ and $L(M_2)$, and the degree to which O favors $L(M_1)$ over $L(M_2)$ depends on the likelihood ratio and $\exp(k_1 - k_2)$. The latter term corrects for an expected overfitting bias, which would otherwise provide the more complex model with an unfair advantage.[12] AIC therefore agrees with implication (a) of (DEGREE), but denies implication (b). This leads us to the following general principle:

---

sider the probability that x is observed in a given interval around x. This probability is equal to the area under the density curve within this interval. If the interval is small, then the area is equal to the density at x times the width of the interval. In order for the area to be invariant under any transformation of x, the density must change whenever the length of the interval changes. So densities, or the differences in densities, fail the requirement of language invariance. On the other hand, the difference of the *probabilities* is invariant, but it is proportional to the length of the small interval around x, which is arbitrary (see Forster, 1995, for further discussion). Therefore the difference measure is caught in a dilemma—it either fails the desideratum of language invariance or it contains an arbitrary multiplicative factor. Fortunately, the arbitrary constant drops out when we take the *ratio* of the likelihoods, or any function of the likelihood ratio, so it is both language-invariant and nonarbitrary. As far as we can see, this class of measures is unique in this regard, at least among likelihood measures.

12. The degree to which O favors H_1 over H_2 may also be a function of the number of data n, even though this is a fact about the data and is not a function of the likelihood ratio. We regard this kind of information as akin to the difference $k_1 - k_2$ because neither of them measures the fit of H_1 and H_2 to the data O.

(DEGREE Prime) The likelihood ratio is the correct way of capturing how the degree of fit between O and H_1 and between O and H_2 influences the degree to which O favors H_1 over H_2.[13]

In the special case of comparing simple hypotheses H_1 and H_2, (DEGREE Prime) reduces to (DEGREE).

The argument for this principle arises out of the *form* of AIC—the fact that AIC can be expressed as O favors $L(M_1)$ over $L(M_2)$ if and only if $P(O|L(M_1))/P(O|L(M_2)) > K$. However, we have to admit that our argument is only as strong as its premises, and not every statistician will agree that AIC stands on firm foundations.[14] In fact, the argument really depends on the premise that model selection should have this form for *some K*. The exact value of K does not matter. As it turns out, almost all model selection criteria in the literature can be expressed in this basic form, including BIC (Schwarz, 1978), variations on AIC (see, e.g., Hurvich and Tsai, 1989), posterior Bayes factors (Aitkin, 1991), and an important class of Neyman-Pearson hypothesis tests.[15] Moreover, all of these criteria apply to a wide variety of statistical applications, including contingency table analysis, regression models, analysis of variance, and time series. There are many independent arguments for (DEGREE Prime).

But is that the end of the debate? One still might dream of a unified perspective from which everything else follows, including a corrected version of Royall's distinction between weak and strong evidence. Akaike's frame-

13. Note that (DEGREE Prime) applies only when the likelihoods of H_1 and H_2 are well defined.

14. For example, it is often charged that AIC is inconsistent (but see Forster, 2000b, for a defense of AIC against this charge). Or it might be maintained that the goal of predictive accuracy is not the primary consideration at hand.

15. There are two cases to consider. In the case of comparing simple hypotheses H_0 and H_1, where H_0 is the null hypothesis, a *best test* of size α is by definition (Hogg and Craig, 1978, 243) a test with a critical region C of size α such that for any other critical region A of size α, $P(C|H_1) \geq P(A|H_1)$. That is, a best test maximizes the probability of rejecting H_0 when H_1 is true. Hogg and Craig (1978, 246) show that for any best test with critical region C, there is a number K such that a data set lies within C if and only if the likelihood ratio of H_1 to H_0 is greater than or equal to K. These are the Neyman-Pearson theorems mentioned in note 10. In the case of comparing a simple hypothesis H_0 against a composite alternative M, a *uniformly most powerful critical region* of size α is by definition (Hogg and Craig, 1978, 252) a region C that provides a best test for comparing H_0 against any point hypothesis in M. So if a uniformly most powerful test exists, the test between H_0 and any representative of M is a likelihood ratio test. In examples in which the assumptions of Akaike's theorem hold (see note 6) and the composite hypothesis has one adjustable parameter, a uniformly most powerful Neyman-Pearson test with $\alpha = .05$ effectively trades off the maximum log-likelihood against simplicity to a degree somewhere between AIC and BIC (Forster, 2000b).

work goes one step beyond Fisher's view that likelihood is fundamental, but one still might wish for more.

6.1   Commentary

Michael Kruse

Likelihood is a key concept for classical, Bayesian, and likelihoodist statistics alike. Yet there is a long-running debate over the proper role of likelihood in statistics, at the center of which is the question of *evidential meaning:*

Q: What does outcome x tell us about the hypotheses we are considering?

"Why Likelihood?" offers an enlightening perspective on this debate. An interesting result, I think, is that it leads us to *reject* one presumption of the debate, viz., that there is some *single* correct concept of evidential meaning.

One answer to Q is that *all* the information in x relevant to evaluating H_1 against H_2 is in the likelihoods, $P(x|H_1)$ and $P(x|H_2)$. If we measure the evidence x for H_1 against H_2 with the *likelihood ratio,* LR $= P(x|H_1)/P(x|H_2)$, we get Forster and Sober's stronger principle DEGREE. To learn what x tells us about H_1 and H_2, look no further than LR.[16]

In contrast, classical statistical methods (e.g., NP testing) treat LR as only *part* of what x tells us about H_1 against H_2. Instead, these methods require that we consider both the actual and *possible* values of LR, the latter of which depend on what *could have* but *actually didn't* happen.

I detect more than a whiff of a primitive philosophical disagreement in this, and not surprisingly, much of the debate consists of appeals to intuitions about what is or is not relevant information.[17] Forster and Sober give us a more productive way to frame the debate by subdividing Q in terms of *inferential aims,* i.e., what we want from those hypotheses. Doing this breaks up Q into distinct questions, three of which are:

Q1   What does x tell us about the *predictive accuracies* of H_1 and H_2?
Q2   What does x tell us about the *probabilities* of H_1 and H_2?
Q3   What does x tell us about the *truth values* of H_1 and H_2?

I thank Elliott Sober for his useful comments.

16. Here we are concerned only with what the *outcome* tells us, not *other* sources of information that may influence our inferences (e.g., prior probabilities).

17. Examples of these are Royall's three points in favor of likelihood that Forster and Sober discuss.

Forster and Sober focus on Q1, arguing that given two composite hypotheses (or models) LIN and PAR, LR = $P(x|L(LIN))/P(x|L(PAR))$ will almost *always* favor PAR—even if LIN is the better model for purposes of prediction.[18] LR, then, is a *biased* source of information about the accuracies of LIN and PAR.

Using LR to make *reliable* judgments about the accuracies of LIN and PAR requires *compensating* for this bias. One class of methods for doing this (including AIC) implies that what x tells us about the accuracies of LIN and PAR depends on both LR *and* the number of adjustable parameters in LIN and PAR. Thus, DEGREE isn't an appropriate answer to Q1.

This is *not,* however, a general argument against DEGREE. If, for instance, we want *coherent* probability assignments, Bayes' theorem allows us to use data to obtain coherent posteriors from coherent priors. But observations affect posteriors *only* through the likelihood function: given a prior distribution for the H_is and data x and y that yield likelihood functions $P(x|H_i) \propto P(y|H_i)$ for all i, then $P(H_i|x) = P(H_i|y)$ for all i. So, DEGREE is the *right* answer to Q2, since LR provides precisely that information in the outcome relevant to maintaining coherence.

A similar point holds for an aim of NP testing, i.e., reliable discrimination between true and false hypotheses. Given this aim, we need to answer Q3. It is well known that in certain situations we can "force" LR to favor H_2 over H_1 if we can sample as long as we like. That is, the *sampling procedure* may guarantee that LR will favor H_2 over H_1 even if H_1 is true. Compensating for *this* bias requires knowing the sampling procedure, so an answer to Q3 will refer to both LR *and* information about how the data were generated.[19]

LR is an important part of each of the answers to Q1–Q3 as a measure of *fit* between data and hypotheses. In *that* sense, the value of LR is vindicated. Yet we also see that *some* inferential aims require appealing to *other* factors in order to use the information in LR properly.

What does this view on the role of likelihood—and, more specifically, LR—have for the prospects of a unified account of evidential meaning? LR emerges as a *tool* whose value derives from our ability to use it to advance our aims. The question of how we ought to *deploy* LR to advance a given aim

18. Here I follow Forster and Sober in identifying the likelihood of a composite hypothesis H with the likelihood of the member of H that best fits x, L(H).

19. As Forster and Sober point out, when H_1 and H_2 are both simple hypotheses, the bias relevant for predictive accuracy drops out. In those *same* situations, the bias relevant to the reliable discrimination of truth and falsity also drops out. This, however, says nothing about what happens once we go *beyond* these special cases.

is *not* one we answer by considering our *intuitions* in particular situations. Rather, it reduces to two relatively well-defined questions: Does LR provide information about hypotheses that, *with respect to our inferential aim,* is biased? And can we *correct* that bias?

How we ought to use LR depends on our inferential aim. One lesson to draw from "Why Likelihood?" then is that disagreements about aims underlie disagreements over the role of likelihood. If the aims of predictive accuracy, coherence, and truth are each distinct from one another, we should *expect* to find a similar diversity in concepts of evidential meaning. The resulting *disunified* cluster of such accounts may be philosophically unsatisfying and leave us wishing for something more. But what more could we really want?

6.2 Commentary

Robert J. Boik

I wish to congratulate Forster and Sober on a thought-provoking paper. Their paper raises many issues that are worthy of discussion, but I will confine my comments to just two. The first issue concerns the relationship between the Akaike information criterion (AIC) and likelihood. Forster and Sober concluded that likelihood receives support from the Akaike framework. I disagree and show that the Akaike framework violates the likelihood principle as well as the law of likelihood. The second issue concerns competing approaches to maximizing predictive accuracy. I argue that if the investigator's goal is to find a model that has maximum predictive accuracy, then the anti-Bayesian approach advocated by Forster and Sober should be abandoned because a superior approach exists. Both issues are related to the Akaike information criterion, so I begin with a brief summary of AIC.

THE AKAIKE INFORMATION CRITERION

Suppose that a process generates n independent data points, $Y_1, \ldots, Y_n$. These random variables can be arranged into an $n \times 1$ vector $\mathbf{Y}$. It is assumed that the process that generated the data can be described by a probability function, say $g(\mathbf{y})$, where $\mathbf{y}$ is a realization of $\mathbf{Y}$. It is assumed that the investigator has some knowledge about g, but the specific form of g is unknown. The investigator would like to use the observed data, $\mathbf{y}$, to find an approximating probability function $\hat{g}$. The approximating function can then be used to predict future observations, $\mathbf{Z}$. One approach to this problem is to use knowledge about g to devise a set of parametric families of distributions to use as potential approximations to g. A specific distribution from the ith approximating family is denoted by $f_i(\mathbf{y}|\boldsymbol{\theta}^{(i)})$, where $\boldsymbol{\theta}^{(i)}$ is a $k_i \times 1$ vector of adjustable parameters.

The investigator would like to choose the "best" among the approximating families of distributions. Akaike (1973) suggested that the Kullback-Leibler (KL) information number (Kullback and Leibler, 1951) be used as a criterion for best. The KL information number is a measure of the distance between a specific approximating distribution and the true distribution. For example, suppose that the ith family has been selected for consideration. The Akaike approach is to approximate $f_i(\mathbf{z}|\boldsymbol{\theta}^{(i)})$ by $\hat{f}_i = f_i(\mathbf{z}|\boldsymbol{\theta}_y^{(i)})$ where $\boldsymbol{\theta}_y^{(i)}$ is the maximum likelihood estimator of $\boldsymbol{\theta}^{(i)}$ and $\mathbf{z}$ is a realization of $\mathbf{Z}$. The subscript on $\boldsymbol{\theta}_y^{(i)}$ serves as a reminder that the estimator is a function of $\mathbf{y}$. Aitchison and Dunsmore (1975) and Geisser (1993) describe alternative choices for the approximating distribution of future observations. Aitchison (1975) called $f_i(\mathbf{z}|\boldsymbol{\theta}_y^{(i)})$ the estimative density because it is constructed by estimating the unknown parameters and then substituting the estimator into the density function. The KL distance between $\hat{f}_i$ and g is

$$
\begin{aligned}
KL(g, \hat{f}_i) &= \int \ln\left[\frac{g(\mathbf{z})}{f_i(\mathbf{z}|\boldsymbol{\theta}_y^{(i)})} \right] g(\mathbf{z})d\mathbf{z} \\
&= \int \ln[g(\mathbf{z})]g(\mathbf{z})d\mathbf{z} - \int \ln[f_i(\mathbf{z}|\boldsymbol{\theta}_y^{(i)})]g(\mathbf{z})d\mathbf{z}.
\end{aligned}
\tag{1}
$$

The KL number is nonnegative and is zero if and only if $\hat{f}_i$ and g are identical (except on sets of measure zero). According to this criterion, the best approximating distribution is the one that minimizes KL information.

The accuracy of two approximating distributions can be compared by taking the difference between their KL distances. The approximator having the smallest KL distance is preferred. The first term on the right-hand side of (1) does not depend on the approximating distribution, so it is of no interest when taking differences between KL distances. The second term is of interest, but in practice it cannot be computed because the true distribution, g, is unknown. Akaike's solution to this problem was to modify the criterion. Instead of searching for the approximator $\hat{f}_i$ that minimizes $KL(g, \hat{f}_i)$,

Akaike suggested that we search for the family of distributions that minimizes the average KL information over all possible realizations of **Y**. An equivalent criterion is to select the family that maximizes predictive accuracy, which can be defined as

$$P_A(g, f_i) = E_Y\left[\int \ln[f_i(\mathbf{z}|\boldsymbol{\theta}_y^{(i)})]g(\mathbf{z})\,d\mathbf{z}\right] = -E[KL(g, \hat{f}_i)] + \text{a constant}, \quad (2)$$

where $E_Y(\cdot)$ is the expected value with respect to $g(\mathbf{y})$, the true distribution of **Y**. Note that predictive accuracy is just the second term on the right-hand side of (1), *averaged over the sample space.*

By means of asymptotic expansions, Akaike showed that if $g(\mathbf{y})$ is contained in the ith family of distributions, then

$$\hat{P}_A(\hat{f}_i) = \ln[f_i(\mathbf{y}|\boldsymbol{\theta}_y^{(i)})] - k_i \qquad (3)$$

is asymptotically unbiased for P_A in (2), where k_i is the number of adjustable parameters in the ith family of distributions. Model selection using Akaike's criterion proceeds by searching for the family that maximizes $\hat{P}_A$ or, equivalently, minimizes AIC $= -2\hat{P}_A$.

BIAS OF AKAIKE'S CRITERION

Forster and Sober incorrectly claim that $\hat{P}_A$ is an unbiased estimator of P_A. Asymptotically, $\hat{P}_A$ is unbiased *if* the family under consideration contains the true probability function g but, in general, $\hat{P}_A$ is biased.

For example, consider a linear model with normal errors. In the ith family, this model contains $k_i - 1$ regression coefficients plus one scale parameter. Sugiura (1978) showed that if the true probability function is $g(\mathbf{y}) = f_i(\mathbf{y}|\boldsymbol{\theta}^{(i)})$, then an unbiased estimator of predictive accuracy is $\hat{P}_A^*(\hat{f}_i) = \hat{P}_A(\hat{f}_i) - \dfrac{k_i(k_i + 1)}{n - k_i - 1}$. That is, the bias of $\hat{P}_A$ in (3) is $k_i(k_i + 1)/(n - k_i - 1)$ and, for fixed k_i, this bias goes to zero as n increases.

If the ith family does not contain g, then the bias of $\hat{P}_A(\hat{f}_i)$ in linear models depends on a noncentrality parameter. This parameter is an index of the difference $\boldsymbol{\mu}_y - \boldsymbol{\mu}_{y,i}$, where $\boldsymbol{\mu}_y$ is the true mean of **Y**, and $\boldsymbol{\mu}_{y,i}$ is the mean of **Y** closest to $\boldsymbol{\mu}_y$ in family i. If g is contained in family i, then the noncentrality parameter is zero; otherwise it is a positive quantity. It can be shown (details are available on request) that the exact bias of $\hat{P}_A$ in the ith family of linear models with normal errors is

$$
\begin{aligned}
\text{Bias}(\hat{P}_A) &= E_Y[\hat{P}_A(\hat{f}_i)] - P_A(g, f_i) \\
&= -\frac{n + 2k_i}{2} + \frac{n}{2}(n + k_i - 1 + 2\lambda_i)e^{-\lambda_i} \sum_{j=0}^{\infty} \frac{\lambda_i^j}{j!(n - k_i - 1 + 2j)},
\end{aligned}
$$

where λ_i is the noncentrality parameter corresponding to the ith family. If the noncentrality parameter remains constant as n increases, then

$$
\text{Bias}(\hat{P}_A) = \frac{k_i(k_i + 1)}{n - k_i - 1} - \frac{2(k_i - 1)\lambda_i}{(n - k_i - 1)(n - k_i + 1)} + O(n^{-3}),
$$

which goes to 0 as n goes to infinity. Presumably, this is the case that Akaike had in mind. In many applications, however, λ_i increases linearly with n. If $\lambda_i = c_i n$, where c_i is a fixed positive constant, then

$$
\text{Bias}(\hat{P}_A) = \frac{k_i(k_i + 1 - 2\lambda_i)}{n - k_i - 1 + 2\lambda_i} + \frac{2n\lambda_i(n + k_i - 1 + 2\lambda_i)}{(n - k_i - 1 + 2\lambda_i)^3} + O(n^{-1}),
$$

which goes to $-2c_i(k_i - 1 + 2c_ik_i)/(1 + 2c_i)^2$ as n goes to infinity. Accordingly, if the noncentrality parameter increases with n, then $\hat{P}_A$ is asymptotically biased. The asymptotic bias can be as large as $-k_i$.

Alternative estimators of predictive accuracy that are asymptotically unbiased whether or not the model under consideration is the correct one have been developed. Sawa (1978) gave an asymptotically unbiased estimator of predictive accuracy for linear models. The Takeuchi information criterion (TIC) (Takeuchi, 1976; Stone, 1977) is asymptotically unbiased for more general models.

MODEL SELECTION AND LIKELIHOOD

Forster and Sober argue that the AIC framework justifies likelihood, at least in part. They identified QUAL and DEGREE as two parts of the likelihood principle (LP), so one might conjecture that it was the LP that was receiving justification. This conjecture is suspect, however, because QUAL and

DEGREE are components of the law of likelihood (LL), rather than of the LP. Before discussing the relationship between AIC and likelihood, it might be prudent to examine the LP and the LL.

The LP states that "the 'evidential meaning' of experimental results is characterized fully by the likelihood function, without other reference to the structure of the experiment" (Birnbaum, 1962, 269). That is, all you need is likelihood. Birnbaum showed that the LP is a consequence of two other principles that are generally accepted by the statistical community; namely, sufficiency and conditionality.

Hacking (1965) thought that the LP went too far in some ways and not far enough in others. First, the LP is deficient because it does not say how to use the likelihood function. Second, the LP is too strong because it uses the likelihood function without reference to the statistical model. Fraser (1963) and others (see Berger and Wolpert, 1984, 47) have criticized the LP on the grounds that the model itself might not be sufficient. Accordingly, the LP should be applied only if the model is believed to capture the entire relationship between the data and the parameters of interest. Third, Hacking argued that belief in the statistical model itself could be uncertain. The investigator should be allowed to formulate alternative models and to evaluate their relative strengths; i.e., model checking should be allowed. Hacking removed these deficiencies by formulating a LL. Royall (1997) attributed the LL to Hacking (1965) and stated the law as follows:

> If hypothesis A implies that the probability that a random variable X takes the value x is $p_A(x)$, while hypothesis B implies that the probability is $p_B(x)$, then the observation $X = x$ is evidence supporting A over B if and only if $p_A(x) > p_B(x)$, and the likelihood ratio, $p_A(x)/p_B(x)$, measures the strength of that evidence. (Royall, 1997, 3)

Royall (2000, 760; 2004 [chapter 5, this volume]) omitted the "only if" condition.

Edwards (1992) argued that it is not necessary to make a distinction between the LL and the LP. He combined them into a single likelihood axiom, stating that

> Within the framework of a statistical model, all the information which the data provide concerning the relative merits of two hypotheses is contained in the likelihood ratio of those hypotheses on the data, and the likelihood ratio is to be interpreted as the degree to which the data support the one hypothesis against the other. (Edwards, 1992, 31)

Joshi (1983) apparently agreed that there is no need to have a LL and a LP. He wrote that the total LP consists of Birnbaum's LP together with the LL, the two parts being kept separate only by convention.

It is apparent that Forster and Sober refer to the LL when they write of QUAL and DEGREE, although they could just as well be referring to Edward's likelihood axiom. In any case, the LL and the axiom agree that within the context of a statistical model, relative evidence is measured by the ratio of likelihoods given the observed data. One implication of the LL as well as the LP is that potential data that might have occurred but did not occur are irrelevant. Accordingly, any evidential procedure that integrates over the sample space (i.e., unobserved data) violates the law, the principle, and the axiom.

AIC AND THE LAW OF LIKELIHOOD

Does the AIC framework justify the LL? I think not, because the AIC framework violates the LL both in theory and in practice. In theory, the AIC framework compares families in terms of predictive accuracies. The justification for $\hat{P}_A$ in (3) is that it estimates predictive accuracy, P_A in (2). Predictive accuracies are useful indices but they do not characterize the evidence that the data provide concerning family i versus family j. In fact, P_A does not even depend on the observed data—P_A is an index of the distance between the true distribution, $g(\mathbf{z})$, and the estimative distribution, $f_i(\mathbf{z}|\boldsymbol{\theta}_y^{(i)})$, *averaged over the sample space* (see equation 2). Predictive accuracy is a property of the hypothesized family of distributions; it is not a property of the data. In practice, P_A is estimated by $\hat{P}_A$ in (3) and $\hat{P}_A$ does depend on the data. Nonetheless, as a measure of evidence, AIC violates the LL. The estimator is obtained by employing asymptotic expansions and then *averaging over the sample space*. The correction term, k_i, is obtained by this averaging process.

Model selection, according to AIC, proceeds by selecting family i over family j if

$$\frac{f_i(\mathbf{y}|\boldsymbol{\theta}_y^{(i)})}{f_j(\mathbf{y}|\boldsymbol{\theta}_y^{(j)})} > e^{k_i - k_j}. \tag{4}$$

Indeed, the left-hand side of (4) is a likelihood ratio. But, as acknowledged by Forster and Sober, it is not the ratio of likelihoods for family i versus family j. Rather, the left-hand side of (4) is the ratio of likelihoods of the most likely member of family i to the most likely member of family j, given the data. It is just one of an infinite number of likelihood ratios that could be computed by choosing specific values for $\theta^{(i)}$ and $\theta^{(j)}$. The LL tells us how to compare f_i to f_j when $\theta^{(i)}$ and $\theta^{(j)}$ are completely specified, but it does not tell us how to compare f_i to f_j when the parameter values are not specified. In particular, the LL does not say that two families should be compared by taking the likelihood ratio of their likeliest members.

Frequentist-based methods of model selection that are consistent with the LL are rare or nonexistent. Ideally, a frequentist would compare likelihood functions evaluated at parameter values that are optimal for the family. That is, evidence about the ith versus the jth families would be examined through the ratio $f_i(\mathbf{y}|\theta_{\text{opt}}^{(i)})/f_j(\mathbf{y}|\theta_{\text{opt}}^{(j)})$, where $\theta_{\text{opt}}^{(i)}$ is the maximizer of $\int \ln[f_i(\mathbf{z}|\theta)] g(\mathbf{z}) \, d\mathbf{z}$ with respect to θ. In practice, this maximization cannot be performed because g is unknown. Accordingly, families cannot be compared in this manner.

BAYESIAN MODEL SELECTION AND THE LAW OF LIKELIHOOD

Forster and Sober state that model selection is the Achilles' heel of the LP. Perhaps it is more accurate to say that the LP is the Achilles' heel of anti-Bayesians. Anti-Bayesians violate the LP. Others can employ Bayesian model selection procedures that are consistent with the LP and the LL. For example, to compare two families, a Bayesian might consider $\theta^{(i)}$ and $\theta^{(j)}$ to be nuisance parameters and use integration to remove them from the likelihood function. Berger, Liseo, and Wolpert (1999) argued that this approach to eliminating nuisance parameters often has advantages over other approaches. The resulting densities, now free of nuisance parameters, can be compared by a likelihood ratio. This ratio is called the Bayes factor (Kass and Raftery, 1995) and can be written as

$$B_{ij} = \frac{\text{posterior odds of model } i \text{ to model } j}{\text{prior odds of model } i \text{ to model } j} = \frac{P(M_i|\mathbf{y})\,P(M_j)}{P(M_j|\mathbf{y})\,P(M_i)} = \frac{f_i(\mathbf{y})}{f_j(\mathbf{y})},$$
$$\text{where } f_i(\mathbf{y}) = \int f_i(\mathbf{y}|\theta^{(i)})m_i(\theta^{(i)})d\theta^{(i)}, \tag{5}$$

$P(M_i|\mathbf{y})$ is the posterior probability of model (or family) i, $P(M_i)$ is the prior probability of model i, and $m_i(\theta^{(i)})$ is a density reflecting prior beliefs about the parameter $\theta^{(i)}$. The simplification in (5) is obtained by expressing

$P(M_i|\mathbf{y})$ as $f_i(\mathbf{y})P(M_i)/f(\mathbf{y})$, where $f(\mathbf{y})$ is the marginal density for $\mathbf{Y}$. Prior probabilities such as $P(M_i)$ need not be specified because they cancel and the Bayes factor is free from their influence.

Note that B_{ij} is a likelihood ratio; it is the likelihood of family i to family j, given the data. Accordingly, model selection based on the Bayes factor does not violate the LL. Bayesians integrate over the parameter space to eliminate nuisance parameters, whereas frequentists integrate over the sample space to eliminate the data. The former approach yields likelihood ratios that characterize the evidence provided by the data. The latter approach yields analytic methods that have long-run optimal properties (e.g., unbiasedness).

Forster and Sober apparently agree that Bayes factors are what we want to examine, but they argue that these ratios cannot be computed because, in general, there is no objective way to choose the prior distribution $m_i(\mathbf{\theta})$. This is a nontrivial issue and might be called the Achilles' heel of Bayesian methods. Furthermore, they correctly argue that the use of improper priors is problematic in this setting because improper priors bring undefined constants of possibly different magnitudes into the numerator and denominator of (5). Kass and Raftery (1995, sec. 5.3) gave some suggestions for solving this problem, but none are entirely compelling.

This problem is partially solved by appealing to asymptotic theory. It can be shown that if the sample size is large, then the Bayes factor depends only weakly on the prior. Specifically,

$$\ln(B_{ij}) = \ln(\hat{B}_{ij}) + O_p(1),$$
$$\text{where } \ln(\hat{B}_{ij}) = \ln\left[\frac{f_i(\mathbf{y}|\mathbf{\theta}_y^{(i)})}{f_j(\mathbf{y}|\mathbf{\theta}_y^{(j)})}\right] - \frac{k_i - k_j}{2}\ln(n), \tag{6}$$

and $\mathbf{\theta}_y^{(i)}$ is the maximum likelihood estimator of $\mathbf{\theta}^{(i)}$ (see Kass and Raftery, 1995, sec. 4). Contrary to the claim of Forster and Sober (in note 3 of their chapter), the approximation in (6) is not based on the use of improper prior distributions. The prior distributions are proper. The prior distribution is not apparent in (6) because it has been absorbed into the $O_p(1)$ term. The correction term, $(k_i - k_j)\ln(n)/2$, is obtained by averaging over the parameter space, whereas the correction term for AIC is obtained by averaging over the sample space.

The absolute error in (6) does not go to 0 as n increases, but the relative error $[\ln(\hat{B}_{ij}) - \ln(B_{ij})]/\ln(B_{ij})$ is $O_p(n^{-1})$ and does go to zero. Accordingly, for sufficiently large samples, the approximate Bayes factor in (6) should be useful for model selection. Kass and Wasserman (1995) showed that for some priors, the approximation error is only $O(n^{-1/2})$.

The approximation in (6) is the basis of Schwartz's (1978) BIC. Schwartz's goal was to select the family having the largest posterior probability, rather than to select the family having the largest predictive accuracy. Schwartz showed that the posterior probability of the ith family is proportional (approximately) to $\exp\{BIC_i\}$, where

$$\text{BIC}_i = \ln[f_i(\mathbf{y}|\boldsymbol{\theta}_y^{(i)})] - \left(\frac{k_i}{2}\right)\ln(n). \tag{7}$$

Unlike Schwartz's BIC, the approximate Bayes factor in (6) does not have a probability-based interpretation. The Bayes factor is a likelihood ratio, rather than a posterior probability. One can use approximate Bayes factors in model selection without assuming that the true distribution belongs to one of the families under consideration and without assigning prior probabilities to the various families.

BAYESIAN MODEL AVERAGING

If one does assign prior probabilities to families of distributions, then Bayes factors can be used to construct predictive distributions. For example, if there are q families, then one could assign $P(M_i) = \tau_i$, where $\tau_i > 0$; and $\sum_{i=1}^{q}\tau_i = 1$. In the absence of other knowledge, each τ_i could be equated to $1/q$. The posterior probability corresponding to the ith family is

$$P(M_i|\mathbf{y}) = \frac{\tau_i B_{i1}}{\sum_{i=1}^{q}\tau_j B_{j1}},$$

and the predictive distribution for a future observation z is $f(z) = \sum_{j=1}^{q}f_i(z)P(M_i|\mathbf{y})$, which can be approximated by

$$\sum_{i=1}^{q} f_i(\mathbf{z}|\boldsymbol{\theta}_y^{(i)}) \frac{\tau_i f_i(\mathbf{y}|\boldsymbol{\theta}_y^{(i)})/n^{k_i/2}}{\sum_{j=1}^{q}\tau_j f_j(\mathbf{y}|\boldsymbol{\theta}_y^{(j)})/n^{k_j/2}}.$$

This predictive distribution represents a weighted average over the q families, whereas the distribution selected by AIC is the single $f_i(\mathbf{z}|\boldsymbol{\theta}_y^{(i)})$ that corresponds to the family having maximum $\hat{P}_A$ in (3). By averaging over families, the Bayesian approach takes model uncertainty into account in a simple way. Kass and Raftery (1995, sec. 6) argued convincingly that this is an important advantage. Wasserman (2000) gave a nontechnical introduction to model averaging. Hoeting et al. (1999) provided a tutorial on Bayesian model averaging. Buckland, Burnham, and Augustin (1997) and Burnham and Anderson (1998, chap. 4) discussed the use of AIC-based weights to do model averaging. It is important to note that averaging over families does not entail averaging over the sample space of the data. Accordingly, Bayesian model averaging does not violate the LL.

THE *FORM* OF MODEL SELECTION METHODS

Forster and Sober note that the *form* of LL-consistent methods such as BIC is the same as that of AIC (compare equations 3 and 7). This is true, but irrelevant. The quantity $\hat{P}_A$ has the *form* of a likelihood ratio because the observed log-likelihood plus a bias correction is an estimator of predictive accuracy. If predictive accuracy could be computed directly, then model selection via AIC would proceed by selecting the family for which P_A in (2) is maximized rather than by selecting the family for which $\hat{P}_A$ in (3) is maximized. The appearance of the likelihood ratio in methods based on different philosophical principles is evidence that the likelihood ratio indeed is important, but it is not evidence about which methods are consistent with the LL. It seems odd to me that anti-Bayesians would point to a Bayesian procedure that is consistent with the LL to bolster support for a frequentist procedure that is not consistent with the LL.

OPTIMAL PREDICTIVE DISTRIBUTIONS

Suppose that one is willing to violate the LL by integrating over the sample space and using predictive accuracy in (2) as a model selection criterion. If high predictive accuracy (low expected KL number) is what the investigator really wants, then the investigator should not use the model selected by AIC. Alternative predictive distributions are superior.

Case I: The Family Contains the True Distribution

Assume that the true density of Y is known to be in the family $f(y|\theta)$, indexed by θ. The investigator wishes to find a predictive density for a vector of future observations Z that will be generated from the same process that gave rise to y. One candidate for this predictive density is the estimative density $f(z|\theta_y)$, where, as before, θ_y is the MLE of θ. Akaike (1973) showed that the expected KL information corresponding to the estimative density is

$$E_Y\{KL[f(z|\theta), f(z|\theta_y)]\} = \frac{k}{2} + O(n^{-1}). \tag{8}$$

A second candidate is the posterior predictive density $f(z|y)$ defined by

$$f(z|y) = \frac{f(z, y)}{f(y)} = \frac{\int f(y|\theta)f(z|\theta)m(\theta)d\theta}{\int f(y|\theta)m(\theta)d\theta}, \tag{9}$$

where $m(\theta)$ is a prior distribution for θ. Using Laplace's method (Tierney and Kadane, 1986), it can be shown (details are available on request) that the expected KL information corresponding to the posterior predictive density is

$$\{E_Y KL[f(z|\theta), f(z|y)]\} = \left(\frac{k}{2}\right)\ln(2) + O(n^{-1}), \tag{10}$$

provided that the prior does not depend on n. A comparison of equations 8 and 10 reveals that, asymptotically, the posterior predictive density has greater predictive accuracy (smaller expected KL information) than does the estimative density and this difference does not depend on the prior.

The advantage of the posterior predictive density can be even larger in finite samples than in infinite samples for certain models. Consider, for example, a regression model under normality. If Jeffreys's (1961) invariant location-scale prior is used for the location and scale parameters, then the posterior predictive distribution is a multivariate t. See Kass and Wasserman (1996, sec. 2) for a summary of Jeffreys's priors. Keys and Levy (1996) showed that the posterior induced by Jeffreys's prior is optimal within a large class of predictive distributions.

Figure 6.2.1 displays the expected KL information numbers for two re-

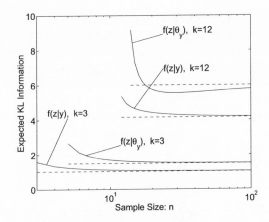

FIGURE 6.2.1 Expected Kullback-Leibler information for Bayesian, $f(z|y)$, and frequentist, $f(z|\theta_y)$, approximating densities when the family contains the true density. Dotted lines represent asymptotic expected KL numbers.

gression models, one having an intercept plus one additional explanatory variable and the other having an intercept plus ten additional explanatory variables. Both models also have a scale parameter. The scale parameter in the estimative density is based on the optimally weighted maximum likelihood estimator (see Keys and Levy, 1996). The expected KL information numbers for $f(z|\theta_y)$ and $f(z|y)$ do not depend on the values of true regression parameters or on the nature of the explanatory variables. Furthermore, these information numbers can be computed without resorting to simulation. A sample size of at least k is required to fit the regression model and a sample size of at least $k + 2$ is required for the expected KL number for the estimative density to be finite. The solid lines in figure 6.2.1 reveal the finite sample advantage of the posterior predictive approach whereas the dotted lines (computed using equations 8 and 10) reveal the asymptotic advantage of the posterior predictive approach over the estimative approach. Aitchison (1975) gave another normal theory example, but without explanatory variables.

The advantage of the posterior predictive approach can be even greater than that illustrated in figure 6.2.1. Suppose that a single observation is obtained from a binomial distribution, $\text{Bin}(n, \theta)$, where n is the number of independent trials and θ is the probability of success on any single trial. It is assumed that $\theta \in (0, 1)$. If the distribution of θ is modeled by a beta distribution, then the posterior predictive distribution is a beta-binomial. The estimative distribution is just another binomial, namely Bin(n, θ^y). If maximizing predictive accuracy is the goal, then one should never use the estimative distribution in this situation because its predictive accuracy is negative infinity and its expected KL number is positive infinity. This occurs because $P(1 \leq Z \leq n - 1|\theta^y) = 0$ whenever $y = 0$ or $y = n$, but the true probability, $P(1 \leq Z \leq n - 1|\theta)$, is never zero. Predictive accuracy of the beta-binomial distribution, on the other hand, is finite regardless of which beta prior is used.

Case II: The Family Does Not Contain the True Distribution

Forster and Sober might object that the above comparisons are not meaningful because, in practice, the true family is never known. A more realistic situation is one in which several approximating families are under consideration, but the true distribution is not contained in any of the families. As an illustration, consider the problem of approximating a complex curve by a polynomial. Figures 6.2.2A and 6.2.2C display the mean responses of $n = 100$ cases. The mean response curves were generated as random mixtures of weighted sine and cosine functions. The data consist of samples of $n = 100$

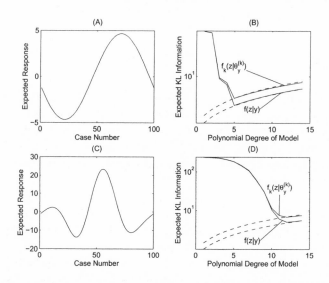

FIGURE 6.2.2 Expected response functions for $n = 100$ cases (A and C). Solid lines in B and D are expected Kullback-Leibler numbers for Bayesian and frequentist approximating densities when the family does not contain the true density. Dotted lines represent asymptotic expected KL numbers.

independent normally distributed random variables having the means displayed in figure 6.2.2A and 6.2.2C. The curves can be approximated by fitting polynomial regression models to the data.

Figures 6.2.2B and 6.2.2D (paired with 6.2.2A and 6.2.2C respectively) display the expected KL numbers for various polynomial models. The posterior predictive and estimative densities are defined as in figure 6.2.1. The expected KL numbers in figures 6.2.2B and 6.2.2D were computed analytically (details are available on request). The solid lines reveal the finite sample expected KL numbers whereas the dotted lines reveal the asymptotic expected KL numbers that would be realized if the true models had been contained in the approximating family. Initially, expected information decreases as polynomial degree ($k - 2$) increases, but it begins to increase when additional polynomial terms add more noise than signal to the fitted model.

The expected KL numbers for the Bayesian and frequentist densities are similar when the polynomial model fits poorly (low degree), but the Bayesian densities have smaller expected KL numbers when the polynomial model fits better (higher degree). Accordingly, figure 6.2.2 verifies that Bayesian posterior predictive densities can be superior to frequentist estimative densities even when the true distributions are unknown and are not members of the approximating family.

SUMMARY

The goal of AIC is to choose a model with good predictive accuracy. The difference between predictive accuracies of two models is the expected log ratio of the predictive (or estimative) densities *averaged over the sample space*. Accordingly, it must be concluded that AIC violates the LL as well as the LP. Thus, Forster and Sober's claim that the LP is supported, in part, by the Akaike framework is suspect.

Bayesian model selection procedures, on the other hand, are consistent with the LL. If sample size is large, then the relevant likelihood ratio (Bayes factor) depends only weakly on the prior distribution and model selection can proceed without specifying the priors.

If one agrees to violate the LL and to compare models according to predictive accuracy, then Bayesian posterior predictive densities are superior to frequentist estimative densities. Furthermore, this advantage appears to extend to the situation in which the true distribution is not contained in one of the approximating families. Accordingly, the approach advocated by Forster and Sober is suspect.

6.3   Rejoinder

Malcolm Forster and Elliott Sober

We find ourselves agreeing with much of what Kruse says in his reply to "Why Likelihood?" and we appreciate his clarification on other points. Boik, on the other hand, believes there is a satisfactory Bayesian solution to the problem of model selection. He also believes that the Akaike solution is flawed. We discuss these points in the following.

INTERPRETATIVE ISSUES

Boik argues that the Akaike information criterion (AIC) "violates" the likelihood principle (which, according to Boik, we should have called the law of likelihood). His argument is that one implication of the law of likelihood is that "potential data that might have occurred but did not occur are irrelevant." "Accordingly, any evidential procedure that integrates over the sample space (i.e., unobserved data) violates the law, the principle, and the axiom." We think this reflects a confusion of the goal that Akaike describes (maximizing predictive accuracy) and the method he identifies in his theorem for achieving that goal (AIC itself). AIC says that there are just two things relevant to estimating predictive accuracy—the actual data and the number of adjustable parameters. The data that one could have obtained but did not is irrelevant to this estimate. Thus, AIC does not violate this implication of the likelihood principle, law, or axiom.

Moreover, Boik says that our claim that the principle "is supported, in part, by the Akaike framework is suspect." We stand by what we claimed— that AIC takes account of something more than likelihood in evaluating models, but that the AIC evaluation of models must agree with the likelihood ordering of those models' likeliest members when the models have the same number of adjustable parameters. Thus, AIC entails that the likelihood principle is true in a special circumstance but not generally.

We noted in our paper that AIC and BIC have the same logical form. Boik finds this point "true, but irrelevant" on the grounds that "the appearance of the likelihood ratio in methods based on different philosophical principles is evidence that the likelihood ratio indeed is important, but it is not evidence about which methods are consistent with the [law of likelihood]." Boik is right that the ubiquity of a quantity does not establish anything about consistency. Our point was simply that the ratio measure specified by (DEGREE) is endorsed not just by AIC, but by other model selection criteria as well.

Boik may have the impression that we think that AIC is the one and only model selection criterion that makes sense. We do not think that this is true. In fact, we say explicitly (Forster and Sober, 1994, sec. 7) that "we do not draw the rash conclusion that Bayesian methodology is irrelevant to Akaike's new predictive paradigm." Forster and Sober (1994) presented Akaike's treatment of model selection as proving that, *by itself*, maximum log-likelihood provides an overestimate of a model's predictive accuracy. Akaike (1973) then showed how the expected bias of that estimate could be corrected by subtracting a penalty term for complexity. However, as every statistician knows, a lack of bias is but one criterion of a "good" estimate; Akaike's theorem has always left open the possibility that there are better methods of maximizing predictive accuracy.

Boik presents the idea of model averaging (see Wasserman, 2000) as a provably better way of maximizing predictive accuracy in some circumstances. We have no objection to this (and did not discuss it in our paper); as Boik notes, this is not the unique prerogative of Bayesian approaches.

THE PROBLEM WITH BAYESIANISM

In footnote 3 of our paper, we criticize the use of improper priors in computing average likelihoods. We did not think of this footnote as a complete statement of our reservations about Bayesianism, so here we want to indicate briefly why we think that there is an additional difficulty.

The version of Bayesianism discussed here is the version that seeks to compare models in terms of the ratio of their posterior probabilities (the Bayes factor). This ratio is a product of two ratios. The first is the ratio of the *likelihoods* of *families* of distributions M_1 and M_2. The second is the ratio of the priors, $P(M_1)/P(M_2)$. However, it is an unalterable fact about the probabilities of nested models that, for example, the probability of (PAR) is greater than or equal to that of (LIN) *relative to any data you care to describe.* No matter what the likelihoods are, there is no assignment of priors and no way of calculating the likelihood ratio consistent with probability theory that can alter the fact that $P(\text{PAR} \mid \text{Data})/P(\text{LIN} \mid \text{Data}) \geq 1$. The reason is that (LIN) is a special case of (PAR) (proof: if LIN is true, then so is PAR, so the probability of PAR being true is greater than or equal to the probability of

LIN being true). How, then, can Bayesians explain the fact that scientists sometimes prefer (LIN) over (PAR)?

Many Bayesians (including Boik) are either unaware of this problem, or fail to believe it, but those who address this problem do so as follows. Instead of (LIN) and (PAR), let us consider (LIN) and (PAR*), where (PAR*) is some subset of (PAR) from which (LIN) has been removed. Since (LIN) and (PAR*) are disjoint, nothing prevents us from ordering their prior probabilities as we see fit. In response, we note that this ad hoc maneuver does not address the problem of comparing (LIN) to (PAR), but merely changes the subject. In addition, it remains to be seen how Bayesians can justify an ordering of priors for the hypotheses thus constructed and how they are able to make sense of the idea that families of distributions (as opposed to single distributions) possess well-defined likelihoods.

Bayesians who proceed with the calculation of the Bayes factor come up with entirely different answers. Schwarz (1978) calculated the likelihood ratio as being the ratio of the sum of two terms—one being the maximum log-likelihood and the other being a penalty term for complexity that is generally greater in magnitude than the one proposed by AIC. On the other hand, Bandyopadhyay, Boik, and Basu (1996) begin from the same assumptions and end up with a simpler result—that the likelihood ratio is the ratio of the maximum likelihoods. How is it possible to get two different answers? We do not have an answer to this question, but instead show how one might get any answer one pleases.

Consider two families of distributions M_1 and M_2 with dimensions k_1 and k_2, respectively. By the probability calculus, the likelihood of a family is the *average* of the likelihoods of the members of the family, where the average is determined in terms of some prior weighting over the members of the family, denoted $P(\theta|M)$, where θ ranges over members of M. If $P(\theta|M)$ is *strictly informationless,* then it is easy to see that $P(\text{Data}|M) = 0$ because almost every θ in M will be very far from the data (have zero likelihood). This means that if we accord equal weight to every distribution in M, the average likelihood of M will be 0. However, it is mathematically possible that a ratio of 0/0 has a finite value, so the fact that the likelihoods are 0 is merely a sign of something strange and not an insurmountable problem in and of itself. The problem is whether there exists a principled way of calculating its value in the case in which the priors are strictly informationless.

As we said in footnote 3 of our paper, the problem arises with "improper" priors, which cannot integrate to 1 (if the prior is 0 then the integral is 0, but if the prior has any finite value, then the integral is ∞). This presents no problem when calculating a posterior distribution, which is provably proportional to the likelihood function, which can then be normalized, but it does create a problem when trying to calculate average likelihoods. This is equally a problem for calculating the *ratios* of average likelihoods as it for calculating average likelihoods.

One approach is to let $P(\theta|M)$ be *approximately* informationless, which avoids the problem of improper priors, and then take the limit to the case of a strictly informationless priors. The simplest method is to divide the members of the family into two subsets—within one subset (which includes the distributions that have nonnegligible likelihood), we let the weights be equal and nonzero; outside this volume, we let the weights be 0. We illustrate this proposal by returning to the examples of (LIN) and (PAR), where the error variance σ^2 is known. For (LIN), we specify a volume V_1 of parameter values for α_0 and α_1 within which the likelihoods are nonnegligible. For PAR, we specify a volume V_2 of parameter values for β_0, β_1, and β_3 with the same characteristic. If we let boldface $\boldsymbol{\alpha}$ and $\boldsymbol{\beta}$ range over curves in (LIN) and (PAR) respectively, the average likelihoods of those families then may be expressed approximately as

$$P(\text{Data}|\text{LIN}) = (1/V_1)\int \ldots \int P(\text{Data}/\boldsymbol{\alpha}, \text{LIN})d\boldsymbol{\alpha},$$
$$P(\text{Data}|\text{PAR}) = (1/V_2)\int \ldots \int P(\text{Data}/\boldsymbol{\beta}, \text{PAR})d\boldsymbol{\beta},$$

where the integration is restricted to the subsets of LIN and PAR with nonzero weights. How are these two likelihoods to be compared? The volume V_1 has two dimensions in parameter space; the volume V_2 has three. Let's assume that the ratio of the integrals has some finite value, which we denote by R. By setting $V_2/V_1 = C$, the likelihood ratio is $R \times C$, which can be any value because C is arbitrary. Except for the last two sentences, most of this material appears in Forster and Sober (1994, sec. 7). Nevertheless, the Bayesian defense (Bandyopadhyay and Boik, 1999) and development (Wasserman, 2000) of the Bayes factor approach to model selection has continued unabated, so all of this needs to be said again.

IS AIC BIASED?

The aim of the Akaike framework is to answer questions about the relationship between the things we can see, such as the maximum likelihoods of models and the number of (independently) adjustable parameters, and things we can't see, like predictive accuracy. Akaike's theorem talks about one such relationship—namely, that AIC provides a unbiased estimate of predictive accuracy. The relationship is best expressed in terms of the space of all possible values of parameters that appear among the models, which we shall refer to as parameter space. Each point in parameter space picks out a particular probabilistic hypothesis. Models are families of such hypotheses and are therefore subsets of points in parameter space. Nested models are nested subsets of points in parameter space. We may always assume that the parameter space is large enough to include a point that represents the distribution that actually generated the data—called the true distribution. A model is true only if it contains the true distribution. The true distribution need not be contained in *any* of the models actually under consideration, and when this is the case, all of the models are false (misspecified).

The method of maximum likelihood estimation picks out a member of each model M, which we denote by $L(M)$. $L(M)$ has a well-defined predictive accuracy that is almost always less that the predictive accuracy of the true distribution. Even when M is true, $L(M)$ may not be the true distribution because of the effect of overfitting. This is a general assumption-free fact. Akaike addressed the problem of deciding how to choose a predictive distribution that is close as possible to the true distribution, where "close" is defined in terms of the Kullback-Leibler discrepancy, or equivalently in terms of predictive accuracy (Forster and Sober, 1994). As Boik points out, there is no reason to restrict one's choice to the maximum likelihood distributions. The Bayesian posterior distribution over the members of each model is also a predictive distribution, as is any average of these, and also averages of the maximum likelihood distributions. Akaike's theorem considers only the maximum likelihood hypotheses.

Akaike's theorem, or at least the proof of the theorem, states that under certain conditions AIC provides an unbiased estimate of the expected predictive accuracy of maximum likelihood hypotheses produced by a model. To explain this, we point out that the proof of the theorem (Akaike, 1973) can be broken into two parts. The first part says that asymptotically for large n, where n is the number of data, under some weak regularity conditions, the differences in the log-likelihoods between two hypotheses (points) in parameter space can be expressed in terms of a norm (a squared distance derived from a multivariate normal probability distribution centered on the point representing the true distribution). Differences in predictive accuracy are also expressed in terms of this norm because differences in predictive accuracy are just expected differences in the log-likelihoods. Under this norm, there exists an orthogonal basis defined on the parameter space such that each model is represented as a vector subspace (since it is the set of all vectors such that certain components are fixed at 0), and the predictively most accurate member of each model is the orthogonal projection of the true distribution onto the subspace. In fact, one can scale the parameters in this new orthogonal basis such that the squared distances defined by this norm are just the squared Euclidean distances in the parameter space. This linearized representation exists for the same reason that any regular non-linear function looks linear when one focuses on a small region of the function, for as the number of data increases, the region in which likelihoods are nonnegligible becomes smaller and smaller.

In our publications, we have referred to the conditions under which the parameter space could be treated as a normed vector space, in this way, as the normality conditions. The assumption that these conditions hold, we have called the normality assumption. Note that this normality assumption makes no reference to whether the error distributions associated with each point hypothesis are normal. In fact, the generality of Akaike's theorem derives from the fact that it applies to any error distribution that satisfies the weak regularity assumptions.

The second part of Akaike's proof shows that *if* the normality assumption is true, then AIC provides an unbiased estimate of predictive accuracy. Boik has correctly pointed out that the assumption that normality holds for finite *n* is a far stronger assumption than the weak conditions for *asymptotic* normality. In fact, one cannot expect it to be true approximately in finite cases, even when the *error* distributions are normal. There is no violation of the part of the theorem that states that AIC provides an unbiased estimate of predictive accuracy *if the normality assumption holds.* The question is whether the normality assumption is true, or approximately true.

The proper measure of the degree to which AIC errs in its estimation of predictive accuracy should be defined in terms of the squared difference between the estimated predictive accuracy and the true predictive accuracy of maximum likelihood hypothesis L(*M*). When normality holds, there is an error of estimation and its probabilistic behavior is described by what we refer to as the error theorem in Forster and Sober (1994, sec. 6).[20] The failure of normality is a different source of error (Boik simply assumes that discrepancies in the biases of estimates of predictive accuracy are not cancelled by other errors, and we assume the same), which we have failed to examine

20. See equation (4.55) in Sakamoto, Ishiguro, and Kitagawa (1986, 77). Similar formulae were originally proven in Akaike (1973).

before. We therefore welcome the opportunity to discuss the error in the special case that Boik describes.

Our disagreement with Boik concerns the way in which he represents the size of these errors. In Forster and Sober (1994), we were careful to define predictive accuracy in a way that does not depend on the size of the sample used to estimate it. This is a general point about the estimation of any quantity. For example, a statistician polls n individuals on their votes in an election and uses the sample *mean* to estimate the proportion of Democratic voters among all voters. The error of the estimation decreases as the sample size increases. One would never argue that the error of estimation increases as the sample size increases because we are really estimating the expected number of Democratic votes in a sample of that size. This is a mistake because the quantity being estimated depends on the sample size. Despite its absurdity, AIC has this peculiar feature. It estimates the expected predictive error *in a set of n data points* from a sample of n observed data points. In order to rid the target quantity of its dependence on n, one should divide by n. This is what we did in defining predictive accuracy (Forster and Sober, 1994), but we were careless about doing so in "Why Likelihood?" When correctly defined, the estimate for predictive accuracy is also divided by n, and the errors of the estimate are divided by n. Boik fails to divide by n in his definition of predictive accuracy, and consequently fails to divide the error of estimation by n.

When one divides by n in the appropriate way, even the worst-case scenario described by Boik does not look so bad. For example, consider the case where Boik claims that there are not only corrections for finite samples, but even in the asymptotic limit when the model is not true (which he calculates for the special case he considers). In reference to this calculation, he says that "the asymptotic bias can be as large as $-k_i$," where k_i is the number of adjustable parameters in the model under consideration. However, the asymptotic bias is actually k_i/n, which is asymptotically zero. Therefore, his claim that there is an asymptotic bias in this example is incorrect. (Similarly, see Forster, 2001, sec. 7, for a rebuttal of the well-known charge that AIC is statistically inconsistent.)

REFERENCES

Aitchison, J. 1975. Goodness of Prediction Fit. *Biometrika* 62:547–554.
Aitchison, J., and I. R. Dunsmore. 1975. *Statistical Prediction Analysis*. Cambridge: Cambridge University Press.
Aitkin, M. 1991. Posterior Bayes Factors. *J. Roy. Stat. Soc.*, ser. B, 53:111–142.

Akaike, H. 1973. Information Theory as an Extension of the Maximum Likelihood Principle. In Petrov, B. N., and F. Csaki, eds., *Second International Symposium on Information Theory.* Budapest: Akademiai Kiado.

Akaike, H. 1985. Prediction and Entropy. In Atkinson, A. C., and S. E. Fienberg, eds., *A Celebration of Statistics.* New York: Springer.

Bandyopadhyay, P., and R. Boik. 1999. The Curve Fitting Problem: A Bayesian Rejoinder. *Phil. Sci.* 66 (supplement): S391–S402.

Bandyopadhyay, P., R. Boik, and P. Basu. 1996. The Curve Fitting Problem: A Bayesian Approach. *Phil. Sci.* 63 (supplement): S264–S272.

Berger, J. O., B. Liseo, and R. L. Wolpert. 1999. Integrated Likelihood Methods for Eliminating Nuisance Parameters. *Stat. Sci.* 14:1–28.

Berger, J. O., and L. Pericchi. 1996. The Intrinsic Bayes Factor for Model Selection and Prediction. *J. Am. Stat. Assn* 91:109–122.

Berger, J. O., and Wolpert, R. L. 1984. *The Likelihood Principle.* Hayward, CA: Institute of Mathematical Statistics.

Birnbaum, A. 1962. On the Foundations of Statistical Inference (with discussion). *J. Am. Stat. Assn* 57:269–326.

Buckland, S. T., K. P. Burnham, and N. H. Augustin. 1997. Model Selection: An Integral Part of Inference. *Biometrics* 53:603–618.

Burnham, K. P, and D. R. Anderson. 1998. *Model Selection and Inference: A Practical Information-Theoretic Approach.* New York: Springer.

Cramér, H. 1946. *Mathematical Methods of Statistics.* Princeton: Princeton University Press.

Edwards, A. W. F. 1987. *Likelihood* (expanded ed.). Baltimore: Johns Hopkins University Press.

Edwards, A. W. F. 1992. *Likelihood* (expanded ed.). Baltimore: Johns Hopkins University Press.

Fisher, R. A. 1938. Comment on H. Jeffrey's "Maximum Likelihood, Inverse Probability, and the Method of Moments." *Ann. Eugenics* 8:146–151.

Fitelson, B. 1999. The Plurality of Bayesian Measures of Confirmation and the Problem of Measure Sensitivity. *Phil. Sci.* 66 (proceedings): S362–S378.

Forster, M. R. 1995. Bayes and Bust: The Problem of Simplicity for a Probabilist's Approach to Confirmation. *Br. J. Phil. Sci.* 46:399–424.

Forster, M. R. 1999. Model Selection in Science: The Problem of Language Variance. *Br. J. Phil. Sci.* 50:83–102.

Forster, M. R. 2000a. Hard Problems in the Philosophy of Science: Idealisation and Commensurability. In Nola, R., and H. Sankey, eds., *After Popper, Kuhn, and Feyerabend.* Boston: Kluwer.

Forster, M. R. 2000b. Key Concepts in Model Selection: Performance and Generalizability. *J. Math. Psych.* 44:205–231.

Forster, M. R. 2001. The New Science of Simplicity. In Zellner, A., H. A. Keuzenkamp, and M. McAleer, eds., *Simplicity, Inference, and Modeling.* Cambridge: Cambridge University Press.

Forster, M. R., and E. Sober. 1994. How to Tell When Simpler, More Unified, or Less *Ad Hoc* Theories Will Provide More Accurate Predictions. *Br. J. Phil. Sci.* 45:1–35.

Fraser, D. A. S. 1963. On the Sufficiency and Likelihood Principles. *J. Am. Stat. Assn* 58:641–647.

Geisser, S. 1993. *Predictive Inference: An Introduction.* New York: Chapman and Hall.

Hacking, I. 1965. *Logic of Statistical Inference.* Cambridge: Cambridge University Press.

Hoeting, J. A., D. Madigan, A. E. Raftery, and C. T. Volinsky. 1999. Bayesian Model Averaging: A Tutorial (with discussion). *Stat. Sci.* 14:382–417.

Hogg, R. V., and A. T. Craig. 1978. *Introduction to Mathematical Statistics.* 4th ed. New York: Macmillan.

Hurvich, C. M., and C. Tsai. 1989. Regression and Time Series Model Selection in Small Samples. *Biometrika* 76:297–307.

Jeffreys, H. 1961. *Theory of Probability.* 3rd ed. New York: Oxford University Press.

Joshi, V. M. 1983. Likelihood Principle. In Kotz, S., N. L. Johnson, and C. B. Read, eds., *Encyclopedia of Statistical Sciences.* 9 vols. New York: Wiley.

Kass, R. E., and A. E. Raftery. 1995. Bayes Factors. *J. Am. Stat. Assn* 90:773–795.

Kass, R. E., and L. Wasserman. 1995. A Reference Bayesian Test for Nested Hypotheses and Its Relationship to the Schwartz Criterion. *J. Am. Stat. Assn* 90:928–934.

Kass, R. E., and L. Wasserman. 1996. The Selection of Prior Distributions by Formal Rules. *J. Am. Stat. Assn* 91:1343–1370.

Keys, T. K., and M. S. Levy. 1996. Goodness of Prediction Fit for Multivariate Linear Models. *J. Am. Stat. Assn* 91:191–197.

Kieseppä, I. A. 1997. Akaike Information Criterion, Curve-Fitting, and the Philosophical Problem of Simplicity. *Br. J. Phil. Sci.* 48:21–48.

Kolmogorov, A. N. 1956. *Foundations of Probability.* Trans. N. Morrison. New York: Chelsea.

Kullback, S., and R. A. Leibler. 1951. On Information and Sufficiency. *Ann. Math. Stat.* 22:79–86.

Lele, S. R. 2004. Evidence Functions and the Optimality of the Law of Likelihood. Chapter 7 in Taper, M. L., and S. R. Lele, eds., *The Nature of Scientific Evidence: Statistical, Philosophical, and Empirical Considerations.* Chicago: University of Chicago Press.

Popper, K. R. 1959. *The Logic of Scientific Discovery.* London: Hutchinson.

Rawls, J. 1971. *A Theory of Justice.* Cambridge: Harvard University Press.

Rosenkrantz, R. D. 1977. *Inference, Method, and Decision.* Dordrecht: Reidel.

Royall, R. M. 1997. *Statistical Evidence: A Likelihood Paradigm.* Boca Raton: Chapman and Hall/CRC.

Royall, R. M. 2000. On the Probability of Observing Misleading Statistical Evidence. *J. Am. Stat. Assn* 95:760–768.

Royall, R. M. 2004. The Likelihood Paradigm for Statistical Evidence. Chapter 5 in Taper, M. L., and S. R. Lele, eds., *The Nature of Scientific Evidence: Empirical, Statistical, and Philosophical Considerations.* Chicago: University of Chicago Press.

Sakamoto, Y., M. Ishiguro, and G. Kitagawa. 1986. *Akaike Information Criterion Statistics.* Dordrecht: Kluwer.

Sawa, T. 1978. Information Criteria for Discriminating among Alternative Regression Models. *Econometrica* 46:1273–1291.

Schwarz, G. 1978. Estimating the Dimension of a Model. *Ann. Stat.* 6: 461–464.

Sober, E. 1988. *Reconstructing the Past: Parsimony, Evolution, and Inference.* Cambridge: MIT Press.

Sober, E. 1999. Instrumentalism Revisited. *Critica* 31:3–38.

Stone, M. 1977. An Asymptotic Equivalence of Choice of Model by Cross-Validation and Akaike's Criterion. *J. Roy. Stat. Soc.,* ser. B, 39:44–47.

Sugiura, N. 1978. Further Analysis of the Data by Akaike's Information Criterion and the Finite Corrections. *Comm. Stat.,* ser. A, 7:13–26.

Takeuchi, K. 1976. Distribution of Information Statistics and a Criterion of Model Fitting. *Suri-Kagaku* (Mathematical Sciences) 153:12–18 (in Japanese).

Tierney, L., and J. B. Kadane. 1986. Accurate Approximations for Posterior Moments and Marginal Densities. *J. Am. Stat. Assn* 81:82–86.

Wasserman, L. 2000. Bayesian Model Selection and Model Averaging. *J. Math. Psych.* 44:92–107.

7 Evidence Functions and the Optimality of the Law of Likelihood

Subhash R. Lele

ABSTRACT

This paper formulates a class of functions, called evidence functions, which may be used to characterize the strength of evidence for one hypothesis over a competing hypothesis. It is shown that the strength of evidence is intrinsically a comparative concept, comparing the discrepancies between the data and each of the two hypotheses under consideration. The likelihood ratio, which is commonly suggested as a good measure for the strength of evidence, belongs to this class and corresponds to comparing the Kullback-Leibler discrepancies. The likelihood ratio as a measure of strength of evidence has some important practical limitations: (1) sensitivity to outliers, (2) necessity to specify the complete statistical model, and (3) difficulties in handling nuisance parameters. It is shown that by using evidence functions based on discrepancy measures such as Hellinger distance or Jeffreys distance, these limitations can be overcome. Moreover, evidence functions justify the use of ratio of conditional, marginal, pseudo or quasi-likelihoods as measures of strength of evidence. It is shown that, provided the model is correctly specified, for a single-parameter case, the likelihood ratio is an optimal measure of the strength of evidence within the class of evidence functions. This result also establishes the connection between the optimality of the estimating functions and the optimality of the evidence functions.

I acknowledge the many interesting discussions I have had with Richard Royall, Jeffrey Blume, and Mark Taper on the topic of statistical evidence. I thank George Casella for pointing out Daniels's (1961) paper.

INTRODUCTION

Hacking (1965, 70) states the law of likelihood as follows:

> If hypothesis A implies that the probability that a random variable X takes the
> value x is $p_A(x)$, while hypothesis B implies that the probability is $p_B(x)$, then the
> observation $X = x$ is evidence supporting A over B if and only if $p_A(x) > p_B(x)$,
> and the likelihood ratio, $p_A(x)/p_B(x)$, measures the strength of that evidence.

Royall (1997) provides several cogent and powerful arguments in favor of
this law. One of the key properties of the likelihood ratio (LR) is that the
probability of strong evidence in favor of the true hypothesis converges to 1
as the sample size increases. That is, eventually, as the information about the
hypothesis increases, we are guaranteed to make the correct choice. In
mathematical terms, it says that, for any choice of K,

$$\lim_{n \to \infty} P\left(\frac{p_A(x)}{p_B(x)} > K \text{ when hypothesis } A \text{ is true} \right) = 1.$$

This is clearly an important assurance. However, as has been emphasized in
ecology (Burnham and Anderson, 1998), and will be discussed later in this
volume (Lindsay, chapter 14; Taper, chapter 15), there is no such thing as the
true model or the true hypothesis. We only have a good or a bad approximation of the true nature of the phenomenon. The above result and the law
of likelihood raise the following question: what happens to the above probability when neither hypothesis is "true," that is, when the true state of nature is something different from either of the two competing hypotheses?
The following simple result provides an insight regarding this issue and
also leads to some important understanding of the law of likelihood itself.

Let X_is be independent, identically distributed random variables following distribution $T(.)$. Let one hypothesis postulate that X_is are independent,
identically distributed random variables following distribution $A(.)$ while
the alternative hypothesis postulate that X_is are independent, identically
distributed random variables following distribution $B(.)$. Let us assume that
neither $A(.)$ nor $B(.)$ equals $T(.)$. There are several ways in which one can
quantify how well the distribution $A(.)$ approximates the distribution $T(.)$.
One such commonly used measure of the similarity or dissimilarity between two probability distributions $f(.)$ and $g(.)$ is defined as:

$$I(f, g) = \int f(x) \log \frac{g(x)}{f(x)} dx.$$

This measure is known as the Kullback-Leibler divergence between $f(.)$ and $g(.)$. It quantifies the dissimilarity between the two distributions. Small values of this measure indicate strong similarity between the two distributions. Notice that $I(f, f) = 0$, that is, the dissimilarity between a distribution and itself is 0. However, note that this is not a symmetric function of f and g, that is, $I(f, g) \neq I(g, f)$. Because it is not symmetric, it is not a "distance." Hence, these dissimilarity measures are usually called "divergence," "discrepancy," or "disparity" measures. For a good discussion of the properties of the Kullback-Leibler divergence, see Burnham and Anderson (1998, chap. 2). In this chapter, I will use the nomenclature "disparity." Let us denote the Kullback-Leibler disparity between $T(.)$ and $A(.)$ by $I(T, A)$. Similarly let $I(T, B)$ denote the Kullback-Leibler disparity between $T(.)$ and $B(.)$. The following result, which is a simple consequence of the law of large numbers, shows that, as the sample size increases, the average log-likelihood ratio converges to the difference in the two disparities.

$$Result\ 1:\ \lim_{n \to \infty} \frac{1}{n} \log \frac{p_A(x)}{p_B(x)} = I(T, B) - I(T, A).$$

It then follows easily that when $I(T, B) > I(T, A)$, that is when probability distribution $A(.)$ is a better approximation of the true distribution $T(.)$, the probability of strong evidence in favor of $A(.)$ converges to 1, as it should. On the other hand, if $I(T, B) < I(T, A)$, the probability of strong evidence in favor of $B(.)$ converges to 1.

One important consequence of this simple result is that one can now interpret the law of likelihood in terms of comparison of the disparities between the data and the two competing hypotheses. In essence, the law of likelihood calculates the likelihood disparity (Lindsay, 1994) between the data and the hypothesized models. One says that the model for which this disparity is smaller is better supported by the data.

The likelihood ratio or the likelihood disparity has several attractive qualities as discussed by Royall (1997), but it also suffers from several shortcomings. For example, it is known to be sensitive to outliers. Further, in order to use the likelihood ratio as strength of evidence, the full distribution of the random variables needs to be specified. In practice, this might be quite difficult. The use of other measures may render the full likelihood unnecessary. For example, quasi-likelihood functions (McCullogh and Nelder, 1989) are useful for statistical inference when only the mean and the variance functions of the random variables are specified. Likelihood-based inference also suffers when there are many nuisance parameters in the model.

Many inference procedures based on conditional likelihood, profile likelihood, pseudolikelihood, and so on are employed to deal with nuisance parameters. Can those "likelihoods" be used to measure strength of evidence? There are situations, for example, in random effects models, hierarchical models, or models of spatial data, where writing the full likelihood is quite cumbersome. Hence, computation of the likelihood ratio will be difficult. However, there are simpler quantities such as the "composite likelihood" (Lindsay, 1988; Heagerty and Lele, 1998; Lele and Taper, 2002) that are used effectively to conduct statistical inference in such situations. Can a composite likelihood ratio quantify strength of evidence? Recently, Lindsay and Qu (2000) have introduced the concept of "quadratic inference functions." Are they quantifying strength of evidence in any sense? It turns out that the common feature to all these practical adjustments to likelihood-based inference is that they all are based on some idea of disparity between the data and the model. In the following, I will attempt to generalize the idea of the likelihood ratio as a measure of strength of evidence to the idea of comparison of the disparities between the data and each of the competing models (hypotheses) as a measure of strength of evidence. If such a generalization is possible, it will open up the possibility of using these practical adjustments to the likelihood-based inference to quantify the strength of evidence.

Such a generalization in terms of disparity measures leads to several interesting possibilities that are important for practicing scientists. For example, one could make the quantification of the strength of evidence robust against outliers; one could measure the strength of evidence when models are only partially specified through the mean and the variance functions; one could measure the strength of evidence for the parameter of interest in the presence of the nuisance parameters; and so on. In fact, there is a close connection between the theory of estimating functions (Godambe, 1960; Godambe and Kale, 1991) and quantification of the strength of evidence. Thus, this should enable us to draw upon significant developments in the theory of estimating functions in the context of nuisance parameters, mixture models, and semiparametric inference.

The outline of this paper is as follows. In the following section, I introduce a class of functions called evidence functions. These functions are based on different disparity measures that quantify the disparity between two given statistical models, fully or partially specified. How do we choose an evidence function from this class? First, I will describe a reasonable optimality criterion based on the probability of strong evidence. Then it will be shown that, within the class of these evidence functions, the log-likelihood ratio has the property of maximizing the asymptotic probability of strong evidence, thus justifying the law of likelihood from the optimality perspective. It will be shown that there are other evidence functions that are equivalent to the likelihood ratio according to this criterion up to the first order. However, for a practicing scientist, these competing evidence functions may be more desirable than the likelihood ratio because of their outlier and model robustness or their ability to deal with nuisance parameters, etc.

EVIDENCE FUNCTIONS

Any function that purports to measure the strength of evidence will be called an "evidence function." In this section, I will discuss some intuitive conditions that an evidence function should satisfy. I will then provide some examples of evidence functions. In the following, the results are stated only for a real-valued parameter space. Generalizations to vector-valued parameter space and nuisance parameters will be discussed in detail elsewhere.

Let Θ denote the parameter space and X denote the sample space. An evidence function purportedly measures the strength of evidence for one parameter value vis-à-vis another parameter value based on the observed data. Thus, the domain of the evidence function should be $X \times \Theta \times \Theta$. It also makes sense that the evidence function has a range that is an ordered field; otherwise, one will not be able to say if one parameter value is better supported than another. Consider a real-valued function of the data and a pair of parameter values, namely, $Ev : X \times \Theta \times \Theta \to R$. Let us call such a function an evidence function.

Given an evidence function, we say that we have strong evidence in favor of θ_1 in comparison to θ_2 if $Ev(X, \theta_1, \theta_2) < -\log K$ for some fixed $K > 1$. We say that we have strong evidence in favor of θ_2 in comparison to θ_1 if $Ev(X, \theta_1, \theta_2) > \log K$ for some fixed $K > 1$ and that we have weak evidence if $-\log K < Ev(X, \theta_1, \theta_2) < \log K$. This is the zone of indifference.

There are some additional conditions that evidence functions should satisfy. I will discuss them in the following.

C1: *translation invariance.* It is reasonable to say that one should not simply be able to add a constant to the evidence function and change the strength of evidence. This is achieved by fixing the value of the evidence function along a curve, say at (θ, θ). The following antisymmetry condition implies translation invariance of the evidence function class: $Ev(X, \theta_1, \theta_2) = -Ev(X, \theta_2, \theta_1)$, that is, it is antisymmetric. This condition also implies that $Ev(X, \theta_1, \theta_1) = 0$.

C2: *scale invariance.* The second invariance corresponds to scaling. One should not be able to simply multiply an evidence function by a constant and change the strength of evidence. The scale invariance can be obtained by the standardized evidence function introduced in the next section.

C3: *reparameterization invariance.* If we have a function ψ, which is a one-one onto mapping of the parameter space, that is, $\psi : \Theta \rightarrow \Psi$, the evidence comparison between (θ_1, θ_2) and between the corresponding points in the transformed space (ψ_1, ψ_2) should be identical. This implies that one cannot simply stretch the coordinate system and change the quantification of the strength of evidence. For example, comparison of the evidence for μ_1 versus μ_2 should be the same as comparing the evidence for $\exp(\mu_1)$ versus $\exp(\mu_2)$.

C4: *invariance under transformation of the data.* The fourth type of invariance the class of evidence functions should satisfy is the transformation of the data invariance. This means that comparison of evidence should remain unaffected by changes in the measuring units. Let $g : X \rightarrow Y$ be a one-one onto transformation of the data. Let the corresponding transformation in the parameter be denoted by $\bar{g}(.)$. Then evidence function should be such that $Ev(X, \theta_1, \theta_2) = Ev(Y, \bar{g}(\theta_1), \bar{g}(\theta_2))$.

Aside from the above conditions, intuitively it is clear that for any function to be considered a reasonable evidence function, the probability of strong evidence in favor of the true hypothesis should converge to 1 as the sample size increases. Toward this goal, some additional regularity conditions are imposed.

The following condition is critical. It says that, on an average, evidence for the true parameter is maximized only at the true value and at no other parameter value.[1] This is similar to the fact that, on an average, log-likelihood is maximized at the true parameter value (Wald, 1949).

R1: $E_{\theta_1}\big(Ev(X, \theta_1, \theta_2)\big) < 0$ for all $\theta_2 \neq \theta_1$.

The next condition stipulates that the weak law of large numbers is applicable to these functions, that is,

R2: $n^{-1}\big(Ev(X, \theta_1, \theta_2) - E_{\theta_1}(Ev(X, \theta_1, \theta_2))\big) \xrightarrow{p} 0,$

provided θ_1 is the true value (or is the best approximating model).

---

1. If the true model is not in the model set, the set must contain a unique best approximating model.

Conditions R1 and R2 imply that, for any fixed $K > 1$, $P_{\theta_1}(Ev(X, \theta_1, \theta_2) <$ $-\log K) \to 1$, that is the probability of strong evidence in favor of the true parameter in comparison to any other parameter converges to 1 as the sample size increases.

Remark: The condition R1 can be replaced by an asymptotic version, namely,

$$R1^*: \liminf_{n \to \infty} \inf_{\theta_2 \neq \theta_1} \left\{ n^{-1} E_{\theta_1}(Ev(X, \theta_1, \theta_2)) \right\} < 0.$$

These conditions are similar to those given in Li (1996), where weak consistency of the estimators obtained from estimating functions is shown. See also Lehmann (1983, chap. 5) for similar results for the log-likelihood.

There are many evidence functions that satisfy these conditions. The data transformation invariance and the parameterization invariance, in some instances, might be satisfied only asymptotically and locally. In the following I will describe some particular examples of evidence functions For the sake of this discussion, I will consider only independent, identically distributed discrete random variables on a finite support.

1) *Log-likelihood-ratio evidence functions.* Let the probability mass function be denoted by $p(x, \theta)$. Consider the log-likelihood-ratio $Ev(X, \theta_1, \theta_2) = \sum_{i=1}^{n} \{\log p(x_i, \theta_2) - \log p(x_i, \theta_1)\}$. This is a real-valued, antisymmetric function. Moreover, it is well known (Rao, 1973, 59) that $E_{\theta_1}\{\log p(x, \theta_2) - \log p(x, \theta_1)\} < 0$. Using the weak law of large numbers, it can be shown that $n^{-1}\sum_{i=1}^{n} \{\log p(x_i, \theta_2) - \log p(x_i, \theta_1)\} \xrightarrow{p}$ $c < 0$. Thus, it easily follows that the probability of strong evidence, $P_{\theta_1}(\sum_{i=1}^{n} \{\log p(x_i, \theta_2) - \log p(x_i, \theta_1)\} < -\log K)$, converges to 1 as the sample size increases.

2. *Disparity-based evidence functions.* Notice that the evidence function considered above relates to the Kullback-Leibler disparity measure, called the likelihood disparity measure by Lindsay (1994). In essence, the above evidence function calculates the likelihood disparity between the data and the hypothesized models. It says that the model for which this disparity is smaller is better supported by the data. However, one can, in principle, use any other disparity measure as well. Such general disparity measures have been used in statistics in the past. For more details, see Beran (1977), Berkson (1980), Read and Cressie (1988), and Lindsay (1994), where such measures are used for the estimation purpose.

Another source for a general survey of divergence measures and their geometry is Rao (1987). I now put them to use to compare evidence for one parameter versus another parameter. I follow the notation introduced in Lindsay (1994).

For the sake of exposition, let us start with a particular disparity measure, the Hellinger distance between two probability mass functions. Let S denote the support, that is the set of values for which the probability is nonzero, for the two probability mass functions. In general, the Hellinger distance between two probability mass functions f and g, both with the same support S, is defined as $HD(f, g) = \sum_{x \in S} [\sqrt{f(x)} - \sqrt{g(x)}]^2$. Let us denote the probability mass functions corresponding to the two hypotheses under consideration by p_{θ_1} and p_{θ_2}. Given the data X, let us denote the empirical probability mass function by p_n. Define the Hellinger distance-based evidence function as $Ev(X, \theta_1, \theta_2) = n\{HD(p_n, p_{\theta_1}) - HD(p_n, p_{\theta_2})\}$.

It is easy to see that this is an antisymmetric function. Notice that condition R2* is satisfied. Standard application of the weak law of large numbers implies that the quantity inside the braces converges to a negative number as the sample size increases, provided θ_1 is the best approximating model. This gives the result that the probability of strong evidence converges to 1 as the sample size increases. It should be emphasized again that when one compares evidence for one parameter against another, all that one does is find which of the two parametric models is closest to the data according to some preselected disparity measure. Comparing evidence can thus be reduced to comparison of disparities between the data and the two models under investigation.

The above discussion can be generalized to a class of disparity measures discussed by Lindsay (1994). Let $\rho(f, g)$ denote the disparity between two models f and g. Then the evidence function based on this disparity measure is given by $Ev(X, \theta_1, \theta_2) = n\{\rho(p_n, p_{\theta_1}) - \rho(p_n, p_{\theta_2})\}$. Under the conditions specified in Lindsay (1994), it is easy to see that the probability of strong evidence converges to 1 as the sample size increases. These evidence functions may have an advantage over the log-likelihood-ratio-based evidence function in terms of robustness in the presence of outliers. Also, there are situations, particularly in the mixture models, for example, Lehmann (1983, 442), where the likelihood is unbounded and hence the log-likelihood-ratio evidence function may not provide the correct answers, whereas Hellinger distance-based evidence functions, being bounded, may be better suited.

3) *Log-quasi-likelihood-ratio.* Many times scientists are unwilling to specify the complete statistical model for the phenomenon under study. Such considerations have given rise to the subject area of estimating functions. Godambe and Kale (1991) is an excellent source of information on

this area. Generalized linear models and quasi-likelihood methods (Mc-Cullogh and Nelder, 1989) are quite familiar to ecologists. Inferences in these situations are based on the specification of only the first two moments, the mean and the variance. A natural question is, can we quantify and compare evidence in such situations? I will show that one can construct evidence functions that are based on only the first two moments, thus allowing such a comparison. This discussion is largely based on the technical results developed by Li (1993), Li and McCullogh (1994), and Li (1996). Suppose that the mean function is μ_θ and the variance function is V_θ. Consider the evidence function given by $Ev(X, \theta_1, \theta_2) = (\mu_{\theta_1} - \mu_{\theta_2})^T V_{\theta_1}^{-1}(X - \mu_{\theta_1}) + (\mu_{\theta_2} - \mu_{\theta_1})^T V_{\theta_2}^{-1}(X - \mu_{\theta_2})$.

It is easy to see that this is antisymmetric and R1 is satisfied (Li, 1996). Moreover, provided the weak law of large numbers is applicable, that is, if $n^{-1}\{(\mu_{\theta_1} - \mu_{\theta_2})^T V_{\theta_1}^{-1}(X - \mu_{\theta_1})\} \xrightarrow{p} 0$ under the true value θ_1, it is easy to see that the probability of strong evidence converges to 1 as sample size increases. This evidence function is closely related to Jeffreys's disparity measure (Jeffreys, 1983).

Intuitively, the result that the probability of strong evidence converges to 1 for this evidence function is obvious, because as the sample size increases the disparity between the data and true parameter becomes 0 whereas the disparity between the data and any other parameter becomes positive. This makes the evidence function 0 only at the true parameter as the sample size becomes large.

Other types of evidence functions may also be constructed based on the composite likelihood (Lindsay, 1988), the profile likelihood (Royall, 1997), potential function (Li and McCullogh, 1994), or the "quadratic inference functions," introduced by Lindsay and Qu (2000).

COMPARISON OF TWO EVIDENCE FUNCTIONS

Now that we have specified a class of evidence functions, can we say something about which evidence function within this class is preferable? In the following I suggest a criterion to compare two evidence functions.

Definition: Given two evidence functions, $Ev1(X, \theta_1, \theta_2)$ and $Ev2(X, \theta_1, \theta_2)$, we say that $Ev1(X, \theta_1, \theta_2)$ is preferable to $Ev2(X, \theta_1, \theta_2)$ if, for all $K > 1$ and for all (θ_1, θ_2), $\dfrac{P_{\theta_1}(Ev1(X, \theta_1, \theta_2) < -\log K)}{P_{\theta_1}(Ev2(X, \theta_1, \theta_2) < -\log K)} > 1$.

This definition says that the evidence function that maximizes the probability of strong evidence in favor of the true parameter uniformly over the

whole parameter space is preferable. However, as with uniformly best estimators, such a uniformly best evidence function may not exist. Moreover,
calculating the probability of strong evidence for a general evidence function and for any sample size is a difficult, if not impossible, task. Instead,
what I propose is an asymptotic criterion. We know that, for most reasonable evidence functions, the probability of strong evidence converges to 1 as
the sample size increases. A similar situation arises when one wants to compare statistical test functions: all reasonable test functions provide consistent tests; that is, for any fixed alternative, the power of the test converges
to 1. There are several different approaches for comparing consistent tests.[2]
One such comparison is based on the Pitman efficiency of a test, where we
let the alternative hypothesis converge to the null hypothesis at a certain rate
and compare the rate at which the power converges to 1. The basic idea is to
make the comparison harder and harder, as the sample size increases. I will
propose a similar approach to comparison of evidence functions. I will let
the alternative hypothesis θ_2 converge to the null hypothesis θ_1 at $n^{1/2}$ rate.
Provided certain regularity conditions are satisfied, one can calculate the
probability of strong evidence for a general evidence function under this
scenario. One can then show that the probability of strong evidence is maximized when the log-likelihood ratio is used as the evidence function. This
will be undertaken in the next section.

OPTIMALITY OF THE LOG-LIKELIHOOD
RATIO EVIDENCE FUNCTION

To facilitate analytical and asymptotic calculations, I will impose some additional regularity conditions on the class of evidence function.

R3: The evidence functions $Ev(X, \theta_1, \theta_2)$ are twice continuously differentiable and the Taylor series approximation is valid in the vicinity of the true
value θ_1.

R4: The central limit theorem is applicable, implying the result that there
exists a function $J(\theta_1)$ such that $0 < J(\theta_1) < \infty$ and

$$n^{-1/2} \left(\frac{d}{d\theta} Ev(X, \theta_1, \theta)|_{\theta_1} \right) \xrightarrow{D} N(0, J(\theta_1))$$

---

2. See Serfling (1980, chap. 10) for a survey.

R5: The weak law of large numbers is applicable, implying the result:

$$n^{-1}\left(\frac{d^2}{d\theta^2}Ev(X, \theta_1, \theta)|_{\theta_1}\right) \xrightarrow{p} -I(\theta_1),$$

where $0 < I(\theta_1) < \infty$. We also assume that function $I(\theta)$ is continuously differentiable up to second order.

STANDARDIZED EVIDENCE FUNCTIONS

In order to compare two evidence functions, we need to standardize them in some fashion.[3] Consider $\tilde{E}v(X, \theta_1, \theta_2) = \dfrac{Ev(X, \theta_1, \theta_2)}{I^{1/2}(\theta_1)I^{1/2}(\theta_2)}$. This will be called a "standardized evidence function."

It is easy to check that the standardized evidence function also belongs to the class of evidence functions. We are now in a position to prove the optimality of the law of likelihood. In the following, we denote θ_1 by θ and $\theta_2 = \theta + \delta n^{-1/2}$ where δ is a real number. Thus, for large sample size, the two parameters are very close to each other, making the distinction between them harder and harder.

Under the regularity conditions, the Taylor series approximation gives

$$\tilde{E}v(X, \theta, \theta + \delta n^{-1/2}) \approx \delta n^{-1/2}\left(\frac{d}{d\theta}\tilde{E}v(X, \theta_1, \theta)|_{\theta_1}\right)$$
$$+ \delta^2 n^{-1}\left(\frac{d^2}{d\theta^2}\tilde{E}v(X, \theta_1, \theta)|_{\theta_1}\right)$$

Hence, from assumptions R4 and R5, we can write the probability of strong evidence as

$$P_\theta(\tilde{E}v(X, \theta, \theta + \delta) < -\log K) \approx \Phi\left\{\left(\frac{J(\theta)}{I^2(\theta)}\right)^{-1/2}\left[\frac{-\log K}{\delta\sqrt{n}} + \frac{\delta}{2}\sqrt{n}\right]\right\}.$$

This result is a slight generalization of the result given in Daniels (1961, equations 3.8 and 3.9), where asymptotic properties for the maximum like-

3. The standardization proposed here is similar to that discussed by Godambe and Kale (1991) in the context of estimating functions.

lihood estimator are studied under only the existence of the first-order derivatives of the log-likelihood. To prove the optimality of the likelihood ratio as a measure of evidence, "an optimal evidence function," we need to show that the above probability is maximized at the Fisher information. First, notice that $\Phi(.)$ is a distribution function and hence is monotonically increasing. Thus, we need to maximize the argument of this function or equivalently minimize $\dfrac{J(\theta)}{I^2(\theta)}$. From the Cramér-Rao lower bound, or a simple application of the Cauchy-Schwartz inequality (Godambe, 1960; Kale, 1962), it follows that $\dfrac{J(\theta)}{I^2(\theta)} \geq I_F^{-1}(\theta)$ where I_F is the Fisher information in the true likelihood with equality obtained for the log-likelihood-ratio evidence function. Thus, the probability of strong evidence is maximized for the likelihood ratio evidence function. This establishes the optimality of the likelihood ratio as an evidence function within the class specified above. In fact, this suggests that an optimal evidence function may be generated from the consideration of optimal estimating functions, provided one can integrate the estimating function properly. Li and McCullogh (1994) discuss the difficulties as well as solutions to the problem of integration of estimating functions to obtain a likelihood-type quantity. Lindsay and Qu (2000) provide another solution to this problem based on the ideas of score tests and quadratic inference functions.

In the following, I observe a few ramifications of the above result.

1. The above result can also be proved under a severely constrained class of evidence function, namely the class that satisfies the condition of information unbiasedness (Lindsay, 1982). As Lindsay points out, these arise out of the consideration of likelihood factors.
2. Evidence functions based on the Hellinger disparity measure are first-order equivalent to log-likelihood-ratio-based evidence functions. However, they are less sensitive to the presence of outliers.
3. Evidence measures based on Jeffreys's disparity can be shown to be optimal in the same sense that maximum quasi-likelihood estimators are optimal. These evidence measures require only partial specification of the model features.
4. In the context of nuisance parameters, evidence functions based on the conditional likelihood ratio can be shown to be optimal.

DISCUSSION

This paper proposes a class of functions that may quantify the strength of evidence called evidence functions.[4] Under certain reasonable regularity conditions, it is proved that the log-likelihood ratio evidence function is the best evidence function. However, there are several shortcomings to the log-likelihood ratio as an evidence function. It is sensitive to outliers in the data and to the specification of the underlying model. Further, it is well known that likelihood-based inference is problematic when there are nuisance parameters. Use of disparity-based evidence measures circumvents the sensitivity to outliers without losing the optimality property (Lindsay, 1994) whereas the use of log-quasi-likelihood evidence functions circumvents the complete model specification problem. Following the results of Godambe (1976) and Lindsay (1982), it is also possible to prove the optimality of the log-conditional likelihood ratio as a measure of the strength of evidence. The nuisance parameter problem may also be handled using the "partially Bayes" approach described in Lindsay (1985). He shows the optimality of the estimating functions obtained using the partially Bayes approach. The results of this paper suggest that the corresponding evidence function should also be optimal. For example, the profile likelihood (Royall, 1997) or the integrated likelihood (Berger, Liseo, and Wolpert, 1999) can be viewed as a particular case of the partially Bayes approach suggested by Lindsay (1985). Such a connection should thus shed some light on the optimality properties of the profile likelihood or the integrated likelihood from the evidential perspective. Mixture models and random effects models consist of another class of problems that might be better handled using disparity-based evidence functions. These and other extensions including the model selection problem discussed by Taper (2004 [chapter 15, this volume]) are worth exploring further.

7.1	Commentary

Christopher C. Heyde

The issues addressed in this paper are important indeed. It is clear that the likelihood ratio will play a key role in comparing specific distributions.

---

4. See Achinstein (2001) and Shafer (1976) for alternative mathematical formulations of a theory of evidence.

What is not so clear is what should be done in the case where explicit choice of distributions is best avoided. This issue will be discussed below.

Doubtless, for reasons of convenience and clarity, the author has usually specialized his discussion to the case of an independent and identically distributed sample. This seems unfortunate, as all the results can be recast in a stochastic process setting with minimal cost or complication. The greater generality afforded by this setting significantly increases the value of the results as a general-purpose scientific tool.

What is needed to achieve the above-mentioned generalization is to note that a log-likelihood is a martingale under modest regularity conditions. Conditions analogous to R1–R5 will hold for most interesting applications, with the R4 and R5 analogs coming from martingale central limit theorems and weak laws of large numbers. All the necessary theory can be found in Hall and Heyde (1980, chap. 6), which provides a detailed discussion of the log-likelihood and its asymptotics in a general setting.

The author concentrates on a single parameter, stating that the vector-valued and nuisance parameter cases will be treated elsewhere. These cases should not be difficult to treat under the assumed conditions of differentiability of the evidence function with respect to the parameter. What is more challenging, however, are nonstandard problems, such as when the parameter set is discrete (as for dealing with the order of an autoregression) or when one has to compare nonnested families of models (for example, different kinds of nonlinear time series, such as bilinear and threshold autoregressive models).

Most of these more challenging problems can probably be treated in a convenient and unified fashion via the formalism of hypothesis testing. See, for example, Heyde (1997, chap. 9) for a framework that should suffice for the task, with appropriate adjustment. Although the ideas of quasi-likelihood estimation have been developed in a setting that mimics the classical likelihood one in requiring continuity of the parameter space and differentiability with respect to the parameter, there is much that can be done outside this setting, for example replacing Fisher information by an empirical version with similar asymptotics. Indeed, quasi-likelihood is a much more flexible methodology than the author acknowledges, as he comments that means and variances need to be specified. In fact, quasi-likelihood is good for things like estimating the mean of a stationary process. A comprehensive treatment of the methodology is provided in Heyde (1997).

The aspirations of the author are considerably similar to those of Lindsay and Markatou in their important forthcoming book.[5] Their methodology is distance based with the intention of providing a framework for model assessment. The general area contains enough scientifically important problems to provide for decades of continuing research. My feeling is that it is of considerable value to view all these topics in terms of information-based comparisons. This, indeed, is the key idea behind likelihood and quasi-likelihood.

7.2   Commentary

Paul I. Nelson

Most members of the statistical community, although seemingly not concerned with patching the cracks in the foundations of their discipline, avidly seek new methods of handling the increasingly complex problems they face. In addition to this natural emphasis on applications, statistical science lacks a broadly applicable, unencumbered conceptual framework that logically flows from a few universally accepted principles. Consequently, paradigm shifts develop very slowly and ultimately win acceptance by persuasion rather than by proof. Bootstrapping, for example, has gained many adherents over the past two decades because in addition to being a very good idea it has been effectively marketed as a solution to otherwise intractable problems. Royall (1997) develops another very good idea, the likelihood principle, and compellingly argues that having observed data $X = x$, the likelihood ratio $L(p_A, p_B) = p_A(x)/p_B(x)$ measures the *evidence* in favor of mass function p_A relative to mass function p_B. For Royall, the likelihood ratio thereby becomes the de facto definition of *evidence*. It is worthy of note that Birnbaum (1962) was able to use only a vague concept of absolute evidence in deriving equivalences among the likelihood, sufficiency, and conditionality principles. Here, Subhash Lele makes a significant addition to Royall's approach by providing a list of properties that reasonable measures of relative evidence should possess. Now, Professor Lele's task is to expand the concept and persuade us to use it.

In the restricted setting of Professor Lele's paper, discrete distributions, and one-dimensional parameters, his evidence functions have great intuitive

5. Editors' note: Dr. Lindsay provides a foretaste of this work in chapter 14.

appeal. Although he states that he will extend these boundaries in future work, I believe that he will run into the same difficulties and need for ad hoc adjustments to basic principles that have limited the application of likelihood-based inference to some interesting but narrow special cases. However, Professor Lele is a forceful advocate of his views, and I look forward to being proven wrong.

The following comments express some of my concerns with the scope and viability of evidence functions:

(1) Hacking's (1965) law of likelihoods, which underlies Lele's approach, can break down when continuous distributions are allowed in the mix, as the following variation on well-known examples illustrates. Let p_A be the normal density with mean 0 and positive standard deviation σ and let p_B assign probability 1 to $\{0\}$. When 0 is observed, $L(p_A, p_B) = p_A(0)/p_B(0)$ can, in the language of Royall (1997), be made to supply arbitrarily large evidence for p_A over p_B by making σ sufficiently small, even though an observation of 0 is perfectly consistent with p_B. Despite p_A being very close, in a practical sense, to p_B for small positive σ, the distributions would still be quite different and their Hellinger distance at a maximum. Although this particular example is artificial, it can easily be tweaked to apply to more realistic situations.

How would Professor Lele define an evidence function in this and the related problem where p_θ is a continuous density function and p_n is the empirical (discrete) mass function? Beran's (1977) approach of fitting a continuous density to p_n when p_θ is continuous introduces the undesirable element of subjectivity in choosing which density estimator to use and is feasible only for large samples.

(2) Condition C4 in the paper, *invariance under transformations of the data*, makes sense only if the family of distributions is closed under the transformation g. If X is binomial based on n trials, for example, invariance of evidence for $g(x) = n - x$ is desirable but a nonissue for $g(x) = \sqrt{x}$. Requiring invariance for all transformations may excessively limit the class of evidence functions.

(3) If one or both of p_A and p_B are composite, even defining an evidence function will require straying far from the basic principles. This problem is even more vexing if nuisance parameters are present.

(4) Inference, in my experience, does not fall neatly into Royall's (1997) three categories: decision, evidence, and belief. Action is often the goal of both evidence and belief. Specifically, Professor Lele does not tell us what the experimenter can or should do with the evidence once it is obtained. The

practical need to cross these boundaries may explain why there is such widespread, simultaneous use of P-values as both evidence and guides to action despite their well-known deficiencies.

7.3   Rejoinder

Subhash R. Lele

THE ROLE OF THE LIKELIHOOD PRINCIPLE IN THE QUANTIFICATION OF THE STRENGTH OF EVIDENCE

Professor Nelson, in the discussion of this paper, and also Professor Cox, in the discussion of Royall (2004 [chapter 5, this volume]) raise the issue of the relevance of stopping rules in the quantification of the strength of evidence. Most scientists intuitively feel that *how* the data were collected should have some bearing on the strength of evidence, although the strict likelihood principle (Berger and Wolpert, 1984) and the use of likelihood ratio as a quantification of the strength of evidence suggests otherwise. I will utilize the standard example commonly invoked to illustrate and support the likelihood principle and show that, under the general evidence function framework, the information in the stopping rule can, indeed should, be incorporated in the strength of evidence.

Consider the experiment where we toss a coin until K heads are observed. In this case, the number of trials is a random variable. Let us assume that there were N trials. As a consequence of this experiment, we obtain a sequence of 0s and 1s. The goal is to quantify the strength of evidence about the parameter $P(\text{head in a single trial}) = p$. There are two ways to model this sequence.

Model 1: We can ignore the fact that the experiment was conducted in such a manner that it terminated after the Kth success and model it as a sequence of i.i.d. Bernoulli trials. Thus, we have $X_i \sim Bernoulli(p)$ with $i = 1, 2, \ldots, N$.

Model 2: We take into account the way the experiment was designed. Under this situation, let Y_i denote the number of failures in between the

I would like to thank Professors Heyde and Nelson for commenting on my paper. Several interesting issues have come up in their discussion and my subsequent thinking.

$(i - 1)$th and the ith success. It is easy to show that this is a geometric random variable with parameter p Thus, we have $Y_i \sim Geometric(p)$ with $i = 1, 2, \ldots, K$.

We consider two parameter values p_1 and p_2. The goal is to compare the strength of evidence for p_1 as compared to p_2.

Kullback-Leibler Evidence Function

Under model 1, the KL evidence function is given by

$$
\begin{aligned}
KLEv(\underline{x}, p_1, p_2) &= \sum_{i=1}^{N} \{\log p(x_i, p_2) - \log p(x_i, p_1)\} \\
&= \sum_{i=1}^{N} x_i \left(\log \frac{p_2}{1 - p_2} - \log \frac{p_1}{1 - p_1} \right) + N \log \frac{1 - p_2}{1 - p_1} \\
&= K \left(\log \frac{p_2}{1 - p_2} - \log \frac{p_1}{1 - p_1} \right) + N \log \frac{1 - p_2}{1 - p_1},
\end{aligned}
$$

whereas, under model 2, the KL evidence function is given by

$$
\begin{aligned}
KLEv(\underline{x}, p_1, p_2) &= \sum_{i=1}^{K} \{\log[(1 - p_2)^{y_i} p_2] - \log[(1 - p_1)^{y_i} p_1] \\
&= K \left(\log \frac{p_2}{1 - p_2} - \log \frac{p_1}{1 - p_1} \right) + N \log \frac{1 - p_2}{1 - p_1}.
\end{aligned}
$$

Thus, we get exactly the same comparison between any two values of the parameter whether we use model 1 or model 2. This is because the likelihood ratio is identical under the two models.

General Evidence Functions

Let us see what happens when we use some other evidence function to quantify the strength of evidence. As an illustration, I will show the calculations for the Hellinger divergence based evidence function. The general form of the HD evidence function is given by:

$$
\begin{aligned}
HDEv(\underline{x}, p_1, p_2) &= \sum_{x \in S} (\sqrt{p_n(x)} - \sqrt{f(x, p_2)})^2 \\
&- \sum_{x \in S} (\sqrt{p_n(x)} - \sqrt{f(x, p_1)})^2
\end{aligned}
$$

This, upon algebraic simplification, can be written as:

$$HDEv(\underline{x}, p_1, p_2) = \sum_{x \in S} \sqrt{p_n(x)}(\sqrt{f(x, p_2)} - \sqrt{f(x, p_1)})$$

Model 1: Here the sample space is $S = \{0, 1\}$. Given the data, $p_n(0) = \dfrac{N - K}{N}$ and $p_n(1) = \dfrac{K}{N}$. Hence the evidence function is given by

$$HDEv(\underline{x}, p_1, p_2) = \sqrt{\left(1 - \frac{K}{N}\right)}(\sqrt{1 - p_2} - \sqrt{1 - p_1})$$
$$+ \sqrt{\frac{K}{N}}(\sqrt{p_2} - \sqrt{p_1}).$$

Model 2: Here the sample space is $S = \{0, 1, 2, \dots\}$. Given the data, $p_n(y) = \dfrac{\sum_{i=1}^{K} I(Y_i = y)}{K}$, and the evidence function is given by

$$HDEv(\underline{y}, p_1, p_2) = \sum_{y=0}^{\infty} \sqrt{p_n(y)}\{\sqrt{p_2(1 - p_2)^y} - \sqrt{p_1(1 - p_1)^y}\}$$

In order to study the behavior of these two situations under the Hellinger divergence-based estimating function, I will consider the asymptotics where K converges to ∞. Recall that evidence functions are defined on the two-dimensional domain. In order to facilitate the visualization, I will plot a slice through the evidence function, where the slice is taken at p_1 being the true value.

It is clear from figure 7.3.1 that the Hellinger distance (HD) based evidence functions under the two models are different. Moreover, the HD evidence function corresponding to model 2 distinguishes any parameter value from the true value more sharply than the evidence function under model 1. It thus stands to reason that if we are going to use an HD evidence function, we should use the HD evidence function under model 2 rather than under model 1. Recall that model 2 correctly describes the true experiment.

In contrast to the Hellinger divergence based evidence functions, the Kullback-Leibler divergence based evidence function comparisons remain the same whether Model 1 or Model 2 is considered to be the true data gen-

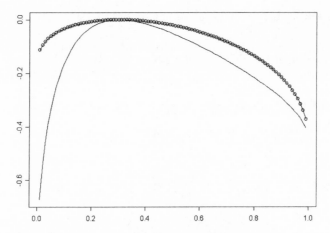

FIGURE 7.3.1 Hellinger distance–based evidence functions and the relevance of stopping rules. The model 1 evidence function is depicted by the line through the circles, and the model 2 evidence function is depicted by the continuous line.

erating mechanism. Thus, stopping rule is irrelevant under the KL evidence function, but it is relevant under the HD evidence function.

One can understand this intuitively by noticing that model 2 explicitly conditions on the fact that the last observation of any observable sequence has to be a success, whereas model 1 ignores this fact. To see that this is important, notice that if the observed sequence had ended in a failure, we will not even bother to entertain model 2 as a candidate model. It gives strong evidence against model 2. However, if a sequence ends with a success, we could conceivably entertain either model. However, if we know how the experiment was conducted, we should not use model 1, because it is a wrong model.[1]

A curious point may be raised here regarding the role of sufficiency and conditioning in statistical inference (Birnbaum, 1962). I wonder if they are strongly related to the use of Kullback-Leibler divergence, or are they more intrinsic than that? It will be interesting to explore the concepts of E-sufficiency (Small and McLeish, 1989) in this context. However, it is clear that if one uses the evidence function framework, the likelihood principle does not seem to play an integral role in the evidential framework.

1. Figure 7.3.1 only compares HD evidence functions under the two models. It does not imply that the HD evidence function under model 2 is better than the KL evidence function.

It should also be noted that the computation of the probabilities of weak and misleading evidence depends on which experiment was conducted and on whether the KL evidence function or the HD evidence function is used.

EVIDENCE FUNCTIONS, COMPOSITE HYPOTHESES, AND NUISANCE PARAMETERS

Professor Nelson raises the issue that the evidential paradigm might be applicable only in simple situations and hence may not be useful in practice. His emphasis is on the issue of handling nuisance parameters and composite hypotheses. I feel this is an important point that can be handled in a unified fashion if one adopts the evidence function ideas rather than a strict likelihood based framework in Royall (1997). Here are some initial ideas.

Evidence functions, in many cases, compare the divergence of the truth from the two hypotheses under consideration. If the two hypotheses are simple hypotheses, the divergences can be defined unambiguously. However, when the hypotheses are composite hypotheses, they each form a collection of many simple hypotheses. Thus, in this case, one has to define divergence of a point (the truth is a single point) from a set (a collection of simple hypotheses). This cannot, in general, be defined unambiguously. In the following, I suggest one possible approach.

Let H denote the composite hypothesis that is a collection of simple hypotheses denoted by h. Given a divergence measure we can compute the divergence between the truth T and any of these simple hypotheses, say $d(T, h)$. Let us define $\rho(T, H, Q) = \int d(T, h)dQ(h)$, where $Q(.)$ is a weight function chosen in such a manner that the resultant integral is a proper divergence measure as well. This is a weighted divergence measure between a point (T) and a set (H). Notice that this definition of divergence depends on two choices, the divergence measure from the truth to a simple hypothesis and the weight measure $Q(.)$. This weight function need not be a probability measure. Based on such a weighted divergence, one can construct the evidence function for comparing two composite hypotheses: $Ev(H_1, H_2) = \rho(T, H_1, Q) - \rho(T, H_2, Q)$. Provided we choose the weight function $Q(.)$ appropriately, this will satisfy the regularity conditions of the evidence functions.

As a simple illustration, consider the case where we have $Y_1, Y_2, \ldots, Y_n$ independent identically distributed random variables following $N(\mu, \sigma^2)$. We want to compare evidence for $\mu = \mu_1$ versus $\mu = \mu_2$. Let us use the profile likelihood function, $PL(\mu; \underline{y}) = \sup_{\sigma^2 > 0} L(\mu, \sigma^2; \underline{y})$, to compare the evidence. One can look upon $PL(\mu_1; \underline{y}) = \sup_{\sigma^2 > 0} L(\mu_1, \sigma^2; \underline{y})$ as a weighted divergence measure with weight 1 given to that value of σ^2 that maximizes $L(\mu_1, \sigma^2)$ and weight 0 for all other values of σ^2. If we were to use the pseudolikelihood (Gong and Samaniego, 1981) to compare evidence; we use $PSL(\mu; \underline{y}) = L(\mu, \sigma^2(\mu); \underline{y})$ where $\sigma^2(\mu) = \dfrac{1}{n}\sum_i (y_i - \mu)^2$. This again corresponds to a weighted divergence measure with weight equal to 1 for $\sigma^2(\mu)$ and 0 for all other values. One could also consider using weighted evidence functions directly instead of going through weighted divergence measures. I have not explored this generalization as yet. My suggestion relates to the discussion by Berger, Liseo, and Wolpert (1999) of the use of integrated likelihood in the context of nuisance parameters.

Another common approach to handling nuisance parameters is based on the use of invariance to eliminate nuisance parameters (Lele and Richtsmeier, 2000). This is also used in the context of analysis of mixed effects where fixed effects are eliminated by invariance (REML) and then likelihood inference is conducted on the transformed data.

To illustrate the power of the use of evidence functions over the likelihood-based approach, let us consider the case of comparing evidence for the common odds ratio from several 2×2 tables. This arises quite commonly in epidemiological studies. I will outline the problem and show that the evidence function approach handles the nuisance parameters quite easily and effectively where the profile and pseudolikelihood approaches have difficulties.

Suppose we want to evaluate the efficacy of a treatment for a particular disease. A commonly used design in this case selects a certain number of patients and divides them in two groups. One group receives the treatment and the other group receives the placebo. We observe how many of the patients survive and how many die in the treatment and control groups. The odds ratio measures the efficacy of the treatment. In practice though, it may be difficult to conduct such a clinical trial in a single hospital. The trial may be replicated in several hospitals and we may have to pool the data from all these hospitals. One of the problems with pooling data from various hospitals is that the efficacy of the treatment may be confounded by the hospital effect.

Let Y_i denote an outcome for patient from the ith hospital. Let

$$\log \frac{P(Y_i = 1|X)}{P(Y_i = 0|X)} = \alpha_i + \beta I_{(X = \text{treatment})}.$$

Here α_i represents the hospital effect and β, the log-odds ratio, represents the treatment effect. A well-known approach to the estimation of the common odds ratio is to use the Mantel-Haenszel estimator. This corresponds to solving an estimating equation $\sum_{i=1}^{K}(n_{11,i}n_{00,i} - \theta n_{10,i}n_{01,i}) = 0$ for θ where $n_{11,i}$ denotes the number of patients who survived after receiving the treatment, $n_{00,i}$ denotes the number of patients who did not survive under placebo, and so on, for the ith hospital. Here, $\theta = \exp(\beta)$ is the odds ratio. Now suppose our goal is to compare evidence for $\theta = \theta_1$ versus $\theta = \theta_2$. Because of the need to estimate too many nuisance parameters, using the profile or pseudolikelihood for this comparison is difficult and not efficient. However, we can easily construct an evidence function by using the above estimating function, as follows:

$$Ev(\theta_1, \theta_2; \underline{y}) = \int_{\theta_1}^{\theta_2} \sum_{i=1}^{K}(n_{11,i}n_{00,i} - \theta n_{10,i}n_{01,i})\,d\theta.$$

We can then use it compare evidence for $\theta = \theta_1$ versus $\theta = \theta_2$. Because the Mantel-Haenszel estimator is known to be highly efficient, one could safely conjecture that the evidence comparisons based on the above evidence function will also be efficient. This clearly shows the effectiveness of the evidence function approach over the standard likelihood-based approach to evidence.

One can also use the conditional likelihood (Godambe, 1976) in the above situation as well as in other cases where there are many nuisance parameters. Once again justifying the use of these different likelihood-type objects for the purpose of evidence comparison is possible under the evidence function setup described in my chapter. Without this general setup, each case will need to be justified individually.

ADDITIONAL ISSUES RAISED BY THE COMMENTATORS

Comparing Discrete and Continuous Random Variables

Professor Nelson raises an issue of comparing discrete and continuous distributions. A simple answer to his question will be the following. Any continuous distribution can be approximated by a multinomial distribution. Once such an approximation is considered, the problem raised by Professor Nelson (his point 1) goes away. Of course, the issue then can be raised about the possible lack of invariance of such a measure to the scheme of discretization. To get around this problem, one could possibly consider the limiting case where the size of the discretization bins converges to 0, the limiting function being the evidence function in the continuous case. A similar issue can be raised against the usual definition of likelihood, and the limiting argument seems to work in that situation (Casella and Berger, 1990). I agree with Professor Nelson that this issue needs further probing.

Invariance under Data Transformations
Professor Nelson thinks that this condition might restrict the class too much. However, intuitively it seems natural that the evidence should be invariant under any one-to-one transformation of the data. I do not see this as being unnatural or terribly restrictive.

Evidence and Decision Making
Professor Nelson has raised an important point: that statistical problems ultimately relate to decision making. Relating evidence to a decision-making process is an important step that should be taken before this framework is useful in practice. I do not now have an answer to this question.

Model Selection
Another important issue that has not been dealt with in this paper is that of model selection (but see Taper, 2004 [chapter 15, this volume]).

Dependent Data Situations
I agree with Professor Heyde that many of these ideas could be extended to dependent data using martingale ideas that he has developed extensively. However, my first goal is to get the initial concepts accepted before extending them to more complex cases.

REFERENCES

Achinstein, P. 2001. *The Book of Evidence.* Oxford: Oxford University Press.

Beran, R. 1977. Minimum Hellinger Distance Estimates for Parametric Models. *Ann. Stat.* 5 : 445–463.

Berger, J. O., B. Liseo, and R. L. Wolpert. 1999. Integrated Likelihood Methods for Eliminating Nuisance Parameters. *Stat. Sci.* 14 : 1–28.

Berger, J. O., and R. L. Wolpert. 1984. *The Likelihood Principle.* Hayward, CA: Institute of Mathematical Statistics.

Berkson, J. 1980. Minimum Chi-Square, not Maximum Likelihood! (with discussion). *Ann. Stat.* 8 : 457–487.

Birnbaum, A. 1962. On the foundations of statistical inference (with discussion). *J. Am. Stat. Assn* 57:269–326.

Birnbaum, A. 1972. More on Concepts of Statistical Evidence. *J. Am. Stat. Assn* 67: 858–871.

Burnham, K. P., and D. R. Anderson. 1998. *Model Selection and Inference: A Practical Information-Theoretic Approach*. New York: Springer.

Casella, G., and R. L. Berger. 1990. *Statistical Inference*. Belmont, CA: Duxbury.

Daniels, H. E. 1961. The Asymptotic Efficiency of a Maximum Likelihood Estimator. In *Proceedings of the Fourth Berkeley Symposium on Mathematical Statistics and Probability*. Berkeley: University of California Press.

Godambe, V. P. 1960. An Optimum Property of Regular Maximum Likelihood Estimation. *Ann. Math. Stat.* 31:1208–1211.

Godambe, V. P. 1976. Conditional Likelihood and Unconditional Optimum Estimating Equations. *Biometrika* 63:277–284.

Godambe, V. P., and B. K. Kale. 1991. Estimating Functions: An Overview. In Godambe, V. P., ed., *Estimating Functions*. Oxford: Oxford University Press.

Gong, G., and F. Samaniego. 1981. Pseudo-Maximum Likelihood Estimation: Theory and Applications. *Ann. Stat.* 9:861–869.

Hacking, I. 1965. *Logic of Statistical Inference*. Cambridge: Cambridge University Press.

Hall, P. G., and C. C. Heyde. 1980. *Martingale Limit Theory and Its Application*. New York: Academic Press.

Heagerty, P., and S. R. Lele. 1998. A Composite Likelihood Approach to Binary Data in Space. *J. Am. Stat. Assn* 93:1099–1111.

Heyde, C. C. 1997. *Quasi-likelihood and Its Application: A General Approach to Optimal Parameter Estimation*. New York: Springer.

Jeffreys, H. 1983. *Theory of Probability*. 3rd ed. New York: Oxford University Press.

Kale, B. K. 1962. An Extension of Cramér-Rao Inequality for Statistical Estimation Functions. *Skand. Aktur.* 45:60–89.

Lehmann, E. L. 1983. *Theory of Point Estimation*. New York: Wiley.

Lele, S. R., and J. T. Richtsmeier. 2000. *An Invariant Approach to Statistical Analysis of Shapes*. London: Chapman and Hall.

Lele, S. R., and M. L. Taper. 2002. A Composite Likelihood Approach to (Co)variance Components Estimation. *J. Stat. Planning Inference* 103:117–135.

Li, B. 1993. A Deviance Function for the Quasi-likelihood Method. *Biometrika* 80: 741–753.

Li, B. 1996. A Minimax Approach to Consistency and Efficiency for Estimating Equations. *Ann. Stat.* 24:1283–1297.

Li, B., and P. McCullogh. 1994. Potential Functions and Conservative Estimating Equations. *Ann. Stat.* 22:340–356.

Lindsay, B. G. 1982. Conditional Score Functions: Some Optimality Results. *Biometrika* 69:503–512.

Lindsay, B. G. 1985. Using Empirical Partially Bayes Inference for Increased Efficiency. *Ann. Stat.* 13:914–931.

Lindsay, B. G. 1988. Composite Likelihood Methods. *Contemporary Math.* 80:221–239.

Lindsay, B. G. 1994. Efficiency versus Robustness: The Case for Minimum Hellinger Distance and Related Methods. *Ann. Stat.* 22:1081–1114.

Lindsay, B. G. 2004. Statistical Distances as Loss Functions in Assessing Model Adequacy. Chapter 14 in Taper, M. L., and S. R. Lele, eds., *The Nature of Scientific Evidence: Statistical, Philosophical, and Empirical Considerations.* Chicago: University of Chicago Press.

Lindsay, B. G., and M. S. Markatou. Forthcoming. *Statistical Distances: A Global Framework for Inference.* New York: Springer.

Lindsay, B. G., and A. Qu. 2000. Quadratic Inference Functions. Technical report. Center for Likelihood Studies, Pennsylvania State University.

McCullogh, P., and J. A. Nelder. 1989. *Generalized Linear Models.* 2nd ed. London: Chapman and Hall.

Rao, C. R. 1973. *Linear Statistical Inference and Its Applications.* 2nd ed. New York: Wiley.

Rao, C. R. 1987. Differential Metrics in Probability Spaces. In Amari, S. I., O. E. Barndorff-Nielson, R. E. Kass, S. L. Lauritzen, and C. R. Rao, eds., *Differential Geometry in Statistical Inference.* Hayward, CA: Institute of Mathematical Statistics.

Read, T. R. C., and N. A. C. Cressie. 1988. *Goodness of Fit Statistics for Discrete Multivariate Data.* New York: Springer.

Royall, R. M. 1997. *Statistical Evidence: A Likelihood Paradigm.* London: Chapman and Hall.

Serfling, R. J. 1980. *Approximation Theorems of Mathematical Statistics.* New York: Wiley.

Shafer, G. 1976. *Mathematical Theory of Evidence.* Princeton: Princeton University Press.

Small, C. G., and D. L. McLeish. 1989. Projection as a Method for Increasing Sensitivity and Eliminating Nuisance Parameters. *Biometrika* 76:693–703.

Taper, M. L. 2004. Model Identification from Many Candidates. Chapter 15 in Taper, M. L., and S. R. Lele, eds., *The Nature of Scientific Evidence: Statistical, Philosophical, and Empirical Considerations.*Chicago: University of Chicago Press.

Wald, A. 1949. Note on the Consistency of Maximum Likelihood Estimate. *Ann. Math. Stat..* 20:595–601.

Mark S. Boyce

OVERVIEW

Although most frequently remembered for his work in astronomy and physics, Galileo is credited with founding the modern experimental method (Settle, 1988). His approach involved iterative formulation of models followed by experiments and careful comparison of results. Galileo insisted that the "Book of Nature is written in the language of mathematics" (Galilei, 1638).

Galileo would have loved this book on the nature of scientific evidence, especially the three chapters in this section, which struggle with the problem of inferring causation. Concepts of evidence as an approach to understanding nature would have satisfied Galileo's persistent passion for linking theory and "cimento" (experiment). Although Galileo worked nearly three hundred years before the discipline of statistics was formalized, one could imagine that his assiduous comparison of data and models might have landed him in a statistics department if he were alive today.

The approach to ecology encompassed within Taper and Lele's "Dynamical Models as Paths to Evidence in Ecology" (chapter 9) is a perfect match to Galileo's vision of science. Taper and Lele's view of ecology follows in the footsteps of three leaders in twentieth-century ecology: G. E. Hutchinson (1978), R. H. MacArthur (1972), and R. M. May (1976). To my mind this view of nature is powerful and exciting because it lays a conceptual framework—the very core of our understanding of the discipline. Likewise, Taper and Lele's chapter fits beautifully into this book's theme of science by relating data to models through evidence. We must first appreciate the conceptual framework from which to select models for evaluation. We need such a framework to know what questions to ask. Applying data to models that do not fit into a larger conceptual framework seems ill fated. Taper and Lele's suggestion that dynamic models be used to determine appropriate design of experiments to elucidate causal pathways exemplifies this approach.

All three chapters in this section focus on ecology. Such a focus helps to examine how the complexities of nature are often far beyond the simple formulations of foundational discussions. Indeed, one can argue that ecosystems are the most complex structures in the universe. Ecosystems are governed not only by all of the laws of physics and chemistry, but also by biology as shaped by natural selection. Then superimposed are all of the complexities of ecological interactions and structures on the landscape and at higher levels of organization (macroecology; see Brown, 1995).

The complexity and scale of ecosystem-level experiments means that some sorts of experiments cannot be replicated feasibly, as developed in "Whole-Ecosystem Experiments: Replication and Arguing from Error" by Miller and Frost (chapter 8). Large-scale ecological phenomena, like large-scale astronomical and geological phenomena, may happen only rarely and there may be no opportunity to manipulate them in ways that allow us to reduce errors as we would in a properly replicated experiment. For example, we are unlikely to see another volcanic eruption on the scale of Mount St. Helens during our lifetimes (Franklin et al., 2000), and the large-scale fires in Yellowstone National Park may recur only every three hundred years (Knight, 1991). We would be naïve not to study such events simply because the treatments cannot be replicated. Instead, we must take advantage of the opportunity to document such events carefully and accumulate data on comparable instances as they happen. Combining results from such investigations using meta-analysis may be one way to gain reliable knowledge of ecosystem-level processes (Arnqvist and Wooster, 1995).

Miller and Frost point out that the ultimate reason for replication is to identify and minimize various sources of error in the system. Perhaps alternative ways exist to isolate and eliminate sources of error. Indeed, if one understands the system sufficiently, we could model the various sources of error. Given the complexity of ecological systems, this may be a risky proposition.

It is appropriate that "Constraints on Negative Relationships: Mathematical Causes and Ecological Consequences" by Brown et al. (chapter 10) appears in the part of the book entitled "Realities of Nature." Here we are confronted with the fascinating implications of the positive semidefinite (PSD) constraint on correlation matrices. Surely most of us considered PSD to be a somewhat arcane property of matrix algebra that became a problem only if we entered our data improperly. In fact, this is an underappreciated constraint that must be considered in interpreting correlation matrices.

Imagine two time series of population counts that have a strong negative correlation approaching $r = -1$. When one species is abundant the other is scarce and vice-versa. Now consider adding a third species to the system. A somewhat disturbing but obvious fact is that this third species *must* be positively correlated with one of the other two. Even if all three species compete, it is simply impossible for three species to show negative correlations in all pairwise interactions.

As I see it, the positive semidefinite constraint on correlation matrices has three implications. First, one might wish to design competition experiments independently in a pairwise fashion because one will not be able to discern competition based on the dynamics among the three species together. Although I couch this in context of competition (as did Brown et al.), this will be the case for any interactions where we expect negative correlations among variables, not only for competitive interactions. Second, we may be able to overcome the PSD limitation of correlation matrices by calculating partial correlation coefficients for those interactions where multiple negative correlations are expected. Third, one must be very cautious about interpreting correlation matrices if multiple negative correlations are expected. For example, the "lack of negatively correlated dynamics among putative competitors in complex multispecies systems cannot be taken as evidence that these species do not in fact compete." Likewise, as Brown et al. point out, one can envisage problems of interpretation in multivariate analysis, e.g., principal components analysis, if multiple negative relationships exist among variables. Here the realities of nature relate to mathematical logic, which of course influences our ability to interpret data logically.

An issue on which authors in this section appear to disagree is over the notion of "common sense" in the analysis of data. Miller and Frost are explicit in their criticism: common sense is used as a catch-all term for a concern with evidential error—experimental errors and errors in the analysis and interpretation of data. Perhaps this is an accurate characterization, if not a trivialization, of common sense. Yet common sense is what keeps many of us in business. Our years of training in ecology allow us to anticipate how an ecological system works, integrating prior knowledge and understanding about other ecological systems. In their chapter, Taper and Lele note that common sense is how we assess whether a model is reasonable, based on our prior knowledge.

However much we would like to rid our attempts to model ecological systems of common sense, I believe that it will retain a crucial role in ecological modeling. Many of the fundamental principles behind models in population ecology are weakly defended by data but are defended because they are so sensible. A fundamental premise in the application of model selection criteria is that "data analysis through model building and selection should begin with an array of models that seem biologically reasonable" (Burnham and Anderson, 1992). Ensuring that we have an array of biologically reasonable alternative models should be the role for us ecologists and our common sense. Perhaps this common sense is as evil as expert opinion in Bayesian statistics. However, there is a fundamental difference because common sense expressed in the choice and development of models can be tested by a confrontation with data. This is not possible with a Bayesian approach to incorporating expert opinion. I am confident that such subjectivity will be an essential element in ecological modeling for the foreseeable future.

The chapters in this section have implications for the tension in ecology between design-based inference and model-based inference. Surely Galileo would have championed model-based inference because it offers rigorous structure. Nevertheless, model-based inference can and should have strong elements of design. And a continuing interplay between experiment and theory will be exactly how we will develop reliable knowledge under Galileo's paradigm of modern science.

REFERENCES

Arnqvist, G., and D. C. Wooster. 1995. Meta-analysis: Synthesizing Research Findings in Ecology and Evolution. *Trends Ecol. Evol.* 10:236–240.

Brown, J. H. 1995. *Macroecology.* Chicago: University of Chicago Press.

Burnham, K. P., and D. R. Anderson. 1992. Data-Based Selection of an Appropriate Biological Model: The Key to Modern Data Analysis. In McCullough, D. R., and R. H. Barrett, eds., *Wildlife 2001: Populations.* London: Elsevier.

Franklin, J. F., D. Lindenmayer, J. A. MacMahon, A. McKee, J. Magnuson, D. A. Perry, R. Waide, and D. Foster. 2000. Threads of Continuity. *Cons. Biol. in Practice* 1:9–16.

Galilei, G. 1638. *Discorsi e dimostrazioni mathematiche intorno, à due nuove scienze, attenenti alla meccanica.* Leiden: Louis Elzevirs.

Hutchinson, G. E. 1978. *An Introduction to Population Ecology.* New Haven: Yale University Press.

Knight, D. H. 1991. The Yellowstone Fire Controversy. In Keiter, R. B., and M. S. Boyce, eds., *The Greater Yellowstone Ecosystem.* New Haven: Yale University Press.

MacArthur, R. H. 1972. *Geographical Ecology: Patterns in the Distribution of Species.* New York: Harper and Row.

May, R. M. 1976. *Theoretical Ecology.* Philadelphia: Saunders.

Settle, T. B. 1988. Galileo's Use of Experiment as a Tool of Investigation. In McMullin, E., ed., *Galileo: Man of Science.* Princeton Junction, NJ: Scholar's Bookshelf.

8 Whole-Ecosystem Experiments: Replication and Arguing from Error

Jean A. Miller and Thomas M. Frost

ABSTRACT

Large-scale ecosystem experiments are fraught with problems both in design and interpretation of their results. However, they are vital for understanding "real" ecological systems that operate in very complex ways over large spatiotemporal scales. Standard statistical analyses cannot be directly applied (Hurlbert, 1984) as these experiments violate most if not all of the underlying assumptions of classical statistics, notably replication. We argue that understanding the types of errors that replication controls allows for better design and interpretation of unreplicated and semireplicated whole-ecosystem experiments. In particular, we argue that ecosystem-size experiments have been and ought to be designed and analyzed by keeping a fundamental focus on errors both quantitative and qualitative. Mayo's error-statistical philosophy of science (Mayo, 1996) provides the tools for such an analysis. Furthermore, her hierarchical approach allows for an "objective" scrutiny of both the experimental design and data in large-scale experiments.

INTRODUCTION: SMALL-SCALE VERSUS LARGE-SCALE EXPERIMENTATION IN ECOLOGY

Experiments have long played a critical part in the development of ecological understanding (e.g., Hairston, 1989; Wilbur, 1997; Resetarits and Bernardo,

Sadly, Tom Frost died before final revisions to our paper were made. I take responsibility for all errors. I consider myself lucky to have met and worked with Tom, however briefly, and am grateful for his enthusiastic introduction to the debates surrounding whole-ecosystem experiments.—JAM

221

1998; Huston, 1999). Most experimental work has been conducted on relatively small scales that lend themselves to replication and the use of standard or classical statistical (e.g., Neyman-Pearson [NP] and Fisherian) testing procedures. This has not been the case, however, for whole-ecosystem experiments. These experiments have helped expand understanding in population, community, and ecosystem ecology (Carpenter et al., 1995) while at the same time attracting a major criticism. Whole-ecosystem experiments are usually conducted using one treatment and one reference system, without the kind of replication that is used in the statistical approaches that are employed in smaller-scale experiments.

The success and importance of small-scale ecological experiments have led some ecologists to argue that only replicated experiments can provide insights into ecological processes. Such small-scale experimental chauvinism is quite shortsighted. Whole areas of scientific knowledge and even whole fields of science involve the use of investigations that either do not lend themselves to replication for practical, financial, or ethical reasons or that revolve around phenomena that are a priori impossible to replicate (e.g., mass extinctions, origin of the universe). Much of geology and Diamond's (1997) assessments of the development of human societies provide clear examples of important information that has been gained without replicated experiments.

Replicated small-scale experiments themselves have important limitations in their applicability to natural ecosystems. A full set of species, life-cycle stages, or ecological processes cannot be readily incorporated into chambers in which experiments are replicated. Thus, their relevance to natural conditions must be considered very carefully (Carpenter, 1999; Schindler, 1998). As S.G. Fisher stated so succinctly, smaller-scale experiments may be much more suited to evaluating the ecology that occurs *in* ecosystems rather than the ecology *of* ecosystems (1997). Whole-ecosystem experiments are essential for evaluating the latter (Carpenter, 1999). Growing awareness of the complexity of ecological systems (Maurer, 1998) has led ecologists to recognize that a range of analytical approaches and tools are needed to understand natural phenomena (Hilborn and Mangel, 1997; Maurer, 1999). An important part of these approaches is using different scales of experiments, from bottles to mesocosms to whole ecosystems (Frost et al., 1995; Drenner and Mazumder, 1999; Huston, 1999).

Numerous important ecological insights have been gained through whole-ecosystem experiments (e.g., Schindler, 1990; Likens, 1992; Carpenter et al., 1995; Carpenter and Schindler, 1998). Even Hurlbert, who authored the seminal paper on pseudoreplication (1984), which decried the frequent lack of replication in smaller-scale experiments where appropriate replication was possible, recognized the importance of unreplicated whole-ecosyst experiments. Concerns have regularly been raised, however, regarding interpretation of whole ecosystem experiments because of their lack of replication (e.g., Stewart-Oaten, Murdoch, and Parker, 1986; Steward-Oaten, Bence, and Osenberg, 1992).

We are *not* arguing that replication on all scales can be replaced or that it is desirable to do so. Large-scale experiments have key advantages for addressing some ecological phenomena but must be used with other experimental scales that incorporate suitable replication. We discuss the use of whole-ecosystem manipulations as a check on laboratory assumptions and consider how to interpret whole-ecosystem experiments given that they *cannot* be evaluated using standard statistical testing procedures.

This evaluation is done by considering why the standard statistical tests are so fruitful to begin with and by searching for means to meet or circumvent their assumptions using other methods when standard procedures are not possible. To this purpose, Mayo's new epistemology of experimental investigations can be used to clarify the issues and suggest solutions (Mayo, 1996). Rather than ignoring or avoiding standard statistical practices, it is necessary to understand the motivation behind particular practices, the rules to determine what types of inferences they license, and why. As we will argue, it is not replication per se but the types of errors for which it controls that are of interest in the design and interpretation of experiments conducted at scales that defy replication.

REPLICATION, BACI DESIGN, AND PSEUDOREPLICATION

We begin by clarifying the meaning of three common concepts used in debates about what can and cannot be learned from whole-ecosystem manipulations.

Replication

Scientists and philosophers attach several meanings to the term "replication." One definition of *replication* in *Webster's Encyclopedic Unabridged Dictionary of the English Language* (1989) is "a copy or the duplication of an experiment, especially to expose or reduce error." Most philosophical and sociological discussions use this meaning of the term. Some authors focus on the first half of this definition and argue that as *exact* duplication of an experiment is an impossible task, experimental results are decided via social negotiation (e.g., Collins, 1985). Others try to justify the claim that replication exposes or reduces error (e.g., Franklin, 1986; Miller, 1997). We are not discussing replication as repetition of an experiment. Instead, as did Hurlbert (1984), we will use *replication* to refer to an experimental design that incorporates multiple experimental (e.g., manipulated) and control (unmanipulated reference) units. We understand *unit* as "any group of things or persons regarded as an entity," to quote *Webster's* again. Specifically, a unit is "one of the individuals or groups that together constitute a whole; one of the many parts, elements, or the like into which a whole may be divided or analyzed." What constitutes the units of any one investigation is determined in reference to three scales (spatial, temporal, and biological aggregation) of manipulation and analysis (Frost et al., 1988). Units are determined in reference to both the spatial and temporal scales at which a manipulation is conducted (e.g., in bottles, small enclosures, or plots up to entire ecosystems, such as ponds or lakes). Time scales may be seconds or centuries, depending on the process under study (e.g., microbial loops or climate changes). The scale of biological aggregation that experimenters are manipulating (e.g., individual organisms, populations, entire ecosystems) is a primary factor in deciding what constitutes an experimental unit. The specific hypothesis, inference, or knowledge that scientists are investigating, as well as the available tools, should determine the appropriate scales to be employed and thus the choice of experimental unit to be manipulated. Manipulations at small scales can be replicated in both senses of the term (repetitions of the experiment with several replicates at the unit of analysis).

Fisher (1947, 22) distinguishes between replication as repetition and replication as enlarging the size of the experiment by increasing the number of experimental units used. He argues that both types of replication provide the same result for increasing the sensitivity of an experiment, especially for detecting differences from the null hypothesis. A null hypothesis "must be exact, that is, free from vagueness and ambiguity, because it must supply the basis of the 'problem of distribution,' of which the test of significance is the solution" (Fisher, 1947, 16). An example of an exact null hypothesis for comparing the death rates of two groups of animals is "the death rates are equal." This does not require specifying exact numerical values for the death rates. In this case, it is evidently the equality rather than any particular values of the death rates that the experiment is designed to test, and possibly to disprove. "The null hypothesis is never proved or established (via significance testing), but is possibly disproved, in the course of experimentation. Every experiment may be said to exist only in order to give the facts a chance of disproving the null hypothesis" (Fisher 1947, 16). Appropriately modeled, null hypotheses can be extremely useful. Often the null hypothesis is that an effect is due to chance so that experimenters can use known chance distributions (e.g., binomial models like coin tossing) to compare their data with. The use and fruitfulness of null hypotheses has been a source of heated controversy in ecology (see Gottelli and Graves, 1996). This controversy does not affect our arguments here. Sensitivity can also be increased qualitatively by increasing the similarity of the units used (Fisher, 1947, 24).

If ecologists want to understand how large-scale complex systems react to a manipulation, replication at the level of experimental units would require manipulating several entire ecosystems. This level of unit replication is a daunting task, as is repeating the experiment. Therefore, experimenters who want to study and manipulate complex ecological interactions at the scale of a whole ecosystem often turn to some kind of BACI design (before-after, control-impact).

BACI Design

BACI studies compare two locations, a control and an intervention site, over an extended period both before and after an experimental manipulation at the intervention site. In such a design, there are only one discrete experimental unit and one control unit, and samples are taken over time in both units (Stewart-Oaten, Murdoch, and Parker, 1986). The BACI design is the most "common type of 'controlled' experiment in field ecology [that] involves a single 'replicate' per treatment" (Hurlbert, 1984, 199). To argue for the sensitivity of this type of experiment, experimenters need to show that the two units (manipulated and control) are very similar (for a simple example of this type of qualitative argument, see Fisher, 1947, 22–25).

The Little Rock Lake (LRL) acidification experiment conducted in northern Wisconsin is an example of this type of design (Brezonik et al., 1993; Frost et al., 1999). The lake was divided into two separate basins by a flexible, inert barrier. Both sides were monitored for one-year prior to the barrier being placed and for another half year after the barrier had been inserted. The utility of the BACI design in inferring causation is augmented by the use of a series of interventions. The treatment side was acidified in steps that approximated 0.5 pH units in three two-year periods (figure 8.1) from a starting pH of 6.1 to a final pH of 4.7 close to the average pH of rain in the region (Watras and Frost, 1989). These three stages represent separate interventions, each of which can be evaluated separately. While not an "active intervention or treatment," the recovery period can be seen as yet another separate intervention in that it provides a changing pH environment and ecosystem responses can be studied during this period (see Frost, Montz, and Kratz, 1998). The experiment was conducted to gain insight into the

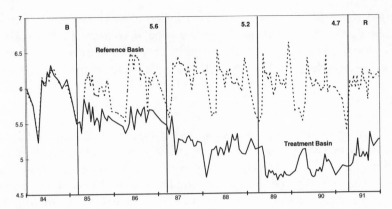

FIGURE 8.1 Surface pH values in the treatment and reference basins of Little Rock Lake. B refers to the baseline period, R to the start of the experiment's recovery period and 5.6, 5.2, and 4.7 refer to the target pH levels that were used during the three two-year acidification periods. Techniques used in this analysis are provided in Brezonik et al. 1993 and Frost et al. 1999.

ecological effects of advancing acidification in seepage lake systems. Many different types of ecological data (chemical and biological responses, etc.) were collected and monitored from this one long-term whole-ecosystem experiment using samples taken every two weeks when the lake was ice free and every five weeks when ice covered. The problem in using such samples (e.g., measurements taken on the same experimental unit through time) as a substitute for replication of experimental units is that events that happened in the unit previously can have a substantial influence on future samples. The samples are not independent. This lack of independence renders many standard statistical procedures inappropriate (cases of pseudoreplication, according to Hurlbert, 1984) if mechanically applied.

Pseudoreplication
Hurlbert coined the term "pseudoreplication" to refer "not to a problem in experimental design (or sampling) per se but rather to a particular *combination* of experimental design (or sampling) and statistical analysis *which is inappropriate for testing the hypothesis of interest*" (Hurlbert, 1984, 190; italics added). This is much too broad a definition. There are many ways to misapply statistical analysis to a host of experimental designs, and many ways to utilize sampling techniques that are inappropriate for testing particular hypotheses. We will restrict the use of the term "pseudoreplication" to refer specifically to cases where the "inappropriateness" comes from violating the assumption of independent replication underlying the use of statistical

tests for inferring that differences between a manipulated and control basin are due to a manipulation. This restriction includes using samples from the same unit through time as independent samples. The errors from violating replication for inferences about "treatment effects" as discussed by Hurlbert (1984, 198), and the solutions suggested by Stewart-Oaten, Murdoch, and Parker (1986) and Stewart-Oaten, Bence, and Osenberg (1992). have important implications for interpreting evidence from BACI experiments above and beyond the use or misuse of statistical tests. We argue that violating replication can be equally or more problematic for the nonstatistical data modeling techniques (e.g., graphing) that Hurlbert suggests to avoid pseudoreplication. An obvious question at this point is, why is replication so desirable? To answer this, we will use Mayo's (1996) approach to assessing methodological rules.

INTRODUCTION TO MAYO'S VIEW ON METHODOLOGICAL RULES

Deborah Mayo, in her *Error and the Growth of Experimental Knowledge,* has suggested that methodological rules are justified to the extent that they ensure testing severity and provide strategies for overcoming particular experimental errors. She argues that rules can be broken if (1) cases where a rule is broken are reported as such and (2) violating the rule does not violate Mayo's concept of severity (see also Mayo, 2004 [chapter 4, this volume). The first situation merely requires that the experimenter be honest about the procedures followed. In one of the cases she analyzes predesignation; this requires that the experimenter state whether the correlation found was pre- or postdesignated when using tests of significance on data (Mayo, 1996, chap. 9). That is, one must be explicit about the data generation process, especially in those cases where common assumptions of a standard practice are violated. The second situation is more difficult for an experimenter to determine and meet in any one particular case. However, the general notion underlying this requirement is straightforward. The intuition underlying severity is captured in the following requirement:

> *Severity requirement:* Passing a test T with evidence e counts as a good test of or good evidence for hypothesis H just to the extent that H fits e and T is a *severe test* of H. (Mayo, 1996, 180)

What criteria do we have for gauging whether a test is severe or not? Mayo (1996, 180) offers two versions of her severity criterion (SC):

1. SC 1a: There is a very high probability that test procedure T would *not* yield such a passing result, if H is false.
2. SC 1b: There is a very low probability that test procedure T would yield such a passing result, if H is false.

Severity captures our commonsense intuitions that a good test should weed out false hypotheses. Hence, it must probe the ways in which the hypothesis, if false, could erroneously be passed. In terms of specific errors, a severe test is one in which, if the error were present, then the test would detect it, with high probability, but not otherwise (see Mayo, 2004 [chapter 4, this volume] for a comprehensive discussion of severity and its informal counterpart, arguing from error). It is important to emphasize here that on her account probability assessments (quantitative or qualitative) always attach to procedures, never to the hypothesis being tested. A hypothesis is either true of false (or, alternatively put, correct or incorrect, an adequate model or not, etc.).

Mayo calls her philosophy "error statistics" because it keeps the centerpiece of Neyman-Pearson (NP) statistics, error probabilities—the probabilities of committing type I and type II errors—central. The strength of standard practice is its ability to measure these probabilities.[1] However, to do that, the rules and underlying assumptions for Neyman-Pearson tests (significance tests, t-tests, etc.) cannot be violated or else the procedure will be unreliable. Mayo (1996) advocates a piecemeal approach to characterize severity independent of the assumptions that justify any particular NP model.[2] Thus, severity can *guide* experiments where NP testing assumptions (predesignation, replication, etc.) are violated, as illustrated below.

Mayo has analyzed the NP rule of predesignation, based on satisfying severity demands (Mayo, 1996, chap. 9). Using Leslie Kish's postdesignation example (Kish, 1970), she argues that methodological rules get their bite by enforcing severity considerations. Kish's example illustrates for signifi-

1. Standard statistical practice here is useful in that two major types of error of concern in making experimental inferences are delineated: we may reject the hypothesis tested when it is true (type I) and we may accept it when it is false (type II) (Neyman and Pearson, 1966, 242).

2. There are also several other differences between Mayo's concept of severity and similar concepts such as power and size in NP testing. For one thing, severity takes into account the actual results of an experiment, which makes it a much more informative concept. See Mayo (1996) for details. See Mayo (1996, chap. 11, sec. 6) for her reinterpretation of rules of acceptance and rejection as inferential rather than behavioral rules.

cance testing how pre- and postdesignation of the hypothesis of interest affects the error probabilities of the test. His example revolved around psychological character traits in children and possible causes (e.g., gradual weaning and high social standards, late bladder training, and little nail biting [Mayo, 1996, 299]). Kish argues that a postdesignated hypothesis, instead of passing at .05 significance level, had an actual significance level of .64. Why the change in the error probability calculation? Mayo analyzed Kish's example using severity considerations. The change in error probabilities is due to the fact that hunting through data for a correlation is an unreliable method, e.g., a poor test, of any hypothesis finally found. This is because, simply due to chance one will find 13 correlations or hypotheses given the size of the data set Kish analyzed even if the null is true. This example illustrates that how the data were generated plays a key role in assessing severity.

Hunting for correlations significantly increases one's chance of finding spurious correlations and investigators must factor this into their results. Mayo does not necessarily condemn breaking predesignation. "What is really being condemned . . . *is treating both predesignated and postdesignated tests alike*" (Mayo, 1996, 296)." In Mayo's account of methodological rules, an experimenter may have the ability to argue that errors resulting from violating an assumption of a test have been taken care of by other methods. As Mayo says, when discussing predesignation:

> From this point of view, the NP admonishment against violating predesignation may be regarded as a kind of warning to the error statistician that additional arguments—possibly outside the simple significance test—may be required to rule out the error at hand. The tests themselves cannot be expected to provide the usual guarantees. This, I suggest, should be seen as an invitation for the error statistician to articulate *other* formal and informal considerations to arrive at a reliable experimental argument. (Mayo, 1996, 298).

Producing such arguments often requires many tests of assumptions, broken or otherwise, and mechanisms. In the case of ecosystem manipulations, assumptions such as replication are not met in the experimental design. Thus, meeting them requires creativity and ingenuity on the part of the investigators if they want to use such tests *or* the "thinking" (i.e., focus on error probabilities) behind them. While Mayo has done this sort of error analysis for use novelty and predesignation, she has not looked at replication (Mayo, 1996, chaps. 8, 9).

WHAT ERRORS DOES REPLICATION CHECK FOR?

Our line of reasoning and interpretation of debates about BACI experiments and pseudoreplication in ecology is in keeping with Mayo's view that methodological rules are to be justified based on insuring severity and checking for specific errors. While serverity is the formal centerpiece of Mayo's epistemology, her concept of "arguing from error" is its informal counterpart.

> It is learned that an error is absent to the extent that a procedure of inquiry with a high probability of detecting the error if and only if it is present nevertheless detects no error. (Mayo, 1996, 445)

The arguments used in the pseudoreplication debates between Hurlbert and others are captured by this latter notion. Hurlbert clearly lays out the two potential errors—he calls them potential sources of confusion—that may arise due to a lack of replication in BACI designs. Replication checks for errors due to natural variability between units and detecting whether a chance event has occurred that may produce effects that could be confused with the effects of a manipulation. The articles by Stewart-Oaten, Murdoch, and Parker (1986) and Stewart-Oaten, Bence, and Osenberg (1992) take up the second element of Mayo's approach to the study of methodological rules—if a particular rule is violated, what other arguments (if possible) are needed to justify the robustness of results for a particular inference in such a case? We will discuss each of these potential errors that can be caused by violating replication and possible ways to circumvent them in turn.

REPLICATION AS A CHECK ON NATURAL VARIATION

According to Hurlbert (1984), "Replication *controls* for the stochastic factor, i.e., among-replicates variability inherent in the experimental material" (191; italics in original). Control is italicized to show that he is using the term in the broader sense he discusses on page 191 as one of the "features of an experimental design that reduce or eliminate confusion" (e.g., experimental error). The first major error of concern in BACI designs that Hurlbert expresses is that the two units are not identical. According to Hurlbert, the validity of comparisons between the manipulated and control unit for treatments effects is justified only if the two units are *identical* not only at the time of manipulation but throughout the experiment with the exception of the treatment effects. According to Hurlbert, "the supposed 'identicalness' of the two plots almost certainly does not exist, and the experiment is not controlled for the possibility that the seemingly small initial dissimilarities between the two plots will have an influence on the . . . [effect under test]" (Hurlbert, 1984, 193). Natural variation assures us of that point. Of course, depending on the size of the treatment effect, this may not be a problem. To make this point in an admittedly exaggerated manner, if one is studying the effect of radiation, even though other cities are not identical to Hiroshima, the effect of a nuclear bomb is pretty unambiguous. Hurlbert also agrees that the size of the expected treatment effect can help compensate for a lack of replication. "When gross effects of a treatment are anticipated or when only a rough estimate of effect is required, or when the cost of replication is very great, experiments involving unreplicated treatments may also be the only or best option" (Hurlbert, 1984, 199–200).

Moreover, even though Hurlbert agrees that significance tests can be validly used to determine whether there is a significant difference between the two locations, nonetheless "the lack of significant differences prior to manipulation cannot be interpreted as evidence of such identicalness. This lack of significance is, in fact, only a consequence of the small number of samples taken from each unit" (Hurlbert, 1984, 200). This is because

> we *know* on first principles, that two experimental units *are* different in probably every measurable property. That is, if we increase the number of samples taken from each unit, and use test criterion (e.g., *t*) values corresponding to an alpha of 0.05, our chances of finding a significant premanipulation difference will increase with increasing number of samples per unit. These changes will approach 1.0 as the samples from an experimental unit come to represent the totality of that unit (at least if the finite correction factor is employed in the calculation of standard errors). (Hurlbert, 1984, 200)

The problem here is that as the number of samples increases, the sampling procedure becomes increasingly sensitive to even minor differences between populations. However, this argument would equally well apply to small-scale experiments, as no set of replicates is ever absolutely identical. While BACI designs cannot use multiple replicates to cancel out noise from random natural variation, this does not mean other reliable arguments for similarity cannot be forwarded to defend the robustness of the results against this particular error. Thus, to place the unobtainable demand for identicalness on BACI designs seems unwarranted.

[handwritten note:] Is the number of [samples] [increases] the sampling procedure is sensitive to minor differences between populations

Rephrasing Hurlbert's First Error of Concern

The real question is not whether the two units are identical—they are not—but whether they are similar enough *in the relevant aspects* that subsequent changes between them are likely to be caused by the manipulation. This is a problem in choosing an appropriate sampling size as well as which features to compare and at what scale of resolution these features will be examined (e.g., choice of resolution at the level of species, genus, etc. in biological experiments; see Frost et al., 1988, 243–48.). This problem is not insolvable in theory. What the relevant features are will depend on background knowledge and other experimental knowledge. The problem then is distinguishing the effect of the impact—if it exists—from the surrounding variability or noise. In Stewart-Oaten, Murdoch, and Parker's view, "only by sampling at many different times, both Before and After Start-up can we hope to estimate the variability due to all sources, both sampling error and random population fluctuations" (1986, 932). They look to an adequate sampling regime before and after a manipulation to control for Hurlbert's first problem (natural variation between locations) with BACI designs.

In setting up the Little Rock Lake (LRL) experiment for testing the effects of increased acidification, LRL was subdivided with a vinyl curtain. Before the curtain was inserted, both basins were monitored for a year to make sure they were similar in a variety of relevant physical, chemical, and biological aspects (e.g., Brezonik et al., 1993; Sampson et al., 1995; Watras and Frost, 1989). After the curtain was inserted, each basin was monitored for six months to try to determine any possible effects on similarity between the basins the separation of them may have induced, as the curtain would have separated previously coupled processes. Also, natural processes are now spatially limited to a much smaller region—approximately half of the area they had available before the curtain was installed.

REPLICATION AS A CHECK ON STOCHASTIC EVENTS
BEYOND NATURAL VARIATION

Hurlbert's second problem is that given the two units are sufficiently similar in the relevant aspects to start with; will they remain that way *over time?* There could be an uncontrolled extraneous event that affects one but not the other plot, thereby increasing the dissimilarity between them that may even, in a worst-case scenario, be falsely attributed to the treatment under test (see Hurlbert, 1984, 191–93). Neither the similarity between the two units nor the size or intensity of the treatment can ameliorate this second situation. We can agree with Hurlbert that replication and interspersion of treatments provide the best insurance against chance events producing such spurious treatment effects, but that does not mean it is the only insurance (method) for detecting and mitigating this particular error. Moreover, Hurlbert's turn to graphical and other nonstatistical techniques as a solution to pseudoreplication only sweeps the problem of violating replication under the carpet.[3] All of the errors Hurlbert raises are errors in the procedure (BACI design) for *generating* data and so would also pose a threat to non-statistical representations of the data as well. Frost et al. (1988, 3) "emphasized simple, primarily graphical, presentations of their data. And have avoided any strictly statistical assessments of the significance of treatment effects . . . due to the fact that whole ecosystem experiment is a contentious area in terms of statistics." They graph biomass (us/L) for both the control and manipulated LRL basins against year (1984–1990). A bar graph is used to show the interannual species turnover rates for each basin (1985–1990). The graph for *K. taurocephala* (Frost et al., 1988, 5) shows massive spikes and increases in the acidified basin. It is visually quite dramatic. However, without replication, these spikes could simply be the result of a stochastic event that occurred only in the manipulated basin at the time of startup. When replication is violated, the errors it controls for are potentially present in the data regardless of how it is presented *unless* the BACI design is supplemented with other procedures for detecting and/or arguments for ruling them out.[4] Turning to graphical presentations to avoid "pseudoreplication" merely makes potential errors less obvious. This is not to deny the fruitfulness of such approaches for modeling data. We only deny that doing so removes the errors for which proper replication controls.

Stewart-Oaten, Murdoch, and Parker (1986, 935–36) conveniently partition the types of stochastic events Hurlbert refers to based on a combination of the spatial and temporal characteristics of their possible effects. They list events with:

1. short-term effects spread over a large area;
2. short-term effects localized to one experimental unit;

3. Note that there are other problems with visual representations. We may have a tendency to see what we expect to see rather than what is really there. The horizontal and vertical scales used may exaggerate or underemphasize important information, etc.

4. Showing that these errors have been ruled out (or not) is the responsibility of the experimenters as well as critics. Possible problems with interpreting their graphs are discussed in the body of Frost et al. (1988).

3. long-term effects spread over a large area; and
4. long-term effects localized to one experimental unit.

Knowledge about possible "artifacts" (i.e., events that are mistakenly attributed to but not caused by the manipulation) is especially critical if one is using samples through time on the two units in lieu of physically independent replicates. This is because an event that happens in a unit may continue to effect subsequent samples from that same unit as the samples are not independent. This is the reason that Hurlbert condemns using samples through time as if they were independent samples as a clear case of pseudoreplication (Hurlbert, 1984,193).

Stewart-Oaten, Murdoch, and Parker (1986) argue that chance events that have effects of short duration are harmless *provided* we sample over a sufficiently long time. The time between samples must be long enough for the process to "forget" its remote past—i.e., "correlation between deviations that are far apart in time is close to zero" (Stewart-Oaten, Murdoch, and Parker, 1986, 932). Sufficiently long sampling periods are needed to check that past events will not build up and influence the next sample so that subsequent samples may be treated as independent. Further, if such short-lived effects occur over a large area, they will also be distinguished from the manipulation effects, as they will also be manifested in the control unit. For the same reason, events with long-lasting effects that occur over a wide area are also, with careful sampling and experimental design, not an insurmountable obstacle to the robustness of results from BACI designs. Stewart-Oaten, Murdoch, and Parker modeled the possible influence of such effects on samples taken through time (1986, 936). They used a Markovian time-series model in a BACI design and argued that, provided samples are adequately spaced in time, their dependency will be weak and analytical results can be justified as robust despite such a violation. They admit that their example model and argument do not justify a claim of strict independence between all observations, only one of low correlation. They further point out that "the independence assumption should also be checked against the data. There are, in fact, powerful tests for doing so." In short, we can turn to sampling procedures and other tests to supplement the BACI design in order to mitigate these three potential sources of error. We are not claiming this is easy to do, only that it is possible in theory to construct such an experimental argument from error. Stewart-Oaten, Murdoch, and Parker still share Hurlbert's concern that there "remains the possibility of a large, long-lasting but unpredictable effect occurring at about the same time as the start-up and affecting one location much more than the other" (Stewart-Oaten, Murdoch, and Parker 1986, 937). Unlike the other three possibilities, the possibility of a long-duration event with localized effects could not be ruled out (i.e., could go undetected) even with a long sampling procedure.

ARGUING FROM ERROR

Stewart-Oaten, Murdoch, and Parker (1986, 937) state that the potential error due to an event with localized, long-lasting effects threatens not only BACI experiments but also other kinds of experiments. This is because "randomized studies share these imperatives in many cases because their results may be due not to the treatment but to the way it is administered, e.g., placebo effects, cage effects in ecological studies, contaminant in drug studies." In short, unknown events (especially those that commence simultaneously with the treatment), potential sources of error, with long-lasting effects that occur in either the manipulated or control unit but not (or to a much lesser extent) in the other unit can affect any experimental design.

However, we disagree with their conclusion that "in both cases [BACI and replicated, randomized designs], possible alternative mechanisms must be proposed (often by the study's critics) and their likelihoods assessed (often rather subjectively)" (Stewart-Oaten, Murdoch, and Parker, 1986, 937). Instead, we suggest both experimenters and critics need to investigate and articulate potential errors in the design of their experiments, paying special attention to data-generating mechanisms and data-modeling techniques. Second, alternative mechanisms brought forth both by the experimenters and their critics must be *tested* and their plausibility argued for based on error considerations (i.e., evidence and severity considerations), not on *subjective* assessments of likelihoods. (See the chapters in this volume by Mayo [chapter 4] and Dennis [chapter 11] for the criticism of subjective assessments in science.)

Probing potential and actual sources of experimental error offers a wealth of local experimental knowledge. As Stewart-Oaten, Murdoch, and Parker (1986, 937) point out, placebo effects and cage effects can be brought to bear as criticisms of small-scale replicated studies. In turn, these criticisms can be tested. If such criticisms pass severe tests, then we have gained important experimental knowledge (e.g., about placebo and cage effects). Frost et al. (1988) offer a suggestion for exploring scaling issues in ecological experiments by the means of such local experimental knowledge about potential errors:

> By means of a graded series of enclosure sizes, it may be possible to derive a relationship between enclosure size and the time scale over which the enclosure exhibits processes that parallel those in a whole lake. Such information could allow for the derivation of scaling laws to directly relate data from experimentally manipulable [and replicable] enclosures to the whole lake community, (Frost et al., 1988, 243)

This type of experimental knowledge would be invaluable for making inferences about ecosystem processes. It is also the type of knowledge we need for making progress with the errors that threaten reliability in unreplicated (pseudoreplicated) experiments.

KEEPING TRACK OF ASSUMPTIONS

To use statistical tools for manipulating our data and drawing inferences, we need to ensure that the assumptions that warrant the use of those tools are met. This, we take it, is Hurlbert's main concern and point about pseudo-replication. We have tried to argue that if these assumptions are not met in the experimental design (e.g., replication), then they must be met elsewhere. If they are successfully met elsewhere (e.g., in the sampling regime), then *sometimes* our results can be robust even if pseudoreplication has occurred. But how does one keep track of the various assumptions in a manner that allows not only us but also others in the scientific community to determine whether they have been accounted for?

MAYO'S PIECEMEAL APPROACH TO EXPERIMENTAL INQUIRY

Understanding the motivation behind a rule and then showing that the goal of that rule has been met are two different tasks. Mayo's severity considerations and argument from error help us get a handle on the first task. Mayo's hierarchy of experiment (a more detailed working out of her models of experiment) is extremely useful for this second task of laying out assumptions and showing whether or not they have been met. Furthermore, severity is difficult, if not impossible, to obtain and assess without using a piecemeal approach in an experiment.

TABLE 8.1 Hierarchy of Models.

Primary Models:
 How to break down a substantive inquiry into one or more local questions that can
  be probed reliably

Experimental Models:
 How to relate primary questions to (canonical)* questions about the particular type
  of experiment at hand
 How to relate the data to these experimental questions

Data Models:
 How to generate and model raw data so as to put them in canonical form
 How to check whether the actual data generation satisfies various assumptions of
  experimental models

 Source: Mayo (1996, table 5.1, p. 130).
 *"Canonical" questions are questions modeled into a standard or simplified experimental
form or using standard known models or mechanisms (e.g., coin-tossing mechanisms are a
canonical form of the binomial distribution, or the spread of arrows or shot around a bull's-eye
target to test for randomness of dispersal). Thus, when testing whether a treatment has an effect
or a phenomenon is acting on our data, we may ask the question experimentally by designing a
null hypothesis of chance where the different known standard (canonical) models of chance can
be used to compare our data. These canonical data models are also used to calculate the probabilities for a quantitative severity calculation.

Mayo uses the term "piecemeal" to indicate that experimental inquires
must be broken down into manageable pieces for investigation in order to
assess severity.

> Severity refers to a method or procedure of testing, and cannot be assessed with-
> out considering how the data were generated, modeled, and analyzed to obtain
> relevant evidence in the first place. I propose to capture this by saying that as-
> sessing severity always refers to a framework of *experimental inquiry.* (Mayo,
> 1996, 11)

Mayo's framework of experimental inquiry is flexible insofar as the models
feed into each other in both directions (see figure 4.1). This relationship is
also maintained in her hierarchy, though it is less visually obvious. Table 8.1,
based on Mayo (1996), explains the roles each model plays in breaking
down an inquiry to make it amenable to severe testing. A whole-ecosystem
experiment is really a complex of experiments. In order to gain reliable information from them, we need to form primary questions/hypotheses that
are specific enough that we can severely test them.

A Piecemeal Approach to Little Rock Lake

First, we need to isolate a specific inference of interest. As tests are directly linked to hypotheses in Mayo's framework, we cannot gauge whether a test is severe unless we know what inference it is testing. The aim behind narrowing our hypotheses is to articulate one that can be severely tested. This process of narrowing hypotheses is dependent on what we want to learn. Scale plays an important role here because certain phenomena occur and can be studied only at particular scales (Frost et al., 1988). However, in order to understand some phenomena, we may not be able to simply reduce all hypotheses to a scale that can allow them to be tested in the laboratory. Thus, simply narrowing our hypotheses will not get us to the point where whole-ecosystem experiments can be dispensed with and the problem of replication would disappear. Experimenters at LRL were able to use "surveys of terrestrial and aquatic ecosystems" to provide the basis from which hypotheses regarding the chemical and biological effects of acidification can be drawn (Sampson et al., 1995) to help narrow hypotheses of interest. In LRL, one specific question of interest was what are the effects of increasing acidification on rotifers, in particular the species *K. taurocephala*. Bioassays were conducted on several species of rotifers from LRL under a variety of food and light conditions. These small-scale laboratory tests provided a source of several "narrowed" testable hypotheses for LRL. Such hypotheses include:

1. Laboratory bioassays can be reliably extrapolated to make predictions in LRL.
2. Bioassays predict either no change or a decrease in rotifer populations in the manipulated basin.
3. Overall rotifer biomass between the two basins will remain equal (approximately).
4. The population of *K. taurcephala*, a species of rotifer, remains the same or decreases in the manipulated basin.
5. Nature assigns interecosystem differences pre- and posttreatment at random.
6. At the time of the manipulation a nonrandom trend or change was observed in rotifer communities (a weaker version of item 3).
7. The pH levels are the primary factor controlling the structure and dynamics of rotifer communities in north temperate lakes.
8. Indirect effects of acidification (e.g., reduction of predators) rather than direct effects cause changes in rotifer populations. (Item 6 is

from Rasmussen et al., 1993; the others are from Carpenter et al., 1989, 1142–1145)

Once we have narrowed down our hypothesis, we then need to translate the primary hypothesis into a testable form, one that will link the primary hypothesis to possible data. This may involve designing an appropriate null hypothesis to test in order to get information about an alternative hypothesis of interest (e.g., number 5 above would be the null hypothesis tested in order to gain evidence for number 6). If the null hypothesis is rejected, mechanistic hypotheses (e.g., numbers 7 and 8) may be forwarded but also need to be tested, often independently of the present experiment.

Given the primary models above for rotifer investigations, there were two experimental models used—bioassays on individual species of rotifers and difference comparisons between the control and manipulated sites in the BACI-designed LRL experiment. Simultaneous sampling through time was conducted on LRL in both the control and manipulated basins. Nearby Crystal Lake (with a near-neutral pH, like the LRL control basin) and Crystal Bog (an acidic environment similar to the manipulated LRL basin) were on a similar sampling regime to provide another control set as background for the LRL manipulation.

Our job is not done once we have chosen our experimental model. Indeed, it has only begun! The model will generate data but we need to model the data in order to get meaningful information out of it. Data are messy. For example, below is how Frost et al. modeled their raw rotifer data for their graphical presentations mentioned earlier.

Samples were processed individually (e.g., by species) but data were subsequently pooled to estimate the mean density of animals throughout the water column. Biomass differences between rotifer populations in LRL control and manipulated basins could then be calculated. Community similarity differences between LRL control and manipulated basins were calculated. These calculations were based on the total of the minimum proportion of total rotifer biomass for each species in either basin of LRL. Rotifer species richness was calculated as the total number of taxa observed within a year. Rates of species turnover indicating the appearance or disappearance of a species in a basin in any one year were calculated. Species turnovers are calculated for each year by determining the number of rotifer taxa that were recorded in a basin at any time within a year but which were then absent during the subsequent year along with the number of taxa that exhibited the opposite pattern, appearing only during the second year of a pair. The total number of appearing and disappearing taxa was then divided by the total number of taxa present at any time in the basin during the first year plus those present during the second year. This number was then multiplied by 100 to obtain the percentage change.

Using the pooled data, four different ways to model it (calculations) could be made. Thus, the same data can be used to answer several different questions depending on how it is modeled.

Experiments produce noise as well as effects of interest. One primary role of data models is to distinguish noise (both chance as well as systematic events whose effects are not of interest) from the event(s)/effect(s) of interest. This is also the level where the assumptions (explicit and implicit) from the experimental design, data generation, and modeling procedures are explicitly listed and checked. Often these assumptions will require independent testing, in which case an entirely new hierarchy of hypothesis, experimental, and data models would be needed. At this level, if they have not already done so, experimenters will need to acknowledge that replication is violated in their experimental design, if that is the case. Any experimental arguments and checks used to cover the errors that violating replication could cause would appear here. This would include arguments for the similarity of both basins in the relevant features, sufficiency of the sampling protocol to be followed, additional monitoring of other nearby bodies of water, other relevant information from previous studies, etc. Questions and results (e.g., from LRL) about the reliability of extrapolating results from replicated and randomized small-scale experiments would appear at this level in their hierarchies of experimental inquiry. Fortunately, we have an arsenal of tools both statistical and nonstatistical to use in this task of gleaning informative (modeled) data from raw, messy data (see, e.g., Franklin, 1986; Galison, 1987; Hacking, 1983; Mayo, 1996; Woodward and Bogen, 1988; Woodward, 1989).

SUBEXPERIMENTS (NESTED HIERARCHIES)

The errors and assumptions in an experimental inquiry can, we believe, be approached effectively by breaking them down in the piecemeal fashion that Mayo advocates and testing for each error one at a time. This often requires using *several* tests. Moreover, each test will come with its own primary, experimental, and data models. A rich and complicated inquiry may require multiple and often nested tests in order to answer or gain information about the original primary question or model. Mayo's hierarchy is a nested hierarchy much like a Russian doll where smaller dolls nest inside larger ones. Testing assumptions and probing errors at the level of the data models will often require developing multiple subexperiments. Each of the five data models used at LRL has its own hierarchy of experimental inquiry to test for possible errors.

Advances in computer simulations and statistical methods for resampling, such as bootstrapping methods (Efron, 1979; Efron and Gong, 1983), have opened up a whole new frontier of quantitative possibilities for testing "what it would be like" if a particular error were committed. This is piece-meal by testing, one at a time, a variety of null models based on canonical models of error. Paleontologists have even less opportunity for physical manipulation and replication than ecologists. As a consequence, paleontologists have developed a large body of literature on computer simulations and bootstrapping techniques (see, e.g., Hubbard and Gilinsky, 1992; Signor and Gilinsky, 1991; Gilinsky, 1991). Several Monte Carlo simulations were run using the rotifer data gathered at LRL as a check on the visual, graphical representations of the data, including a new data model—randomized intervention analysis (RIA). However, as a new model for determining whether a change occurred after the manipulation at LRL, its reliability had to be checked. One way to check that a new technique is working is to compare its results with other successful techniques or models. RIA's results were compared to the results of another statistical technique for time series analysis, autoregressive integrated moving average (ARIMA) models (Rasmussen et al., 1993; Box and Jenkins, 1976). In this case, Rasmussen et al. (1993, 154) reported that "conclusions from RIA for the Little Rock Lake *K. tauro-cephala* parallel those we found [using ARIMA models]."

Subexperiment: RIA Tests on Rotifer Data
One of the data models for estimating P-values for rotifer data at LRL is randomized intervention analysis (RIA). This data model requires its own hierarchy of models.

Primary model. RIA is used to determine whether a change has occurred following experimental manipulation of an ecosystem. Null hypothesis: "nature assigns inter-ecosystem differences to pre/post treatment at random. More precisely, the null model states that all possible permutations of the data have an equal probability of being observed" (Carpenter et al., 1989, 1143).

Data model. Using the same rotifer ·data, the series of parallel observations from both the manipulated and control basins both before and after the manipulation that Frost et al. had used, Carpenter et al. modeled it differently: "A time series of interecosystem differences is then calculated, and from these are calculated mean values for the premanipulation and post-manipulation difference, D(pre) and D(post), respectively. The absolute value of the difference between D(pre) and D(post) is the test statistic" (Carpenter et al., 1989, 1143).

Experimental model. Carpenter et al. (1989, 1143–44) give an excellent description of the experimental model and caveats about its interpretation (italics added): "*P*-values can be approximated by Monte Carlo methods. To do this, differences are randomly assigned to times before or after the manipulation, regardless of their position in the actual sequence. Many such permutations are generated, and the resulting frequency distribution of [D(PRE) − D(POST)], where D(·) is a statistic of interest, is examined. The proportion of values of [D(PRE) − D(POST)] that exceeds the observed value is the approximate *P*-value for a one-sided test. The *P*-value indicates the probability of a test statistic as or more extreme than that observed occurring under the null hypothesis. If this *P*-value is low, then one concludes that a non-random change in the inter-ecosystem difference occurred following the manipulation." One potential error in their experimental design (RIA) is that autocorrelation in the data could affect its results (estimated *P*-values). They could rule out nonnormality and heterogeneous variances as affecting the test results because "RIA uses randomization to derive an error distribution from the data itself" (Carpenter et al., 1989, 1143). Therefore, they needed to find out what effect autocorrelation would have as their samples were not independent. This required another subexperiment with its own hierarchy of models within the RIA subexperiment itself. (See what we mean about Russian dolls!)

Simulating Errors in Estimated *P*-values in RIA
Primary model. Null hypothesis: depending on sample size autocorrelation has no effect on *P*-values determined by RIA. (The hypothesis could also state that any effect would be greater than or less than some preset boundary.) *Experimental model.* Using a Monte Carlo simulation to represent the LRL experiment, they ran RIA on a sequence of random numbers. "The pre-manipulation sample consisted of $N/2$ normally distributed random numbers with mean $\overline{X}$ and standard deviation s. The post manipulation sample consisted of $N/2$ normally distributed random numbers with mean $\overline{X} = ms$, where m is the specified manipulation effect." They then repeated the procedure, but this time "the sequences of random numbers were autocorrelated" (Carpenter et al., 1989, 1144). Data models are the same as above for RIA it-

self. Using 36 simulated experiments, they determined the effects of sample size on autocorrelated data in RIA. Frost et al. (1999) also checked for the effect of autocorrelation using both an autoregressive model and a moving average model against which to check the results of RIA.

Summary of RIA Hierarchies

RIA was one of the data models used to determine whether the null hypothesis should be accepted or rejected (first hierarchy in this section). However, the samples used were not independent and thus investigators needed to determine the effect of autocorrelation and sample size on the reliability of using RIA (second hierarchy in this section). They used a Monte Carlo simulation to see what effect the error autocorrelation, if present, would have on estimated P-values using RIA. What effect does autocorrelation have on the results from modeling data using RIA?

> In all of the autocorrelated simulations examined, the true P-value was .05 if the P-value from RIA was less than .01. Therefore, as a conservative rule of thumb, the P-value from RIA should be .01 to reject the null hypothesis when the time series is autocorrelated. If the time series is autocorrelated and P from RIA is between .01 and .05, then the results are equivocal. . . . Autocorrelation was not a severe obstacle to interpretation of our data. Overall, about one-third of the time series were autocorrelated, but in most of these cases the P-values were so high or low that the results of RIA were unequivocal. Autocorrelation caused equivocal results in 3% of the cases we examined. (Frost et al., 1999, 1149)

Thus, as in Mayo's look at violating predesignation in psychology, violating independence also effects the severity assessments using t-tests (the test that Carpenter et al. used). In particular, to achieve a P-value of .05, using RIA one must have a P-value of 0.01 or less. Furthermore, P-values between .01 and .05 are equivocal. One cannot reject the null at the .05 level using RIA and satisfy severity, as the results of RIA are equivocal at that level. While the bar is raised in order to reject the null using RIA, nonetheless, the results do support Stewart-Oaten, Murdoch, and Parker (1986), who argued on theoretical grounds that autocorrelation may be only a minor problem in the analysis of time series of differences between experimental and reference ecosystems given sufficient time between samples. However, Carpenter et al. emphasize that while RIA tests whether the same changes or trends occurred in both basins after the manipulation, it does not prove that the manipulation caused the change between basins. Determining causality, they state, rests "on ecological rather than statistical arguments" (Carpenter et al., 1989, 1143). RIA does provide support that the visually dramatic spikes in the graphically presented data are not an artifact of graphing techniques. They show a real effect, but not that the manipulation caused the effect.

ERRORS, ASSUMPTIONS, AND REPLICATION: A BRIEF SUMMARY OF ROTIFER *K. TAUROCEPHALA* EXPERIMENTS

All experiments are, to varying degrees, simplifications. A major assumption underlying smaller-scale experiments is that their results can be extrapolated to natural systems or reality. That is, we assume our simplification or reduction of parameters maintains the relevant aspects of the natural situations or phenomena under study in a complex system. Large-scale experiments test this assumption by running a manipulation in nature. Gonzalez and Frost (1994) offer an example of the importance of this type of check. Laboratory assays on the effects of pH on reproduction under varied food conditions and survivorship without food indicated that the rotifer *K. taurocephala* would experience either no change or a decline in abundance with increasing acidification. In the LRL acidification experiment, however, *K. taurocephala* exhibited a marked increase in abundance when pH was lowered. This increase seems to be due to a reduction of its predators at lower pH. This result suggests that care must be taken in extrapolating results from the laboratory to a natural system. (See Schindler, 1998, for a comprehensive discussion of ecosystem manipulations as checks on smaller-scale experiments.)

Can we provide arguments and evidence that this rotifer response is not due to some localized event with long-term effects that occurred only in the manipulated basin of LRL? Yes. First, we have observational data from another pair of local bodies of water to supplement data from both our control basin and experimental basin (although these are not replicates in Hurlbert's sense). Crystal Lake, a pH-neutral lake, and Crystal Bog, with high acidity, were both monitored alongside LRL and showed a similar relative abundance of *K. taurocephala* in the acidic environment. Second, the experimental argument from LRL is further bolstered by similar representation in other ecosystem acidification experiments, notably from the NTL-LTER program. These two *independent* lines of evidence bolster the inferences at LRL because it would seem *very unlikely,* given their similar responses in nearby neutral and acidified waters, that *K. taurocephala* flourished contrary to laboratory expectations based on some event that was *unique* to LRL's experimental basin. Thus, there is strong evidence that this potential source of error either did not occur *or* if it did occur, did not affect LRL results for rotifers. *While this is a qualitative argument, it is a strong argument based on statistical thinking, that is, a focus on error, if not significance testing.* However, even with these two lines of independent evidence, the replicated laboratory studies showed that pH alone could not be directly responsible for this flourishing and this raises a real problem for inferring that acidification was the direct cause of *K. taurocephala's* success in LRL. Clearly an acidic *environment* plays a role, but the specific causal mechanism(s) *in* that environment has not been demonstrated simply based on abundance studies. This contradiction between lines of evidence illustrates another reason why ecosystem (environmental) studies are so crucial for understanding ecological phenomena. Large-scale experiments not only provide evidence for specific hypotheses but are themselves a source of new hypotheses.

Data from LRL on the shortening of spine length on *K. taurocephala,* as well as the observed decline in their major predator, suggest that the decline in predators was a direct cause in their flourishing in LRL's experimental basin.

Can we now legitimately infer that the indirect effects of acidification on predation more than offsets direct acid effects on the rotifer *K taurocephala?* No, to do so would be tantamount to changing the question under test. One inference tested at LRL was whether our results from the laboratory experiments could be reliably extrapolated to a more complex ecological system. The results from Little Rock Lake (both graphical and Monte Carlo simulations) suggest that the extrapolation was not reliable. The LRL study was designed to test a variety of inferences about the effects of increasing acidification on rotifer populations. It was not designed to *test* indirect effects on rotifers by decreased predation, and thus experimenters did not design their tests to see how that hypothesis could be in error. Thus while LRL rotifer studies have generated a new hypothesis (the alternative mechanism of which Stewart-Oaten et al. spoke), on the philosophy of experiment we are espousing, it would be premature and unwarranted to claim that LRL and the other independent studies have tested this decreased predation hypothesis severely. Instead, we would recommend that more observations from a variety of sources (natural experiments like Crystal Lake and Crystal Bog, small-scale laboratory experiments with replication, computer simulations, and remodeling the LRL data) be used to test this hypothesis by probing both it and the methods used to generate it for possible errors.

CONCLUSIONS: THE USE OF STATISTICAL TESTS
AND ERROR STATISTICAL THINKING

What is the attraction of using statistical tests, even where their use is highly problematic or unwarranted? One answer is the supposed rigor that quantitative analysis gives to results. However, numbers are only as reliable as the procedures that produce them. Statistical tests are very useful for analyzing data provided their assumptions are met. It is up to the investigator to ensure that assumptions are met and to report *how* they were met. However, statistical tests are not the only tests that must meet assumptions in order to produce reliable data for a particular inference. Hurlbert suggests that if assumptions cannot be directly met in the experimental design, then graphical representation of the data should be substituted for statistical analysis. This is a reasonable suggestion. Unfortunately, graphical representation, which can be both honest and extremely useful, can be just as or even more misleading than misapplying statistical tests, where at least the assumptions are generally well known.[5] Looking *only* at the graphs, they *appear* to be incontrovertible proof of the effect of decreasing lake pH on *K. taurocephala* (abundance was plotted against pH level). In the discussion of these graphs, further evidence (e.g., loss of major predators combined with shortening spines and RIA analysis) suggested that such a relationship was probably indirect and more likely due to changes in the food web caused by acidification. Simply forgoing statistical tests when their assumptions are not directly met in the experimental design does not solve the very real problems with providing evidence for manipulation effects in unreplicated ecosystem experiments.

Attempting to use statistical tests forces experimenters to deal with possible sources of errors that had they not attempted to use these tests in the first place they may well have been unaware of. For example, we are compelled to think replication of experimental and control units matters for detecting and mitigating the possibility of a stochastic event and the effects of natural variation between units. However, rather than throwing up our hands in despair or turning to alternative presentations of data simply in order to avoid problematic assumptions, we feel such violations of common methodological rules should be a call to find and articulate other quantita-

5. Stewart-Oaten et al. (1986, 930) make a similar point: "Since their assumptions are frequently implicit, these graphs and tables usually provide a less reliable basis for conclusions, rather than a more reliable one."

tive and qualitative arguments[6] *guided* by statistical reasoning with its focus on error probabilities. Let us be clear, we are not championing the misuse of statistical tests when their assumptions are violated. We are urging experimenters to try to meet (or approximate) these assumptions in order to avoid the errors they were designed to overcome with other means when possible and to take into account how violations of such assumptions/methods affects the severity or reliability of their tests and to report this. This style of error reasoning should also guide criticism of experiments. It is also important to emphasize that we are not suggesting that nonstatistical methods, tests, and representations of data be dispensed with in any way. They can provide extremely reliable and powerful experimental arguments and evidence for a hypothesis. We are emphasizing that one limitation on these methods is that the methods in and of themselves may not force experimenters to engage in statistical thinking with its focus on error probabilities. We also pointed out that this focus on error and arguing from error not only leads to more reliable inferences but leads to the growth of local ecological and methodological knowledge (e.g., about cage effects). The lesson we have tried to illustrate by looking at the Little Rock Lake acidification ecosystem experiment is that while it is not always legitimate to use statistical tests, it is always legitimate and good to use statistical reasoning with its focus on error probabilities. However, this is a demanding and complicated task that requires great ingenuity and creativity on the part of investigators. We believe that Mayo's error-statistical epistemology of experiment provides excellent guidance for this often daunting task.

Deborah Mayo (1996) has reinterpreted classic frequentist statistics into a much more general framework that she calls error statistics to indicate the continuing centrality and importance of error probabilities and *error-probabilistic reasoning* in testing hypotheses. By using a piecemeal approach to experiment, she has been able to characterize her concept of severity independent of the specific assumptions in any one Neyman-Pearson or Fisherian test. Her generalization of statistical reasoning above and beyond any one statistical test provides a consistent and coherent approach to testing and assessing both quantitative *and* qualitative evidence and hence can be directly applied to whole-ecosystem experiments. Using her error-statistical approach, we argued that even though replication was violated in

6. E.g., Monte Carlo simulations to test for effects of autocorrelation, similarity arguments, independence arguments, that is, arguments based on independent evidence from other manipulations and observational data (or natural experiments), etc.

the LRL experiment (a case of pseudoreplication), the potential errors that replication would have controlled for were adequately met by alternative methods and arguments and could be reliably ruled out. Given the evidence from LRL on the marked increase in *K. taurocephala* in combination with laboratory studies on the detrimental effects of acidification on it, we could reliably infer that its flourishing must be due to an indirect effect of the acidification of the lake. However, based on the standards of severe testing, it would be premature and unwarranted to infer that decreased predation was the causal mechanism behind this phenomenon. This is because the LRL experiment was not designed to test this specific inference and to probe the ways in which it would be an error to assert it. More work would need to be done at all levels of testing (small-scale to large-scale, remodeling of data, observational evidence, etc.) and other arguments from error made before we would be warranted in making this particular inference. This example also underscores the fact that whole-ecosystem experiments can be excellent sources for generating new hypotheses. This seems to us especially fruitful given the large spatial and temporal scales and the degree of complexity at which ecological phenomena operate in the world. Because, when all is said and done, ultimately, we want to test inferences about the real world, large-scale and complex as it is. The error-statistical approach to experiment, we believe, will help us succeed in this task.

8.1   Commentary

William A. Link

INTRODUCTION

Miller and Frost (subsequently MF) discuss "how to design and interpret . . . experiments given that standard statistical practice (based on assumptions of replication, randomization, independence, etc.) is inapplicable in any direct way."[7] This is a tall order. I understand and have sympathy for their motivation—scientists simply do not always have the luxury of being able to carry out randomized and replicated experiments.

In such cases, model-based analysis can be an attractive, indeed the only

7. Editors' note: This quote was deleted from the final version of the Miller and Frost chapter; see discussion in the rejoinder.

alternative to design-based analysis. As Edwards (1998, 324) says, "There is a place for both types of inference in environmental and ecological resource assessment, especially in view of the fact that so many situations disallow a strict design-based approach." Still, model-based analysis is generally regarded as inferior in inferential value, as evidenced by Edwards's subsequent complementary assessment of a model-based analysis: "[the authors] seem to use model-based inference only when necessary, and in these instances they use it as conservatively as possible."

The position embraced by MF is that we can learn from data collected in less than optimal fashion. I agree with them, that it is simply incorrect to argue "that only replicated experiments can provide *insights* into ecological processes" (my emphasis). Insights can be gained with far less rigor.

The "error-statistical approach" proposed by Mayo (1996) and enthusiastically supported by MF seems to me a mixture of model-based analysis, good sense, and common sense, with just a touch of nonsense. The commonsense part of it may be summarized as follows: Carefully choose a model, with clearly articulated premises. When the premises are not testable, argue for their reasonableness; when the premises *are* testable, do so. Assess the effects of model misspecification. Use your head.

I agree entirely; this is a good course of action. The conclusions reached will be useful for induction and retroduction (Romesburg, 1981), both of which are appropriate tools in scientific inquiry, especially if used to generate testable hypotheses in the hypothetico-deductive method (Box, 1976; Nichols, 1991). The resulting inference, though, is weak, especially compared to one obtained under the hypothetico-deductive paradigm of scientific inquiry. No amount of epistemology-craft can alter the basic facts.

MF say "it is not replication per se but the types of errors for which it controls that are of interest in the design and interpretation of experiments." The message is that if we're smart enough to identify sources of confounding, or confident enough that their effects are small or nonexistent, then we can manage without replication, randomization, and other such niceties. We may simply assert that we trust our model and proceed with our inquiry. Such reasoning justified quota sampling and led to the classic failure of three major polls in predicting the outcome of the 1948 Dewey-Truman presidential election—a textbook example of what can go wrong with model-based analysis.

I have analyzed an ugly data set or two, myself, and always tried to follow the commonsense course of action described by MF. When I'm done, I report the results as conditional on the model and squirm a bit when pressed as to how much confidence I can put in the model itself. What MF have extracted from Mayo's new epistemology, apparently, is that I need no longer squirm.

At issue is the confidence one can place in a model-based analysis. In this essay, I discuss confidence in models, emphasizing the need for careful evaluation of its source.

THE PROTOTYPICAL DESIGNED EXPERIMENT

Chapter 2 of Fisher's *The Design of Experiments* (1935) illustrates "the principles of experimentation . . . by a psycho-physical experiment." "A lady declares that by tasting a cup of tea made with milk she can discriminate whether the milk or the tea infusion was first added to the cup" (Fisher, 1971, 11; Box, 1976, reports that the example was based on Fisher's own experience of spurned hospitality). A test is proposed: the lady is presented with eight cups of tea and required to identify which four had the tea poured into the milk and which four were mixed (horrors!) in the reverse order.

To be on the safe side, the eight drinks are randomly assigned to teacups and tasted in random order. As Fisher notes in presenting this classic example of a designed experiment, "whatever degree of care and experimental skill is expended in equalizing the conditions, other than the one under test, which are liable to affect the result, this equalization must always be to a greater or less extent incomplete and, in many important cases, will certainly be grossly defective" (Fisher, 1971, 19).

The appeal of this designed experiment lies in the certainty with which a null distribution can be calculated. Assuming that the lady in question is completely incapable of distinguishing the two preparations, the randomization guarantees a priori that the number of cups correctly identified is a hypergeometric random variable; in particular, that there is 1 chance in 70 that she will correctly distinguish and identify the two groups of four cups. The appeal of designed experiments lies in the confidence one can place in models for the data they produce.

CONFIDENCE IN MODELS

By a model, I mean a pair $\{P, \pi\}$, where P is a probabilistic description of phenomena[8] ϕ, and π is a set of premises about ϕ, implying P.

---

8. Here, I am not using the word "phenomena" in the special sense given by Miller and Frost, which distinguishes "phenomena" and "data," but rather in the wider sense in which data are one type of phenomenon.

Data sets for ecological applications are frequently observational. Associated models usually involve uncertain dependencies on measured covariables and may depend on untestable premises. In this context, models are rightly regarded as abstractions: π is a simplification of reality, P is of necessity an approximation; Burnham and Anderson (1998) go so far as to label the phrase "true model" an oxymoron. They're likely right, at least with regard to the collection of models we might assemble to describe a particular observational data set. Their sentiment echoes that of Box (1976, 792), who said that "the statistician knows, for example, that in nature, there never was a normal distribution, there never was a straight line, yet with normal and linear assumptions, known to be false, he can often derive results which match, to a useful approximation, those found in the real world."

Let us allow for the moment that such a thing as "truth" exists (rather than acceding to the intellectual anarchy and hairsplitting of Pontius Pilate syndrome).[9] Then there is a true P, say P_ϕ, of which the modeled P may be only an approximation. Our confidence in a model is a subjective measure of the validity of π as descriptive of ϕ, expressed in terms of the effect of using P as a surrogate for P_ϕ.

Using the tools of randomization and replication, a scientist guides the occurrence of phenomena in such a way as to allow the construction of models in which a great deal of confidence can be placed. The hypergeometric distribution used to assess the tea-drinking lady's claim is not an approximation, but truth, provided the null hypothesis is true.

Thus I would modify Box's dictum that "all models are false, but some are useful" to "all analyses are model-based—some you have more confidence in." What distinguishes models is the credibility of the assumptions needed to justify their use.

Confidence Based on Randomization
Randomized experiments afford a good deal of confidence in their models, but it is a worthwhile exercise to examine the source of the confidence. Consider Fisher's test of the tea-drinking lady's sensory acuity.

Ideally, there would be no need for randomization. The lady's claim was not qualified with respect to the quality of the china used, nor with regard to the order of presentation; there is however an implicit ceteris paribus—all other things being equal. We want irrelevant features of the experiment to neither enhance nor diminish her performance.

---

9. "What is Truth?" (John 18:38).

The fact is that if ceteris ain't paribus, randomization won't make it so. If the cups are not the same, presenting them in random order will not make them the same. The results will depend in greater or lesser measure on what is essentially an irrelevancy: the randomization.

We could envision, as an extreme case, some sort of unobservable characteristic of four cups that would force the lady, every time, to choose those four as the ones prepared with milk first, regardless of her ability under ceteris paribus. Then, the outcome of the experiment would be determined entirely by the randomization; the tasting—the part of the experiment that actually involves the lady—would have no part in producing the data used to assess the lady's claim.

Fisher, under the heading of "sensitiveness," says: "it is in the interests of the sensitiveness of the experiment, that gross differences . . . should be excluded . . . not as far as possible, but as far as is practically convenient" (Fisher, 1971, 24). On the other hand, Fisher says, experimental refinements are not required: "the validity of the experiment is not affected by them" (Fisher, 1971, 24). In modern statistical parlance, lost sensitivity due to "gross differences" is loss of power when the null hypothesis is false; the "validity of the experiment" is the truth of the distributional assumptions, when the null hypothesis is true.

I believe that the cost of our confidence in the validity of the experiment is greater than simply accepting the possibility of low power. There is an additional cost, of allowing our conclusions to be determined by a component of variation that has nothing to do with the question at hand, as I shall now explain.

Let X denote an outcome and let R denote a randomization; our evaluation of the randomized experiment is based on $P(X)$, which is the average value, across randomizations, of $P(X|R)$. Let $\mathbf{I}(X)$ be an indicator of the occurrence of X (1 if X occurs, 0 otherwise). Then Fisher's "validity of the experiment" is simply that the rate parameter for the Bernoulli trial $\mathbf{I}(X)$, $P(X) = E(\mathbf{I}(X))$, be correctly specified.

The experiment has two stages, the randomization and the tasting. We may therefore identify components of variation for the Bernoulli trial $\mathbf{I}(X)$. The double expectation theorem for variances yields

$$\mathrm{Var}(\mathbf{I}(X)) = \mathrm{Var}(E(\mathbf{I}(X|R)) + E(\mathrm{Var}(\mathbf{I}(X|R)),$$

or equivalently,

$$P(X)[1 - P(X)] = \mathrm{Var}\{P(X|R)\} + E\{P(X|R)[1 - P(X|R)]\}$$

The first term on the right-hand side is the component of variation associated with the randomization. We may define "randomization dependence" as the proportion of the total variation represented by this component, viz.,

$$\theta(X) = \frac{\text{Var}(P(X|R))}{P(X)(1 - P(X))}.$$

Clearly, $0 \leq \theta(X) \leq 1$, with $\theta(X) = 0$ representing the desirable case of ceteris paribus, when the randomization is unnecessary, and $\theta(X) = 1$ representing the pathological and undesirable case where the result of the experiment is completely determined by the randomization.

Fisher's "validity of the experiment" is an a priori assessment, before the randomization, averaged across potential outcomes of which we will have a sample of size 1. I can set up the experiment here in my office as I write, using my four wretched coffee mugs and four Styrofoam cups. I can carry out the randomization right now, flipping coins and using random number generators to specify the order in which the lady is to taste the specimens. Once I have done so, $P(X|R)$ is fixed; I will have determined, in some measure, the outcome of the experiment, without the lady even being present. In particular, $100\theta(X)\%$ of the variation in $I(X)$ is attributable solely to the randomization.

AN ADMONITORY TALE

Andy Brand Raisin Bran Inc.'s quality control team is required to test whether raisins are being added to the mix at the proper rate. It is known from long experience that when the equipment producing the cereal is operating properly, the number of raisins in an A-ounce sample is a Poisson random variable, with mean βA. The QC team invents a device that is intended to sample A^* ounces of cereal, and count the number of raisins present; the number A^* was calculated so that the probability of getting 0 raisins in a sample is 5% if the cereal is being produced according to specifications.

The only problem with the QC device is that it has trouble sampling precisely A^* ounces of cereal; it is determined that the sample size is RA^*, with R uniformly distributed over the range 0.50 to 1.97. The company statistician is called in. After a few quick calculations, she reports that $P(\text{no raisins}|R)$ ranges from .003 to .224 depending on R, but that fortunately the average value of $P(\text{no raisins}|R)$ is .05.

Her employer, baffled by the subtleties of randomization, fires her and the entire QC team, even despite protestations that θ(no raisins) is only 6.7%.

* * *

One has to question whether "validity" has been achieved while papering over irrelevant variation. There is something analogous to the almost universally despised contrivance of using a randomized decision rule to obtain exact level decision rules: the desired outcome is attained, but at the cost of irrelevant variation.

My argument does not call into question the soundness or elegance of Fisher's proposed experiment, nor the important role of randomization. It does, however, highlight the importance of controlling for sources of variation in a scientific inquiry. This can be done, and should be done, by appropriate modeling. Uncontrolled differences not only decrease "sensitivity" (power), but also increase randomization dependence. To rely too heavily on randomization is to obtain model confidence in the presence of irrelevant variation.

I conclude this section with a comment on the relative importance assigned to randomization in Bayesian and frequentist philosophies. It is interesting to compare a frequentist's blanket statement that "randomization is irrelevant to Bayesian inference" (Dennis, 1996, 1,096) with that of a leading group of Bayesian analysts:

> A naive student of Bayesian inference might claim that because all inference is conditional on the observed data, it makes no difference how those data were collected. This misplaced appeal to the likelihood principle . . . [leads to an] incorrect view [that] sees no formal role for randomization in either sample design or surveys. (Gelman et al., 1998, 190)

Gelman et al.'s compelling presentation of "ignorable designs" (designs under which data collection considerations can be overlooked in Bayesian analysis) points out that when no covariates are observed, randomization is a necessity; nevertheless, "distinguishing information is usually available in some form" (Gelman et al., 1998, 221) that will enhance a model-based analysis.

CONFIDENCE UNDER MAYO'S EPISTEMOLOGY

Confidence in a model-based analysis can be built up by testing of model premises (normality, independence, etc.). This is the backbone of Mayo's "argument from error":

> It is learned that an error is absent to the extent that a procedure of inquiry with a high probability of detecting the error if and only if it is present nevertheless detects no error. (Mayo, 1996, 445) [10]

In other words, we can be confident in model premises that have been tested and not rejected under conditions guaranteeing high power and low type I error rates. It is worth noting, however, that tests of model premises are one of the least appropriate uses of the hypothesis-testing paradigm. Hypothesis tests are attempts at proof by contradiction: assume what you're attempting to disprove, and if the results are inconsistent with the assumption, reject the assumption. However in testing model premises the burden of proof changes: the analyst specifies H_0 (e.g., "normality"), and if fortune smiles and the evidence is not overwhelmingly against the null, the analyst treats the assumption as established truth. Reading MF, I have a nagging suspicion that Mayo accepts null hypotheses that have not been rejected.

Readers of MF may share some of my discomfort with Mayo's epistemology, at least as presented by MF. Consider:

> It is not enough that mistakes are logically possible, since we are not limited to logic. Unless one is radically skeptical of anything short of certainty, specific grounds are needed for holding that errors actually occur in inquiries, that they go unnoticed and that they create genuine obstacles to finding things out. (Mayo, 1996, 4)

I read this statement repeatedly,[11] wondering what we have besides logic. The philosophy seems to shift the burden of model uncertainty away from the scientist. This is especially surprising, given that the ideas presented are supposed to be "an alternative epistemology to Bayesianism that does not

10. Editors' note: This quote was deleted from the final version of the Miller and Frost chapter; see discussion in the rejoinder.
11. Editors' note: This quote was deleted from the final version of the Miller and Frost chapter; see discussion in the rejoinder.

rely on subjective priors."[12] (By the way, Bayesian analysis need not rely on subjective priors, and when it does, the sensitivity of its conclusions to changes in the prior should always be reported.)

CONCLUSION

All analyses are model-based. What distinguishes models is the credibility of the assumptions needed to justify their use.

This being the case, it makes sense to carefully evaluate the source of our confidence. Randomization has its purpose in Bayesian or frequentist analysis: to put the final polish on model validity. It is not a good idea to rely too heavily on randomization for model validity: as Fisher (1971, 49) says, "In inductive logic . . . an erroneous assumption of ignorance is not innocuous."

MF rightly encourage us not to despair if standard statistical practice (based on assumptions of replication, randomization, independence, etc.) is inapplicable in any direct way. As Edwards (1998, 324) says, "There is a place for . . . [model-based] inference in environmental and ecological resource assessment." On the other hand, assumptions are still assumptions, no matter how many "severe" tests they have "passed."

8.2	Commentary[13]

Charles E. McCulloch

Miller and Frost (MF) make the important point that whole-ecosystem experiments are essential to our understanding of large-scale phenomena but are terribly difficult to conduct, according to accepted statistical principles for asserting cause and effect. They also argue (correctly in my opinion) that Hurlbert's viewpoint on pseudoreplication is overly strident and should be

---

12. Editors' note: This quote was deleted from the final version of the Miller and Frost chapter; see discussion in the rejoinder.

13. Editors' note: The commentary presented here discusses an earlier version of the Miller and Frost chapter. In response to these commentaries by McCulloch and by Link, Miller has substantially restructured the chapter to increase clarity. Miller discusses these changes in her rejoinder.

relaxed. After all, lack of replication rarely prevents a perfectly valid regression analysis.

However, I disagree with one of their major premises: that lack of replication is the main reason we cannot assert cause and effect from whole-ecosystem experiments. I assert this by considering three conceptual experiments.

1. Suppose we had a pair of lakes, one manipulated and one unmanipulated, *known to be identical in all ways except for the manipulation.* Then, any differences between them could be attributed to the manipulation, despite the lack of replication. Even as a conceptual experiment, this is not very useful, since the point of MF is to make inferences in the presence of error.

2. Suppose we had a manipulated and unmanipulated lake that might additionally differ due to the natural variability among lakes (in some population of lakes) but *with the degree of variability known.* It is easy to construct a statistical test that, if it rejects the null hypothesis of no difference, can be used to attribute cause to the manipulation. This shows that a primary role of replication is merely to allow an empirical estimate of error.

3. What if we have replication but the manipulated and unmanipulated lakes differ in a way other than just the manipulation? For example, suppose we wish to attribute absence of a species of fish in Adirondack lakes to the "natural" manipulation of acidification. The more acidified lakes also tend to be higher in elevation, carrying with it a host of other confounding factors. Even replication taken to its fullest extent of measuring every lake and measuring each lake so many times so as to be error-free would not allow us to assert cause and effect. Bottom line: replication does not eliminate bias (often due to confounding), and that is the bugaboo of asserting cause and effect. Instead, it reduces variability and allows empirical estimates of the size of that variability.

The above example is observational in nature, and hence it is easier to understand the possible presence of confounding factors. However, the same is true of experiments with actual manipulations but without the presence of randomization. Lack of randomization allows the introduction of confounding and hence bias, which obscures causal inference. I feel that randomization is a more fundamental notion than replication for this purpose.

There is much practical advice in MF about how to strengthen the inference of causation. I think, however, it will be just as fruitful to return to the ideas of inference as expounded in the introductory chapters of Box, Hunter, and Hunter (1978), the ideas of quasi-experimental design as described in Campbell and Stanley (1981) (see also James and McCulloch, 1994), and the adaptation of modern statistical approaches to the inference of cause and effect from observational studies (e.g., Rosenbaum, 1995).

8.3   Rejoinder

Jean A. Miller

PRELIMINARY REMARKS

First, when Link wonders "what we have besides logic" to work with, he was referring to a specific quote we had from Mayo (1996) that he reproduced in his commentary. It immediately became clear upon reading his commentary that we had taken the quote out of context and that probably no one other than a professional philosopher could add the appropriate context, and so it was cut from our chapter. Logic, in the context of this quote, referred only to a very narrowly defined type of logic—deductive logic [14] and not to logic or reasoning in a general manner of applying rational thought and evidence to support an argument. The problem with philosophical applications of deductive logic to science is that it does not take into account how the data were generated but takes evidence as a given; see Mayo's rejoinder (chapter 4, this volume). Debates about whole-ecosystem experiments revolve around how data are generated; thus this specific type of logic is uninformative for such discussions.

Second, we agree with both commentators that there are many excellent books and methods for using statistical tests for a variety of evidence—experimental and observational. Our aim, which their criticism has helped me

Sadly, Tom Frost died before we received McCulloch's and Link's commentaries. I use the first person plural when referring to our paper and views that Tom expressed and the first person singular for all other comments. I take responsibility for all mistakes.

I thank Chuck McCulloch and Bill Link for their useful and insightful commentaries on our paper. I will first deal with some minor concerns that I revised the paper to account for. I will spend the rest of this commentary discussing their main concern about the role of randomization versus replication for causal inference.

14. See chapter 2.1 for a philosophically technical definition of deductive logic.

to greatly focus, was and is to provide a more general and philosophical treatment of why such tools are needed and the types of arguments needed to ground the use of them in whole-ecosystem experiments. I would add to their suggestions two books by the econometrician Aris Spanos (1986, 1999) on "misspecification testing." In general, the statistical tests that Hurlbert discusses are ones on which the probability model of normality, independence and identically distributed samples is assumed. Misspecification testing is an approach for testing the assumptions of a model outside of the model (e.g., testing one model whose use is justified based on the assumptions of normality, independence, and identically distributed samples using another model that does not share *all* of the same assumptions). As misspecification tests lie outside rather than within the boundaries of a postulated statistical model being used, they can be used to check for violations of the assumptions of the initial model. The Monte Carlo simulations on the effects of autocorrelation provide an example of this.

Third, originally we had made some comparisons of our approach with the Bayesian approach but have since cut those sections as redundant. The chapters in this volume by Mayo (chapter 4) and Dennis (chapter 12) make the same points and more than adequately express our views.[15] When Link discusses how Bayesian approaches can and do take into account how the evidence was gathered, e.g., randomization, then it would seem that Bayesians are accepting basic frequentist criteria and thus there is no conflict at that point. Where Bayesians and frequentists differ after this point, we recommend that the reader see the chapters by Mayo and Dennis to decide whether Bayesianism adds anything substantial to the frequentist position. It is important to note here, however, that for frequentists probability attaches to the *procedure* used to produce the data, not to the inference. For frequentists, "the chances are in the sampling procedure, not in the parameter" (Freedman et al., 1991, 351). This is in direct contrast to Bayesian approaches that assign probability to the inference itself (e.g., its truth status). Thus, Link's "confidence in models" as a "subjective measure of the validity" needs to be clearly distinguished from the frequentist concept of confidence intervals. Confidence intervals represent the frequency (probability) with which an observed event will happen within the context of the procedure employed.

Finally, we were overly optimistic in our first draft, and I took our commentators' chiding to heart; they influenced the revision that is published in this volume. However, I am still optimistic in that I feel their commentaries

---

15. See also Reckhow (1990) for a Bayesian approach to ecosystem experiments.

illustrate the usefulness of the very approach that we suggest by focusing on the properties of particular methods for detecting and controlling specific sources of error. I will illustrate this in the rest of this commentary when discussing randomization, which was the major criticism (and rightly so) of our paper. I view their criticism as conducting the very type of appraisal of methodological rules that we have been arguing for and support.

AN (OVERLY) SIMPLE RESPONSE TO QUESTIONS CONCERNING RANDOMIZATION IN ECOSYSTEM EXPERIMENTS

Hurlbert and many others focus on the absence of replication as foreclosing the possibility of a causal analysis in ecosystem manipulations. A quick and oversimplified explanation is that without replication at the level of experimental units, it simply is impossible to randomize, and therefore replication is the real problem. Even if we realize that randomization assumes replication, this response does not adequately uncover the rationale for desiring randomization in and of itself. This response is inadequate, as we have learned nothing new about randomization as a methodological rule, e.g., what errors it controls.

RANDOMIZATION TO CONTROL EXPERIMENTER BIAS

Hurlbert states the differences between replication and randomization:

> Replication reduces the effects of "noise" or random variation or error, thereby increasing the *precision* of an estimate of, e.g., the mean of a treatment or the difference between two treatments. Randomization eliminates possible bias on the part of the experimenter, thereby increasing the *accuracy* of such estimates. (Hurlbert, 1984, 192)

By randomizing experimental and control units, experimenters cannot consciously or unconsciously let this choice be influenced by untested biases. For example, in trying out the efficacy of a new drug, bias could enter if a pharmaceutical company chose the healthiest members of the test population to give the drug to and gave the weakest members the placebo. Such a treatment assignment seems to stack the cards in favor of the drug's efficacy and this seems an unreliable test of the drug (i.e., it lacks severity). There are many methods (double-blind trials, having others repeat an experiment, etc.) besides randomization for eliminating possible experimenter bias. This is the most intuitively obvious role for randomization.

We could randomize our ecosystem manipulation without replication. Given a list of the 10,000 lakes in Wisconsin, we could find a method for randomly choosing one to experiment on. We could pull names out of the hat. Now experimenter bias can be ruled out, but intuitively this seems a very unreliable procedure. To understand why this is the case, we need to more fully examine the role of randomization in experimental inquiries. First, however, I will briefly define "cause" and "confounding factors," as these notions lie at the heart of why randomization plays a crucial role in causal inferences, as pointed out by McCulloch and Link.

COUNTERFACTUAL DEFINITION OF CAUSE AND CONFOUNDING FACTORS

The entire notion of cause is a landmine of controversy and a field in its own right. See Greenland, Robins, and Pearl (1999, sec. 1) for a quick historical overview of cause and the development of *counterfactual notions of cause* starting in the eighteenth century with the work of David Hume. In our chapter, we basically accept the standard counterfactual notion for determining cause and effect expounded in their paper.[16] Briefly, if some object or event that we label "cause" does not occur, then the object or event that we call the "effect" also will not occur. The underlying causal argument that needs to be sustained in the Little Rock Lake experiment is that if the experimental basin had not been acidified, then particular effects that did occur in it (or changes of a certain magnitude) would also not have occurred.

Greenland, Robins, and Pearl (1999) also provide a short yet excellent discussion on *confounding* factors—the major concern in justifying causal inferences—based on Mill's "informal conceptualization" as Greenland, Robins, and Pearl explain:

> Mill's requirement suggests that a comparison is to be made between the outcome of his experiment . . . and what we would expect the outcome to be if the

16. See also Woodward (1997) for a discussion of counterfactual reasoning in scientific explanation. The reader may wish to see Dawid (2000) for an attack on counterfactual concepts of cause and responses to Dawid by Pearl (2000), Robins and Greenland (2000), Rubin (2000), Shafer (2000), and Wasserman (2000).

agents we wish to study had been absent. If the outcome is not that which one would expect in the absence of the study agents, his requirement insures that the unexpected outcome was not brought about by extraneous circumstances. If, however, those circumstances do bring about the unexpected outcome, and that outcome is mistakenly attributed to effects of the study agent, the mistake is one of confounding (or confusion) of the extraneous effects with the agent effects. . . . The chief development beyond Mill is that the expectation for the outcome in absence of the study exposure is now most always explicitly derived from observation of a control group that is untreated or unexposed. (Greenland, Robins, and Pearl, 1999, 31–32).

Confounding factors (e.g., Hurlbert's sources of confusion, Link's ceteris paribus conditions) can be seen as bias on the part of nature rather than the experimenter and unknown (unknowable even) to the experimenter. The impetus for having a control basin is to represent what ought to have happened in the acidified basin had it not been acidified. Some amount of change or variation through time would be expected in both basins but not beyond a certain magnitude or of certain types. Of course, working out what is "reasonable" variation between the two basins is where a lot of work and arguing from error come into play, as we have discussed in our paper (e.g., similarity arguments).

Both our commentators feel that lack of randomization is equally, if not more, problematic for causal inference than replication: "randomization is a more fundamental notion than replication for this purpose" (McCulloch)." This is because "lack of randomization allows the introduction of confounding and hence bias, which obscures causal inference." Link agrees, quoting Fisher: "whatever degree of care and experimental skill is expended in equalizing the conditions, other than the one under test, which are liable to affect the result, this equalization must always be to a greater or less extent incomplete, and in many important cases will certainly be grossly defective." Why is this the case? McCulloch highlights the problem for causal inference even with replication:

> Even replication taken to its fullest extent of measuring every lake and measuring each lake so many times so as to be error-free would not allow us to assert cause and effect. *Bottom line: replication does not eliminate bias (often due to confounding) and that is the bugaboo of asserting cause and effect.* Instead, [replication] reduces variability and allows empirical estimates of the size of that variability (italics added).

Thus, randomization as a methodological rule seems to get its bite from its ability to check or control confounding factors. But how does it do this?

INTERPLAY BETWEEN REPLICATION AND RANDOMIZATION

Hurlbert quotes from Fisher to demonstrate the interplay between replication and randomization:

> With respect to testing, the "main purpose [of replication], which there is no alternative method of achieving, is to supply an estimate of error [i.e., variability] by which the significance of these comparisons is to be judged . . . [and] the purpose of randomization . . . is to guarantee the validity of the test of significance, this test being based on an estimate of error made possible by replication" (Fisher, 1971, 63–63).

Hurlbert explains, quoting Fisher again, that the validity of using significance testing is guaranteed by randomization because randomization "guarantees that, on the average, 'errors' are independently distributed." As Hurlbert points out, this is important because "a lack of independence of errors prohibits us from knowing [the true] α, the probability of a type I error" (Hurlbert, 1984, 192, citing Fisher, 1926, 506). It is this role of guaranteeing the independent distribution of errors that makes randomization such a powerful tool in experimental inquiries. This role is also at the heart of McCulloch's and Link's discussions of randomization. It will be useful here to discuss the difference between precision and accuracy to better understand this second role that randomization plays.

PRECISION VERSUS ACCURACY

Fisher argues that replication increases precision but randomization increases accuracy. A classic example of this difference is illustrated by target shooting. If all my arrows cluster closely together in one spot but far from the bull's-eye, my shooting is very precise but inaccurate as an estimate of the location of the center of the target. On the other hand, if my arrows are equally distributed (e.g., the distribution is due to chance, not to a systematic error on my part) around the bull's-eye, taking the average of them will provide an accurate (i.e., good) estimation for the location of the bull's-eye.

This is because the errors associated with each shot ought to cancel out, e.g., too far up, too far down, to the left, etc. For accuracy, we want assurance that the errors associated with each shot are independent of each other so as to cancel out. Ideally, we want an experiment that is both accurate and precise. The closer the shots are to the bull's-eye, the smaller the errors that will be attached to each shot (i.e., variability between shots). But a more random spread of the arrows around the bull's-eye leads to a more accurate estimate of the true center because the error from each shot will offset the error from another shot. One can be very precise but totally off the mark for an explanation. Being very precise about a wrong answer or mechanism does not seem to be what we are aiming for in causal inferences.

CONFOUNDING FACTORS, CETERIS PARIBUS CONDITIONS, AND RANDOMIZATION

Let us look more closely at our commentators and Fisher's discussion of explicitly introducing chance via randomization into experimental designs to control the possibility of confounding. Link explains the experimental situation in regards to randomization succinctly in the context of Fisher's lady drinking tea.

> Ideally, there would be no need for randomization. The lady's claim was not qualified with respect to the quality of the china used, nor with regard to the order of presentation; there is however an implicit ceteris paribus—all other things being equal. *We want irrelevant features of the experiment to neither enhance nor diminish her performance* (italics added).

Harking back to our target-shooting example, without randomization, we have no guarantee that all other things are equal. But precisely what "other things" are we concerned with here? What needs to be equalized is the variety of possible confounding factors or errors from each replicate.

Fisher is very explicit on *how* randomization "checks" confounding factors in the ceteris paribus clause:

> The element in the experimental procedure which contains the essential safeguard is that the two modifications of the test beverage are to be prepared "in random order." *This, in fact, is the only point in the experimental procedure in which the laws of chance, which are to be in exclusive control of our frequency distribu-*

tion, have been explicitly introduced. The phrase "random order" itself, however, must be regarded as an incomplete instruction, standing as a kind of shorthand symbol for the full procedure of randomization, by which the validity of the test of significance may be guaranteed against corruption *by the causes of disturbance which have not been eliminated.* (Fisher, 1947, 19; italics added)

Note that the errors being randomized are those left over *after* all known possible differences have already been ruled out (e.g., type of china used, strength of the infusion, type of tea used, temperature, etc., which had already been predetermined for each cup (19–20). Only after all steps have been taken to ensure as best as possible uniformity in the units (teacups) except for the order of milk and water, randomization of the cups is to be done last!

The explicit introduction of chance at the last stage of the design ensures, in this case, that:

If, now, after the disturbing causes are fixed, we assign, strictly at random, 4 out of 8 cups to each of our experimental treatments, then every set of 4, whatever its probability of being so classified, will certainly have a probability of exactly 1 in 70 of *being* the 4, for example, to which the milk is added first. However important the causes of disturbance may be, even if they were to make it certain that one particular set of 4 should receive this classification, the probability that the 4 so classified and the 4 which ought to have been so classified should be the same. (Fisher, 1947, 20)

Fisher's point is that randomization does not eliminate confounding factors—it does not. What it does do is make it highly improbable (in his example 1 in 70) that the sources(s) of confounding would appear solely in one of the groups. However, this probability can be calculated only by explicitly introducing chance into an experimental design.

FORCIBLY INTRODUCING CHANCE INTO AN EXPERIMENT

Randomization will not make ceteris paribus conditions equal. Each cup or set of cups can be influenced by some factor that we have not controlled for. And there will always be such factors; at least we can never rule them out. Randomization, as a methodological rule, has strength insofar as it introduces a chance distribution into the ceteris paribus conditions. Thus any errors associated with ceteris paribus are randomized. In the tea experiment, there is now only a 1 in 70 chance that any set of cups will be the cups; thus there is only a 1 in 70 chance that a confounding factor would affect the cups so chosen and bias the results. As in the target-shooting example, randomization does not make the errors disappear. Instead, the randomized design ensures the independent distribution of the errors so that, in the end, these errors either cancel out or the probability that the procedure would be open to them (e.g., open to confounding) can be determined.

Link, however, feels that even randomization cannot overcome the possibility of confounding factors determining the results of the test:

> The fact is that if ceteris ain't paribus, randomization won't make it so. If the cups are not the same, presenting them in random order will not make them the same. The results will depend in greater or lesser measure on what is essentially an irrelevancy: the randomization.
>
> We could envision, as an extreme case, some sort of unobservable characteristics of four cups that would force the lady, every time, to choose those four as the ones prepared with milk first, regardless of her ability under ceteris paribus. Then, the outcome of the experiment would be determined entirely by the randomization; the tasting—the part of the experiment that actually involves the lady—would have no part in producing the data used to assess the lady's claim.

For the sake of argument, what if four of the cups had some "unobservable" and unknown confounding factor? What are the chances that a random assignment of cups to treatment and control would assign all four with the unobservable force choice factor to the manipulated (e.g., milk first) group—1 in 70. Thus, even if the lady were forced by unknown forces to choose those cups, the chances that the four cups would be "the four" and that her response would be consistent with her purported ability would be low. This is because the chance that the treatment and the unobserved factor would both be present in the exact same four sets of cups is very low. What would show up though is a discrepancy between her expected results and the observed results unless she was merely guessing. Systematic errors (forced choice cups) also falls under ceteris paribus conditions and thus should also be equally distributed among the cups if the treatment given them remains randomized. Even if she did have the skill and answered correctly to the forced choice with the milk-first group, she would have been incorrect on the "forced" taste difference when there ought to have been none (e.g., the tea-first cups). If the lady felt afterwards that the test was not "fair," then, as she is making the claim, she would need to suggest that there was some systematic bias in the experiment.

The lady could insist that the experiment could be repeated using different cups. Replicating the experiment while varying the physical elements used could help experimenters discern if any of these elements play a role in her discerning abilities. Let's consider a more plausible example of this concern with ceteris paribus conditions. Suppose she drove to London for the test and the city water incapacitated her ability. In that case, she would have had to claim that during this test, she did not have the ability and could only guess. Being honest, if she really had the ability, she should notice its loss in the new circumstances (e.g., the difference between expected and observed outcomes). This, of course, can be tested by conducting the test in the normal situation in which she asserts that she can tell a difference. By conducting the experiment on home ground as it were, we have another way of ruling out ceteris paribus conditions. One may, at that point want to test whether it is something in the local water that makes her discrimination possible. Nonetheless, even if the water contributed to her ability, she would have demonstrated her ability, albeit only in special circumstances (e.g., only with local water). One can imagine an entire tea-tasting program being set up varying the types of water, cups, and teas used, etc. Moreover, we would want to conduct a double-blind trial so that whoever was serving the tea and recording the lady's responses could not tip her off. We don't want another Hans the counting horse incident!

CRITICISMS WITH CAUSAL INFERENCES
EVEN IN RANDOMIZED EXPERIMENTS

Like Stewart-Oaten, Murdoch, and Parker (1986, on cage and placebo effects), Greenland, Robins, and Pearl (1999, 32) also argue in their article that "confounding is also possible in randomized experiments, because of systematic elements in treatment allocation, administration, and compliance and because of random differences between comparison groups." Link's forced-choice cups seem to me another example of these types of problems. However, by exploring such errors, much can be learned as they are phenomena in their own right, as we suggested in our chapter. While randomization may not be an absolute guarantee against bias and that ceteris is paribus, it is a very strong method for countering bias and providing a calculated distribution for errors in the ceteris paribus clause.

EVALUATING REPLICATION AND RANDOMIZATION
AS METHODOLOGICAL RULES

Replication controls for natural variation. Randomization controls for experimenter bias, and confounding factors in the ceteris paribus clause. If we
were to rewrite our paper, randomization would play a larger role in our discussion. However, replication is a necessary assumption in randomization
and thus seems to me the more fundamental notion. Having multiple replicates provides for the independent distribution of errors by randomization.
An implicit assumption of ours, and I believe of others in the pseudoreplication debates, is that were replication possible, experimenters would, of
course, randomize treatments as well and so the errors each was designed to
detect and control for are lumped together under replication in discussions.

The main focus of the pseudoreplication debates over causal inference is
the specter of a localized stochastic event with long-term effects influencing
one treatment to a greater degree than another as a potential cause of the
difference between treatments rather than the experimental manipulation
per se. However, using independent evidence from similar manipulations
and observations of "natural experiments" (e.g., naturally occurring bodies
of water that resemble the experimental set-up like Crystal Lake and Crystal
Bog) suggests that in the LRL experiment the ceteris paribus clause holds.
While these other lines of evidence are not replicates nor randomized in a
strict sense of those terms, they do provide for a qualitative argument from
error insofar as it seems unlikely that the differences noted at LRL were due
to something unique (to the ceteris paribus condition not holding) in either
basin. The strength of this qualitative argument against both experimenter
bias and broken ceteris paribus conditions lies in the independence of these
multiple lines of evidence. The ceteris paribus conditions at each site are
surely different and independent of one another and when the results are
compared should to some extent cancel out. While purely qualitative, this
line of reasoning seems in tune with Fisher's injunction that randomization
works to the extent that it "guarantees that, on the average, 'errors' are independently distributed" (Hurlbert, 1984, 192, citing Fisher, 1926, 506). Neither replication nor randomization solves all problems with making causal
inferences.

For example, in controlled, replicated, and randomized laboratory studies acidification was shown to be detrimental to rotifer species under a variety of light and food conditions. Predation was not a factor in the laboratory tests and hence falls under its ceteris paribus clause. This evidence in
light of the evidence from LRL and other ecosystem experiments and ob-

servations led to the suggestion that the flourishing of *K. taurocephala* was due to decreased predation—an indirect cause of acidification and similar to the concerns about ceteris paribus clauses discussed in our chapter and this rejoinder over an ability to make causal inferences given the complexity of real ecosystems. We suggested that the decreased predation hypothesis itself needs to be tested as it was not the hypothesis LRL was designed to test and hence no account of the ways in which data from LRL would be unreliable for testing this hypothesis had been probed (i.e., it has not been severely tested).

The point we have tried to make in our chapter and that I hope I have made here is not that using other methods and arguments in place of or to approximate proper replication and randomization in some way makes it valid to apply tests of significance to the LRL data. It does not; however, in trying to use standard tests, (e.g., significance tests) where assumptions like replication and randomization are explicit can alert investigators to potential errors in their design. The point is, with or without significance tests, these errors from violating replication and randomization remain unless other means can be found to detect and circumvent them. We have tried to show that by understanding the properties of these methodological rules for detecting and controlling specific types of errors, we can better design experiments and model and interpret data using other methods and arguments both quantitative and qualitative in an attempt to approximate them and hence increase the reliability of whole ecosystem experiments for making inferences. Moreover, Mayo's piecemeal approach to error that we have been championing also highlights and suggests the ways in which the growth of experimental knowledge (e.g., effects of scale) can be developed for understanding complex interactions/manipulations in ecology. The types of arguments we discussed are similar to the more rigorous statistically valid causal arguments used in small scale experiments in that their focus is on detecting and curtailing the effects of procedural errors.

A not uncommon claim in the pseudoreplication literature is that causal arguments in ecology will have to rely on ecological rather than statistical arguments, however, the term 'ecological arguments' is rather vague and could mean anything from speculative theory and arguments to the best explanation to conformity with limitations and mechanisms in other sciences like physics, chemistry, and biology. In our chapter and here, ecological arguments are interpreted as qualitative (and quantitative) experimental arguments from error. If, as Link pointed out, we don't want another Dewey-Truman disaster, then the reliability of the procedures used to generate data for testing a hypothesis must be appraised honestly and uncertainties duly noted. Mayo's piecemeal approach to severe testing and arguing from error gets to the heart of the problem—how to assess whether or not the processes that were used to generate and model the data are reliable for the inference being made. Perhaps the greatest strength of Mayo's approach for ecological phenomena lies in its qualitative aspects, her informal argument from error, which provides guidelines for designing and assessing similarity arguments, arguments from the independence of lines of evidence, the role of experimental assumptions, etc. based on an ability to detect and control for specific types of errors for making particular types of inferences using, often very complicated experimental contexts. Decisions and policies need to be made, often immediately (e.g., controlling acid rain). For making inferences about real ecosystems especially to inform environmental policies multiple types of experimental arguments are needed at all scales of manipulation to rule out the errors at any one scale.

REFERENCES

Box, G. E. P. 1976. Science and Statistics. *J. Am. Stat. Assn* 71:791–799.

Box, G. E. P, W. G. Hunter, and J. S. Hunter. 1978. *Statistics for Experimenters.* New York: Wiley.

Box, G. E. P., and G. M. Jenkins. 1976. *Time Series Analysis: Forecasting and Control.* Rev. ed. San Francisco: Holden-Day.

Box, G. E. P., and G. Tiao. 1975. Intervention Analysis with Applications to Economic and Environmental Problems. *J. Am. Stat. Assn* 70:70–79.

Brezonik, P. L., J. G. Eaton, T. M. Frost, P. J. Garrison, T. K. Kratz, C. E. Mach, J. H. McCormick, J. A. Perry, W. A. Rose, C. J. Sampson, B. C. L. Shelley, W. A. Swenson, and K. E. Webster. 1993. Experimental Acidification of Little Rock Lake, Wisconsin: Chemical and Biological Changes over the pH Range 6.1 to 4.7. *Can. J. Fish. Aquat. Sci.* 50:1101–1121.

Burnham, K. P, and D. R. Anderson. 1998. *Model Selection and Inference: A Practical Information-Theoretic Approach.* New York: Springer.

Campbell, D. T., and J. Stanley. 1981. *Experimental and Quasi-experimental Designs for Research.* Boston: Houghton Mifflin.

Carpenter, S. R. 1999. Microcosm Experiments Have Limited Relevance for Community and Ecosystem Ecology: Reply. *Ecology* 80:1085–1088.

Carpenter, S. R., S. W. Chisholm, C. J. Krebs, D. W. Schindler, and R. F. Wright. 1995. Ecosystem Experiments. *Science* 269:324–327.

Carpenter, S. R., J. J. Cole, T. E. Essington, J. R. Hodgson, J. N. Houser, J. F. Kitchell, and M. L. Pace. 1998. Evaluating Alternative Explanations in Ecosystem Experiments. *Ecosystems* 1:335–344.

Carpenter, S. R., T. M. Frost, D. Heisey, and T. K. Kratz. 1989. Randomized Intervention Analysis and the Interpretation of Whole-Ecosystem Experiments. *Ecology* 70:1142–1152.

Carpenter, S. R., and D. W. Schindler. 1998. Workshop on Ecosystem Manipulation. *Ecosystems* 1:321–322.

Collins, H. 1985. *Changing Order: Replication and Induction in Scientific Practice.* London: Sage.

Dawid, A. P. 2000. Causal Inference without Counterfactuals. *J. Am. Stat. Assn* 95: 407–424.

Dennis, B. 1996. Discussion: Should Ecologists Become Bayesians? *Ecol. Appl.* 6: 1095–1103.

Dennis, B. 2004. Statistics and the Scientific Method in Ecology. Chapter 11 in Taper, M. L., and S. R. Lele, eds., *The Nature of Scientific Evidence: Statistical, Philosophical, and Empirical Considerations.* Chicago: University of Chicago Press.

Diamond, J. 1997. *Guns, Germs, and Steel: The Fates of Human Societies.* New York: Norton.

Drenner, R. W., and A. Mazumder. 1999. Microcosm Experiments Have Limited Relevance for Community and Ecosystem Ecology: Comment. *Ecology* 80:1081–1085.

Edwards, D. 1998. Issues and Themes for Natural Resources Trend and Change Detection. *Ecol. Appl.* 8:323–325.

Efron, B. 1979. Bootstrap Methods: Another Look at the Jackknife. *Ann. Stat.* 7:1–26.

Efron, B., and G. Gong. 1983. A Leisurely Look at the Bootstrap, the Jackknife, and Cross-Validation. *Am. Stat.* 37:36–48.

Fisher, R. A. 1926. The Arrangement of Field Experiments. *J. Ministry Agric. (London)* 33:503–513

Fisher, R. A. 1947. *The Design of Experiments.* 4th ed. London: Oliver and Boyd.

Fisher, R. A. 1971. *The Design of Experiments.* 9th ed. New York: Hafner.

Fisher, S. G. 1997. Creativity, Idea Generation, and the Functional Morphology of Streams. *J. N. Am. Bentholog. Soc.* 16:305–318.

Franklin, A. 1986. *The Neglect of Experiment.* Cambridge: Cambridge University Press.

Freedman, D., R. Piscani, R. Purves, and A. Adhikari. 1991. *Statistics.* 2nd ed. New York: Norton.

Frost, T. M., D. L. DeAngelis, S. M. Bartell, D. J. Hall, and S. H. Hurlbert. 1988. Scale in the Design and Interpretation of Aquatic Community Research. In Carpenter, S. R., ed., *Complex Interactions in Lake Communities.* New York: Springer.

Frost, T. M., S. R. Carpenter, A. R. Ives, and T. K. Kratz. 1995. Species Compensation and Complementarity in Ecosystem Function. In Jones, C. G., and J. H. Lawton, eds., *Linking Species and Ecosystems.* New York: Chapman and Hall.

Frost, T. M., P. K. Montz, and T. K. Kratz. 1998. Zooplankton Community Responses during Recovery from Acidification: Limited Persistence by Acid-Favored Species in Little Rock Lake, Wisconsin. *Restor. Ecol.* 6:336–342.

Frost, T. M., P. K. Montz, T. K. Kratz, T. Badillo, P. L. Brezonik, M. J. Gonzalez, R. G. Rada, C. J. Watras, K. E. Webster, J. G. Wiener, C. E. Williamson, and D. P. Morris.

1999. Multiple Stresses from a Single Agent: Diverse Responses to the Experimental Acidification of Little Rock Lake, Wisconsin. *Limnol. Oceanogr.* 44: 784–794.

Galison, P. 1987. *How Experiments End.* Chicago: University of Chicago Press.

Gelman, A. B., J. B. Carlin, H. S. Stern, and D. B. Rubin. 1998. *Bayesian Data Analysis.* Washington: Chapman and Hall.

Gilinsky, N. L. 1991. Bootstrapping and the Fossil Record. In Gilinsky, N. L., and P. W. Signor, eds., *Analytical Paleobiology.* Knoxville, TN: Paleontological Society.

Gonzalez, M. J., and T. M. Frost. 1994. Comparisons of Laboratory Bioassays and a Whole-Lake Experiment: Rotifer Responses to Experimental Acidification. *Ecol. Appl.* 4:69–80.

Gottelli, J. J. and G. R. Graves. 1996. Null Models in Ecology. Washington: Smithsonian Institution.

Greenland, S., J. M. Robins, and J. Pearl. 1999. Confounding and Collapsibility in Causal Inference. *Stat. Sci.* 14:29–46.

Hacking I. 1983. *Representing and Intervening.* Cambridge: Cambridge University Press.

Hairston, N. G., Sr. 1989. *Ecological Experiments: Purpose, Design, and Execution.* Cambridge: Cambridge University Press.

Hilborn, R., and M. Mangel. 1997. *The Ecological Detective: Confronting Models with Data.* Princeton: Princeton University Press.

Hubbard, A., and Gilinsky, N. 1992. Mass Extinctions as Statistical Phenomena: An Examination of the Evidence Using Chi-Square Tests and Bootstrapping. *Paleobiology* 18(2): 148–160.

Hurlbert, S. H. 1984. Pseudoreplication and the Design of Ecological Field Experiments. *Ecol. Monographs* 54:187–211.

Huston, M. A. 1999. Microcosm Experiments Have Limited Relevance for Community and Ecosystem Ecology: Synthesis and Comments. *Ecology* 80:1088–1089.

James, F. C., and C. E. McCulloch. 1994. How to Strengthen Inferences about Causes of Trends in Bird Populations. *Bull. Ecol. Soc. Am.,* vol. 75, no. 2, part 2.

Kish, L. 1970. Some Statistical Problems in Research Design. In Morrison, D., and R. Henkel, eds., *The Significance Test Controversy.* Chicago: Aldine.

Likens, G. E. 1992. *The Ecosystem Approach: Its Use and Abuse.* Oldendorf/Luhe, Germany: Ecology Institute.

Maurer, B. A. 1998. Ecological Science and Statistical Paradigms: At the Threshold. *Science* 279:502–503.

Maurer, B. A. 1999. *Untangling Ecological Complexity: The Macroscopic Perspective.* Chicago: The University of Chicago Press.

Mayo, D. G. 1996. *Error and the Growth of Experimental Knowledge.* Chicago: University of Chicago Press.

Mayo, D. G. 2004a. An Error-Statistical Philosophy of Evidence. Chapter 4 in Taper, M. L., and S. R. Lele, eds., *The Nature of Scientific Evidence: Statistical, Philosophical, and Empirical Considerations.* Chicago: University of Chicago Press.

Mayo, D. G. 2004b. Rejoinder. Chapter 4.3 in Taper, M. L., and S. R. Lele, eds., *The Nature of Scientific Evidence: Statistical, Philosophical, and Empirical Considerations.* Chicago: University of Chicago Press.

Miller, J. A. 1997. Enlightenment Error and Experiment: Henry Cavendish's Electrical Researches. Master's thesis, Virginia Tech.

Neyman, J., and E. S. Pearson. 1966. *Joint Statistical Papers of J. Neyman and E. S. Pearson.* Berkeley: University of California Press.

Nichols, J. D. 1991. Science, Population Ecology, and the Management of the American Black Duck. *J. Wildlife Management* 55:790–799.

Pearl, J. 2000. Comment on Dawid. *J. Am. Stat. Assn* 95:428–431.

Petersen, J. E., C.-C. Chen, and W. M. Kemp. 1997. Scaling Aquatic Primary Productivity Experiments under Nutrient- and Light-Limited Conditions. *Ecology* 78: 2326–2338.

Rasmussen, P. W., D. M. Heisey, E. B. Nordheim, and T. M. Frost. 1993. Time-Series Intervention Analysis: Unreplicated Large-Scale Experiments. In Scheiner, S. M., and J. Gurevitch, eds., *Design and Analysis of Ecological Experiments.* New York: Chapman and Hall.

Reckhow, K. H. 1990. Bayesian Inference in Non-replicated Ecological Studies. *Ecology* 71:2053–2059.

Resetarits, W. J., and J. Bernardo, eds. 1998. *Experimental Ecology: Issues and Perspectives.* Oxford: Oxford University Press.

Robins, J. M., and S. Greenland. 2000. Comments on Dawid. *J. Am. Stat. Assn* 95: 431–435.

Romesburg, H. C. 1981. Wildlife Science: Gaining Reliable Knowledge. *J. Wildlife Mgmt.* 45:293–313.

Rosenbaum, P. R. 1995. *Observational Studies.* New York: Springer.

Rubin, D. B. 2000. Comments on Dawid. *J. Am. Stat. Assn* 95:435–438.

Sampson, C. J., P. L. Brezonik, T. M. Frost, K. E. Webster, and T. D. Simonson. 1995. Experimental Acidification of Little Rock Lake, Wisconsin: The First Four Years of Chemical and Biological Recovery. *Water Air Soil Poll.* 85:1713–1719.

Schindler, D. W. 1990. Experimental Perturbations of Whole Lakes as Tests of Hypotheses Concerning Ecosystem Structure and Function. *Oikos* 57:25–41.

Schindler, D. W. 1998. Replication versus Realism: The Need for Ecosystem-Scale Experiments. *Ecosystems* 1:3231–3334.

Schindler, D. W., K. H. Mills, D. F. Malley, D. L. Findlay, J. A. Shearer, I. J. Davies, M. A. Turner, G. A. Linsey, and D. R. Cruikshank. 1985. Long-Term Ecosystem Stress: The Effects of Years of Experimental Acidification on a Small Lake. *Science* 228:1395–1401.

Shafer, G. 2000. Comments on Dawid. *J. Am. Stat. Assn* 95:438–442.

Signor, P., and N. Gilinsky. 1991. Introduction to Gilinsky, N. L., and P. W. Signor, eds., *Analytical Paleobiology.* Knoxville, TN: Paleontological Society.

Spanos, A. 1986. *Statistical Foundations of Econometric Modeling.* Cambridge: Cambridge University Press.

Spanos, A. 1999. *Probability Theory and Statistical Inference.* Cambridge: Cambridge University Press.

Stewart-Oaten, A., J. R. Bence, and C. W. Osenberg. 1992. Assessing Effects of Unreplicated Perturbations: No Simple Solutions. *Ecology* 73 : 1396–1404.

Stewart-Oaten, A., W. W. Murdoch, and K. R. Parker. 1986. Environmental Impact Assessment: "Pseudoreplication" in Time? *Ecology* 67 : 929–940.

Wasserman, L. 2000. Comments on Dawid. *J. Am. Stat. Assn* 95 : 442–443.

Watras, C. J., and T. M. Frost. 1989. Little Rock Lake (Wisconsin): Perspectives on an Experimental Ecosystem Approach to Seepage Lake Acidification. *Arch. Environ. Contam. Toxicol.* 18 : 157–165.

Wilbur, H. M. 1997. Experimental Ecology of Food Webs: Complex Systems in Temporary Ponds. *Ecology* 78 : 2279–2302.

Woodward, J. 1989. Data and Phenomena. *Synthese* 79 : 393–472.

Woodward, J. 1997. Explanation, Invariance, and Intervention. *Phil. Sci.* 64 (proceedings): S26–S41.

Woodward, J., and J. Bogen. 1988. Saving the Phenomena. *Phil. Rev.* 97(3): 303–352.

9 Dynamical Models as Paths to Evidence in Ecology

Mark L. Taper and Subhash R. Lele

ABSTRACT

As scientists we are interested in understanding how nature works. Many statistical techniques and models concentrate on association rather than causation. It is important that we move from exploration and description to explanation. Dynamical models are useful for incorporating explicitly causal pathways in the statistical models. Consequently, dynamical models help in the design of experiments to differentiate among causal pathways. Dynamical models often generate unexpected predictions that can lead to very strong inferences on the underlying processes of nature.

MOVING FROM EXPLORATION TO EXPLANATION

Natural scientists like to understand how nature works. Usually this quest begins with exploration of empirical patterns in nature (Maurer, 2004 [chapter 2, this volume]). This search may involve a visual exploration of dependencies of variables using computer programs and high-speed graphics that allow rotation of data in three dimensions and sometimes animation to simulate the fourth dimension. The associations between variables that are seen can be tested for statistical significance using various techniques such as the Monte Carlo randomization procedures (Manly, 1997). These associations, although suggestive, do not necessarily reveal causality. Most of the statisti-

This paper is based on a presentation given by Ted J. Case at the first meeting of the "Evidence Project" working group at the National Center for Ecological Analysis and Synthesis in the fall of 1999. We have benefited from many further interactions with Dr. Case on the subject, and we thank him for sharing views.

cal applications in science deal with the certification that the observed pattern has not arisen due to pure random processes. For example, testing for complete spatial randomness in an observed point pattern is a common theme in many spatial analyses in the ecological literature. This exorcising of causality from statistical thinking (and hence its application in science) is a feature that runs deep in the discipline of statistics. Pearson (1911) forcefully argued for focusing statistics on correlation rather than on causation: "Beyond such discarded fundamentals as 'matter' and 'force' lies still another fetish amidst the inscrutable arcana of modern science, namely, the category of cause and effect" (Pearson, 1911).

While observed associations may not *prove* anything about causality, they may *suggest* mechanistic hypotheses that might explain the observed patterns. To move from exploration to explanation it is often useful to develop a dynamical model.

DYNAMICAL MODELS

A dynamical model is simply a model concerned with change. Elucidating causation requires a dynamical model at a very fundamental level. An agent of some kind influences a situation and causes a change. This is a dynamical description. Even modern statistics is reembracing, if somewhat cautiously, its ability to address causation beyond the bounds of randomized experiments (Rubin, 1991; Stone, 1993; Pearl, 1995; Barnard et al., 1998; Freedman, 1999; Greenland, Robins, and Pearl, 1999). At the center of this modern approach are explicit dynamical models of cause and effect based on an acceptance of Hume's counterfactual definition of causation (Hume, [1748] 1999; but see Dawid, 2000). The construction of a dynamical model hones our intuition about the possible mechanisms that could produce the pattern and helps us determine if, at least in principle, the pattern is logically consistent with the suspected processes.

A dynamical model usually begins with certain axioms and assumptions that are safely taken for granted. It also abstracts nature by explicitly or implicitly deciding to ignore some variables and processes that are thought to be irrelevant to the problem at hand. Finally, these assumptions are combined into deductions about nature by the application of some logical machinery. In ecological systems there are typically many potential interacting variables influencing the phenomenon of interest. To ensure that our logic does not go astray, we require a structured formal logical system. This is usually mathematics, where it may take the form of differential equations, difference equations, or partial differential equations as descriptions of these processes in ecology. However, a dynamical model may also be a computer simulation model, a block diagram, a physical model, or even a verbal model. A dynamical model is not necessarily a deterministic model. One can incorporate environmental noise and randomness in deterministic dynamical models by turning them into stochastic models. For example, many population dynamics models, such as the logistic growth model, started as deterministic differential equation models. However, they have been modified to incorporate randomness and turned into stochastic differential equations. Similar modifications have been made to species interaction models such as the famous Lotka-Volterra equations. A dynamical model, when formed as the basis for statistical models, helps refine our thinking about the forces acting on the system, their underlying mechanisms, the time course of the response, and the critical variables to measure. In this sense, a dynamical model arises *in advance of* a statistical model. In the following, we explore the use of dynamical models, deterministic or stochastic, as paths to evidence in ecology. The main idea is that dynamical models are more likely to lead to the understanding of causation than simple statistical association models.

STUDYING THE SYSTEM OF COMPETING SPECIES USING DYNAMICAL MODELS

When trying to learn about causes in nature, it is usually beneficial to construct models that represent biological mechanisms. The natural world is complex and a model must reduce this complexity by focusing on the essential components that are minimally necessary and sufficient to understand the problem at hand. Yet even with this abstraction, the model may include many variables that interact to produce surprising and unanticipated consequences.

With many interacting species, the net result of the removal of one species on the abundances of the others can be very difficult to predict without a careful mathematical model. Indirect effects in community ecology refer to effects between two species mediated wholly in terms of changing population densities of intermediary species (Holt, 1977; Bender, Case, and Gilpin, 1984; Roughgarden and Diamond, 1986; Schoener, 1993). Indirect effects are also referred to as "interaction chains" (Wootton, 1993). Indirect effects pass from one species to another via the density changes in one or more intermediary species in the food chain. A colorful example of an indirect effect was offered by Charles Darwin (1859) in the *Origin of Species:*

Humblebees [bumblebees] alone visit red clover (Trifolium pratense), as other bees cannot reach their nectar ... The number of humblebees in any district depends in a great degree on the number of field-mice, which destroy their combs and nests ... Now the number of mice is largely dependent, as every one knows, on the number of cats ... Hence it is quite credible that the presence of a feline animal in large numbers in a district might determine, through the intervention first of mice and then of bees, the frequency of certain flowers in that district! (Darwin, 1859, 74)

The community interaction that Darwin describes involves a single food chain and the indirect positive effect of cats on bumblebees is intuitive. When food webs contain multiple species at each trophic level, however, the indirect effects may be somewhat counterintuitive. For example, consider the simplest possible case of just three resource competitors with the idealized niches shown in the top of figure 9.1. The important feature of this sketch is that two pairs of competitors (species 1 and 2, and species 2 and 3) have large overlaps in resource use and thus compete strongly while the other pair (species 1 and 3) overlaps less and thus competes more weakly.

The biology implicit in figure 9.1 can be represented by a system of dynamical equations called the Lotka-Volterra competition equations:

$$\frac{dN_1}{dt} = \frac{r_1 N_1}{K_1}(K_1 - N_1 - \alpha_{12}N_2 - \alpha_{13}N_3)$$

$$\frac{dN_2}{dt} = \frac{r_2 N_2}{K_2}(K_2 - \alpha_{21}N_1 - N_2 - \alpha_{23}N_3)$$

$$\frac{dN_3}{dt} = \frac{r_3 N_3}{K_3}(K_3 - \alpha_{31}N_1 - \alpha_{32}N_2 - N_3),$$

where N_i, K_i, and r_i are the population size, the ecological carrying capacity, and intrinsic growth rate of species i, and α_{ij} is the coefficient of competition of species j on species i.

An α matrix (matrix of competition coefficients) consistent with figure 9.1 is:

$$\alpha = \begin{bmatrix} 1 & 0.5 & 0.063 \\ 0.5 & 1 & 0.5 \\ 0.063 & 0.5 & 1 \end{bmatrix}.$$

If all three species have the same $Ks = 100$, the equilibrium densities are $N_1^* = N_3^* = 88.81$ and $N_2^* = 11.19$.

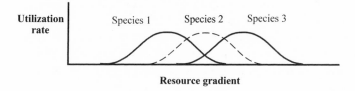

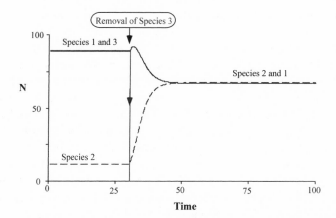

FIGURE 9.1 Top: Resource utilization curves for three hypothetical competitors. The overlap between curves is roughly proportional to interspecific interaction strength in the Lotka-Volterra competition equations. Bottom: A simulation of the three-competitor model. Species 3 is removed at time 30. This results in a lower equilibrium density for species 1 (at 66.7) as well as a higher equilibrium density for species 2 (also at 66.7).

What is the effect of removing species 3 on the equilibrium densities of species 1 and 2? The removal of species 3 will result in a decrease in species 1's equilibrium density and an increase in that of species 2. This is because in the context of all three species, the indirect effect of species 3 on species 1 is beneficial because species 3 depresses species 2, which is a stronger competitor with species 1 than is 3. Such interactions are represented in human terms in the saying "my enemy's enemy is my ally." A numerical simulation verifies this expectation (figure 9.1, bottom). If the species follow the Lotka-Volterra competition equations, the exact response can be predicted by simply examining the sign of term in the inverse of the α matrix (Bender, Case, and Gilpin, 1984; Case, 1999).

Notice that the immediate response of species 1, upon removal of species 3, was an increase in numbers. The rate of this increase is proportional to the direct effect of the species 3 on 1, which is $\alpha_{13} = 0.0625$). However, through the indirect effect of species 3 on species 2, species 1's trajectory turns around and ultimately reaches a lower equilibrium than before the removal of species 3. Thus, by tracking actual population dynamics and by comparing with the dynamical model, a greater understanding of the working of the community can be gained than by simply inspecting end points.

In short, our simple expectations for species removal experiments based on two-species systems can lead us astray when three or more species are present because the indirect effects get mingled with the direct effects. This is not an isolated example. Yodzis (1988) found that reversals in sign between the direct and total interactions for species pairs were the rule rather than the exception in multispecies models of communities. This means that we have to be cautious in interpreting the results of perturbation experiments (Bender, Case, and Gilpin 1984). As an example, consider a meta-analysis geared to summarizing published results from perturbation experiments with the expectation that interspecific competition is revealed by an increase in density upon removal of the competitor (e.g., Gurevitch et al., 1992). This expectation ignores the important issue of potential indirect effects. In the example above, the removal of species 3 causes a decrease in the density of species 1, even though they are direct competitors. No amount of additional replication would make this result disappear. This is not a problem of experimental noise but rather the lack of a logical distinction between direct and indirect effects. Study of the dynamical model helps us make this important distinction and realize that it is possible to separate the direct and indirect effects based on their differential temporal patterns of response. A species in the Lotka-Volterra model always shows an immediate increase in population density following the removal of one of its direct competitors. This initial increase may be followed by either an increase or decrease from the starting level depending on the indirect effects. Any meta-analysis dealing with the effects of species removal on remaining species densities should at a minimum distinguish between short-term results and long-term results; otherwise no matter how much data is brought to bear on the issue, the interpretation of the results will be muddled. Thus, it is the dynamical model that gives the critical insights to make the necessary distinctions and thus helps us design appropriate experiments. A simple statistical association approach in such a situation will lead us astray. For example, two-dimensional marginal distributions of the species densities do not provide enough information about the three-dimensional dynamical system (see Brown et al., 2004 [chapter 10, this volume] for a similar discussion).

DYNAMICAL MODELS AS GUIDES TO EXPERIMENTAL DESIGNS

An important function of dynamical models in the unveiling of nature is their potential guidance in designing good experiments. Examination of dynamical systems models of community interaction, such as those discussed in the previous section, led to the realization that two kinds of ecological experiments are necessary to understand community interactions. These were termed "pulse" and "press" experiments by Bender, Case, and Gilpin (1984). In pulse experiments, observations are made of the short-term dynamical response of the species in a community to a perturbation, while in press experiments, observations are made of the new species equilibria that result when a perturbation is maintained indefinitely.

As a more detailed example of the impact of dynamical models on the design of ecological experiments, consider the scientific question of whether two predators feeding on a single prey species interact, modifying their foraging behavior for success. To answer the scientific question, we must ask a statistical question.

In one set of exclosures in nature, we exclude predator species 1; in another set, we exclude predator species 2; in still another treatment, we exclude both predators; and finally, we set up controls where both predators are allowed access to the prey (e.g., Wootton, 1993). We'd like to know whether, on the one hand, the combined predation rate when both predators are allowed access to prey can be modeled as additive effects based on the results from single-predator experiments so that their combined effects are substitutable to some extent, or whether, on the other hand, predators interact in some way so that their combined effects are supra-additive or multiplicative. However, testing the additivity hypothesis requires some knowledge about the dynamics of the individual predator-prey interactions. To make the point, we will contrast two overly simplistic hypotheses for these dynamics. In both models, we assume that during the time frame of the experiment, there is no prey recruitment or movement into or out of exclosures, so that prey numbers are only changed by mortality. We also assume that predator numbers are uncontrolled and unknown.

In the first model, each prey individual has some fixed probability d_i of being eaten by all individuals of predator species i per unit time, so the change in numbers of prey (N) follows an exponential decay. The loss of prey to predator 1 alone is modeled as

$$\frac{dN}{dt} = -d_1 N.$$

Under the assumption that the two predators act independently (i.e., with no interaction modification), the combined death rate of the prey when both predators are allowed access to prey is

$$\frac{dN}{dt} = -d_1 N - d_2 N. \tag{1a}$$

The exclosures are initially stocked with N_0 prey and the remaining prey are measured at time T. By solving the differential equation (1a), we have the expectation under the null hypothesis that the numbers of prey are

$$N(T) = N_0 \exp(-T(d_1 + d_2)). \tag{1b}$$

The death rates d_1 and d_2 may be determined over the same time interval in the single-predator experiments, providing a basis for a statistical test for the additivity of predator effects.

Now, contrast this with an alternative dynamical model in which a constant number of prey are consumed per predator per unit time, regardless of prey numbers. This might be the case if the number of predators is small, and they are easily satiated by a single meal. Consequently, regardless of prey numbers, a fixed number are eaten each day, for this model, the prey death rate would not follow (1a) but rather

$$\frac{dN}{dt} = -d_1 - d_2. \tag{2a}$$

At the close of the experiment at time T, the expectation under the null hypothesis for this dynamical model is that prey numbers at time T are

$$N(T) = N_0 - T(d_1 + d_2), \tag{2b}$$

where again d_1 and d_2 are also determined over the same time interval T in the single-predator experiments and then these rates are used to test the assumption of the additivity of the two species' effects as modeled in 2b. The two models make very different predictions about the time course of prey depletion. Under model 1b, prey population size declines exponentially with time, while under model 2b, population decline is linear.

Now here's the rub. Without a priori knowledge of which model is being followed (1a or 2a), an interaction modification, as detected by the application of test 1b, may not be one as detected by 2b, and vice-versa. Without knowing whether the prey dynamics actually follows 1a or 2a or some other unknown function, we cannot conclude much at all about the presence or absence of interaction modifications. To experimentally get this required information requires that prey survival be determined at more frequent intervals in the single-species experiments. This will allow one to reveal the actual time course of numerical change in prey numbers and to discern between models 1a, 2a, and other possibilities. Thus, dynamical models can be used to determine a priori what experimental information will be most useful in distinguishing hypotheses.

The number of models that could potentially be used to describe a set of interacting species may be large, but usually only a few of these models will be biologically reasonable (but see Taper, 2004 [chapter 15, this volume], for a discussion of when this is not the case). The "reasonableness" of a model can assessed using two criteria: first, commonsense judgment and prior knowledge about the species and their population dynamics; and second, how well the suspected single-species and two-species models fit the observations in those experiments compared to alternative models. For example, consider a system in which species interact such that their per capita growth rates are affected in an additive manner (such as in equation 1a), but which are being modeled as if these effects were multiplicative. This multiplicative model will simply yield a poor fit to the changing species densities in the two-species experiments. The investigator will find a better fit by adopting the additive model. Hopefully, this test of alternative models will yield a clear winner so that it can be used as the null model for systems with more than two species in a test for interaction modifications. The questions of what "fit" is (see Lindsay, 2004 [chapter 14, this volume]), how to best quantify it, and how to relate it to evidence in science (Lele, 2004 [chapter 7, this volume]) are thorny and important issues, and they occupy the bulk of this book. However, if one's goal is scientific understanding, one must never forget that it is the models under comparison that are preeminent.

Some of the most potent models in ecology are those that develop appropriate statistical tests directly from explicit assumptions about the underlying dynamical processes and assumptions about reasonable bounds on critical parameters. Randomness is an integral component of all real processes. This includes both the biological processes we wish to understand and the observations by which we obtain data. A vital step in forging the link between theory and data is the explicit incorporation of stochastic elements into the very structure of dynamical models as well as into the description of the observation process (Ives, 2000). Recent studies that analyze temporal and spatial variation in (unmanipulated) population density from long-term census data include Cushing et al. (1998), Dennis, Kemp, and Taper (1998), and Lele, Taper, and Gage (1998). While alternative models and alternative parameter combinations can always yield conflicting explanations for the same data (or, in statistical jargon, yield models that are non-identifiable), at least the model assumptions and parameters that are most pivotal will be explicit. This can be followed by new experiments tailored to resolve processes and decide among competing explanations.

In summary, linking data gathering to a dynamical model forces the ecologist to specify the dynamics that are being assumed and aids in the planning of experimental designs that allow the development of statistically valid tests of those model assumptions. To show that the dynamical model is consistent with the observed dynamics generally will require information on the time course of population growth, not just two snapshots of initial numbers and final numbers at the end of the experiment. Dynamical models often raise predictions not originally considered in the construction of the model. These new predictions may suggest new experiments and observations. In this way, dynamical models serve the useful function of statistically linking data to theory in a consistent manner. Hand (1995) chides the statistical community for being too caught up with the models and the data and not concerned enough with the "problem." What is clear from this discussion is that *concern for process-based dynamical models is concern for the scientific "problem."*

DYNAMICAL MODELS AND UNANTICIPATED CONNECTIONS

As mentioned above, a useful feature of dynamical models is that they often suggest unanticipated connections between variables or new predictions about the phenomenon under study—connections that were not the original focus of the model. These serendipitous connections can suggest new features of nature to explore and measure and thus provide additional ways of testing the model and discerning among competing explanations. Tests of unanticipated predictions represent strong support for the processes embodied in a model. William Whewell, in his important treatise on philosophy of science (1858), placed "consilience of inductions" above goodness of fit and predictions of facts of the same kind in his hierarchy of confirmation. Thus, the fit of observational or experimental data to these serendipitous predictions is in fact a stronger support than a model's mere fit to the data that inspired it or even the support gained when the model fits new observations of the original type well (see also Maurer, 2004 [chapter 2, this volume], and Scheiner, 2004 [chapter 3, this volume]).

As an example, consider experimental tests for "higher-order interactions" or "interaction modifications" in ecology. These occur when the presence of one species modifies the strength of interaction between two other species. One particularly interesting type of interaction modification is where one species' mere presence "intimidates" others, such that they alter their behavior and thus the strength of their interaction (Beckerman, Uriarte, and Schmitz, 1997). These are typically rapid changes in the interactions between species—*not* mediated by numerical changes in population size of intervening species in the food chain (as with indirect effects), but simply due to changes in the behavioral reaction of a species due to another species. For example, Peacor and Werner (1997) found that the presence of predacious dragonfly larvae, even if caged and restricted from eating prey, caused small tadpoles to reduce their foraging effort and that this in turn led to a reduction in the predation rate of these tadpoles by another predator. Similarly, Beckerman, Uriarte, and Schmitz (1997) glued shut the jaws of spiders so that they could not kill or consume grasshopper prey. Nevertheless, the mere presence of such spiders reduced grasshopper foraging activity, leading to decreased rates of exploitation of grass by the grasshoppers. A dynamical model may be able to incorporate such mechanisms much more easily and realistically than simple statistical association models.

SUMMARY REMARKS

The goal of a modeler is to provide a mechanism sufficient to reproduce the pattern of interest. One way that a dynamical model may fail is by not capturing or reproducing the essential features of nature that it attempts to explain. Of course, even if the model can reproduce the natural pattern, this does not necessarily imply that the forces being modeled are the forces producing the pattern in nature. You may have simply found a happy coincidence. You may be right but for the wrong reasons (Dayton, 1973). Thus, the second way a model may fail is by reproducing the result, but only by invoking mechanisms that either are not operating in nature or are operating but overridden by more potent forces. A process is assumed to act in some simple way, and it does not. Also possibly, the assumptions may individually be correct but additional complicating factors may be operating that have been neglected in the model's structure, and these factors are sufficiently important to override the behavior of the system. The art of model building is then to distill nature down to its essential features, adding no more complexity than is needed to explain the phenomenon of interest, to discern among alternative explanations, and to not include any process that is contradicted by known facts.

One of the central points of this volume is that we learn about nature by selecting among descriptions of nature (models) based on evidence for those models in data. Clearly, our inferences can be no more interesting, no better, and no more subtle than our models. This book focuses on evidential procedures for comparing models; as a consequence, most of the models described here are simple so as not to cloud the issues. But it is important to recall that the generation and analysis of models is a critical step in the scientific process. If we compare trivial models our science will be trivial.

9.1   Commentary

Steven Hecht Orzack

Understanding ecological process is the central focus of this paper. It is the claim of Taper and Lele that such understanding can come only (or at least most easily) when one employs dynamical models. They define such models as those that are "concerned with change." The contrast they make is between such models and statistical models, those that simply describe associations between all or most variables. This is an important distinction, and it is easy to agree with Taper and Lele that dynamical models can be an excellent means to elucidate the truth. After all, we do not want to simply *describe*, say, species abundance or distribution patterns; instead, we seek to *explain* these features of ecosystems and their changes by analysis of the biological essentials of the situation.

To this extent, the contrast Taper and Lele draw is of use; however, it is of little use by itself. The reason is that any attempt to understand, say, a population's dynamics must involve dynamical *and* statistical thinking, and the most difficult issue is deciding what mix of the two to employ. What I mean by this can be illustrated by consideration of a specific model, such as the stochastic analysis of extinction dynamics described by Lande and

I thank Greg Dwyer, Matthew Leibold, and Cathy Pfister for comments and advice. This work was partially supported by NSF awards SES-9906997 and EIA-0220154.

Orzack (1988). This model allows one to make short-term and long-term predictions about the extinction behavior of a population as determined by its life history. The use of the word "its" reflects the issue at hand. This model ignores genetic differences among individuals in regard to life history traits; yet these may affect population dynamics in many instances. The same oversight is a feature of many ecological models, another example being the stochastic analysis of reproductive effort evolution by Orzack and Tuljapurkar (2001). To this extent, these and many other models are nondynamical because they rely upon a point value for any given vital rate of the life history even though we know that there may often be some temporally changing distribution of values within the population.

For this reason, these models are statistical because as Levins (1966, 427–28) stated when discussing his view of the generic structure of mathematical models in population biology, "many terms enter into consideration only by way of a reduced number of higher-level entities." This means that at a certain level we do not seek an explanation; instead, we rely upon static description. Yet these models are also dynamical because we seek to *explain population dynamics* as a function of a temporally changing environment. These two models illustrate a general point: all dynamical models are a mix of the statistical and the dynamical. (However, one can create a purely statistical model.)

So the real issue is not whether one should use a purely dynamical model, since such a thing does not exist. Instead, the real issue is knowing what mix of static description and dynamical explanation is feasible and meaningful to use in any given modeling effort. At present, this is still mainly a matter of tradition, rather than a matter of something more "scientific." I share a hope with many others that information criteria will be used more and more to guide investigators in choosing between more and less complicated models (see Taper, 2004 [chapter 15, this volume].) Here, a more complicated model would be one with more dynamical detail and less static description. Information criteria have the virtue that the basis for a choice of a certain level of complexity is at least clearly delineated, as opposed to remaining something more ineffable. Of course, such criteria (and any other similar analytical tools) are only as good as the biological insights and knowledge that the investigator brings to bear on the analysis. Ideally, there is some back-and-forth between intuition and the interpretation of these criteria over time.

In lieu of using such criteria in a judicious and biologically informed way, it is easy to fall prey to generic a priori claims about the world, which ultimately hinder understanding. Most common among ecologists is the belief that the real world is "always" more complex than one can understand. Even if one were not to use information criteria to determine how complex a particular explanation needs to be, one can show that such attitudes about how the world "must be" can be wrong. This can be illustrated in the context of the analysis of direct and indirect competitive effects discussed by Taper and Lele. As they note, it can be a serious mistake to ignore indirect effects when trying to understand the ecological consequences of perturbing a network of competing species. Removing a competing species may *inhibit* a competitor, because the now-absent competitor competed more strongly with a third species, which now can compete more strongly with the remaining species. By itself, this is an interesting and important insight, but it does not imply that a more complicated model, a dynamical model, is always or usually necessary for understanding species interactions. One reason is that indirect effects can sometimes be understood in terms of the concatenation of pairwise competitive effects, while in other instances they cannot. This point is well illustrated in the work of Wootton (1993), who demonstrated that several species of bird predators enhance acorn barnacle numbers at an intertidal locality in the Pacific Northwest of the United States because they consume limpets, an important barnacle competitor. This is a multispecies interaction readily understood in terms of pairwise effects. (The same is true of the bumblebee example discussed by Taper and Lele.) On the other hand, the similar colors of barnacles and limpets result in an enhancement of limpet numbers because they make it more difficult for bird predators to consume limpets. Here, the effect of bird predation on limpet abundance can be understood only by understanding the effect of barnacles on limpets. This is a multispecies interaction not readily understood in terms of pairwise effects. The overall point is that understanding "complexity" does not always require a more complex model.

There is another, more pragmatic reason why invoking dynamics and complexity can be a trap. Pursued to its logical end, the attitude implies that in any analysis one needs data on, for example, all of the pairwise competitive interactions between the species present before one can know whether indirect effects are important. But there is no reason to stop there. After all, even higher-order interactions might well affect community dynamics in such a way that indirect effects act against direct effects. The point is that there is no closure to the problem; there are always more levels to include in one's analysis (unlike in physics, where one can infer that quantum uncertainty is not due to ignorance about underlying processes.)

Of course, information criteria can play some role in providing biological closure in an analysis like this. They may indicate that there is no significant evidence for the existence of certain kinds of interactions. However, even if there is such support, it can remain an open question as to whether such interactions need to be understood in a particular attempt to understand ecological processes.

Consider this question in the context of the analysis of competitive removal experiments. Taper and Lele mention the work of Gurevitch et al. (1992), who looked for competition by determining whether a species removal resulted in the numerical increase of another species. The consequences of indirect effects are omitted. Because the web of competitive interactions could be such that species removal results in a decrease of another species (as described above), the use of Gurevitch et al.'s criterion to assess the prevalence of competition would seem to be incorrect. Yet the overall consequence of using this criterion is simply a matter of conservatism in testing. In lieu of an indication that indirect effects act in the same direction as direct effects (such that the indirect effects of a species removal results in the long-term *increase* of a noncompeting species), one sees that Gurevitch et al.'s criterion relies upon the simplest but most unambiguous evidence for competition. It excludes indirect competitive effects of the kind described by Taper and Lele. This makes it harder to find evidence for competition; however, it does not make it impossible. All other things being equal, a comparison of the kind outlined by Gurevitch et al. would provide a lower bound for the frequency of competitive interactions. This is hardly the worst possible situation, at least if one wants to be conservative in making an overall judgment about the general importance of competition in nature.

The prediction that Gurevitch et al. used also has the additional virtue of being less tied to particular expectations stemming from the use of language such as "competitor." After all, a long-term increase in one species' abundance after the removal of a second species is an unambiguous indication of competition. But the interpretation of the dynamical consequences of indirect effects is not so clear. Should species with indirect effects of the kind Taper and Lele describe be regarded as competitors, just like species with a direct effect on one another? The answer to this question is far from obvious.

While the use of dynamical analyses in efforts to understand ecological processes can often result in greater understanding, this fact by itself does not motivate a claim that such analyses are usually necessary. What counts are results that lead to true understanding, no matter how they are obtained. *Any* modeling approach can be a means to this end.

9.2   Commentary [1]

Philip M. Dixon

Most of the chapters in this book focus on statistical evidence. How does a set of observations provide evidence about one (or more) statistical models? Although there are many different philosophical interpretations of evidence, all of them include a quantitative model and data. The model may be simple (two groups have the same mean) or complex, but in all cases the discussion about evidence is a discussion about logically valid conclusions about the model or its parameters. This chapter by Taper and Lele reminds us that a discussion of evidence is only one part of the real issue: how do we use data to understand ecological concepts? Evidence about models (or model parameters) is only evidence about ecological concepts if the model is an adequate approximation of the concepts.

Taper and Lele emphasize some of the difficulties translating ecological concepts into models and parameters. In both of the authors' major examples (competition in multispecies systems and higher-order interactions in a predator-prey system), simple dynamical (differential equation) models are used translate concepts into expected results. In both cases, the results are more complex than they "should" be. The competition example demonstrates that removal of a competitor can decrease (not increase) the abundance of a target species in a multispecies system. The predator-prey example demonstrates that conclusions about interactions depend on the form of the dynamical model (i.e., whether predator numbers influence prey death rate or prey per capita death rate). In either case, focusing on statistical models and evidence may miss the appropriate ecological concept. Part of the problem is the overlapping, but not equivalent, vocabulary of ecology and statistics. Although there are quite a few examples of words with different meanings in the two subjects (e.g., density, population, scale, hypothesis), one of the most problematic words is interaction.

Ecological interactions and statistical interactions are not equivalent. A

1. Editors' note: This commentary is based on an earlier version of chapter 9. Revision of the chapter introduced new issues not considered in the commentary.

statistical interaction is a nonadditive component of a model. A change in the rest of the model (e.g., adding a new predictor variable or transforming the response) may introduce, eliminate, or change the magnitude of a statistical interaction. In Taper and Lele's second example, the statistical model depends on the form of the dynamical model, which determines whether $N(t)$ should be log transformed. In many cases, the amount of statistical interaction depends on whether or not the data are transformed. However, not all statistical interactions can be eliminated by changing the model (Box, Hunter, and Hunter, 1978, 219). Also, the data can be used to find a transformation that minimizes the interaction, which usually leads to a model that is simpler to describe and easier to interpret (Box and Cox, 1964). A statistical interaction, a particular sort of pattern in the data, may not describe the ecological concept unless it is clear what sort of model is appropriate.

Dynamical models are one tool to explore the relationship between ecological concepts and patterns in data. Other sorts of models may also indicate counterintuitive consequences of ecological concepts, e.g., the complex interplay among competition, environmental heterogeneity, and spatial scale (Garrett and Dixon, 1997).

Like any model, dynamical models simplify reality. This simplification affects both the interesting parts of the model (the relationship between ecological concepts and model parameters) and the uninteresting parts of the model (e.g., nuisance parameters and variation). Other sorts of models may better describe the "uninteresting" parts.

Two examples illustrate the value of considering additional types of models, especially the stochastic models briefly mentioned by Taper and Lele. Deterministic dynamical models predict the mean response, but they do not say anything about patterns of variability about that mean. Fitting a statistical model requires specifying a function for the mean and a function for the variability around the mean (i.e., characteristics of the error). If you assume constant variance around the mean, most field data sets don't fit a logistic population growth model very well. However, a stochastic version of logistic population growth predicts that the variability will increase with the population size (Dennis and Costantino, 1988). This model fits field data much better. Deterministic differential equation models also assume that parameters are constant. The reality is that ecological populations live in a seasonally changing world. Sometimes it is possible to simplify a model and ignore seasonality, but the debate over chaos in measles dynamics illustrates the potential dangers (Grenfell et al., 1994).

The purpose of the modeling, to connect ecological concepts to measurable patterns in data, is more important than details of the modeling technique. There is no substitute for careful thinking about the system. Used appropriately, dynamical modeling is one way to explore those connections.

9.3	Rejoinder

Mark L. Taper and Subhash R. Lele

To facilitate discussion of various points raised by Drs. Dixon and Orzack, we define more clearly what we mean by static models and dynamical models. Further, we describe how stochasticity enters into these classes of models.

a. Dynamical models are models that are explicitly *temporal* in nature. They model how a system under investigation changes over time.
b. Static models describe the equilibrium towards which the dynamical model converges as time progresses. This could be either a spatial or a nonspatial equilibrium state.
c. We can further classify dynamical and static models each into two classes: deterministic and stochastic. Stochastic models use statistical models, through the incorporation of noise term (additive or nonadditive) to model the unaccounted features of the system along with inherent uncertainty of the natural phenomena.

Thus, we have four classes of models: dynamical and deterministic, dynamical and stochastic, static and deterministic, and static and stochastic.

Professor Dixon makes an important point: many of the frequently used dynamical models are deterministic. They provide information only about a system's average behavior and are mute regarding the variability around it. We agree with his comments. We suggest in our chapter that an important step in linking theory to data is the explicit incorporation of random elements into the structure of a dynamical model. Such stochastic-dynamical models can be used to predict variability as well as central tendency. We did not emphasize this point in our paper and would like to thank him for raising it explicitly.

Dixon also discusses the distinction between statistical interaction and ecological interaction. We do not agree with his assessment that statistical and ecological interactions are distinct. We feel that if there is truly an ecological interaction, the statistical model should reflect it. Transformation of the model might seem to remove such an interaction term; however, if interpreted broadly, it should reappear when one back-transforms it to the observation scale. It is a question of proper modeling rather than statistical wizardry.

Professor Orzack wonders if our claim that pairwise analysis of species interactions may not be adequate may itself be inadequate. Are more levels of complexity needed? Should one keep on adding higher-order interactions ad infinitum? The extent of the mix and/or the number of interactions is related both to the biology of the system and to the purpose of the study. If the addition of an interaction term potentially can change the decisions significantly, we ought to add it; if the decisions are invariant to the addition, then we should not bother to add it. Certainly, in the system described by Wootton (1993), several kinds of nonlinearities (higher-order interactions) play important roles in the community's ecology. It is not likely that either adequate predictions or adequate explanations will be possible without these added complexities. An important theme found in many chapters of this book (e.g., Lele, chapter 7; Lindsay, chapter 14; Taper, chapter 15; Taper and Lele, chapter 16) is that analysis should be tailored to the problem. The construction and selection of models is in our mind an essential part of an analysis.

Orzack also raises the issue of the extent of mix between stochastic and dynamical components. Both these issues strongly relate to model selection (Taper, chapter 15) and the concept of model adequacy (Lindsay, chapter 14). Stochastic versions of dynamical models have a built-in mechanism for including much marginal complexity. Processes that cannot be ignored but need not be modeled explicitly can be subsumed in the noise terms. Several chapters in this volume point out that some of the noise in our models represents deterministic processes that are either unknown or that the modeler wishes to suppress.

Orzack claims the Gurevitch et al. (1992) criterion for the prevalence of competition in ecological systems will be conservative because in some cases the removal of one species may cause the decrease of another through indirect interactions rather than through direct competition. We believe this claim is incorrect. As in the well-known phenomenon of "apparent competition" (Holt, 1977), indirect effects can lead to increases in the density of some species after the removal of a species with no direct competitive interaction. Without knowing the relative frequency of types of indirect interactions, whether the criterion is conservative or liberal cannot be inferred. Making assumptions about the relative frequency of type of indirect interactions seems to be a tenuous step. Uncertainty regarding the frequency of even direct interactions, let alone indirect ones, is indicated by the need to perform a meta-analysis.

The Gurevitch et al. approach does not *exclude* indirect effects, as Orzack states. It simply does not distinguish indirect effects from direct effects. However, meaningful conclusions can be drawn with the Gurevitch approach about the frequency with which ecological press experiments with the removal of one species result in the long-term increase in a putative competitor. One can even operationally define such results as competition. However, we should realize from our understanding of dynamical models that this definition would sum over direct and indirect pathways. Further consideration of dynamical models of community ecology would indicate that even this definition would be far from unambiguous. Very few ecological removal experiments are continued for sufficiently long times for the resulting communities to reach equilibrium (Yodzis, 1988).

Implicit in Orzack's comments is that static patterns may be enough for understanding ecological behavior. It is true that much can be learned from static patterns in nature; however, we should note that different dynamical systems can give rise to the same static patterns, similar to the fact that the stationary distribution of a stochastic process does not characterize its transition distributions. Thus, if our decisions are based on the transition distributions, we need to model those explicitly. One can then use the stationary distribution to estimate the relevant parameters, assuming they are estimable using the stationary distribution alone. If they are not estimable, we are forced to sample the process in such a fashion that transitions are observed. Thus, thinking about dynamics explicitly is essential for drawing proper inferences as well as for designing proper experiments (observational or manipulative). We may be able to extract the information we need from static patterns without observations of transitions, but we will not be able to know this without first studying the patterns potentially stemming from the alternative models under consideration. Dynamics should not be shoved under the rug in the name of simplicity or parsimony.

Both dynamical and statistical components of ecological models are important. An appropriate mix depends on the goals of modeling. One should not rely solely on the statistical models for quantifying the strength of evidence; the strength of evidence is a function of both the statistical component (intrinsic variability and unaccounted for interactions) and the dynamical component (spatiotemporal changes in the system and causal links). Static models are also useful to quantify strength of evidence, provided they reveal important features of the underlying dynamical model.

Dr. Orzack is not alone in the ecological literature in his interest in the static results of species removal experiments, nor are we alone in our skepticism. To pursue this topic further here would distract from the thrust of this volume; instead, we direct interested readers to a recent and insightful review by Peter Abrams (2001).

We would like to thank Drs. Dixon and Orzack for providing interesting and important comments on our paper. These discussions have helped us to clarify several key points inadequately discussed in our chapter.

REFERENCES

Abrams, P. A. 2001. Describing and Quantifying Interspecific Interactions: A Commentary on Recent Approaches. *Oikos* 94:209–218.

Barnard, J., J. Du, J. L. Hill, and D. B. Rubin. 1998. A Broader Template for Analyzing Broken Randomized Experiments. *Soc. Methods Res.* 27:285–317.

Beckerman, A. P., M. Uriarte, and O. J. Schmitz. 1997. Experimental Evidence for a Behavior-Mediated Trophic Cascade in a Terrestrial Food Web. *Proc. Nat. Acad. Sci.* 94:10735–10738.

Bender, E. A., T. J. Case, and M. E. Gilpin. 1984. Perturbation Experiments in Community Ecology: Theory and Practice. *Ecology* 65:1–13.

Box, G. E. P., and Cox, D. R. 1964. An Analysis of Transformations (with discussion). *J. Roy. Stat. Soc.,* ser. B, 26:211–252

Box, G. E. P., W. G. Hunter, and J. S. Hunter. 1978. Statistics for Experimenters. New York: Wiley.

Brown, J. H., E. J. Bedrick, S. K. Morgan Ernest, J.-L. E. Cartron, and J. F. Kelly. 2004. Constraints on Negative Relationships: Mathematical Causes and Ecological Consequences. Chapter 10 in Taper, M. L., and S. R. Lele, eds., *The Nature of Scientific Evidence: Statistical, Philosophical, and Empirical Considerations.* Chicago: University of Chicago Press.

Case, T. J. 1999. *An Illustrated Guide to Theoretical Ecology.* New York: Oxford University Press.

Cushing, J. M., R. F. Costantino, B. Dennis, R. A. Desharnais, and M. Henson-Shandelle. 1998. Nonlinear Population Dynamics: Models, Experiments, and Data. *J. Theoret. Biol.* 194:1–9.

Darwin, C. 1859. *On the Origin of Species by Means of Natural Selection, or Preservation of Favored Races in the Struggle for Life* (1st ed.). London: Murray.

Dawid, A. P. 2000. Causal Inference without Counterfactuals (with commentary). *J. Am. Stat. Assn* 95:407–448.

Dayton, P. K. 1973. Two Cases of Resource Partitioning in an Intertidal Community: Making the Right Prediction for the Wrong Reason. *Am. Nat.* 107:662–670.

Dennis, B., and R. Costantino. 1988. Analysis of Steady State Populations with the Gamma Abundance Model: Applications to Tribolium. *Ecology* 69:1200–1213.

Dennis, B., W. P. Kemp, and M. L. Taper. 1998. Joint Density Dependence. *Ecology* 79:426–441.

Freedman, D. 1999. From Association to Causation: Some Remarks on the History of Statistics. *Stat. Sci.* 14:243–258.

Garrett, K. A., and Dixon, P. M. 1997. Environmental Pseudointeraction: The Effects of Ignoring the Scale of Environmental Heterogeneity in Competition Studies. *Theoret. Popul. Biol.* 51:37–48.

Greenland, S., J. M. Robins, and J. Pearl. 1999. Confounding and Collapsibility in Causal Inference. *Stat. Sci.* 14:29–46.

Grenfell, B. T., A. Kleczkowski, S. P. Ellner, and B. M. Bolker. 1994. Measles as a Case-Study in Nonlinear Forecasting and Chaos. *Phil. Trans. Roy. Soc. London,* ser. A, 348:515–530.

Gurevitch, J., L. L. Morrow, A. Wallace, and J. S. Walsh. 1992. A Meta-analysis of Competition in Field Experiments. *Am. Nat.* 140:539–572.

Hand, D. J. 1995. Discussion of the paper by Chatfield. *J. Roy. Stat. Soc.,* ser. A, 158:448.

Holt, R. D. 1977. Predation, Apparent Competition, and the Structure of Prey Communities. *Theoret. Popul. Biol.* 12:197–229.

Hume, D. [1748] 1999. *An Enquiry Concerning Human Understanding,* ed. Tom L. Beauchamp. Oxford and New York: Oxford University Press.

Ives, A. R. 2000. Stochasticity and Statisticians in Environmental Biology. *Trends Ecol. Evol.* 15:485–486.

Lande, R., and S. H. Orzack. 1988. Extinction Dynamics of Age-Structured Populations in a Fluctuating Environment. *Proc. Nat. Acad. Sci.* 85:7418–7421.

Lele, S. R. 2004. Evidence Functions and the Optimality of the Law of Likelihood. Chapter 7 in Taper, M. L., and S. R. Lele, eds., *The Nature of Scientific Evidence: Statistical, Philosophical, and Empirical Considerations.* Chicago: University of Chicago Press.

Lele, S. R., M. L. Taper, and S. Gage. 1998. Statistical Analysis of Population Dynamics in Space and Time Using Estimating Functions. *Ecology* 79:1489–1502.

Levins, R. 1966. The Strategy of Model Building in Population Biology. *Am. Sci.* 54:421–431.

Lindsay, B. G. 2004. Statistical Distances as Loss Functions in Assessing Model Adequacy. Chapter 14 in Taper, M. L., and S. R. Lele, eds., *The Nature of Scientific Evidence: Statistical, Philosophical, and Empirical Considerations.* Chicago: University of Chicago Press.

Manly, B. F. J. 1997. *Randomization, Bootstrap, and Monte Carlo Methods in Biology.* 2d ed. New York: Chapman and Hall.

Maurer, B. A. 2004. Models of Scientific Inquiry and Statistical Practice: Implications for the Structure of Scientific Knowledge. Chapter 2 in Taper, M. L., and S. R. Lele, eds., *The Nature of Scientific Evidence: Statistical, Philosophical, and Empirical Considerations.* Chicago: University of Chicago Press.

Orzack, S. H., and S. Tuljapurkar. 2001. Reproductive Effort in Variable Environments, or Environmental Variation *Is* for the Birds. *Ecology* 82:2659–2665.

Peacor, S. D. and E. E. Werner. 1997. Trait-Mediated Indirect Interactions in a Simple Aquatic Food Web. *Ecology* 78:1146–1156.

Pearl, J. 1995. Causal Diagrams for Empirical Research. *Biometrika* 82:669–710.

Pearson, K. 1911. *The Grammar of Science.* 3rd ed. London: A. and C. Black.

Roughgarden, J., and J. M. Diamond. 1986. Overview: The Role of Species Interactions in Community Ecology. In Diamond, J. M., and T. J. Case, eds., *Community Ecology.* New York: Harper and Row.

Rubin, D. B. 1991. Practical Implications of Modes of Statistical Inference for Causal Effects, and the Critical Role of the Assignment Mechanism. *Biometrics* 47:1213–1234.

Scheiner, S. M. 2004. Experiments, Observations, and Other Kinds of Evidence. Chapter 3 in Taper, M. L., and S. R. Lele, eds., *The Nature of Scientific Evidence: Statistical, Philosophical, and Empirical Considerations.* Chicago: University of Chicago Press.

Schoener, T. W. 1993. On the Relative Importance of Direct versus Indirect Effects in Ecological Communities. In Kawanabi, H., J. Cohen, and I. Iwasaki, eds., *Mutualism and Community Organization.* Oxford: Oxford University Press.

Stone, R. 1993. The Assumptions on Which Causal Inference Rest. *J. Roy. Stat. Soc.,* ser. B, 55:455–466.

Taper, M. L. 2004. Model Identification from Many Candidates. Chapter 15 in Taper, M. L., and S. R. Lele, eds., *The Nature of Scientific Evidence: Statistical, Philosophical, and Empirical Considerations.* Chicago: University of Chicago Press.

Taper, M. L., and S. R. Lele. 2004. The Nature of Scientific Evidence: A Forward-Looking Synthesis. Chapter 16 in Taper, M. L., and S. R. Lele, eds., *The Nature of Scientific Evidence: Statistical, Philosophical, and Empirical Considerations.* Chicago: University of Chicago Press.

Whewell, W. 1858. *Novum Organon Renovatum.* London.

Wootton, J. T. 1993. Indirect Effects and Habitat Use in an Intertidal Community: Interaction Chains and Interaction Modifications. *Am. Nat.* 141:71–89.

Yodzis, P. 1988. The Indeterminacy of Ecological Interactions as Perceived through Perturbation Experiments. *Ecology* 69:508–515.

10 Constraints on Negative Relationships: Mathematical Causes and Ecological Consequences

James H. Brown, Edward J. Bedrick, S. K. Morgan Ernest, Jean-Luc E. Cartron, and Jeffrey F. Kelly

ABSTRACT

One problem in inferring process from pattern is that offsetting and indirect effects, nonlinearities, and other difficulties in complex systems may prevent a clear pattern from being generated even though the hypothesized process is operating. One such complication is the constraint on the magnitudes of correlations and covariances when some of the relationships are negative. The positive semidefinite (PSD) condition requires that each of the eigenvalues and each of the principal minors of a correlation matrix be nonnegative, thereby placing limits on the possible values of the correlation coefficient. We use simple ecological examples to illustrate how processes that can generate strongly negative correlations in a simple two-variable system can be constrained from generating such clear patterns when there are multiple variables. The PSD condition is more than a statistical curiosity; it reflects real constraints on the magnitudes and patterns of interactions in ecological networks and other complex systems.

INTRODUCTION

Consider the following hypothetical example. We have three competing species; call them 1, 2, and 3. In separate experiments, each of the three pairs (1-2, 2-3, 1-3) exhibits perfectly negatively correlated population dynamics, so that the cross-correlation coefficients for the time series with zero lag,

J.H.B. and S.K.M.E. thank NSF Grant DEB-9707406, and S.K.M.E. thanks NSF Grant DGE-9593623 for support.

which are identically equal to the Pearson product moment correlation coefficient, r, are all equal to -1. Now, compete all three species in the same experiment. What will be the outcome? All three correlation coefficients cannot be -1, and the range of possible values is constrained: if two of the values are -1, the third must be 1; and if all three values are negative and equal, they can be no less than $-.5$. If we competed 11 species in one competition experiment, and the correlations were all equal and negative, then r is no less than $-.1$. So r^2, the proportion of variation explained by each pairwise interaction, would be no greater than .01. The message is that it may be difficult to infer the existence of negative interactions even when the mechanisms that give rise to them are operating.

Much of scientific inference is an effort to learn about mechanisms or processes from patterns in data. Regardless of whether one analyzes data from an unmanipulated system or from a controlled experiment, inferences about processes must come from patterns of relationships. In ecology, the processes are typically interactions between organisms and components of their biotic or abiotic environments. The interactions generate patterns of relationships among variables. The signs and magnitudes of the relationships are often quantified as correlations. As illustrated by the above examples, when there are relationships among multiple variables and some of them are negative, the magnitudes of the correlations may be mathematically constrained. This is a property of correlation matrices known as the positive semidefinite (PSD) condition. This results from the requirement that all eigenvalues of the correlation matrix be greater than or equal to 0 and that each leading principal minor is also at least 0. The PSD condition is not a mathematical or statistical artifact. It is a powerful constraint on the pattern of correlation that can be exhibited by a system with multiple negative interactions.

There are many cases in ecology and other disciplines where the PSD condition can be expected to affect scientific inference. Most ecological structures and dynamics reflect the outcome of interactions among multiple components or variables. Whenever some of these relationships are negative, the PSD condition may constrain the size of observed correlations. Examples include networks of interactions among multiple species that include competition, predation, parasitism, and disease; and tradeoffs among multiple traits in the evolution of life histories or other responses to environmental variation. Since the components of a system that an ecologist chooses to study are typically only a subset of the potentially interrelated variables, the PSD condition can affect the magnitudes of correlations even when the negative relationships are not explicitly included in the analysis. Finally, the PSD condition can influence the apparent explanatory power of multivariate techniques, as will be discussed later.

MATHEMATICAL FRAMEWORK

The PSD condition is a formalization of the necessary condition that the variance of a random variable must be positive when the random variable is a linear combination of other variables. Some notation is needed to define the PSD condition on a correlation matrix. Suppose that we have data on p variables or features, denoted $X_1, X_2, \ldots, X_p$. The (sample) correlation matrix for these variables is denoted by the $p \times p$ symmetric matrix

$$
R_1 = \begin{bmatrix} r_{11} & r_{12} & \cdots & r_{1p} \\ r_{21} & r_{22} & \cdots & r_{2p} \\ \vdots & \vdots & \vdots & \vdots \\ r_{p1} & r_{p2} & \cdots & r_{pp} \end{bmatrix}.
$$

The ith row and jth column element of R_p is r_{ij}, the correlation between the X_i and X_j. The diagonal elements are 1: $r_{ii} = 1$. The off-diagonal elements satisfy the symmetry condition $r_{ij} = r_{ji}$ for $i \neq j$. The p subscript on R_p is used here to identify the dimension of the correlation matrix.

There are several equivalent definitions of PSD, of which we give three. The definitions assume that the matrix in question is symmetric. (1) R_p is PSD if each eigenvalue of R_p is nonnegative. (2) R_p is PSD if each of the p leading principal minors of R_p is nonnegative. The leading principal minors are the determinants of the matrices $R_1, R_2, \ldots, R_p$, where R_i is the upper submatrix of R_p that gives the correlation matrix of $X_1, X_2, \ldots, X_i$. (3) Let $c_1, c_2, \ldots, c_p$ be given constants. Then R_p is PSD if $\sum_{i=1}^{p} \sum_{j=1}^{p} c_i c_j r_{ij} \geq 0$ for all possible choices of $c_1, c_2, \ldots, c_p$. The equivalences among these definitions of PSD and definitions for eigenvalues and determinants are given by Rao (1973).

The third condition is easy to understand statistically. Suppose the variables $X_1, X_2, \ldots, X_p$ have been standardized to have mean 0 and variance 1. Then the correlation matrix R_p is also the (sample) variance-covariance matrix of $X_1, X_2, \ldots, X_p$. Thus, for any set of constants $c_1, c_2, \ldots, c_p$, the (sample) variance of the linear combination $Y = c_1 X_1 + c_2 X_2 + \cdots + c_p X_p$ satisfies (Rao, 1973)

$$\text{var}(Y) = \sum_{i=1}^{p} \sum_{j=1}^{p} c_i c_j r_{ij} = \sum_{i=1}^{p} c_i^2 + \sum_{i \neq j} c_i c_j r_{ij}. \tag{1}$$

For R_p to be a correlation matrix, equation (1) must give a nonnegative value for the variance of every linear combination of $X_1, X_2, \ldots, X_p$. If there exists at least one set of constants for which $\sum_{i=1}^{p} \sum_{j=1}^{p} c_i c_j r_{ij} < 0$, then R_p is not a correlation matrix. We note that the PSD condition is both necessary and sufficient for R_p to be a correlation matrix. That is, every correlation matrix R_p is PSD and furthermore, if R_p is symmetric, has diagonal elements of 1, and PSD, then there exist variables $X_1, X_2, \ldots, X_p$ for which R_p is the correlation matrix.

To illustrate a simple constraint on correlations that the PSD condition imposes, suppose that all the off-diagonal correlations are identical, and equal to r. With Y defined to be the sum $Y = X_1 + X_2 + \cdots + X_p$, equation (1) gives $\text{var}(Y) = p + p(p - 1)r = p\{1 + (p - 1)r\}$.

For this variance to be nonnegative, we must have $1 + (p - 1)r \geq 0$ or $r \geq -1/(p - 1)$. The bound $r \geq -1/(p - 1)$ is implied by the single linear combination with $c_1 = c_2 = \cdots = c_p = 1$. However, it can be shown that this lower bound on r is the worst-case scenario, that is (1) holds for all $c_1, c_2, \ldots, c_p$ if and only if $r \geq -1/(p - 1)$. So, for $p = 3$, the most negative value possible for r is $r = -.5$. For $p = 11$, the most negative value possible is $r = -.1$. The upper bound on r is 1, regardless of p.

This paper is concerned primarily with limits on pairwise correlations among $p = 3$ variables, for which the correlation matrix is

$$R_3 = \begin{bmatrix} 1 & r_{12} & r_{13} \\ r_{12} & 1 & r_{23} \\ r_{13} & r_{23} & 1 \end{bmatrix}.$$

This expression makes explicit use of the symmetry assumption by giving only the unique entries in the matrix. To understand the constraints that the PSD condition imposes, suppose r_{13} and r_{23} are given. Then, what values of r_{12} are feasible? To answer this question, we consider the second characterization of PSD matrices. With $R_1 = [1]$ and

$$R_2 = \begin{bmatrix} 1 & r_{12} \\ r_{12} & 1 \end{bmatrix},$$

the determinants, or principal minors, of R_3, are $\det(R_1) = 1$, $\det(R_2) = 1 - r_{12}^2$ and

$$\det(R_3) = 1 - (r_{12}^2 + r_{13}^2 + r_{23}^2) + 2r_{12}r_{13}r_{23}. \tag{2}$$

Each of the determinants must be nonnegative for R_3 to be a correlation matrix. So, for fixed values of r_{13} and r_{23} we can calculate the lower bound, r_{12}^L, and upper bound, r_{12}^U, on r_{12}. We must have $-1 \leq r_{12} \leq +1$ for $\det(R_2) \geq 0$ (the "usual" constraint on a single correlation) and $r_{12}^L \leq r_{12} \leq r_{12}^U$ for $\det(R_3) \geq 0$, where $r_{12}^L = r_{13}r_{23} - \sqrt{1 - r_{13}^2 - r_{23}^2 + r_{13}^2r_{23}^2}$ and $r_{12}^U = r_{13}r_{23} + \sqrt{1 - r_{13}^2 - r_{23}^2 + r_{13}^2r_{23}^2}$ are the solutions to the quadratic equation $1 - (r_{12}^2 + r_{13}^2 + r_{23}^2) + 2r_{12}r_{13}r_{23} = 0$. For example, if $r_{13} = r_{23} = -.707$, or $r_{13} = r_{23} = +.707$, then $0 \leq r_{12} \leq 1$. If $r_{13} = -.60$ and $r_{23} = .60$, then $-1 \leq r_{12} \leq .280$.

We can use the bounds on r_{12} to support the following general statements about pairwise correlations among three variables. If each pairwise correlation is negative, then not all of the correlations can be extremely strong. Further, if two of the pairwise correlations are strongly positive, the third cannot be strongly negative. However, two strongly negative correlations can be accompanied by a strong positive correlation. Additional emphasis on these points is given throughout the paper.

A REAL EXAMPLE: RELATIONSHIPS AMONG LATITUDE, CLUTCH SIZE, AND BODY SIZE IN OWLS

The constraints on the observed pattern of variation imposed by the PSD condition can be illustrated by the relationships among latitude, clutch size, and body size in owls of the family Strigidae (Cartron, Kelly, and Brown, 2000). In endothermic vertebrates, these relationships have typically been studied in pairwise fashion, and three common patterns of variation have been identified: a positive correlation between clutch size and latitude, a positive correlation between body size and latitude (Bergmann's rule), and a negative allometry between clutch size and body size. Allometry refers to relationships between biological traits and body size that can be characterized by power laws, so they are usually treated statistically by log-transforming the variables.

Cartron, Kelly, and Brown (2000) analyzed covariation among the three variables, using a dataset for 77 owl species. They expected to find the relationships shown in figure 10.1a. They actually obtained the pattern of correlations shown in figure 10.1b: strongly positive between clutch size and latitude, $r = .57$; moderate and negative between clutch size and body size, $r = -.33$; weakly positive and insignificant between body size and latitude,

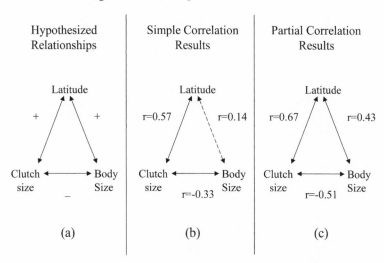

FIGURE 10.1 Hypothesized and actual correlations between latitude, clutch size, and body size for Strigid owls. Bold lines indicate significant ($p < .05$) correlations while dashed lines indicate nonsignificant correlations.

$r = .14$ (where $N = 77$ in all cases). So these results provide at best weak evidence for the expected patterns, and hence for the operation of the mechanistic processes that have been hypothesized to generate the strong pairwise correlations observed in previous studies. When partial correlation analysis was used to hold the influence of the third variable mathematically constant, all three relationships were highly significant and in the predicted direction (figure 10.1c): clutch size and latitude, $r = .67$; clutch size and body size, $r = -.51$; body size and latitude, $r = .43$. The inferences that would be drawn from these two analyses would obviously be quite different. Only the latter provides strong evidence supporting all three of the predicted relationships. It suggests that the mechanisms that underlie these patterns in other endothermic vertebrates are also operating in owls.

The first analysis, which consisted of simple pairwise correlations, does not give such an accurate picture. It fails to support Bergmann's rule, the positive relationship between body size and latitude, and provides only limited support for the negative allometry of clutch size. The reason is that with one negative and two positive correlations, all three cannot simultaneously be strong due to the PSD condition. As shown above, in the case of three variables, 1, 2, and 3, given the correlations r_{13} and r_{23} between vari-

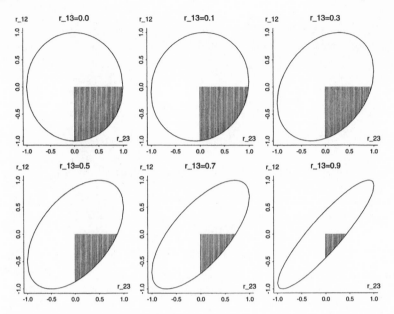

FIGURE 10.2 Enclosed region gives permissible value of r_{12} for given values of $-1 \leq r_{23} \leq 1$ and selected given values of r_{13}. The shaded area gives permissible values of $r_{12} \leq 0$ for given values of $r_{23} \geq 0$ and $r_{13} \geq 0$.

ables 1 and 3 and 2 and 3, respectively, the limits on the possible values for the correlation r_{12} are given by the endpoints of the interval $r_{13}r_{23} \pm \sqrt{1 - r_{12}^2 - r_{13}^2 + r_{13}^2 r_{23}^2}$, as shown in figure 10.2. The shaded area in the figure delineates the range of possible values when one correlation is negative and two are nonnegative.

Cartron, Kelly, and Brown (2000) performed additional analyses of the data for owls. Phylogenetically independent contrasts did not materially change the interpretation. At the generic level, even partial correlation analyses provided strong support for only one or two, but never three, of the relationships predicted in figure 10.1a in different genera. The general message, however, remained unchanged: the patterns depicted in figure 10.1c and the mechanisms that produce them appear to occur in owls, but the evidence is obscured because of the PSD property of correlation matrices. For example, if an investigator "tested" for the existence of Bergmann's rule by regressing body size against latitude, there would be little evidence for this pattern in the family as a whole and most of the genera.

BROADER IMPLICATIONS

As implied by the preceding example, whenever some of the relationships are negative, the PSD condition can place severe limits on patterns of correlation and complicate ecological inference. The PSD condition is due to properties of correlation matrices: all eigenvalues and each leading principal minor must be at least 0. Note, however, that these properties are not just mathematical or statistical artifacts. They powerfully constrain not only the relationships among variables, but also the operation and outcome of the mechanisms that produce these relationships. In the case of the owls, response to selection for increased body size at high latitudes is limited so long as clutch size is positively correlated with latitude and negatively correlated with body size (presumably due to other selective pressures). In the case of three coexisting competing species, it is simply impossible for the population dynamics of all three pairs to be strongly negatively correlated. One way to view this is to realize that the direct negative interaction between any given pair is opposed by a net positive effect of the negative interaction between the other two pairs (figure 10.3). The outcome of the three-way interaction must reflect a balance between these direct and indirect effects.

Because competition is a negative interaction, inferences about the process will be constrained by the PSD condition whenever there are three or more competing species. In the 1970s, experimental studies of protists and fruit flies asked whether the outcomes of pairwise interactions could predict the outcomes of multispecies interactions (Vandermeer, 1969; Gilpin, Carpenter, and Pomerantz, 1986). These experiments focused on which species "won" by driving the others extinct and concluded that the hierarchy of exclusion was generally transitive and predictable. Had the experiments been designed so that multiple species were able to coexist, however, it is almost certain that their dynamics would have been complicated and relatively unpredictable. This is certainly the case for the dynamics of coexisting, competing species in the field. For example, several desert rodent species, which exclusion experiments have shown to compete strongly, do not exhibit negatively correlated population dynamics when they occur together in unmanipulated systems, and some even exhibit positive correlations in population fluctuations (Gilpin, Case, and Bender, 1982; Brown and Heske, 1990; Valone and Brown, 1996). The lack of negatively correlated dynamics among putative competitors in complex multispecies systems cannot be taken as evidence that these species do not in fact compete.

A

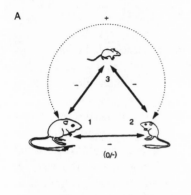

FIGURE 10.3 Bold arrows indicate the correlations between species in pairwise interactions. The dotted arrow indicates the indirect effect of species 1 on species 2 mediated through species 3. Signs in parentheses indicate how the indirect interaction could affect the correlation sign of the pairwise interaction if the third species was not taken into account. A: When all three species exhibit strong negative pairwise interactions, the indirect effects could reduce or negate the direct negative interaction between two species. B: When two pairs of species have negative correlations and the other pair have a positive correlation, the indirect interaction does not weaken the direct interaction.

B

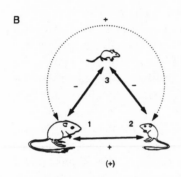

Other cases where the PSD condition must apply occur in macroecology. For example, in analyses using many species within large taxonomic groups, population density tends to be negatively related and area of geographic range to be positively correlated with body size, but there is also a positive correlation between population density and geographic range (e.g., Hanski, 1982; Brown, 1984, 1995; Brown and Maurer, 1987). Constraints imposed by the PSD condition seem to be expressed in two ways. First, none of the pairwise relationships is typically very strong. In fact, while the expected correlations are usually statistically significant, the patterns of variation are complex and better described by "constraint envelopes," drawn to characterize the boundaries of the distributions, than by linear regressions fitted through the entire datasets (Brown and Maurer, 1987; Brown, 1995). Second, the positive relationship between abundance and distribution is often most clearly seen when analyses are restricted to closely related, ecological similar species. This has the effect of holding body size relatively constant, because "closely related, ecological similar species" also tend to be similar in size.

Appreciation of the implications of the PSD condition should lead to caution in interpreting ecological relationships. Many ecological systems are composed of multiple interacting units, and some of the interactions are typically negative. Inferring process from pattern often necessarily involves interpreting correlative data. This applies, for example, to both experimental and nonmanipulative studies of ecological communities composed of multiple species and of local populations composed of multiple interacting individuals. We call attention to two situations that are particularly relevant.

First, all of the interacting units may not be included in the analysis. Almost always data are available only for a simplified subset of the possible interactors. Thus, for example, studies in community ecology typically focus on some small number of species. Although these species may on average interact more strongly with each other than with other species in the community, there is no assurance that all significant negative interactions are confined to the subset selected for study. It is easy to see therefore, how, as a consequence of the PSD condition, the components not included in the analysis could influence the structure and dynamics of the selected subsystem. It would be easy to miss or misinterpret the roles of important mechanisms. For example, many experimental field studies of competition remove one species and monitor the response of one or a very small number of other species. Sometimes, no compensatory increase is observed, and the investigators infer the species in question do not compete (Schoener, 1974; Connell, 1975). An alternative hypothesis, however, is that other members of the community that are not monitored compensate for the removal. Note that such a response does not necessarily mean the species that were hypothesized to compete do not in fact do so.

The second caution applies to explicitly multivariate analyses (see also Koyak, 1987). With increasing frequency ecologists have used two general classes of multivariate techniques, usually for somewhat different purposes. On the one hand, procedures such as principal components analysis (PCA), principal factor analysis (PFA), canonical discriminant analysis (CDA), and canonical correlation analysis (CCA) are used primarily as tools to describe the associations among variables in terms of a smaller number of uncorrelated variables. While these methods characterize the structures of correlation and covariance matrices, they can give quite different results depending upon which variables are included in the analysis, especially when some of them are negatively correlated. On the other hand, methods such as path analysis and other forms of structural equation modeling (SEM) are often used to characterize the structure of interaction networks and to choose among alternative models (e.g., Mitchell, 1992, 1994; Wootton, 1994; Grace and Pugesek, 1997; Siemann, 1998). These techniques are of some utility in identifying important pathways of direct and indirect interaction (but see Smith, Brown, and Valone, 1997, 1998; Grace and Pugesek, 1998). When SEM is used to estimate the relative strengths of particular processes or interactions, such inference can be limited by the problem of negative correlations. Both of the above classes of multivariate techniques have variations that utilize either correlation or covariance matrices. The PSD condition also applies to covariance matrices, but the constraints are not as clearly understood because they depend on the variances of the individual characteristics.

To the extent that ecology endeavors to infer process from pattern, this effort is complicated by the inherent complexity of ecological systems. Populations, communities, and ecosystems are all composed of multiple units that interact with each other and with their abiotic environment. These interactions are both positive and negative. Consequently, both the structure and dynamics of these systems exhibit complex correlation structures. The PSD constraint on correlation matrices is only one, but an important one, of the factors that makes the ecologist's exercise of inferring causation from correlation particularly challenging.

10.1   Commentary

Robert D. Holt and Norman A. Slade

A fundamental axiom of scientific inference is that correlation need not imply causation. One message of the interesting contribution by Brown et al. is that "the observation of weak (or no) correlation need not imply no causation" (with apologies for the triple negation), particularly when multiple factors interact in a causal nexus. We agree with essentially all the points raised by the authors. We have a few specific thoughts about the content of their paper, but will mainly focus our remarks on the more general issue of "constraints" in the pursuit of scientific knowledge.

A SPECIFIC COMMENT

Some readers may have the impression that the authors are advocating the use of partial correlations. Though this technique has its uses, one should be

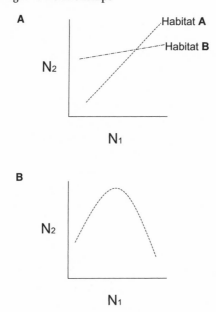

FIGURE 10.1.1 Misleading aspects of correlation as a measure of the strength of association. A: Perfect correlation, but different strength of effect. B: No correlation, but strong nonlinear association.

aware of its limitations. If one measures a correlation between variables X and Y, with no measurement error, and later discovers data about the correlation between those two variables and a third variable, Z, then this discovery does not change the initial estimate of a correlation between X and Y. This structural independence is not true for partial correlations; adding variables to a data set can often change the magnitude and even direction of partial correlation between two variables (for a concrete example in community ecology, see Carnes and Slade, 1988). Thus, there is a kind of robustness in correlations as descriptors of pattern, which is absent for partial correlations.

If the goal of scientific investigation is to go beyond pattern to discern generative processes, it is important to recognize that simple correlations are constrained along the lines sketched by the authors, and moreover that any given correlation can arise from a multitude of underlying causes, leading to the potential for error in causal interpretation. Using correlations as measures of strengths of association between variables can be misleading in any case, for reasons other than those discussed in Brown et al. Figure 10.1.1 shows several simple examples. In Figure 10.1.1a, we show the relationship between the abundances of two species in each of two locations. At each location, the correlation is identically unity. Yet if it is actually the case that the abundance of species 1 impacts the abundance of species 2, clearly the strength of the effect is stronger at location A than at location B. In Figure 10.1.1b, there is no overall correlation between the two variables, yet they clearly have a strong nonlinear relationship. We suspect that most ecologists are interested in the functional relationships between variables, which cannot be assessed with a purely correlational approach. Correlational techniques are almost inescapable, particularly in exploring new areas of inquiry, but they should probably be viewed largely as tools for generating hypotheses, which will then be addressed in other ways.

GENERAL REMARKS

The effects discussed by Brown et al. have to do with constraints in sets of correlational measures of association. Constraints arise in science in various ways. All of science comprises a web of belief, and some philosophers (e.g., Quine and Ullian, 1978) detect a whiff of empirical contingency even in the austere domain of abstract mathematical truths. Nonetheless, to make any headway in any particular scientific investigation one takes certain aspects of the world as given in order to examine other aspects. We find it useful to distinguish three grades of constraints, in decreasing order of universality, stringency, and durability: necessity; hard contingency; and soft contingency. The Brown et al. paper is concerned with the first of these, but we feel a broader consideration of the nature of constraints in inferential reasoning is useful to this volume, and more broadly to the effective pursuit of the scientific enterprise.

NECESSARY CONSTRAINTS

Consider first necessary relationships among factors. Philosophically, "necessity" refers to definitions or to mathematical, deductive relationships. The authors show that because a matrix composed of correlation coefficients has a mathematical property known as "positive semidefinite" there can be constraints bounding the magnitudes of correlation coefficients in multivariate settings, particularly when all or most of the correlation coefficients are negative. So, knowing the values of certain matrix elements predetermines (at least to a degree) the possible magnitudes of other matrix elements. This follows not from facts about the natural world, but from the meaning and mathematical definition of "correlation coefficient".

Similar necessary constraints arise throughout science. For instance, in dynamical systems whose properties are described by a system of differential equations, if there are physical or biological units involved (e.g., mass, time, species), the physical units on the left side of the equations must match the physical units on the right side of the equations. So, in the basic equation of Newtonian mechanics, $F = ma$, "force" has the units [mass][distance]/[time^2], and in the basic equation of population dynamics, $dN/dt = rN$, where N is population size, the intrinsic growth rate r has units 1/[time]. Dimensional analysis can be used to reach surprisingly robust conclusions about natural phenomena. Okubo (1980) for instance used dimensional analysis to show in a few lines of reasoning that the speed of the wavefront of an invasive species moving into homogeneous terrain must be proportional to $\sqrt{rD}$, where r is intrinsic growth rate and D is a diffusion coefficient. Brute-force analysis of reaction-diffusion dynamical models leads to the same qualitative result, but only after much more difficult analyses (though such analyses are necessary to derive the proportionality constant).

At Princeton University, according to an apocryphal (one hopes) story circulating among the students in the physics department, a brilliant graduate student convinced himself that he had the key to the unified "theory of everything" and buried himself in the library for several years of hard analysis. One day he decided to show his work to his advisor and emerged into the light of day with reams of paper covered with symbols. His advisor's first comment was "This can't be right." After staring at the equations for a while, the advisor then sadly said, "This can't even be wrong." The student had set up equations that were dimensionally incorrect, and so were not just incorrect, but nonsensical from the get-go. Ignoring necessary constraints can lead to completely invalid inferences about the world.

The phenomenon discussed by Brown et al. is similar to this but more insidious. One could doubtless find examples in the ecological literature of the use of correlation matrices where interpretations of each pairwise interaction or relationship were made without due recognition of the necessary interdependence that arises in multivariate correlation analyses.

HARD CONSTRAINTS

Many important constraints are not necessary but can be viewed as similarly rigid and unquestioned, at least in a given domain of inquiry. In particular, the nature of physical laws and the values of physical parameters are discovered, not defined. Some physical laws (e.g., conservation of mass) are accurate only in certain domains. To an ecologist, however, such facts about the world can be treated as absolute constraints, within which biological processes occur.

This seems obvious, but the conscious recognition of hard constraints can be rather useful. Consider for instance competition. Holt, Grover, and Tilman (1994) examined mechanistic models of competition between two consumer species for a single limiting resource (e.g., nitrogen), and attacked by a predator. This is a four-dimensional system, the analysis of which can be dauntingly complex. However, by assuming a closed system one can apply the principle of conservation of mass, so that all variables in the system (when measured in comparable units) must add up to a fixed nutrient mass. In effect, using the principle of conservation of mass permitted Holt, Grover, and Tilman (1994) to reduce the dimensions of the system from four to three. In the original four-dimensional phase space, the dynamics of the system were constrained to lie on a three-dimensional plane. Additional such constraints (e.g., one for each distinct nutrient) further reduce the possible dimensionality of the system's dynamics, with each fresh constraint reducing the potential "surface" of movement by one dimension. Recognizing the existence of such material constraints greatly facilitates analysis. Comparable insights with many important implications are currently emerging from the explicit incorporation of chemical stoichiometry into community ecology (Sterner and Hessen, 1994)

SOFT CONSTRAINTS

Finally, many constraints arise over short to long time scales from contingent facts of earth history and biological processes. One can find many different kinds of "soft constraints" in ecology and evolutionary biology. Part of the process of intellectual maturation in these disciplines has revolved around an increasing appreciation of the need to be explicit about constraints (e.g., in behavioral ecology, see Krebs and Davies, 1997, 10). We now briefly touch on several soft constraints that are in our opinion particularly important in ecology and evolution.

The results in Holt, Grover, and Tilman (1994) alluded to above were derived by assuming a closed arena for competition. But the inorganic nutrient pool may actually be influenced by spatial processes, coupling the pool with an external environment (e.g., nitrogen deposition, leaching). Does this matter? Grover (1997) shows that this violation of assumptions need not vitiate the conclusions of Holt, Grover, and Tilman (1994), provided the system reaches equilibrium. Assuming equilibrium in effect imposes a constraint on a system, a constraint that arises because dynamical forces push the state of a system towards a particular configuration (e.g., a plane of potential equilibria). In like manner, Charnov (1995) assumed that populations are stationary (with births equaling deaths) in order to derive a startling variety of predictions about regularities in the life histories of vertebrates. Holt and Gomulkiewicz (1997) analyzed adaptive evolution in a stable "sink" population (one in a habitat where fitness is less than 1, but extinction is prevented by recurrent immigration). They showed that a demographic constraint could prevent adaptation by natural selection; specifically, to be retained by selection, mutants had to have absolute fitnesses greater than unity, regardless of the fitness of the immigrant (ancestral) type.

Assumptions of stationarity, stability, or equilibria are often not literally true; populations do go extinct, and explode, and in general vary through time in their abundance, and shift in genetic composition as well. But an assumption of "near-equilibrium" may nonetheless be close enough to the truth to be viewed as a "soft" contingent constraint, within which other dynamical processes (e.g., selection) may occur. For instance, consider the study of evolutionary dynamics. If ecological processes effectively equilibrate population size to a fixed carrying capacity, it simplifies analysis of evolutionary dynamics to assume a constant population size, N, and focus solely on changes in gene frequency. This is, in fact, the standard protocol in classical population genetics. Conversely, to understand short-term population dynamics, it is often reasonable to assume that the properties of organisms are fixed and concentrate on changes in numbers.

Abstractly, whenever one takes a dynamical system and reduces its dimensionality by taking particular entities that are potentially dependent variables, making them into constant or time-varying parameters, one has imposed a soft constraint upon system behavior. We contend that this is an essential intellectual maneuver for grappling with the behavior of complex systems and often does correspond reasonably well to reality. A clear understanding of system properties in the face of soft constraints provides yardsticks for understanding the consequences of breaking those constraints.

In some circumstances, different constraints lead to similar outcomes. Brown et al. show that with equivalent competition among n species, the magnitude of the correlation coefficient necessarily declines with increasing n. Similar effects can emerge from dynamical models of species interactions. For instance, consider a model of shared predation leading to indirect competition among prey (from Holt, 1984). Assume that a predator feeds equivalently upon n prey species. Each prey in the absence of the predator exhibits logistic growth with equal r and K, and they do not directly compete. At equilibrium, the predator reduces total prey numbers to N. This is a "soft" constraint, because the system does not have to be at equilibrium (for the particular model in question [Holt, 1984], the equilibrium is globally stable, so it should be reasonable to assume that the system will "usually" be near equilibrium). Because of the assumed symmetry in prey growth, each prey species equilibrates at an abundance of N/n. If one prey species is removed, again the prey equilibrate, to $N/(n-1)$. The proportional change in a focal prey species abundance is $[N/(n-1) - N/n]/(N/n) = 1/(n-1)$. This quantity, which measures pairwise indirect competition, declines with increasing n. The conclusion that interspecific interactions will seem weaker when more species are simultaneously present thus seems to emerge from several different kinds of constraints.

Another example of a biological constraint was used by Ronald Fisher (1930) to examine the evolution of sex ratio: in sexual species all individuals must have a mother and a father. This fact, when used in evolutionary arguments, predicts the evolution of equal investment in the two sexes during reproduction if near relatives do not mate and compete. This again is a contingent constraint, because it depends upon the fact that there are just two sexes in the vast majority of living organisms. This particular constraint is embedded deeply enough in the organization of life that it has almost the "hard" status of a physical constraint.

The final, very broad class of soft constraints we wish to touch upon are those referred to loosely as "historical" or "phylogenetic" constraints. One reason introduced rabbits could wreak such havoc on the vegetation of Australia was that these plant species had evolved on a continent without a history of exposure to rabbits or rabbitlike herbivores. The constraint here reflects the contingent history of a place; the water barrier between Australia and Asia prevented species in the former continent from experiencing certain selective agents in their evolutionary history. Another very important class of constraints comes from considering the evolutionary dynamics of lineages themselves. Evolution depends upon variation, but variation always arises via mutation against the background of a phenotype and genotype generated by evolution. Not everything is possible. The range of potential variation is constrained by the kind of organism in question, and the developmental pathways it has inherited from its ancestors, along with its environment (e.g., historical factors of stabilizing selection, continuity in environments experienced by a given lineage). In general, small changes in developmental pathways are more likely than large changes. This is a soft constraint on evolution because it is not impossible for mutants or large or unique effects to arise or for species to persist through episodes of environmental change that facilitate large evolutionary changes. One way to view the history of life is that it involves the interplay of constraint breaking (e.g., the development of flexible behaviors and learning) and constraint generation (e.g., metabolic dependence upon oxygen) (Rosenzweig, 1995; Joel S. Brown, personal communication).

Brown et al. convincingly show that necessary constraints arising from the PSD properties of correlation matrices may make it particularly difficult to infer causation from correlation, given the multivariate causal complexity of biological systems. We agree, and further suggest that the explicit recognition of constraints—including necessary constraints, hard constraints, and soft constraints—should be an essential part of the conceptual repertoire of scientists, both to avoid unwitting error (the main point of the Brown et al. paper) and to assist in the challenge of understanding the structure and dynamics of complex biological systems in general.

10.2	Commentary

Steve Cherry

Brown et al. discuss how a mathematical property (positive semidefiniteness, or PSD) of covariance and correlation matrices imposes constraints on the scientific inferences one can make about complex ecological systems, particularly in the presence of negative relationships. They go much further and claim that the PSD property constrains "not only the relationships among variables, but also the operation and outcome of the mechanisms that produce these relationships." I found the discussion interesting but ultimately was unconvinced by their arguments.

Throughout their chapter, Brown et al. discuss the use of linear models and in particular the use of correlation analysis to identify important ecological relationships among pairs of variables in a multivariate setting. Correlation measures the strength of the linear relationship between two quantitative variables, and that is all it measures. It provides no information on causal mechanisms and, as is well known, can lead to incorrect conclusions

I thank Mark Taper and Robert Boik for helpful comments and discussion.

about the cause of linear relationships in the presence of confounding variables (spurious correlations). Thus, as Brown et al. point out the existence of a positive correlation between the dynamics of two species does not imply the absence of competition. The existence of a negative correlation does not imply the presence of competition.

Holt and Slade note that partial correlation analysis has limitations. I agree and add two more caveats. One is that partial correlations are correlations. They possess the same mathematical properties and constraints that all correlations do. The second caveat concerns the conclusion that the partial correlation analysis of the owl data provided strong evidence for the correctness of Bergman's rule. This is a somewhat misleading claim because the hypothesis that body size is positively correlated with latitude is not the same as the hypothesis that body size is positively correlated with latitude conditional on clutch size. Within the context of linear regression Brown et al. are evaluating two different models.

It is also worth noting that many of the problems Brown et al. discuss (e.g., the effects of confounding variables and spurious results in multivariate analysis) are not confined to the analysis of quantitative data. For example, one can easily imagine settings in which a hypothesized relationship between two quantitative variables (e.g. latitude and body size of owls) is masked by a third qualitative variable. Another example is Simpson's paradox in the analysis of two-way tables. In such cases, the PSD property of correlation matrices is not even an issue.

Brown et al. start with a simple thought experiment. Three competing species are compared in three separate experiments. In experiment 1, species 1 and 2 compete in the absence of species 3; in experiment 2, species 1 and 3 compete in the absence of species 2; and in experiment 3, species 2 and 3 compete in the absence of species 1. In each case, the population dynamics are perfectly negatively correlated providing evidence that the species have negative impacts on one another. Brown et al. argue that these relationships cannot be observed when all three species are allowed to compete together because if two of the pairs are perfectly negatively correlated, the third pair must be perfectly positively correlated. They write, "The message is that it may be difficult to infer the existence of negative interactions even when the mechanisms that give rise to them are operating." It is certainly true that the three pairs of variables cannot all have correlations of -1, and one can use the PSD property to show why this cannot occur. The problem can also be simply illustrated by noting that if $cor(X_1, X_2) = -1$ (implying that $X_1 = k_1 X_2 + a_1$ for some negative constant k_1 and constant a_1) and $cor(X_1, X_3) = -1$ (implying that $X_1 = k_2 X_3 + a_2$ for some negative constant k_2 and constant a_2) then $cor(X_2, X_3) = 1$ because $cor(X_2, X_3) = cor(k_1X_2 + a_1, k_2X_3 + a_2) = cor(X_1, X_1)$.

There is a problem with this example, however. If the three variables mean the same thing in each experiment and if all measurements are made the same way under identical circumstances, then geometrically incompatible linear relationships in an experiment involving all three species do not become geometrically compatible in separate experiments involving the three distinct pairs. This is true even if the example is made more realistic by supposing that the negative correlations are all between $-.55$ and $-.65$. If strong geometrically incompatible pairwise linear relationships are seen in separate experiments, then those experiments differ from one another and from the single experiment involving all three species in some fundamental way. The mechanisms that led to the negative interactions in the separate experiments will not be operating in the same way in the single experiment and may not be operating at all.

The limitations of correlation analysis discussed by Brown et al. are real, but the PSD property is blameless. Holt and Slade actually place the blame where it belongs in their discussion of necessary constraints. The problem results from the mathematical definition of variance, covariance, and correlation. A number of mathematical properties arise as a result of those definitions. Variances cannot be negative. A variable cannot be negatively correlated with itself. Covariance and correlation matrices must be positive semidefinite because the variance of any linear combinations of random variables must be positive.

Our view of reality can be constrained by these mathematical properties. Ecological processes leading to the existence of competitive interactions among three species can exist, but if an investigator attempts to identify those interactions by looking for negative pairwise correlations, then the attempt is doomed to produce weak evidence at best. The mathematical properties of correlation place constraints on what correlation analysis can reveal about the relationships. Do the mathematical properties constrain "the operation and outcome of the mechanisms that produce these relationships"? Brown et al. believe they do, arguing, for example, that for owls "response to selection for increased body size at high latitudes is limited so long as clutch size is positively correlated with latitude and negatively correlated with body size." But owls were evolving for millions of years before variance, covariance, and correlation were ever defined and the mathematical properties of those measures of variability and association have nothing to do with the operation of selective pressures in owls.

The hypotheses we formulate, the data we collect, how we collect those data, and the mathematical and statistical tools we use to analyze and model data all constrain our view of how ecological processes operate, but they do not constrain the operation of the processes.

| 10.3 | Rejoinder |

James H. Brown, Edward J. Bedrick, S. K. Morgan Ernest,
Jean-Luc E. Cartron, and Jeffrey F. Kelly

We generally agree with the comments on our chapter by Cherry and by Holt and Slade (H&S). However, their comments bring up issues that warrant response.

Both commentaries raise statistical issues related to correlation analysis, and in particular to our comparison of results obtained using linear correlation and partial correlation analyses. We suggested that the partial correlations provided additional, stronger evidence for relationships between latitude and body size and clutch size in owls, because the mechanisms operating on the three variables cannot all be strong due to the positive semidefinite (PSD) constraint. We agree that the pairwise correlation coefficients describe the actual magnitude of the linear relationships among the variables. But it is of interest to ask whether the patterns and mechanisms, well documented in other studies of latitudinal variation (see references in Cartron, Kelly, and Brown, 2000), are not also present in the owl case, but obscured by the relationships among the three variables.

We are very much aware that it is difficult to infer causation from correlation or mechanistic process from statistical pattern. Indeed, our chapter was intended to call attention to one class of problems. We and many other ecologists are not content to describe patterns. We view the search for and documentation of patterns as the first, inductive phase of a two-part research program. If the patterns are interesting, they can serve as inspiration for the second, inductive phase: the erection and evaluation of mechanistic hypotheses. Our chapter and the commentaries suggest the importance of a third phase. Once a process or mechanism has been discovered, it is desirable to determine when and how strongly it operates in nature.

In both of our examples, interspecific competition and the joint influence of variables associated with latitude on clutch size and body size, there is considerable evidence from earlier studies, not only for the existence of the patterns, but also for the mechanisms underlying them. Consider the case of competition, which has been repeatedly documented in laboratory and field experiments. These experiments, which most frequently have been conducted on pairs of species, hold constant or control the influence of other variables. When this is done and population growth of the two species is limited by some shared, homogeneous resource, competition is usually observed—in fact, it is typically so strong that one species drives the other to extinction. But does competition operate in nature, and what are the consequences? For decades (from Hutchinson, 1961, to Huisman and Weissing, 1999) ecologists have grappled with the seeming paradox that species with similar requirements frequently coexist in nature but usually exclude each other in controlled laboratory experiments.

In the field, in contrast to the experiments, species typically do not interact in a pairwise fashion, isolated from other entities and interactions. Instead, each species is a part of a complicated network of interactions that include positive and negative, direct and indirect effects of multiple species and abiotic variables. This kind of complexity leads to complicated nonlinear (in the general sense) patterns and processes, but the study of such complex systems is in its infancy. How can we obtain evidence for the role of competition or some other process in such a system? Our chapter was intended to point out one class of problems, but not to provide definitive solutions. We showed why it is important to recognize the implications of the PSD condition. Partial correlation analysis can sometimes be informative, suggesting that the PSD constraint may be limiting the observed magnitude of covariation. But partial correlation is not a powerful general solution to the problem of detecting patterns and inferring processes in complex dynamic systems with many variables. Cherry and H&S in their commentaries and Taper and Lele (2004 [chapter 16, this volume]) bring up other, related issues, such as the nonlinearities that arise when new variables are added to a simple two-dimensional system (see also Huisman and Weissing, 1999). These are promising beginnings, but there is a long way to go.

Cherry is unconvinced by our claim that the PSD not only is a statistical property of covariance and correlation matrices, but also reflects a constraint on the way that networks of interactions are organized in nature. We agree that we have not proven our claim, but we will continue to advocate it. Most mathematics and statistics were originally developed for practical applications, to describe relationships in the natural world. Indeed, by characterizing these relationships symbolically, mathematics implies that they hold universally, not just for a single example. One should always question whether a mathematical statement or model can justifiably be applied to a particular case, based on whether the variables and parameters in the math-

ematics represent a sufficiently realistic abstraction of the parameters and
processes in the application. In the present example of the PSD, such scrutiny will typically focus on whether the application meets the assumption
of the statistical model: e.g., linear relationships, normal distributions, and
so on. As noted above, the commentaries of Cherry and H&S correctly point
out problems in applying linear parametric statistics, which make such assumptions, to complex systems, where they may not hold.

   This is exactly the point that we wanted to make about the PSD. The
magnitudes of negative relationships that can be observed in isolated two-
dimensional systems and be described by linear parametric models cannot
exist in complex networks of many interacting components. While the
same mechanisms may be operating, they cannot have the same outcome.
Because of the PSD, the magnitudes of covariances and correlations are constrained. Does this mean that there are limits on the strength of competition
and other negative interactions in such networks? We think that it does, but
this is a deep issue that warrants much attention. It is at the heart of scientific, statistical, and philosophical issues about evidence and inference.
Cherry is correct that species were competing and owls were affected by
variables associated with latitude long before statisticians invented covariance and correlation matrices. But we will stick with our assertion that the
PSD points to a real and powerful constraint on the magnitude of negative
interactions in complex networks.

   H&S use our chapter as a springboard to make some general points about
the kinds of constraints that affect the structure and dynamics of ecological
systems. We agree that the term "constraint" has been applied fairly loosely
to a wide variety of phenomena. Constraints do indeed come in many colors—or flavors. H&S's classification and definitions are a useful step in the
ongoing effort to place the variety of constraints in a conceptual framework
that includes rigorous operational definitions. Elsewhere, Brown (Brown
and Maurer, 1987; Brown, 1995; Guo, Brown, and Enquist, 1998) has also
considered the kinds of ecological constraints and their implications. Following these treatments, we would distinguish between "absolute" and
"probabilistic" constraints. Our probabilistic constraints seem to correspond
to H&S's category of "soft" constraints. Such constraints reflect the operation of inherently stochastic processes that affect the likelihood but not the
certainty of obtaining parameter values within some specified range. An example is the effect of body size on the minimum area of geographic range
(Brown and Maurer, 1987). Species of large body size are unlikely to be confined to small geographic areas, but this is not impossible, and indeed some

endangered species of large body size currently occupy very small geographic ranges.

There is less correspondence in the way that the more powerful constraints are treated. Our category of absolute constraints seems to encompass H&S's categories of both "necessary" and "hard" constraints. Indeed, we see problems in the way that H&S try to distinguish between their two categories. H&S suggest that "many important constraints are not necessary but can be viewed as *similarly rigid and unquestioned,* at least in a given domain of inquiry" (emphasis added). They use the example of applications to ecology of physical laws, such as chemical stoichiometry. Currently many such applications are qualitative. We would respond by suggesting that hopefully the applications will become more quantitative, and that eventually they will take the form of rigorous mathematical models. Then the "hard" constraints will have been shown to be "necessary" or "absolute." Brown and Maurer (1987), Brown (1995), and Guo, Brown, and Enquist (1998) offer examples, showing how power functions that define absolute constraints due to allometry or competition can be used to predict limits on parameter values.

This takes us back to the relationship between correlation and causation, pattern and process. Empirical patterns, such as the statistical distributions of macroecological data, can provide clues to processes and inspire mechanistic hypotheses. We would reserve the term "constraints" for mechanisms that impose limits. We look forward to the day when these mechanisms can be characterized quantitatively with mathematical models. When this is done, we suspect that most limiting processes will fall into the categories of being either absolute (hard) or probabilistic (soft). There will still remain, however, challenging problems of inferring when and how these processes operate in complex ecological systems in nature. Continuing to focus on constraints may prove to be especially productive, because it may be easier to obtain evidence for processes when they set limits than when they contribute to the complex behavior within these boundaries.

REFERENCES

Brown, J. H. 1984. On the Relationship between Abundance and Distribution of Species. *Am. Nat.* 124:255–279.

Brown, J. H. 1995. *Macroecology.* Chicago: University of Chicago Press.

Brown, J. H., and E. J. Heske. 1990. Temporal Changes in a Chihuahuan Desert Rodent Community. *Oikos* 59:290–302.

Brown, J. H., and B. A. Maurer. 1987. Evolution of Species Assemblages: Effects of Energetics and Species Dynamics on the Diversification of the North American Avifauna. *Am. Nat.* 130:1–17.

Carnes, B. A., and N. A. Slade. 1988. The Use of Regression for Detecting Competition with Multicollinear Data. *Ecology* 69:1266–1274.

Cartron, J.-L. E., J. F. Kelly, and J. H. Brown. 2000. Constraints on Patterns of Covariation: A Case Study in Strigid Owls. *Oikos* 90:381–389.

Charnov, E. L. 1995. *Life History Invariants: Some Explorations of Symmetry in Evolutionary Biology.* Oxford: Oxford University Press.

Connell, J. H. 1975. Producing Structure in Natural Communities. In Cody, M. L., and J. M., eds., *Ecology and Evolution of Communities.* Cambridge, MA: Belknap.

Fisher, R. A. 1930. *The Genetical Theory of Natural Selection.* Oxford: Clarendon.

Gilpin, M. E., M. P. Carpenter, and M. J. Pomerantz. 1986. The Assembly of a Laboratory Community: Multispecies Competition in *Drosophila.* In Diamond, J., and T. J. Case, eds., *Community Ecology.* New York: Harper and Row.

Gilpin, M. E., T. J. Case, and E. E. Bender. 1982. Counter-intuitive Oscillations in Systems of Competition and Mutualism. *Am. Nat.* 119:584–588.

Grace, J. B., and B. H. Pugesek. 1997. A Structural Equation Model of Plant Species Richness and Its Application to a Coastal Wetland. *Am. Nat.* 149:436–460.

Grace, J. B., and B. H. Pugesek. 1998. On the Use of Path Analysis and Related Procedures for the Investigation of Ecological Problems. *Am. Nat.* 152:151–159.

Grover, J. P. 1997. *Resource Competition.* London: Chapman and Hall.

Guo, Q., J. H. Brown, and B. J. Enquist. 1998. Using Constraint Lines to Characterize Plant Performance. *Oikos* 83:237–245.

Hanski, I. 1982. Dynamics of Regional Distribution: The Core and Satellite Species Hypothesis. *Oikos* 38:210–221.

Holt, R. D. 1984. Spatial Heterogeneity, Indirect Interactions, and the Structure of Prey Communities. *Am. Nat.* 124:377–406.

Holt, R. D., and R. Gomulkiewicz. 1997. How Does Immigration Influence Local Adaptation? A Reexamination of a Familiar Paradigm. *Am. Nat.* 149:563–572.

Holt, R. D., J. P. Grover, and D. Tilman. 1994. Simple Rules for Interspecific Dominance in Systems with Exploitative and Apparent Competition. *Am. Nat.* 144:741–771.

Huisman, J., and F. J. Weissing. 1999. Biodiversity of Plankton by Species Oscillations and Chaos. *Nature* 402:407–410.

Hutchinson, G. E. 1961. The Paradox of the Plankton. *Am. Nat.* 95:137–145.

Koyak, R. A. 1987. On Measuring Internal Dependence in a Set of Random Variables. *Ann. Stat.* 15:1215–1228.

Krebs, J. R., and N. B. Davies. 1997. *Behavioural Ecology.* 4th ed. Oxford: Blackwell.

Mitchell, R. J. 1992. Testing Evolutionary and Ecological Hypotheses Using Path Analysis and Structural Equation Modeling. *Funct. Ecol.* 6:123–129.

Mitchell, R. J. 1994. Effects of Floral Traits, Pollinator Visitation, and Plant Size on *Ipomopsis aggregata* Fruit Reproduction. *Am. Nat.* 143:870–889.

Okubo, A. 1980. *Diffusion and Ecological Problems: Mathematical Models.* Berlin: Springer.

Quine, W. V., and J. Ullian. 1978. *The Web of Belief.* 2nd ed. New York: McGraw-Hill.

Rao, C. R. 1973. *Linear Statistical Inference and Its Applications* 2nd ed. New York: Wiley.

Rosenzweig, M. L. 1995. *Species Diversity in Space and Time.* Cambridge: Cambridge University Press.

Schoener, T. W. 1974. Resource Partitioning in Ecological Communities. *Science* 185: 27–39.

Siemann, E. 1998. Experimental Testing of Effects of Plant Productivity and Diversity on Grassland Arthropod Diversity. *Ecology* 79:2057–2070.

Smith, F. A., J. H. Brown, and T. J. Valone. 1997. Path Analysis: A Critical Evaluation Using Long-Term Experimental Data. *Am. Nat.* 149:29–42.

Smith, F. A., J. H. Brown, and T. J. Valone. 1998. Path Modeling Methods and Ecological Interactions: A Response to Grace and Pugesek. *Am. Nat.* 152:160–161.

Sterner, R., and D. O. Hessen. 1994. Algal Nutrient Limitation and the Nutrition of Aquatic Herbivores. *Ann. Rev. Ecol. Syst.* 25:1–29.

Taper, M. L., and S. R. Lele. 2004. The Nature of Scientific Evidence: A Forward-Looking Synthesis. Chapter 16 in Taper, M. L., and S. R. Lele, eds., *The Nature of Scientific Evidence: Statistical, Philosophical, and Empirical Considerations.* Chicago: University of Chicago Press.

Valone, T. J., and J. H. Brown. 1996. Desert Rodents: Long-Term Responses to Natural Changes and Experimental Manipulations. In Cody, M. L., and J. A. Smallwood, eds., *Long-Term Studies of Vertebrate Communities.* San Diego: Academic Press.

Vandermeer, J. H. 1969. The Competitive Structure of Communities: An Experimental Approach with Protozoa. *Ecology* 50:362–370.

Wootton, J. T. 1994. Predicting Direct and Indirect Effects: An Integrated Approach Using Experiments and Path Analysis. *Ecology* 75:151–165.

PART 4

SCIENCE, OPINION, AND EVIDENCE

Mark L. Taper and Subhash R. Lele

OVERVIEW

What types of information should public decisions be based on? Does expert opinion have a place in the scientific process? Can expert opinion help us understand the world better than we would using only documented facts? Would we make better decisions with expert opinion than without? These questions are critical in many scientific endeavors but especially in ecology, where there is a paucity of hard data but a superfluity of expert opinion. This section attempts to address these questions as they relate to statistical analysis and decision making.

Brian Dennis (chapter 11) discusses the influence of postmodernism and relativism on the scientific process and in particular its implications, through the use of subjective Bayesian approach, in statistical inference. He argues that subjective Bayesianism is "tobacco science" and that its use in ecological analysis and environmental policy making can be dangerous. He claims that science works through replicability and skepticism, with methods considered ineffective until they have proven their worth. He argues for the use of a frequentist approach to statistical analysis because it corresponds to the skeptical worldview of scientists.

Daniel Goodman (chapter 12), on the other hand, argues that the Bayesian approach is best for making decisions. He argues from the betting strategist's point of view and claims that one needs to put probabilities on various hypotheses. Consequently, the Bayesian approach is highly appropriate. However, he joins Dennis in disapproving of the subjective aspects of Bayesianism. He suggests, as an alternative, using related data to create "objective" priors. Goodman grounds his suggestions with thoughts on the philosophical and mathematical underpinnings for such an approach, which is very closely related to the empirical Bayes approach as well as meta-analysis, although these connections need to be investigated further.

Subhash Lele (chapter 13) agrees with Brian Dennis that subjective Bayesianism could be dangerous to science. However, he believes that along with subjectivity, there is often much useful information in the opinions of experts. He suggests a way to utilize this information fruitfully in statistical and scientific inference. There are two major obstacles to using the Bayesian approach: the difficulty of eliciting useful information from the scientists, and the possibility of "bad" information. Perhaps instead of eliciting information on the form of prior distributions of model parameters, one could elicit expert guesses for possible observations with the attributes of real data. This is operationally easier and, more importantly, allows for "calibration" of the expert opinion. There are two interesting features that come out of this. First, it is possible to utilize intentionally misleading expert opinion fruitfully. Second, one can estimate the informativeness of the expert. Sometimes it is even better to *not* use expert opinion if all that it is going to do is add noise to the observed data.

Given the contentious nature of the topic, commentaries on these papers provide interesting counterpoint to the ideas in the paper. They work hard at being good apologists for the Bayesian viewpoint, thus providing balance to the strong viewpoints in the main papers. We cannot expect the resolution of such a long-standing controversy; however, these chapters and commentaries together further our understanding of the role of expert opinion in scientific inference.

11   Statistics and the Scientific Method in Ecology

Brian Dennis

ABSTRACT

Ecology as a science is under constant political pressure. The science is difficult and progress is slow, due to the variability of natural systems and the high cost of obtaining good data. Ecology, however, is charged with providing information in support of environmental policy decisions with far-reaching societal consequences. Demand for quick answers is strong, and demand for answers that agree with a particular point of view is even stronger.

The use of Bayesian statistical analysis has recently been advocated in ecology, supposedly to aid decision makers and enhance the pace of progress. Bayesian statistics provides conclusions in the face of incomplete information. Bayesian statistics, though, represents a much different approach to science than the frequentist statistic studied by most ecologists. The scientific implications of Bayesian statistics are not well understood.

I provide a critical review of the Bayesian approach. I compare, using a simple sampling example, Bayesian and frequentist analyses. The Bayesian analysis can be cooked to produce results consistent with any point of view because Bayesian analysis quantifies prior personal beliefs and mixes them with the data. In this, Bayesian statistics is consistent with the postmodern view of science, widely held among nonscientists, in which science is just a system of beliefs and has no particular authority over any other system of beliefs. By contrast, modern empirical science uses the scientific method to identify empirical contradictions in skeptics' beliefs and permit replication and checking of empirical results. Frequentist statistics has become an indispensable part of the scientific method.

I also undertake a critical discussion of statistics education in ecology. Part of the potential appeal of Bayesian statistics is that many ecologists are confused about frequentist statistics, and statistical concepts in general. I identify the source of confusion as arising from ecologists' attempts to learn statistics through a series of non-calculus-based statistical methods courses taken in graduate school. I prescribe a radical change in the statistical training of ecological scientists that would greatly increase the level of confidence and facility with statistical thinking.

INTRODUCTION

Science in the Crosshairs

Tobacco company scientists argue that there is no evidence that smoking tobacco is harmful. Biblical creation scientists argue that the evidence for evolution is weak. Institutes paid for by industry are staffed by degreed scientists whose job is to create in the minds of politicians and the public the illusion of major scientific disagreements on environmental issues.

We are awash in a sea of popular postmodernism. Fact in the postmodern view is just strongly held belief (Anderson, 1990). Native Americans, according to tribal creation stories, did not originally cross over from Asia, but arose independently and originally in the Americas. The U.S. government is concealing dark secrets about the existence of extraterrestrial life, secrets that became partly exposed after a crash near the town of Roswell, New Mexico. Crystals have health and healing properties. One's personality and tendencies are influenced by the positions of solar system objects at the moment of birth. O. J. Simpson did not kill Nicole Brown and Ron Goldman.

The postmodern outlook is not confined to popular culture, but permeates intellectual life as well. Humanities disciplines at universities have abandoned the traditional empirical view of science (Sokal and Bricmont, 1998). Feminists claim that the methods and requirements of science are biased against females. According to feminist scholars, if females formed the reigning power structure, science would be more cooperative in occupation and more tolerant of multiple explanations. "Science studies" historians focus on questionable behavior of well-known scientists toward colleagues, in order to expose science as a subjective power struggle. Multicultural philosophers portray science as just another of many legitimate ways of knowing, its successes due primarily to the dominance of European culture. Scientists' writings are deconstructed by literary theorists in order to reveal how the scientists were trapped by the prevailing cultural mental prisons. Political polemics from the intellectual left and right reveal disbelief and disrespect for established scientific knowledge.

Ecology

The questioning of science and the scientific method continues within the science of ecology as well. Ecology has become a highly politicized science. Once a quiet backwater of biology, ecology burst into high public profile in the early 1970s (Earth Day, April 1970, was a watershed event) with the emergence of popular concern about environmental issues. After passage of the National Environmental Policy Act, the Endangered Species Act, the National Forest Policy and Management Act, and many other federal and state laws, ecologists and the findings of ecology suddenly had great influence in the lives of people everywhere. The scientific information from ecology, coupled with the environmental laws, forced constraints on peoples' behavior and economic activity.

Many ecological topics, from evolution to conservation biology to global climate change, hit people close to home. As a result, scientific signals are often masked or distorted by political noise. An ecological discovery that has impact on human conduct will often have a debunking campaign mounted against it by moneyed interests. Government agency scientists are sometimes muzzled and have their findings reversed by the pen stroke of a political appointee. Natural resource departments at state universities are pressured by special interest groups. Radio talk show hosts set themselves up as authoritative spokespersons on environmental topics.

Among practicing, credentialed ecologists, the science itself is quite contentious. The topic intrinsically attracts many participants, and the competition for admission to programs, jobs, journal space, grants, and recognition is fierce. Severe scientific and political infighting surfaces during position searches at university departments. Anonymous peer reviews can be ignorant and vindictive. Resources for curiosity-driven research are scarce; ecological research is funded more and more by agencies and companies with particular agendas. Pressures mount from environmental decision makers for definitive answers. In this postmodern cacophony, how can a healthy ecological science thrive?

In fact, some ecologists in the past couple of decades have questioned whether the Popperian hypothetico-deductive approach and the collection of inquiry devices known as the scientific method are too constraining for ecology. Good empirical data in ecology have often been too slow in coming or too difficult and expensive to collect, and scientific progress in ecology has seemed painfully slow. The calls for relaxed scientific guidelines have come from two main sources. First, some "theoretical ecologists" have sought scientific respect for their pencil, paper, and computer speculations on ecological dynamics. Their mathematical models, however, frequently play the role of "concepts" rather than "hypotheses," due to the lack of connections to data and the lack of widely accepted ecological laws with which to build models. As a result, theoretical ecologists have called for judging mathematical models under different criteria than scientific hypotheses would be judged. Articles in the November 1983 issue of *The American Naturalist* debated this question, among others, within the context of community ecology.

Second, applied ecologists and social decision makers have often viewed the beetles-and-butterflies focus of ecological natural history research to be an unaffordable luxury. Academic ecology research is an exotic world of Galapagos birds, Caribbean lizards, jungle orchids, and desert scorpions; it is slow, intellectual, and, to some onlookers, produces few useful generalities. Yet ecology also deals with vital topics within which major social decisions must be made, regardless of the amount of evidence available. Answers in the form of "best judgments" by experts are needed, and fast. For example, will breaching the Columbia watershed dams save salmon, or not?

A partial reading list for a seminar course on the scientific method in ecology might include Connor and Simberloff (1979, 1986), Saarinen (1980), Hurlbert (1984), Strong, et al. (1984), Hairston (1989), Underwood (1990), Peters (1991), Schrader-Frechette and McCoy (1993), Dixon and Garrett (1993), and the entire November 1983 issue of *The American Naturalist.*

Inevitably, bound up in this question about ecological science are concerns about statistical practice. The lack of true replication in many ecological experiments exposed by Hurlbert (1984) was a shocker. The lack of attention to power in many ecological studies has also been criticized (Toft and Shea, 1983; Peterman, 1990). Distribution-free statistical methods have been advocated (Potvin and Roff, 1993), but some of those arguments have been challenged (Smith, 1995; Johnson, 1995). Stewart-Oaten (1995) criticized the tendency of ecologists to view statistics as a set of procedural rules for data analysis. The widespread misinterpretation of statistical hypothesis testing has inspired much discussion (Simberloff, 1990; Underwood, 1990; Yoccoz, 1991; Johnson, 1999). The lack of statistical connections between "nonlinear dynamics" models and ecological data was criticized (Dennis et al., 1995). Specific ecological topics, such as the prevalence of density-dependent population regulation, have spawned their own statistical literature (see Dennis and Taper, 1994).

Bayesian Statistics
According to some (Reckhow, 1990; Ellison, 1996; Johnson, 1999), there is a statistical solution to many of ecology's ills, and that solution is Bayesian. Bayesian statistics is remarkably different from the variety of statistics

called frequentist statistics that most of us learned in college. Bayesian statistics abandons many concepts that most of us struggled (with mixed success) to learn: hypothesis testing, confidence intervals, P-values, standard errors, power. Bayesians claim to offer improved methods for assessing the weight of evidence for hypotheses, making predictions, and making decisions in the face of inadequate data. In a cash-strapped science charged with information support in a highly contentious political arena, the Bayesian promises are enticing to ecological researchers and managers alike.

But is there a price to pay? You bet. Bayesians embrace the postmodern view of science. The Bayesian approach abandons notions of science as a quest for "objective" truth and scientists as detached, skeptical observers. Like postmoderns, Bayesians claim that those notions are misleading at best. In the world of Bayesian statistics, truth is personal and is measured by blending data with personal beliefs. Bayesian statistics is a way of explicitly organizing and formulating the blending process.

There is an enormous literature on Bayesian statistics. A glance at the titles in any current statistics journal (say, *Journal of the American Statistical Association* or *Biometrika*) might convince a casual onlooker that the world of statistics is becoming Bayesian. The Bayesian viewpoint is indeed gaining influence. The burgeoning literature, however, tends to be highly mathematical, and a scientist is right to question whether the attraction is mathematical instead of scientific. Actually, frequentism is alive and well in statistics. Introductory textbooks and courses remain overwhelmingly frequentist, as do canned computer statistics packages available to researchers. Frequentist and Bayesian statisticians waged war for many years, but the conflict quieted down around 1980 or so, and the two camps coexist now in statistics without much interaction.

Ecology, however, represents fertile, uncolonized ground for Bayesian ideas. The Bayesian-frequentist arguments, which many statisticians tired of twenty years ago, have not been considered much by ecologists. A handful of Bayesian papers have appeared in the ecological literature (see Ellison, 1996, and other papers featured in the November 1996 *Ecological Applications*). Their enthusiastic exposition of Bayesian methods, and their portrayal of frequentism as an anachronistic yoke impeding ecological progress, has attracted the attention of natural resource managers (Marmorek, 1996).

The Bayesian propagule has arrived at the shore. Ecologists need to think long and hard about the consequences of a Bayesian ecology. The Bayesian outlook is a successful competitor, but is it a weed?

I think so. In this paper, I attempt to draw a clear distinction for ecologists between Bayesian and frequentist science. I address a simple environmental sampling problem and discuss the differences between the frequentist and the Bayesian statistical analyses. While I concur with Bayesians regarding critiques of some of the imperfections of frequentism, I am alarmed at the potential for disinformation and abuse that Bayesian statistics would give to environmental pressure groups and biased investigators. At the risk of repeating a lot of basic statistics, I develop the sampling example rather extensively from elementary principles. The aim is to amplify the subtle and not-so-subtle conflicts between the Bayesian and frequentist interpretations of the sampling results. Readers interested in a more rigorous analysis of the scientific issues in the frequentist/Bayesian debate are urged to consult Mayo's (1996) comprehensive account.

One thing has become painfully clear to me in twenty years of extensive teaching, statistical consulting, reviewing, and interacting in ecology: ecologists' understanding of statistics in general is abysmally poor. Statistics should naturally be a source of strength and confidence for an ecologist, no matter how empirically oriented. Unfortunately, it is all too frequently a source of weakness, insecurity, and embarrassment. The crucial concepts of frequentism, let alone Bayesianism, are widely misunderstood. I place the blame squarely on ecologists' statistical educations, which I find all wrong. In a later section of this paper, I offer some solutions to this problem. Ecologists, whether Bayesian or frequentist, will be better served by statistics with a radical revision of university statistics coursework.

WHAT IS FREQUENTISM?

Nature cannot be fooled. —Richard Feynman

Suppose a reach of a stream is to be sampled for Cu pollution. A total of 10 samples will be collected from the reach in some random fashion, and Cu concentration (μgL^{-1}) will be determined in each sample.

The purpose of the samples is to estimate the average concentration of Cu in the water at the time of sampling. The sampling could be a part of an ongoing monitoring study, an upstream/downstream before/after study, or similar such study.

Frequentist statistics involves building a probability model for the observations. The modeling aspect of statistics is crucial to its understanding and proper use; however, the noncalculus statistics methods courses taken by ecologists-in-training tend to emphasize formulas instead of models. I, therefore, develop the modeling aspect in more detail than is customary in frequentist analyses, so that the approach may be contrasted properly with the Bayesian way.

We might suppose that the observations of Cu concentration could be modeled as if they arose independently from a normal distribution with mean μ and variance σ^2. Applied statistics texts would word this model as follows: each sample is assumed to be drawn at random from a "population" of such samples, with the population of Cu values having a frequency distribution well approximated by a normal curve. The population mean is μ, and the population variance is σ^2. Mathematical (i.e., postcalculus) statistics texts would state: the observations, $\{X_i\}$, $i = 1, 2, \ldots, n$ are assumed to be independent, identically distributed normal (μ, σ^2) random variables. Regardless of the wording and symbology, the important point is that a probability model is assumed for *how the variability in the data arose*. The analyses are based on the model, and so it will be important to evaluate the model assumptions somehow. If the model is found wanting, then proper analyses will require construction of some other model.

We suppose the observations are drawn; their numerical values are $x_1, x_2, \ldots, x_n$ $(n = 10)$. Symbols are used for the actual values drawn, so that the subsequent formulas will be general to other data sets; the lowercase notation indicates fixed constants (sample already drawn) instead of uppercase random variables (sample yet to be drawn). The distinction is absolutely crucial in frequentist statistics and is excruciating for teachers and students alike. (Educators in ecology should be aware that non-calculus-based basic statistics textbooks, in response to the overwhelming symbol allergies of today's undergraduates, have universally abandoned the big X–little x notation and, with it, all hope that statistics concepts are intended to be understood). Let us suppose that the investigator has dutifully calculated some summary statistics from the samples, in particular the sample mean $\bar{x} = (\sum_{i=1}^{n} x_i)/n$ and the sample variance $s^2 = [\sum_{i=1}^{n}(x_i - \bar{x})^2]/(n - 1)$, and that the resulting numerical values are $\bar{x} = 50.6$, $s^2 = 25.0$. The probability model for the observations is represented mathematically by the normal distribution, with probability density function (pdf) given by

$$f(x) = (\sigma^2 2\pi)^{-1/2} \exp[-(x - \mu)^2/(2\sigma^2)], \quad -\infty < x < \infty.$$

This is the bell-shaped curve. The cumulative distribution function (cdf) is the area under the curve between $-\infty$ and x and is customarily denoted $F(x)$. The probabilistic meaning of the model is contained in the cdf; it is the probability that a random observation X will take a value less than or equal to some particular constant value x:

$$F(x) = P[X \le x] = \int_{-\infty}^{x} f(u)\,du.$$

Again, the lower- and uppercase Xs have different meanings. The constants μ and σ^2 are "parameters." In applied statistic texts, μ and σ^2 are interpreted respectively as the mean and variance of the population being sampled. Bear in mind that the population here is the collection of all possible samples that could have been selected on that sampling occasion. It is this potential variability of the samples that is being modeled in frequentist statistics.

The likelihood function is an essential concept to master for understanding frequentist and Bayesian statistics alike. The pdf $f(x)$ quantifies the relative frequency with which a single observation takes a value within a tiny interval of x. The whole sample, however, consists of n observations. Under the independence assumption, the product $f(x_1)f(x_2)\cdots f(x_n)$ quantifies the relative frequency with which the whole sample, if repeated, would take values within a tiny interval of $x_1, x_2, \ldots, x_n$, the sample actually observed. The product is the "probability of observing what you observed" relative to all other possible samples in the population. Mathematically, the random process consists of n independent random variables $X_1, X_2, \ldots, X_n$. The product $f(x_1)f(x_2)\cdots f(x_n)$ is the joint pdf of the process, evaluated at the data values.

The joint pdf of the process, evaluated at the data, is the *likelihood function*. For this normal model, the likelihood function is a function of the parameters μ and σ^2. The relative likelihood of the sample $x_1, x_2, \ldots, x_n$ depends on the values of the parameters; if μ were 100 and σ^2 were 1, then the relative chance of observing sample values clustered around 50 would be very small indeed. Written out, the likelihood function for this normal model is

$$\begin{aligned} L(\mu, \sigma^2) &= f(x_1)f(x_2)\cdots f(x_n) \\ &= (\sigma^2 2\pi)^{-n/2} \exp\left[-(2\sigma^2)^{-1} \sum_{i=1}^{n} (x_i - \mu)^2 \right]. \end{aligned}$$

An algebraic trick well known to statisticians is to add $-\bar{x} + \bar{x}$ inside each term $(x_i - \mu)$ in the sum. Squaring the terms and summing expresses the likelihood function in terms of two sample statistics, $\bar{x}$ and s^2:

$$L(\mu, \sigma^2) = (\sigma^2 2\pi)^{-n/2} \exp\{-(2\sigma^2)^{-1}[(n-1)s^2 + n(\mu - \bar{x})^2]\}. \tag{1}$$

Only the numbers $\bar{x}$ and s^2 are needed to calculate the likelihood for any particular values of μ and σ^2; once $\bar{x}$ and s^2 are in hand, the original data values are not required further for estimating the model parameters. The statistics $\bar{x}$ and s^2 are said to be *jointly sufficient* for μ and σ^2.

Because likelihood functions are typically products, algebraic and computational operations are often simplified by working with the log-likelihood function, $\ln L$:

$$\ln L(\mu, \sigma^2) = -(n/2)\ln(2\pi) - (n/2)\ln \sigma^2 - (n - 1)s^2/(2\sigma^2)$$
$$- n(\mu - \bar{x})^2/(2\sigma^2). \tag{2}$$

Modern frequentist statistics can be said to have been inaugurated in 1922 by R. A. Fisher, who first realized the importance of the likelihood function (Fisher, 1922). Fisher noted that the likelihood function offers a way of using data to *estimate* the parameters in a model if the parameter values are unknown. Subsequently, J. Neyman and E. S. Pearson used the likelihood function to construct a general method of statistical hypothesis testing; that is, the use of data to select between two rival statistical models (Neyman and Pearson, 1933). More recently, the work of H. Akaike (1973, 1974) launched a class of likelihood-based methods for model selection when there are more than two candidate models from which to choose.

Two cases for inferences about μ will be considered: σ^2 known and σ^2 unknown. The "σ^2 known" case is obviously of limited practical usefulness in ecological work. However, it allows a simple and clear contrast between the Bayesian and frequentist approaches. The "σ^2 unknown" case highlights the differences in how so-called nuisance parameters are handled in the Bayesian and frequentist contexts and also hints at the numerical computing difficulties attendant on the use of more realistic models. I concentrate on point estimates, hypothesis tests, and confidence intervals.

Known σ^2

We assume that σ^2 is a known constant, say, $\sigma^2 = 36$.

Fisher (1922) developed the concept of *maximum likelihood* (ML) estimation. The value of μ, call it $\hat{\mu}$, that maximizes the likelihood function (eq. 1) is the ML estimate. The ML estimate also maximizes the log-likelihood function (eq. 2). It is a simple calculus exercise to show that eq. 2 is maximized by $\hat{\mu} = \bar{x}$. Thus, a point estimate for μ calculated from the sample is $\hat{\mu} = 50.6$.

ML point estimates were shown by Fisher (1922) and numerous subsequent investigators to have many desirable statistical properties, among them: (a) asymptotic unbiasedness (statistical distribution of estimate approaches a distribution with the correct mean as n becomes large); (b) consistency (distribution of the parameter estimate concentrates around the true parameter value as n becomes large); (c) asymptotic normality (distribution of the parameter estimate approaches a normal distribution, a celebrated central limit theorem–like result); and (d) asymptotic efficiency (the asymptotic variance of the parameter estimate is as small as is theoretically possible). These and other properties are thoroughly covered by Stuart and Ord (1991). Deriving these properties forms the core of a modern Ph.D.-level mathematical statistics course (Lehmann, 1983).

These statistical properties refer to behavior of the estimate under hypothetical, repeated sampling. To illustrate, the whole population of possible samples induces a whole population of possible ML estimates. In our case, to each random sample $X_1, X_2, \ldots, X_n$ there corresponds a sample mean $\overline{X}$, a random variable. The frequency distribution of the possible estimate values is the *sampling distribution* of the ML estimate. The sampling distribution plays no role in Bayesian inference, but is a cornerstone of frequentist analyses.

When reading statistics papers, one should note that the "hat" notation for estimates (e.g., $\hat{\mu}$) is frequently used interchangeably to denote both the random variable $(\overline{X})$ as well as the realized value $(\overline{x})$. This does not create confusion for statisticians, because the meaning is usually clear from context. However, the distinction can trip up the unwary. A quick test of one's grasp of statistics is to define and contrast μ, $\overline{x}$, $\overline{X}$, and $\hat{\mu}$ (ornery professors looking for curveballs to throw at Ph.D. candidates during oral exams, please take note).

In our example, the sampling distribution of the ML estimate is particularly simple. The exact sampling distribution of $\overline{X}$ is a normal distribution with a mean of μ and a variance of σ^2/n. The independent normal model for the observations is especially convenient because the sampling distribution of various statistics can be derived mathematically. In other models, such as the multinomial models used in categorical data analysis or the dependent normal models used in time series analysis, the sampling distributions cannot be derived exactly and instead are approximated with asymptotic results (central limit theorem, etc.) or studied with computer simulation.

A *statistical hypothesis test* is a data-driven choice between two statistical models. Consider a fixed reference Cu concentration, μ_0, that has to be maintained or attained, for instance, $\mu_0 = 48$. One position is that the reference concentration prevails in the stream, the other position is that it does not. The positions can be summarized as two statistical hypotheses; these are H_0: the observations arise from a normal (μ_0, σ^2) distribution, and H_1: the observations arise from a normal (μ, σ^2) distribution, where μ is not restricted to the value μ_0. In beginning statistics texts, these hypotheses are often stated as H_0: $\mu = \mu_0$, H_1: $\mu \neq \mu_0$. A decision involves two possible errors, provided the normal distribution portion of the hypotheses is viable. First, H_0 could be true but H_1 is selected (type I error); second, H_1 could be true but H_0 is selected (type II error). Both errors have associated *conditional* probabilities: α, the probability of erroneously choosing H_1, given H_0 is true, and β, the probability of erroneously choosing H_0, given H_1 is true. Both of these error probabilities are set by the investigator. One probability, typically α, is set arbitrarily at some low value, for instance .05 or .01. The corresponding hypothesis assumed true, H_0, is termed the *null* hypothesis. The other probability is controlled by the design of the sample or experiment (sample size, etc.) and the choice of test statistic. The hypothesis assumed true, in this case H_1, is the *alternative* hypothesis.

Several important points about statistical hypothesis tests must be noted. First, α and β are not the probabilities of hypotheses, nor are they the unconditional probabilities of committing the associated errors. In frequentist statistics, the probability that H_0 is true is either 0 or 1 (we just do not know which), and the unconditional probability of committing a type I error is either α or 0 (we do not know which). In the frequentist view, stating that "H_0 has a 25% chance of being true" is meaningless with regard to inference.

Second, the simpler hypothesis, that is, the statistical model that has fewer parameters and is contained within the other as a special case, is usually designated as the null hypothesis, for reasons of mathematical convenience. The sampling distributions of test statistics under the null hypothesis in such situations are often easy to derive or approximate.

Third, statistical theory accords no special distinction between the null and alternative hypotheses, other than the difference by which the probabilities α and β are set. The hypotheses are just two statistical models, and the test procedure partitions the sample space (the collection of all possible samples) into two sets: the set for which the null model is selected, and the set for which the alternative is selected.

Fourth, the concordance of the statistical hypothesis with a scientific hypothesis is not a given, but is part of the craft of scientific investigation. Just because an investigator ran numbers through PROC this-or-that does not mean that the investigator has proved anything to anyone. The statistical hypothesis test can enter into scientific arguments in many different ways, and weaving the statistical results effectively into a body of scientific evidence is a difficult skill to master. Ecologists who have become gun-shy about hypothesis testing after reading a lot of hand-wringing about the misuse of null hypotheses and significance testing will find the discussions of Underwood (1990) and Mayo (1996) more constructive.

In our stream example, the hypothesis test is constructed as follows. The likelihood function under the null hypothesis is compared to the maximized likelihood function under the alternative hypothesis. The likelihood function under the null hypothesis is eq. 1, evaluated at $\mu = \mu_0 (= 48)$ and $\sigma^2 = 36$. The maximized likelihood function under the alternative hypothesis is eq. 1, evaluated at $\mu = \hat{\mu} = \bar{x} = 50.6$ and $\sigma^2 = 36$. The likelihood ratio statistic, $L(\hat{\mu}, \sigma^2)/L(\mu_0, \sigma^2)$, or a monotone function of the likelihood ratio such as

$$G^2 = -2 \ln[L(\mu_0, \sigma^2)/L(\hat{\mu}, \sigma^2)]$$

forms the basis of the test. High values of the test statistic G^2 favor the alternative hypothesis, while low values favor the null. The decision whether to reject the null in favor of the alternative will be based on whether the test statistic exceeds a *critical value* or cutoff point. The critical value is determined by α and the statistical sampling distribution of the test statistic.

A well-known result, first derived by S. S. Wilks (1938), provides the approximate sampling distribution of G^2 for many different statistical models. If the null hypothesis is true, then G^2 has an asymptotic chi-square distribution with 1 degree of freedom, under hypothetical repeated sampling. (The number of degrees of freedom in Wilks's result is the number of independent parameters in H_1 minus the number of independent parameters estimated in H_0, or $1 - 0 = 1$ in our example.) Using this result, one would reject the null hypothesis if G^2 exceeded $\chi^2_\alpha(1)$, the $100(1 - \alpha)$th percentile of a chi-square (1) distribution ($\chi^2_{.05}(1) \approx 3.843$). Because our example involves observations from the mathematically convenient normal distribution, the sampling distribution can be calculated exactly. Letting $Z = (\bar{X} - \mu_0)/\sqrt{\sigma^2/n}$, the expression for G^2 can be algebraically rearranged (using the $+\bar{X} - \bar{X}$ trick again; the upper case $\bar{X}$ reminds us that hypothetical repeated sampling is being considered): $G^2 = Z^2$. Because Z has a standard normal distribution, the chi-square result for G^2 is exact (square of a standard normal has a chi-square (1) distribution). The decision to reject can be based on the chi-square percentile, or equivalently on whether $|Z|$ exceeds $z_{\alpha/2}$, the $100(1 - \alpha/2)$th percentile of the standard normal distribution ($z_{.025} \approx 1.960$).

For the stream example, the attained value of Z is $z = (50.6 - 48)/\sqrt{36/10}$ ≈ 1.37. For $\alpha = .05$, the critical value of 1.96 is not exceeded. We conclude that the value $\mu_0 = 48$ is a plausible value for μ; there is not convincing evidence otherwise. The P-value, or attained significance level, is the probability that Z for a hypothetical sample would be more extreme than the attained value z under the null model. From the normal distribution, $P[|Z| > 1.37] = P \approx .17$. If the test statistic has exceeded the critical value, then also P will be less than α.

Confidence intervals and hypothesis tests are two sides of the same coin. A confidence interval (CI) for μ can be defined in terms of a hypothesis test: it is the set of all values of μ_0 for which the null hypothesis $H_0: \mu = \mu_0$ would not be rejected in favor of the alternative $H_1: \mu \neq \mu_0$. A CI can be considered a set of plausible values for μ. The sets produced under hypothetical repeated sampling would contain the true value of μ an average of $100(1 - \alpha)\%$ of the time. The form of the interval is here

$$(\bar{x} - z_{\alpha/2}\sqrt{\sigma^2/n}, \bar{x} + z_{\alpha/2}\sqrt{\sigma^2/n}).$$

The realized CI for the stream example, with $\alpha = .05$, is

$$(50.6 - 1.96\sqrt{36/10}, 50.6 + 1.96\sqrt{36/10}) = (46.9, 54.3).$$

Note that under the frequentist interpretation of the interval, it is not correct to say that $P[46.9 < \mu < 54.3] = 1 - \alpha$. The interval either contains μ or it does not; we do not know which. The concept of a CI can be likened to playing a game of horseshoes in which you throw the horseshoe over a wall that conceals the stake. Your long-run chance of getting a ringer might be 95 percent, but once an individual horseshoe is thrown, it is either a ringer or it is not (you just do not know which).

There are *one-sided* hypothesis tests, in which the form of the alternative hypothesis might be $H_1: \mu \geq \mu_0$ (or $\leq$), and associated one-sided confidence intervals (see Bain and Engelhardt, 1992). The one-sided test or CI might be more appropriate for the stream example, if, for instance, the data are collected to provide warning as to whether an upper level, μ_0, of Cu concentration has been exceeded.

Unknown σ^2

Realistic modeling studies must confront the problem of additional unknown parameters. Sometimes the whole model is of interest, and no particular parameters are singled out for special attention. Other times, as in the stream example, one or more parameters are the focus, and the remaining unknown parameters ("nuisance parameters") are estimated out of necessity.

The parameter σ^2 in the normal model is the perennial example in the statistics literature of a nuisance parameter. That the estimate of μ becomes more uncertain when σ^2 must also be estimated was first recognized by W. S. Gosset (Student, 1908). The problem of nuisance parameters was refined by numerous mathematical-statistical investigators after the likelihood concept became widely known (Cox and Hinkley, 1974, is a standard modern reference).

In frequentist statistics, a leading approach is to estimate σ^2 (or any other nuisance parameter) just as one would estimate μ. The approach has the advantage of helping subsidiary studies of the data; for instance, in a monitoring study (such as the stream example), one might have an additional interest in whether or not σ^2 has changed. In this case, σ^2 is not really a nuisance, but rather an important component of the real focus of study: the model itself.

For estimation, the likelihood function (eq. 1) is regarded as a joint function of the two unknowns, μ and σ^2. The ML estimates of μ and σ^2 are those values that jointly maximize the likelihood (eq. 1) or log-likelihood (eq. 2). A simple calculus exercise sets partial derivatives of $\ln L(\mu, \sigma^2)$ with respect to μ and σ^2 simultaneously equal to 0. The resulting ML estimates are $\hat{\mu} = \bar{x}$, $\hat{\sigma}^2 = \dfrac{(n-1)}{n} s^2$. Note that the ML estimate of σ^2 is not the sample variance (which contains $n - 1$ in the denominator). An estimate that adjusts for a small-sample bias is $\tilde{\sigma}^2 = s^2$. The ML estimate of σ^2, however, has smaller mean-squared error in small samples; $\hat{\sigma}^2$ and $\tilde{\sigma}^2$ are virtually identical in large samples.

Hypothesis tests and confidence intervals for μ again revolve around the likelihood ratio statistic. With additional unknown parameters in the model, the statistic compares the likelihood function maximized (over the remaining parameters) under the null hypothesis H_0: $\mu = \mu_0$, with the likelihood maximized (over all the parameters including μ) under the alternative hypothesis H_1: $\mu \neq \mu_0$. When $\mu = \mu_0$, the value of σ^2 that maximizes $\ln L(\mu, \sigma^2)$ (eq. 2) is

$$\hat{\sigma}_0^2 = \frac{(n-1)}{n} s^2 + (\mu_0 - \bar{x})^2 = \frac{1}{n} \sum_{i=1}^{n} (x_i - \mu_0)^2.$$

The (log-) likelihood ratio statistic is

$$G^2 = -2 \ln \left[\frac{L(\mu_0, \hat{\sigma}_0^2)}{L(\hat{\mu}, \hat{\sigma}^2)} \right],$$

where in the brackets is the ratio of the null and alternative likelihoods, evaluated at the ML estimates. With some algebraic rearrangement (the $+\bar{x} - \bar{x}$ trick again), G^2 becomes

$$G^2 = -n \ln \left(\frac{\hat{\sigma}^2}{\hat{\sigma}_0^2} \right) = -n \ln \left[\frac{\sum_{i=1}^n (x_i - \bar{x})^2}{\sum_{i=1}^n (x_i - \bar{x})^2 + n(\bar{x} - \mu_0)^2} \right]$$

$$= -n \ln \left(\frac{1}{1 + \dfrac{t^2}{n-1}} \right),$$

where

$$t = \frac{\bar{x} - \mu_0}{\sqrt{s^2/n}}$$

is recognized as Student's t-statistic. The hypothesis test can be based on the asymptotic chi-square (1) sampling distribution of G^2, or better yet, on the known exact distribution of $T = (\bar{X} - \mu_0)/\sqrt{S^2/n}$. One rejects $H_0: \mu = \mu_0$ in favor of $H_1: \mu \neq \mu_0$ if $|T|$ exceeds $t_{\alpha/2,n-1}$, the $100[1 - (\alpha/2)]$th percentile of the Student's t-distribution with $n - 1$ degrees of freedom.

For the stream example with $\mu_0 = 48$, the attained value of T is $t = (50.6 - 48)/\sqrt{25/10} \approx 1.64$. For $\alpha = .05$, the critical value of $t_{.025,9} \approx 2.262$ is not exceeded by $|t|$, and we conclude that the value $\mu_0 = 48$ is a plausible value for μ. The P-value for the test is obtained from Student's t-distribution (9 degrees of freedom): $P[|T| > 1.64] = P \approx .14$.

Confidence intervals, as before, can be defined by inverting the hypothesis test. The values of μ_0 for which H_0 is not rejected—that is, for which $|T| \leq t_{\alpha/2,n-1}$—make up the interval

$$(\bar{x} - t_{\alpha/2,n-1}\sqrt{s^2/n}, \bar{x} + t_{\alpha/2,n-1}\sqrt{s^2/n}).$$

This constitutes a $100(1 - \alpha)\%$ CI for μ. The interval also represents a *profile likelihood* CI. For a range of fixed values of μ_0, $L(\mu_0, \sigma^2)$ is maximized (over σ^2 values) and compared to $L(\hat{\mu}, \hat{\sigma}^2)$ (the maximized likelihood under the model H_1). The set of μ_0 values for which $G^2 \leq k$, where k is some fixed constant, is a profile likelihood CI. In the above interval, $k = n \ln\left[1 + \dfrac{(t_{\alpha/2,n-1})^2}{n-1}\right]$, the critical value of the likelihood ratio test using the exact Student's t-distribution. In nonnormal models, k is typically a percentile of the chi-square distribution used to approximate the sampling distribution of G^2. Frequently for such models, repeated numerical maximizations are necessary for calculating profile likelihood intervals.

The stream example gives a 95% CI of

$$(50.6 - 2.262\sqrt{25/10}, \, 50.6 + 2.262\sqrt{25/10}) = (47.0, 54.2).$$

This CI represents a range of plausible values for μ, taking into account the uncertainty of estimation of σ^2. The word "plausible" has not yet been commandeered by Bayesians, and I use it here in its perfectly good English meaning. The word is descriptive of the interpretation of a CI as a hypothesis test inversion; no lurking Bayesian undertones need be inferred.

Much of standard introductory statistics, in the form of t-tests, tests of independence in contingency tables, analysis of variance, and regression, can be understood in the context of the above concepts. In particular, normal linear models (analysis of variance and regression) are formed by allowing μ to be reparameterized as

$$\mu = \beta_0 + \beta_1 r_1 + \beta_2 r_2 + \cdots + \beta_m r_m,$$

where $\beta_0, \beta_1, \ldots, \beta_m$ are unknown parameters and $r_1, r_2, \ldots, r_m$ are values of covariates (indicator variables or predictor variables).

Model Evaluation
Once the point estimates are calculated, tests are performed, and confidence intervals are reported, the job is not done. The estimates and tests have valid sample space statistical properties only if the model is a reasonable approximation of how the original data arose. *Diagnostics* are routine checks of model adequacy. They include examining residuals (in this case, $x_i - \hat{\mu}$) for approximate normality via normal quantile-quantile plots or tests, tests for outliers or influential values, and graphical plotting of model and data. The model implicit in the statistical analysis is to be questioned, and if found wanting, some other model might be necessary.

Such model checking, it must be noted, involves sample space properties of the model. If the correct model is being used, observations of the process are expected to be in *control,* that is, within the usual boundaries of model-predicted variability. A process out-of-control is indicated by wayward observations and calls for further investigation. This is a standard principle of *quality control,* which involves the systematic and routine use of statistical models to monitor variability and is used in virtually all modern manufacturing processes (Vardeman, 1994).

WHAT IS BAYESIANISM?

> *What he and I are arguing about is different interpretations of data.* —Duane Gish, in an evolution/creation debate

The concepts of frequentism revolve around hypothetical repetitions of a random process. The probabilities in a frequentist problem are probabilities on a sample space. The quantity α, for instance, is the probability that the sample will land in a particular region of sample space, given that a particular model describes the process.

In Bayesian statistics, sample space probabilities are not used. Instead, probability has a different meaning. Probability in Bayesian statistics is an investigator's personal measure of the *degree of belief* about the value of an unknown quantity such as a parameter.

Let us again turn to the example problem. We have a sample of 10 observations of Cu concentration in a stream. We assume that these observations can be modeled as if they arose independently from a normal distribution with a mean of μ and a variance of σ^2. Again, we treat separately the cases of σ^2 known and σ^2 unknown.

Known σ^2

There is only one unknown parameter, μ. The Bayesian formulates his/her beliefs about the value of μ into a *prior probability distribution.* The prior distribution has pdf denoted by $g(\mu)$ and cdf given by $G(\mu) = \int_{-\infty}^{\mu} g(v)dv$.

The form of $g(\mu)$ must be specified completely by the investigator. There are various ways to do this. One way is to "elicit" such a distribution by determining the odds the investigator would give for betting on various values of μ.

The subsequent formulas work out algebraically if we assume that the form of $g(\mu)$ is a normal pdf with a mean of θ and a variance of τ^2, with the values of θ and τ^2 to be elicited. However, more complicated distributional forms nowadays are possible to implement in practice. Skewed gamma-type distributions or curves fitted to the investigator's odds declarations can be used.

Our investigator, in this example, works for the mining company upstream. This investigator would give one-to-three odds that the Cu concentration is below 18.65, and three-to-one odds that the Cu concentration is below 21.35. If the 25th and the 75th percentiles of a normal distribution are set at 18.65 and 21.35 respectively, then solving $G(18.65) = .25$, $G(21.35) = .75$ simultaneously gives $\theta \approx 20$, $\tau^2 \approx 4$.

It is worth pausing a moment to reflect on the prior. It is not claimed that μ is a random variable. Indeed, μ is a fixed quantity, and the objective is to estimate its value. Rather, μ is an *unknown* quantity, and *personal beliefs about μ can be represented as if they follow the laws of probability.* This is because the odds that the investigator would give for the value of μ increase smoothly from 0 to $+\infty$ as the value of μ increases from $-\infty$ to $+\infty$. Such increase and range are the precise properties of a cdf written in terms of odds: $G(\mu)/[1 - G(\mu)]$.

Data, in the Bayesian view, modify beliefs. The data enter the inference through the likelihood function. The likelihood function is as central to Bayesian inference as it is to frequentist inference. However, its interpretation is different under the two outlooks.

In Bayesian statistics, the likelihood arises as a conditional probability model. It is the joint pdf of the process, evaluated at the data, *given values of the unknown parameters.* In other words, the set of beliefs about all possible values of μ and all possible outcomes of the data production process are contained in a joint pdf, say $h(x_1, x_2, \ldots, x_n, \mu)$. The likelihood function (eq. 1) is the conditional pdf of $x_1, x_2, \ldots, x_n$, given μ:

$$L(\mu, \sigma^2) = (\sigma^2 2\pi)^{-n/2} \exp\{-(2\sigma^2)^{-1}[(n-1)s^2 + n(\mu - \bar{x})^2]\}$$
$$= h(x_1, x_2, \ldots, x_n | \mu).$$

The frequentist simply regards the likelihood as a function of possible values of μ, with no underlying probability attached to the μ values.

What is sought in Bayesian statistics is the probability distribution of beliefs after such beliefs have been modified by data. This distribution is known as the *posterior* distribution and is the distribution of μ given the data, $x_1, x_2, \ldots, x_n$. Bayes' theorem in probability is a mathematical result about joint and conditional probability distributions that is not in dispute between frequentists and Bayesians. In the present context, the theorem is used to write the pdf of μ given $x_1, x_2, \ldots, x_n$, denoted $g(\mu|x_1, x_2, \ldots, x_n)$, in terms of the likelihood function and the prior distribution:

$$g(\mu|x_1, x_2, \ldots, x_n) = Ch(x_1, x_2, \ldots, x_n|\mu)g(\mu).$$

The quantity C is a *normalization constant* that causes the area under $g(\mu|x_1, x_2, \ldots, x_n)$ to be equal to 1. It is

$$C = \frac{1}{\int_{-\infty}^{+\infty} h(x_1, x_2, \ldots, x_n|\mu)g(\mu)d\mu}.$$

The calculation of C is the mathematical and computational crux of Bayesian methods. Obtaining C is algebraically straightforward for the forms of the prior and the likelihood selected in our stream example. The quantity μ appears quadratically in the exponential function in the product $h(x_1, x_2, \ldots, x_n|\mu)g(\mu)$, and so C is related to the integral of a normal distribution. The end result is that the posterior pdf is that of a normal distribution

$$g(\mu|x_1, x_2, \ldots, x_n) = (\tau_1^2 2\pi)^{-1/2} \exp[-(\mu - \theta_1)^2/(2\tau_1^2)],$$

where the mean θ_1 is

$$\theta_1 = \left(\frac{\sigma^2}{\sigma^2 + \tau^2 n}\right)\theta + \left(\frac{\tau^2 n}{\sigma^2 + \tau^2 n}\right)\bar{x},$$

and the variance τ_1^2 is

$$\tau_1^2 = \frac{\sigma^2\tau^2}{\sigma^2 + \tau^2 n}.$$

The mean θ_1 is a weighted combination of the prior mean, θ, and the sample mean of the data, $\bar{x}$. As the sample size increases, the weight on the prior mean decreases, approaching 0 in the limit as $n \to \infty$.

The point estimate of μ in Bayesian statistics is usually taken to be the expected value of the posterior distribution: θ_1. This estimate can be regarded as a *prediction* of the value of μ. The posterior distribution, like the prior, represents degree of belief. The prior prediction was θ, and the posterior prediction θ_1 quantifies how the prior prediction has been modified by the advent of the data. In our stream example, the ten data points changed the prior prediction of $\theta = 20$ into the posterior prediction of $\theta_1 \approx 36.1$. The variance of the posterior is $\tau_1^2 \approx 1.89$.

One should note that the Bayesian point estimate of μ is *biased* in the frequentist sense. If hypothetical repetitions of the sampling process were imagined (for the same Bayesian with the same prior), the frequency distribution of the Bayesian's estimates would be off-center from μ. If we denote by Θ_1 the sample-space random version of the point estimate θ_1, then the expected value of Θ_1 over the sample space is

$$E(\Theta_1) = \left(\frac{\sigma^2}{\sigma^2 + \tau^2 n}\right)\theta + \left(\frac{\tau^2 n}{\sigma^2 + \tau^2 n}\right)E(\bar{X})$$

$$= \mu + \left(\frac{\sigma^2}{\sigma^2 + \tau^2 n}\right)(\theta - \mu).$$

The amount of bias is seen to be the difference $\theta - \mu$ (bias in the prior) multiplied by the Bayesian weight. In the stream example, if the null hypothesis ($\mu = 48$) were true, the amount of bias in the Bayesian estimate is about -13.3; this Bayesian's long-run frequency distribution of estimates would be centered at a distance more than twice the standard deviation ($\sigma = 6$) below the true value of μ.

The posterior distribution also yields to the Bayesian information about the uncertainty with which to regard the prediction. One way to summarize this uncertainty is the Bayesian credible interval, formed by taking an interval containing $100(1 - \alpha)\%$ of the probability in the posterior density. The smallest such interval is the *highest probability region* (HPR). The HPR is analogous to the confidence interval of frequentist statistics but has a much different interpretation. The Bayesian asserts that there is a 95% chance that μ is within a given 95% HPR, because probability represents belief on a parameter space (all possible values of μ). The frequentist cannot assert that there is a 95% chance that μ is within a given 95% CI, because probability to a frequentist represents long-run frequency on a sample space (all possible outcomes of the sample). With our normal model, the HPR region is the interval centered at the posterior mean, θ_1, containing $100(1 - \alpha)\%$ of the area under the posterior density:

$$(\theta_1 - z_{\alpha/2}\sqrt{\tau_1^2}, \theta_1 + z_{\alpha/2}\sqrt{\tau_1^2}).$$

In the stream example, the 95% HPR is (33.4, 38.8), which is quite different from the 95% confidence interval (46.9, 54.3) obtained under the frequentist approach. However, as the sample size becomes large and the data swamp the prior beliefs, the HPR in this normal-based example converges rapidly to the confidence interval. In other words, the Bayesian and the frequentist will report essentially the same interval estimate for μ if good data are available. While this asymptotic behavior of HPRs is typical for standard statistical models, it is somehow not a comforting point of agreement for Bayesians and frequentists, in that the interpretation of the two intervals is so different. Also, for some models and circumstances, the rate of convergence of the Bayesian HPR to the frequentist CI is alarmingly slow (see discussion of σ^2 unknown, below).

A key aspect of Bayesianism is adherence to the *likelihood principle.* The principle states that sample space probabilities are irrelevant to inferences about unknown parameters. The data only influence the inferences through the likelihood function. This principle is embodied in the posterior density, $g(\mu|x_1, x_2, \ldots, x_n)$. All inferences about μ are contained in the posterior density and are phrased in terms of probabilities on parameter space. Only the data actually observed appear in the posterior (via the likelihood function); no hypothetical data, such as a critical value for $\bar{x}$, or probabilities of hypothetical data, such as P-values or type I and II error probabilities, are considered in the conclusions about μ.

Bayesians are adamant on this point (Lindley, 1990; Berger and Berry, 1988). Type I and II error probabilities and P-values are probabilities of "data that didn't happen," and Bayesians question what relevance such quantities could possibly have for conclusions about a parameter.

The use of sample space probabilities in frequentist statistics has surprising, and to Bayesians, undesirable consequences. Foremost is the dependence of the statistical conclusions on the *stopping rule* of the experiment. For instance, were the stream samples drawn sequentially, one by one, until some threshold high or low value of $\bar{x}$ was attained? Or were simply ten samples drawn? Or did the investigator actually draw eleven samples but drop one jar accidentally?

Unknown σ^2

Bayesians claim that the treatment of nuisance parameters within the Bayesian framework is one of the key advantages of their approach. Let us examine how this claim operates in practice.

With σ^2 unknown, the concept behind the Bayesian analysis is straightforward. The posterior distribution for μ, represented by the pdf $g(\mu|x_1, x_2, \ldots, x_n)$, is still sought. First though, beliefs about μ and σ^2 must be summarized in a joint prior distribution for μ and σ^2. So represented, the beliefs are entered into the mix with the likelihood function (eq. 1), and the posterior distribution for μ is then obtained (at least in principle) with Bayes' theorem.

The joint prior pdf for μ and σ^2, denoted $g(\mu, \sigma^2)$, is that of a bivariate continuous distribution. The distribution would presumably be defined for positive real values of σ^2 and real (or positive real) values of μ. A joint distribution in general would contain some correlation between μ and σ^2. Rarely, however, can any dependence of beliefs about μ on those about σ^2 be acknowledged or elicited. Consequently, the form often proposed for the joint pdf is a product of univariate prior pdfs for μ and σ^2:

$$g(\mu, \sigma^2) = g_1(\mu)g_2(\sigma^2).$$

Here $g_1(\mu)$ is a pdf for μ (such as the normal pdf in the σ^2-known case above), and $g_2(\sigma^2)$ is a pdf for σ^2. The form of $g_2(\sigma^2)$ selected by the investigator could be a gamma, reciprocal gamma, lognormal, or other distribution on the positive real line. The product form of $g(\mu, \sigma^2)$ assumes (or implies) that the prior information about μ is independent of that of σ^2.

With the elicited joint prior in hand, the analysis proceeds via Bayes' theorem. The joint posterior pdf for μ and σ^2 is proportional to the product of the prior pdf and the likelihood function

$$g(\mu, \sigma^2|x_1, x_2, \ldots, x_n) = C_1 h(x_1, x_2, \ldots, x_n|\mu, \sigma^2)g_1(\mu)g_2(\sigma^2).$$

The likelihood function is again written as $h(x_1, x_2, \ldots, x_n|\mu, \sigma^2)$ to emphasize its role as a conditional pdf. The constant C_1 is the normalization constant given by

$$C_1 = \frac{1}{\int\int h(x_1, x_2, \ldots, x_n|\mu, \sigma^2)g_1(\mu)g_2(\sigma^2)d\mu\, d\sigma^2},$$

where the integrals are over the ranges of μ and σ^2 in the prior pdfs. Some remarks about the daunting process of obtaining C_1 are given below. In principle, the joint posterior pdf $g(\mu, \sigma^2|x_1, x_2, \ldots, x_n)$ contains all the beliefs about μ and σ^2, updated by the data $x_1, x_2, \ldots, x_n$. Moreover, the nuisance parameter σ^2 is vanquished by integrating it out of the joint posterior to get the posterior marginal distribution for μ:

$$g(\mu | x_1, x_2, \ldots, x_n) = \int_0^\infty g(\mu, \sigma^2 | x_1, x_2, \ldots, x_n) d\sigma^2.$$

This posterior pdf for μ reflects all beliefs about μ after the advent of the data. The pdf could be used, for instance, to obtain an HPR for μ, just as was done in the σ^2-known case above.

The technical difficulties with the analysis reside in evaluating the multiple integrals for C_1 and in integrating out σ^2 to get the marginal posterior for μ. For nearly all forms of prior distributions $g_1(\mu)$ and $g_2(\sigma^2)$, the integrals must be performed numerically. Up until the middle 1980s, the lack of symbolic results for the integrals were the death knell for Bayesianism, because methods for reliably evaluating multidimensional integrals were poorly developed. However, clever simulation methods were devised for these integrals; the methods exploit the fact that the integrals are essentially expected values of functions with respect to the prior distributions. Papers on Bayesian analyses in the statistics literature subsequently exploded in number, starting in the late 1980s. The simulation methods have been, for a decade, a part of the hidden culture of statisticians, described tersely or implicitly in dense mathematical terms in the statistics literature, but are now receiving excellent expositions for broader scientific audiences (for instance, Robert and Casella, 1999; Press, 2003). Investigators must be warned, however, that the numerical methods at present involve heavy computer programming efforts, postcalculus statistics knowledge, and sometimes days of computer time; the methods are not ready yet for routine use by busy laboratory or field scientists. For example, the developers of the WinBUGS software, one of the best present-day entrance points for Bayesian calculations, state on their current website (as of spring 2003) that "WinBUGS aims to make practical MCMC methods available to applied statisticians." Certainly though, software development is rapidly catching up to the Bayesian ideas, and scientists who are not "applied statisticians" can expect soon to be able to join in the fun of drawing conclusions in the absence of, or even in spite of, good data.

What if the investigator does not really have, or is unwilling to admit, any prior beliefs about σ^2? Bayesian writers have proposed "uninformative priors" for such situations. Use of these priors has also been advocated for situations in which investigators disagree about the prior information and require a relatively "neutral" prior for mediation (Lee, 1989). However, there are different approaches to specifying neutral priors. One is the *maximum entropy* approach (Jaynes, 1968). The investigator in the maximum entropy approach specifies only numerical summaries of the prior distribution, such as the mean alone or both the mean and the variance. The prior is then the distribution that maximizes the "entropy content" (expected value of $-\ln g_2(\sigma^2)$) of the prior, while retaining the numerical summaries. If the mean of the prior for σ^2 is fixed at ϕ, for example (and the range is taken to be the positive real line), the maximum entropy criterion yields an exponential distribution for σ^2 with pdf $g_2(\sigma^2) = (1/\phi) \exp(-\sigma^2/\phi)$. Another approach is that of the *uniform prior:* the beliefs about σ^2 are taken to have a uniform distribution. This type of prior is sometimes called an *improper prior* because it is not integrable over the entire range of the parameter (here, the positive real line). Actually, the uniform distribution for σ^2 is taken to range properly from 0 to, say, some large unspecified number γ. The prior pdf (a constant, $1/\gamma$) is then integrable, and the value of γ, if large, turns out to affect the calculations about μ only negligibly.

A conceptual problem with uninformative priors is that ignorance about σ^2, expressed in an uninformative prior distribution for σ^2, does not translate into ignorance about a function of σ^2, say $\ln(\sigma^2)$. For instance, if σ^2 has a uniform distribution on the interval from 0 to γ, then the distribution of $\ln(\sigma^2)$ is nonuniform. A uniform prior distribution for σ^2 leads to a different posterior distribution for μ than when a uniform prior for $\ln(\sigma^2)$ is used. This disparity has motivated some Bayesians to investigate how to choose the scale upon which their ignorance is to be expressed. Textbook discussions of such investigations gravitate to scales that allow convenient algebra, i.e., scales for which the problematic integrals noted above can be evaluated symbolically (e.g., Lee, 1989).

So that some numerical results might be displayed for the stream example without having to refer to a workstation, let us employ such a scale for σ^2. Suppose the prior distribution for $\ln(\sigma^2)$ is taken to be a uniform distribution on some large, unspecified interval of the real line. Then, from the transformation rule for distributions (Rice, 1995), the prior distribution for σ^2 has a pdf of the form $g_2(\sigma^2) = C_2/\sigma^2$, where C_2 is a constant. This prior is improper on the entire positive real line, but again it will be thought of as ranging from 0 to some large but unspecified upper value. The joint prior distribution for μ and σ^2 becomes, assuming independence of beliefs about the two parameters, the product of marginal prior pdfs:

$$g(\mu, \sigma^2) = C_2(\sigma^2\sqrt{\tau^2 2\pi})^{-1} \exp[-(\mu - \theta)^2/(2\tau^2)].$$

Substituting this joint prior into the expression for the posterior distribution for μ and σ^2 above, and using the normal likelihood function (eq. 1) for $h(x_1, x_2, \ldots, x_n|\mu, \sigma^2)$, one obtains

$$g(\mu, \sigma^2|x_1, x_2, \ldots, x_n) = C_1(\tau^2)^{-1/2}(2\tau)^{-(n+1)/2}(\sigma^2)^{-(n+2)/2}$$
$$\times \exp\left\{-\frac{1}{2\sigma^2}(n-1)s^2 - \frac{1}{2\sigma^2}n(\mu - \bar{x})^2 - \frac{1}{2\tau^2}(\mu - \theta)^2\right\}.$$

This posterior pdf for μ and σ^2 is a dome-shaped function reflecting the joint beliefs about the two parameters after the advent of the data. The nuisance parameter is now eliminated in an act that is to Bayesians conceptually as well as algebraically symbolic. The terms in the posterior involving σ^2 are (thanks to our selection of prior) in the form $(\sigma^2)^{-\alpha}\exp(-b/\sigma^2)$. The form is like a reciprocal gamma pdf and yields $b^{-(\alpha-1)}\Gamma(\alpha - 1)$ when integrated over the positive real line. Thus

$$g(\mu|x_1, x_2, \ldots, x_n) = \int_0^\infty g(\mu, \sigma^2|x_1, x_2, \ldots, x_n)d\sigma^2$$
$$= C_3\left[\frac{(n-1)s^2}{2} + \frac{n(\mu - \bar{x})^2}{2}\right]^{-n/2}\exp\left[-\frac{(\mu - \theta)^2}{2\tau^2}\right]$$
$$= C_3\left[1 + \frac{t^2}{(n-1)}\right]^{-n/2}\exp\left[-\frac{(\mu - \theta)^2}{2\tau^2}\right].$$

Here $t = \sqrt{n}(\bar{x} - \mu)/s$ is the t-statistic that would be used by frequentists to test a particular value of μ as a null hypothesis. The posterior distribution for μ is seen to be the prior normal pdf for μ weighted by the pdf of a Student's t-distribution. Rather awkwardly, the normalization constant C_3 cannot be evaluated symbolically, and so to obtain a point estimate or HPR, the workstation must be booted up even for this simple illustrative example. Fortunately, the numerical integration for one dimension is straightforward.

The 95% HPR resulting for the stream example (using, as will be recalled, $\bar{x} = 50.6$, $s^2 = 25.0$, $n = 10$, $\theta = 20.0$, $\tau^2 = 4.0$) is approximately (17.3, 25.3). Recall that the frequentist 95% confidence interval based on the Student's t-distribution was (47.0, 54.2). In the previous case, in which σ^2 was taken as a known constant ($\sigma^2 = 36$), the HPR (33.4, 38.8) was considerably closer to the corresponding frequentist confidence interval (46.9, 54.3). Apparently, when σ^2 is unknown, the preponderance of weight remains in this example on the prior; the HPR is remarkably insensitive to the data. While the frequentist results take at face value the estimate, s^2, of σ^2, the Bayesian results in the face of lack of knowledge about σ^2 contain an inherent preference for beliefs about μ over data. Indeed, one can ask just how much evidence is necessary for the Bayesian here to start noticing the data. What if the values of $\bar{x}$ and s^2 from the stream had resulted from larger sample sizes? Suppose the value of n is increased in the posterior pdf for μ above, while keeping all the quantities fixed at the same values. At $n = 40$, prior beliefs are still heavily weighted: the HPR is (21.8, 30.0). A strangely sudden change of heart occurs between 60 and 65 observations. At $n = 60$, the posterior pdf for μ has developed a prominent "shoulder" near $\mu = 50$, and the upper end of the HPR has started to reach upward; the HPR is (26.1, 40.1). The frequentist 95% confidence interval for $n = 60$ is by contrast only two units wide: (49.3, 51.9). At $n = 63$, the posterior pdf for μ is bimodal, one peak influenced by the prior, and one peak influenced by the data; the HPR is (28.8, 47.1). By $n = 65$, the data peak has grown taller than the prior peak, and the HPR is (31.2, 48.3). By $n = 70$, the laggard lower end of the HPR has finally entered the 40s; the HPR is (42.7, 49.0).

If $\mu = 48$ were cause for alarm, the frequentist scientist would have detected this state of affairs with as few as 10 observations. It would take at least 65 observations before our Bayesian, to whom beliefs are evidence on equal footing with data, would sound a Cu pollution warning.

DISCUSSION

Beliefs
It should be evident from the above example that Bayesian and frequentist statistics arise from different views about science. In Bayesian statistics, beliefs are the currency traded among investigators. Beliefs are evidence. Data are used to modify beliefs.

"Beliefs" is not necessarily a four-letter word. In Bayesian statistics, nothing precludes the prior density being "rationally" constructed based on commonsense information. Indeed, investigators frequently encounter situations in which a parameter is not completely unknown. In our stream example, it is known for a fact that the mean Cu concentration is not a negative quantity. Is it not possible to account for such knowledge in the analysis?

Frequentists account for such information by building more realistic statistical models. The normal distribution is a mathematical approximation. Real Cu concentrations cannot be negative. In the stream example, if the normal approximation is adequate, negative Cu concentrations are wildly improbable, and a negative value would result in an extremely bad model. However, a different distribution model might be required, if, for instance, some concentrations are bunched near zero. Even a new, more realistic model is always subject to questioning, via model evaluation procedures.

Frequentists also use prior information in designing surveys and experiments. For instance, the optimal allocation of samples in a stratified sampling scheme depends on knowing the variances within each stratum (Scheaffer, Mendenhall, and Ott, 1996). In addition, selecting a sample size in experimental design depends on the desired power, which in turn depends on the effect size and the variance (Ott, 1993). The investigator must have some prior information about the effect size and variance for the design to achieve the desired power.

To the frequentist, though, such prior information is regarded as *suspect*. This is a major departure point from the Bayesians. Frequentists build on prior information tentatively, using sample space variability properties constantly for checking the reliability of the knowledge. Prior information is placed in the likelihood function itself, and is thereby vulnerable to empirical challenge.

Bayesians will cry foul at my rough handling of the Bayesian analysis in the stream example. The selection of the prior mean of $\theta = 20$ by the mining company scientist seems lopsided and biased, cynically calculated to come to a predefined conclusion. The scientist is supposed to formulate the prior distribution based on real costs and consequences of being wrong. What if the scientist were forced to deliver on the bet implied by the prior? It must be remembered, however, that the scientist is employed by the mining company.

First, the real cost to the scientist comes not from being far from the truth, but rather from defeat. The two are not necessarily concordant. The scientist has a real financial stake in defending the mining company's view. If this scientist is not willing, some other scientist will gladly step in and cook the numbers.

Second, the scientist can be wrong, or even deceived. It is quite possible that the scientist's beliefs were genuine. During previous monitoring of the stream, say, Cu concentrations might indeed have hovered around μgL^{-1}, and the scientist had no reason to believe that today's samples would be different. In this hypothetical scenario, the company had a pollution event and failed to inform its monitoring group. Either way, the scientist's beliefs do nothing but contaminate the data analysis. They add no legitimate information to the estimate of Cu concentration.

In fact, there is often a conflict of interest between a scientist's beliefs and the truth. In an ideal world, the scientist who is most successful in discovering truth will be the most successful in building a scientific career. While this circumstance is fortunately common, the real world of scientific careers admits additional complexities (Sindermann, 1982). Scientists gain reputations for being advocates of certain theories. Laboratories and research programs grow from mining particular techniques or points of view. A scientist's career is measured in the form of socially warranted visibility: jobs, papers, research grants, citations, and seminar invitations. Young graduate students know one career syndrome well: stubborn adherence of older scientists to old-fashioned explanations and quick dismissal by such scientists of newer ideas before even understanding them. Senior scientists know another career syndrome well: rapid study and advocacy by younger scientists of fashionable new hypotheses that contradict established doctrine and are beyond the frontiers of available data. Everyone in science knows of investigators who took wrong turns toward untenable hypotheses and then spent whole careers defending the hypotheses with contrived arguments. To an individual scientist with a career to build, maintain, and defend, victory, rather than truth, is often the objective.

Scientific Method

Is science just a postmodern "way of knowing" after all? At the level of the individual scientist, it would certainly seem so, given all the explicit and implicit social pressures. Science in the postmodern view is a belief system, and scientists achieve success only by participating in a socially warranted system of thought and action, which changes from place to place and year to year.

Bayesianism, through the incorporation of personal beliefs into statistical analyses, accepts the postmodern view of science. A scientist's acceptance or rejection of a hypothesis is a decision made in light of beliefs influenced by costs or utilities. To the Bayesian, science is improved by explicitly stating, organizing, and acting on beliefs. A scientist summarizes his/her prior beliefs into a probability distribution and modifies those beliefs in a controlled and systematic way with data. Observers are free to quibble with the scientist's prior, or use their own priors and come to their own conclusions. Consensus of beliefs will supposedly emerge as data become more available and priors become diluted. However, the process for this emergence is not clear, for, human nature being what it is, priors will inevitably become more opinionated in the face of growing data. Fundamentally, at the heart of it all, the interpretation of results is in terms of beliefs. In Bayesianism, beliefs are sanctioned, not repudiated.

Modern science, though, has been wildly successful despite the imperfect humans that make up the ranks of scientists, and, incidentally, despite the almost complete absence of Bayesianism in day-to-day scientific life. The postmodern claim that science is socially constructed reality is an intellectual fraud (Sokal and Bricmont, 1998). Hydrogen atoms and the speed of light are the same in India, Alaska, and the Andromeda galaxy. It is true that scientists, and groups of scientists, often come to the wrong conclusions. It is the *process* that is responsible for the enormous gains in understanding we have attained, in ecology and in other disciplines. Our understanding does not just jump from one fashionable paradigm to another; it *improves*. Science is like a river that flows mostly forward, but with slow pools and backcurrents here and there. It is the collective process of empirical investigation, involving weeding out of untenable notions and careful checking of working hypotheses, that makes progress possible. The invisible empirical hand of Galileo, the Adam Smith of science, promotes the emergence of reliable knowledge.

Bayesians and postmoderns alike miss the fundamental idea of science. Science is not about beliefs; science is about skepticism. Science is not about prediction, estimation, making decisions, data collection, or data interpretation. Scientists engage in these activities, but these activities do not constitute science. Science, rather, is about constructing convincing explanations and acquiring reliable knowledge. A "convincing" explanation is one that a reasoned skeptic is forced, by logic and evidence, to accept as at least a serious contender for the true explanation. A "reliable" result can be reproduced by others, who can rely on them for building further explanations. Scientific arguments are aimed at reasoned skeptics. "Reasoned" means open to acknowledging evidence that might contradict prior points of view.

The *scientific method* is a series of logical devices for eliminating or reducing points of reasoned skepticism. One premise of the scientific method is that human judgment is inherently flawed. This is because reasoned skeptics might validly argue in any situation that a scientist's personal beliefs are suspect. Successful scientists seek to counter that criticism by adopting investigative methods that eliminate conscious or unconscious biases. Frequentist analyses are an important tool in the scientific method.

Frequentism accepts only a portion of the postmodern critique. To the frequentist, the actions and behaviors of individual investigators are indeed mired in beliefs. However, to the frequentist, the methods of statistical analysis are set up to discount those beliefs as much as possible. The assumption that a scientist's judgments are not to be trusted has a long history in frequentist statistics and is expressed in the concepts of design-based sampling, replication, randomization, experimental design, unbiased estimation, model diagnostics, and explicit stopping rules.

Frequentist statistics adheres to the principles of the scientific method. Experimental subjects are selected at random. Observations are sampled at random. Variability of the process under study is carefully controlled and modeled, so that future investigators can replicate and check the work. In frequentist hypothesis testing, the skeptic's null hypothesis is *assumed* to be true, but unlike the Bayesian's prior, the assumption is just an argumentative device. The assumption is then found to be tenable or untenable under the data. The statistical models used by the investigator are suspect and must have demonstrated reliability and usefulness for future investigators. By the continual modeling of and referral to sample space variability of a data production process, frequentism can not only show that some hypotheses are untenable in a classic Popperian way, but can also establish that other hypotheses are operationally reliable and can serve as the bases for future studies (Mayo, 1996).

Evidence

Bayesians claim that scientists long for numerical measures of evidence. If only one could attach, in some reliable way, a number to a hypothesis, indicating the relative weight of evidence for that hypothesis as opposed to others, then scientific conclusions would be clearer and more helpful to policy decisions. Why must we avoid doing what seems natural, that is, stating that the chance hypothesis A is true is Q%?

The answer is that the number is scientifically meaningless, and the price is too high. In Bayesian analyses, the evidentiary number cannot exist except in the personal belief system of the investigator. Neither priors nor likelihood functions can be empirically challenged in the Bayesian scheme, and so personal beliefs are always present to some degree in conclusions. With the postmodern foot in the door, the way is opened for limitless political pressure to influence the weight of evidence. Bayesian statistics might seem like a shot in the arm for a stalled science, but Bayesian science unfortunately fails to convince.

The evidentiary number in Bayesian statistics is likelihood, modified by beliefs. Is it possible to eliminate the belief considerations, while retaining likelihood? Various investigators through the years have proposed statistical approaches that accept the likelihood principle but reject the use of priors (Edwards, 1972; Royall, 1997). The relative weight of evidence for hypothesis A over hypothesis B is determined by comparing their likelihoods under these schemes.

The likelihood principle eliminates consideration of any sample space events, other than the actual data outcome, as evidence. But likelihood has little absolute meaning by itself, without appeal to sample space properties. One cannot determine from the statement "ln L_A = -43.7" whether A is a viable model or not. One cannot determine from the statement "ln L_A − ln L_B = 5.8" whether model A is unquestionably better than model B. The viability of a model depends on a variety of things, for instance, on whether it *fits*. A difference in log-likelihoods as big as 5.8 might easily be within the range of variability expected by chance. Without analyses based on hypothetical sample space events, these possibilities cannot be addressed.

Also, the likelihood principle eliminates consideration of stopping rules. Whether the sample size was sequentially determined or fixed, the evidence is the same under the likelihood principle. Unfortunately, we cannot then determine whether or not the investigator's results are unusual under a particular experimental protocol, and consequently we cannot question the likelihood upon which the investigator's conclusions are based. The stopping rule dependence exists because we do not trust the scientist: we insist upon the option of repeating the study to as close a degree as possible.

Finally, the current likelihood-only analyses are developed only for simple hypotheses, that is, for statistical models with no estimated parameters (Royall, 1997). Nuisance parameters are simply not treated (but it is a topic under active study and new developments are emerging; see Royall, 2000). The practical realities of real scientific problems strongly suggest that likelihood-only methods are not yet ready for prime time.

It should be clear by now that evidence in science is not and should not be a single number. Evidence is a structure of arguments, in which each structural piece survives continual and clever empirical challenges.

Tobacco Company Science

There is a class of scientific-appearing people that I call unreasoned skeptics. Unreasoned skeptics do not accept the tenets of the scientific method. They view science as an activity of data interpretation either in light of prior beliefs or to maximize certain utilities. Money, power, and influence are the objects of the scientific game. In this they have a decidedly postmodern outlook. Unreasoned skeptics include tobacco company scientists and biblical creation scientists. It is perhaps fortunate that these professional debaters tend not to know much about statistics, for I fear that they would find Bayesian statistics well suited for their sponsored disinformation campaigns.

EDUCATION

> I have taken eighteen credits of statistics classes, but I still do
> not understand statistics. —*Ph.D. student in wildlife*

The distressed student confessed this to me toward the end of a long, rigorous program of graduate study. The student had taken a succession of graduate statistics "methods" courses, such as regression, analysis of variance, experimental design, nonparametric statistics, and multivariate statistics, virtually the entire "service" offering of the university statistics program, and had worked hard and received near-perfect grades. Yet, the student felt that the subject was still a mystery. My impression, based on twenty years of teaching, research, and consulting in ecological statistics, is that this student's confusion about statistics is not an isolated case, but rather represents the norm in the life sciences. What this student's case illustrates is the sad fact that *the "applied" courses insisted upon for their students by life science educators are designed to perpetuate the confusion.* One can take statistics methods courses until the cows come home and be no nearer to understanding statistics than one is to understanding quantum mechanics.

In this last section of my essay, I offer a prescription for change. It might seem like a digression, but I contend that it is a crucial part of the Bayesian/ frequentist problem. Many ecologists have never really been comfortable with statistical concepts (e.g., they think that a P-value is the probability of a hypothesis, etc.), and this discomfort can be exploited by polemicists.

Ecologists are ill-served by their statistics education. For a science in which statistics is so vital, it is paradoxical that statistics is such a source of insecurity and confusion. It is as if the subject of statistics is a big secret. Ecologists are given glimpses and previews of the subject in their methods courses, but the subject itself is never revealed. Shouldn't ecologists instead be trained to wield statistical arguments with strength and confidence?

The topic of statistics could hardly be more important to a science than it is in ecology. First, ecologists are routinely confronted by nonstandard data. The random mechanisms and sampling schemes encountered in ecology often are not well described by the statistical models underlying off-the-shelf statistical methods. I find it ironic that ecologists spend a fair amount of time and journal space arguing about statistics; quantitatively oriented ecologists even teach statistics and attempt to invent new statistical methods. These are tasks for which ecologists (without the education I discuss below) are by and large untrained. With methods courses, one never learns the foundational principles from which statistics methods arise; one merely learns the methods that have already arisen. No amount of methods courses and no amount of familiarity with computer packages can compensate for this gap in understanding. In particular, jury-rigged attempts to transfer off-the-shelf analyses to nonstandard situations can result in embarrassment and frequently is the subject of useless controversy.

Second, ecological systems are stochastic. Stochastic models are rapidly becoming an integral part of the very theories and concepts of ecology. Yet confusion about stochastic models has often marred published ecological discourse. For instance, the debates on density dependence vs. density independence, a staple in the ecological literature since the 1950s, continue to feature mathematically incorrect statements about persistence, autocorrelation, and statistical tests (see discussion by Dennis and Taper, 1994).

I propose that ecologists take fewer statistics courses. No, that is not a typo. The core of an ecology graduate student's statistical training should be a one-year course sequence in mathematical statistics. The standard "math-stat" course offered at most colleges and universities is an upper division undergraduate course (which usually can be taken for graduate credit). It is where the secret is revealed, and by the way is where statisticians commence training their own students. With this course sequence, statistics will be a source of strength and confidence for any ecologist. Though the math-stat sequence is a tough challenge, the ecologist will be rewarded by needing far fewer methods courses in their educations. The usual math-stat sequence, incidentally, gives balanced coverage to both the frequentist and the Bayesian approaches without developing the scientific issues to any great degree (Bain and Engelhardt, 1992).

Proper preparation for math-stat is essential. Statistics is a postcalculus subject, and that is the heart of the educational problem. There is no way around this fact. The reduced number of methods courses during the training of a Ph.D. will have to be partially compensated by thorough undergraduate calculus preparation. This does not mean "business calculus" or even "calculus for life sciences." This means genuine calculus, taken by scientists and engineers. Genuine calculus should be considered a deficiency in graduate admissions; the one-year math-stat sequence can be undertaken in the first or second year of graduate training. It will help if math-stat is not the first exposure to statistics. The student who has had the usual undergraduate basic statistics course will be continually amazed in math-stat to discover how unified and powerful the concepts of statistics are. This basic curriculum—calculus, introductory statistics, and math-stat—can be followed by selected methods courses of particular interest to the student.

One pleasant surprise to a student taking this curriculum will be that the methods courses are quite easy for someone with math-stat training. Textbooks on many statistics topics become readily accessible for self-study. Imagine reading statistics as one reads biology! Another, perhaps more disconcerting surprise will be that the student will be sought out *constantly* in his/her department by faculty and other grad students as a source of statistical help. The demand for statistics consulting can be overwhelming. But that hints at a third surprise, a very pleasant one indeed: command of statistics greatly enhances employability.

Ecologists-in-training who are particularly interested in theoretical ecology, or who are simply interested in strong prospects for gainful employment as scientists, might consider acquiring a graduate degree in statistics. Our M.S. statistics graduates from the University of Idaho wave at their faculty advisers from their lofty tax brackets.

Ecologists struggling with the Bayesian/frequentist issues will ultimately benefit from greater statistical understanding in general. Statisticians are not going to solve the scientific issues of ecological evidence; ecologists must do that to their own satisfaction. But at their current command level of statistics, ecologists are not ready for the task.

CONCLUDING REMARKS

Ecology is a difficult science because of the large variability in ecological systems and the large cost of obtaining good information. It is difficult also because of the sensitive political nature of its scientific questions and the pressing demand for quick answers. However, if ecologists yield to the Bayesians' call for watering down the standards of evidence, the end result will be tobacco company science, not science.

| 11.1 | Commentary |

Charles E. McCulloch

Even as a dyed-in-the-wool frequentist, I must cry foul with Brian Dennis's rough treatment of Bayesians. Bayesians do test hypotheses and do calculate quantities very similar to (and, in many ways, *better* than) confidence intervals, *P*-values, and standard errors. To say that "Bayesian statistics provides conclusions in the face of incomplete information" is to obfuscate the fact that frequentists always use incomplete data. And his stream example is patently slanted towards a frequentist treatment.

I will elaborate on two of these points. Bayesians can calculate the probability that a hypothesis is false. This has a much more straightforward interpretation than the awkward definition of a frequentist *P*-value and is what many scientists would like to calculate.

Dennis's stream example is smack in the middle of the frequentist domain: it is easy to envision repeated sampling (even indefinitely), and there is only a single, easily estimated source of variability. But what is he to do in devising a scientifically defensible conservation plan for an endangered species in the one location in which it remains? We might have some decent mortality data, only a little fecundity data (except perhaps in related species), and only expert opinion on survival if the habitat becomes fragmented. How do we combine these different sources of information and variability in a framework in which the frequentist paradigm is inappropriate? It is a hard problem but not one that should cause us to bury our heads in the sand.

Even in settings more amenable to frequentist inference, the Bayesian paradigm can be attractive. In an area of interest to me, namely maximum likelihood estimation of variance components, how do we deal with setting confidence intervals? Even the large sample distribution of the estimates is an awkward mixture of point masses (as estimates hit the boundary of zero) and chi-square distributions with unknown mixing fractions. And this doesn't even begin to deal with the small sample distribution, which cannot be written down in closed form. The Bayesian approach, together with modern computing power, straightforwardly handles this problem, even in small and moderate-sized samples.

I must, however, agree with Dennis on the difficulties of specification of prior distributions. In my mind, this is an Achilles' heel of the Bayesian approach. This is especially true in high-dimensional problems, in which the joint distribution of twenty-five parameters might be needed. As Dennis emphasizes, the specification is in many cases not innocuous and must be treated with care.

That said, prior distributions sometimes capture bona fide expert opinion, not just *personal* beliefs, and are a synthesis of data and experience. Dennis also minimizes much of the current research on default, objective, and robust priors, all of which are intended to deal with the difficulty of personalistic priors. Also, ethical Bayesians invariably do sensitivity analyses to assess the influence of their prior specification.

I agree with Dennis in being extremely skeptical of the Bayesian approach in ecology. But let us not try to blame the ills of postmodern society and the ignorance of science by the general populace on the Bayesians. And we should keep in mind the situations in which the Bayes paradigm is attractive and likely to be effective.

Regarding his suggestion as to how to educate ecologists, I say "get with the program." Statistics minors at Cornell (a popular minor for ecologists) have, for over twenty years, been required to take a theory of probability and a theory of statistics course. To this is added a matrix algebra or modeling course, as well as the usual yearlong data analysis and methods sequence. It makes an ecosystem of difference.

| 11.2 | Commentary |

Aaron M. Ellison

I comment on three substantive issues raised by Dennis: (1) the scientific method; (2) differences between Bayesian and frequentist statistics; and (3) statistical education for ecologists. I draw inspiration for this essay from the writings of Neil Postman (e.g., Postman and Weingartner, 1969; Postman, 1995), Paul Feyerabend (1975, 1978, 1987), and David Orr (1992), who, being firmly rooted in the postmodernist camp, provide valuable insights about education, science, and society.[1]

This essay was written during a week of fieldwork in southern Oregon and northern California, supported by a grant from the Packard Foundation. The Darlingtonia fens of the Siskiyou and Klamath Mountains and the coastal redwood forests reminded me why I am a scientist and an ecologist.

1. Dennis's critique of Bayesian inference is motivated in large measure by a strong aversion to the lapping waves of "popular postmodernism" on the shoals of science and the impact of feminist theory in the sciences. In reality, Dennis is concerned about the distinction between *relativism* and *objectivism* within the scientific method (see Feyerabend, 1987, for a historically informed, philosophical review). *Postmodernism* is a reaction to *modernism* and denies an unbridled faith in progress (see Cahoone, 1996). Dennis further muddies the waters by lumping *deconstructionism*, a method of literary criticism invented by Paul de Man (1989) and Jacques Derrida (1967) with postmodernism. Although many postmodernists are also relativists (see Gross and Levitt, 1994) and engage in literary (or scientific) deconstruction, it is possible to be an objective scientist within a postmodern framework, just as it is

Scientists construct convincing explanations and acquire reliable knowledge. We do this through rigorous application of "the scientific method," a device for eliminating or reducing points of reasoned skepticism. All scientists share a common vision that objective truth is "out there," but because there is a diversity of reasons that we become scientists, we approach that truth in many ways (Feyerabend, 1975, 1978). In the end, however, the "collective process of empirical investigation, involving weeding out of untenable notions and careful checking of working hypotheses" is independent of frequentist or Bayesian (or other) methods (Galileo had neither at his disposal). That this process leads to *progress* is a core belief of modernism (see footnote 1). Straitjacketing ourselves to one method, as advocated by Dennis, reminds us of an observation by Heisenberg (1958, 58): "We have to remember that what we observe is not nature itself, but nature exposed to our methods of questioning." Such straitjacketing even may slow progress, if in fact progress exists. "Tobacco company science" can benefit by having only one acceptable method of scientific inquiry. To our collective detriment, many political decisions hinge on the distinction between $P = .05$ and $P = .06$, even in the face of more than adequate scientific evidence.

It is worth noting that "science" (and the science of ecology) is not equivalent to truth. Rather (to extend a concept developed by Postman, 1995), science is a specialized language and method that we employ to learn about truth (objective reality) and to explain errors (arising from nonscientific, objectively false versions of reality). Similarly, statistics is not a science; it is a specialized language and tool that we employ to speak about science and to describe scientific results. Thus, statistics is a language used to interpret another language (science) to learn about truth and explain error. As a plethora of human languages enriches our social lives, so a multiplicity of statistical languages can enrich our understanding of science and bring us closer to understanding reality.

"THE" SCIENTIFIC METHOD

Ecologists are preoccupied with the hypothetico-deductive, falsificationist method laid out by Popper (1959), viewed through the lens of statistics developed by Fisher (1922) and Neyman and Pearson (1928), and placed on a

---

possible to be a relativist within a modernist framework, and neither need engage in deconstructionism. Einstein and Heisenberg were successful relativist-modernists, whereas I consider Bayesians to be objectivist-postmodernists.

pedestal by Platt (1964), Strong (1980), Simberloff (1980), and Peters (1991). When published, the papers by Strong (1980) and Simberloff (1980) were useful and important antidotes to decades of fuzzy thought and storytelling masquerading as ecological science (although there is much useful data in those old papers), but subsequent ossification of hypothetico-deductivist methods in ecology shares much with the spread of gospel by religious disciples blindly following their master. Advances in modern science, including ecology, owe much to the professed use of the hypothetico-deductive, falsificationist method. However, in many (most?) instances, it is applied post hoc to ideas conceived inductively and tested with experiments deliberately designed to address the favored "alternative" hypothesis. We should not forget that induction, which Popper attempted to toss into the dustbin of scientific history, is still widely used in "real" mathematics and undergirds most (ecological) modeling activities,[2] activities that Dennis does not consider science. This is indeed odd, as models are convincing explanations based on reliable knowledge and often generate falsifiable hypotheses. Feyerabend (1975, 1978), Howson and Urbach (1993), and Schrader-Frechette and McCoy (1993) provide useful antidotes to knee-jerk hypothetico-deductivism and are recommended reading for all ecologists who want to become more than just producers and curators of ANOVA tables.

THE LANGUAGE OF SCIENCE:
FREQUENTIST AND BAYESIAN STATISTICAL INFERENCE

I have discussed elsewhere in detail some of the contrasts between frequentist and Bayesian inference (Ellison, 1996; see Howson and Urbach, 1993, for a readable review and Earman, 1983, for a thorough vetting of the philosophical arguments concerning hypothetico-deductivist and Bayesian methods). I only emphasize here that frequentist statistics answer the question "how probable are my data, given my (null) hypothesis?" (i.e., $P(\text{data} \mid H)$), whereas Bayesian inference answers the question "how probable is my hypothesis (null or alternative(s)), given my data?" (i.e., $P(H \mid \text{data})$). Far from asserting that frequentist methods are "an anachronistic yoke impeding ecological progress," I suggested that Bayesian inference had utility both in ecological research (where it would allow us to use effectively *unbiased* results

2. Nicholls, Hoozemans, and Marchand (1999) is a recent example with real-world policy implications.

from previous research,[3] as in the "macroecological" approach advocated by Brown and Maurer, 1989, and Brown, 1999, among others) and, more importantly, in expressing degrees of uncertainty in a policy- and decision-making framework. My interest in this topic initially grew out of teaching a course in decision theory, where Bayesian inference is used extensively (e.g., Berger, 1985; Smith, 1988; Chechile and Carlisle, 1991).

Fundamentally, we scientists seek to understand the world around us. We do this by drawing objective conclusions from observations and experiments that also allow for accurate predictions of future events. Frequentist statistics are very useful for drawing objective conclusions from observations and experiments but are not well suited for prediction and inference. Bayesian methods are useful for prediction and inference, given objective, *unbiased* data from careful observations, controlled experiments, appropriate null hypotheses, carefully constructed priors, and, when necessary, formal elicitation of expert opinion (unlike, in all respects, the "cooked" example Dennis presents).[4] For example, whereas clinical trials of new drugs are grounded firmly in frequentist methods, medical diagnosis and consequent improvements in health care and longevity owe much to the principles of "specificity" and "sensitivity" that emerge directly from Bayes' theorem (e.g., Rosner, 2000). Like clinicians who are responsible for the health and well-being of their patients, ecologists would be well served by using all available tools to understand the mechanics of ecological systems (frequentist analysis) and to predict their responses to future conditions (model selection with or without Bayesian *or* frequentist inference; see, for example, Burnham and Anderson, 1998).

Until recently, ecologists, like statisticians, have had only asymptotically accurate, frequentist methods available in their statistical toolkit. Increasing

3. Editors' note: see Goodman 2004 (chapter 12, this volume) for an extensive discussion on this process.

4. Dennis's example examining copper concentrations is cooked in at least two ways. First, the "expert opinion" of the investigator employed by the mining company is clearly biased. Second, and statistically more importantly, Dennis has changed the point-null hypothesis from that presented in the frequentist example ($\mu_0 = 48$) to that presented in the Bayesian example ($\mu_0 = N(20, 4)$). Dennis equates the expert prior with the null hypothesis when in fact, the Bayesian also would be testing the same null hypothesis as the frequentist. There are other, known problems with testing point null hypotheses using Bayesian inference (see Lee, 1997, 124 f), notably that it is impossible to put a continuous prior probability density on a point estimate (Lee, 1997, 126). However, it is possible to put a prior probability density on a very small interval around the point estimate, as in μ ($\mu_0 - \delta, \mu_0 + \delta$). The data, and a properly identified prior on a point-null hypothesis, would clearly have helped to identify the mining scientist as biased.

computational speed allowed for the development and use of exact statistical tests, many of which do not require knowledge of underlying distributions. Further increases in the power of desktop computers, and improved methods of numerical analysis and approximation, have meant that Bayesian software can now be implemented more easily (e.g., Pole, West, and Harrison 1994; Cook and Broemeling, 1995; Albert. 1996). Not surprisingly, given these tools, a careful examination of assumptions of, and similarities and differences between, Bayesian and frequentist inference has led to a greater diversity of approaches in the major statistical journals. Contrary to Dennis's assertion, practitioners[5] of frequentist and Bayesian inference interact and discuss areas of overlap and disagreement (e.g., Pratt, 1965; Smith, 1984; Efron, 1986, 1998; Casella and Berger, 1987; Lewis and Berry, 1994; Samaniego and Reneau, 1994; Hill, 1968; Robbins, 1964; Moore, 1997; Gopalan and Berry, 1998). It makes sense for ecologists to adopt the same degree of pluralism. I suggest that ecologists adopt good statistical practice and use whatever method is appropriate to the question ($P(\text{data} \,|\, H)$ or $P(H \,|\, \text{data})$) —frequentist, exact, likelihood, information-theoretic, or Bayesian.

STATISTICAL EDUCATION FOR ECOLOGISTS

Dennis argues that statistics is a postcalculus subject, and therefore all ecologists should have one year of calculus prior to entry into graduate school, wherein they should take a standard math-stat course in statistical theory. In principle, I agree, but I suspect that this blanket prescription would not accomplish very much. "Real" undergraduate calculus generally is poor preparation for anything other than the follow-up advanced calculus courses, and undergraduates do not readily transfer their calculus skills to other courses (e.g., ecology, physics, statistics).

Statistics (and mathematics) is neither truth nor science; rather, it is a specialized language that has been developed to speak about science. Statistical education (I explicitly avoid the term "training," which, as Orr, 1992, has pointed out, is what we do to animals) therefore should be about understanding the language of statistics and understanding what limitations that language places on our understanding and interpretation of scientific re-

---

5. Neither should be called adherents to "frequentism" or "Bayesianism." These terms conjure up visions of religion and politics (pick your favorite "ism"), neither of which has a place in objective modernist or postmodernist science.

sults. Of course, to develop that understanding, it helps to know what statistics are, how they are used, how they are interpreted, and why society reveres them when it tells us what we want to know and disparages them otherwise (contrast the front section of any issue of *USA Today* with Huff's classic *How to Lie with Statistics* [1954]). Substitute "ecologists" for "society" in the last sentence, and we have the kernel of statistical education for ecologists. This has been broadly recognized by statisticians, who have developed curricula addressing these goals not only narrowly within statistics classes, but also across disciplinary boundaries, at both high school and university levels (e.g., Mosteller, 1988; Cobb and Moore, 1997; Nolan and Speed, 1999; Roberts, Scheaffer, and Watkins, 1999). Ecologists need to catch up to this trend, not form a rear guard protecting the gates of a decaying castle. We also need this education so that we know how to provide unbiased, objective information to the individuals and groups we have charged with making ecological and environmental policy decisions. Otherwise, we will have nothing left to study.

11.3   Rejoinder

Brian Dennis

I appreciate the comments of Charles E. McCulloch and Aaron M. Ellison, but I take issue with the points they raise.

CEM states that "Bayesians do test hypotheses and do calculate quantities very similar to (and, in many ways, *better* than) confidence intervals, *P*-values, and standard errors." I did not claim that Bayesians do not calculate quantities analogous to confidence intervals and standard errors; in fact, I calculated some of those Bayesian things in the paper (*P*-values, of course, are forbidden to Bayesians because of the likelihood principle). The concern I have raised is that such quantities calculated by Bayesians contain their subjective beliefs and are thereby unconvincing to a skeptical observer. Saying that the Bayesian quantities are better does not make them better.

CEM states: "To say that 'Bayesian statistics provides conclusions in the face of incomplete information' is to obfuscate the fact that frequentists always use incomplete data." It is not clear what he means here: perhaps that frequentists do not trust the "data" represented by their beliefs? When frequentists use incomplete data, which to them often means low sample size,

a proper analysis informs them and anyone else of the limits to our knowledge of the subject at hand. Is this not a virtue? The Bayesian and frequentist analyses differ the most when data are incomplete.

CEM claims (after describing a tough, long-standing problem in estimation arising from a roguish likelihood function): "The Bayesian approach, together with modern computing power, straightforwardly handles this problem, even in small and moderate-sized samples." This gets at one of the main
reasons for the recent attraction to Bayesian methods in ecology: when the
information contained in data about an unknown quantity is spotty, the
Bayesian approaches give "answers," whereas the frequentist approaches
merely indicate that the information contained in the data about the unknown quantity is spotty. There is absolutely no question that the reader's
prior beliefs make great "smoothers" that can tame the most wrinkled likelihoods. The problem is, so do mine. Bayesians "handle" these problems, but
do they do it convincingly?

CEM agrees with me on the difficulties of specifying prior distributions,
the "Achilles' heel" of Bayesian statistics, according to him.[6] A more fundamental criticism I raised, though, concerns our overall lack of guidance
about how to do convincing science under Bayesian rules. This includes
problems beyond specification of priors, such as how to evaluate models (for
instance, goodness of fit) under the likelihood principle. The knee is the
Achilles' heel of the leg, so to speak.

According to CEM, "prior distributions sometimes capture bona fide expert opinion, not just *personal* beliefs, and are a synthesis of data and experience." I absolutely agree. But how are we to know when the priors are not
personal? It is precisely here, in the formation of an individual's opinion,
where the postmodern critique of science scores a touché. Any given expert's opinion is too easily deconstructed. An expert says that forty-nine previous studies indicate that dietary supplement A reduces the risk of health
threat B? What about the one methodologically correct study that the expert
omitted, which indicates the opposite (and which was not funded by the
dietary supplement industry)? What possible relevance does that expert's
opinion have to the results of the data at hand? Science works because it is
a process by which many admittedly biased investigators can achieve standardization and reliability. Nature is the referee.

Resource management in the face of scant information, such as the conservation problem posed by CEM, is an area in which I fear the Bayesian ap-

6. These difficulties should not be minimized; see for instance Anderson (1998).

proaches could do major damage. The scientific problems in this area are politicized enough; any scientific signal in the data risks being obscured by the noise over beliefs and utilities. For instance, the recent Bayesian "prospective analysis" in the Plan for Analyzing and Testing Hypotheses (PATH), which recommended, in the face of scant evidence, breaching dams as the best method to recover the dwindling stocks of salmon in the Snake River, did not enhance the credibility of science in the public arena (see Mann and Plummer, 2000).

I have not blamed Bayesians for the ills of postmodern society and the ignorance of science by the general populace, as CEM implies. Bayesian statistics for the most part has yet to be discovered by either society's special interests or postmodern scholars. I have commented instead on the apparent similarities of their worldviews.

CEM says, to my call for ecologists to take mathematical statistics, "get with the program." I say "hear, hear." Of course, a probability and mathematical statistics course sequence is standard for statistics minors (and majors) at the University of Idaho, Cornell, and scores if not hundreds of other colleges and universities. I suspect, though, that CEM paints too rosy a picture of ecological statistics education at Cornell. For instance, as of spring 2003, Cornell undergraduate biology majors (which includes "ecology and environmental biology" as an area of emphasis) are required to take two semesters of approved college mathematics courses, including one semester of calculus, according to the Cornell online catalogue. For most biology students that calculus course is, unless Cornell biology students are very different from those in the rest of the U.S.A., a one semester survey course ("Math 106" at Cornell). Unfortunately, the minimal prerequisite listed at Cornell for the probability and mathematical statistics courses is the *third* semester of a real calculus sequence (Math 111, 112, and 213 at Cornell). Those prerequisites for math-stat are standard across the country; the third course, multivariable calculus, is needed for multivariate distributions. Thus, Cornell undergraduate biology majors, like biology majors all over the U.S.A., are *structurally* unprepared for ever being able to understand statistics. Graduate ecology students, of course, generally reflect undergraduate biology preparation and cannot take math-stat without taking at least a year to remedy undergraduate calculus deficiencies. While some farsighted Cornell grad students in ecology and evolutionary biology no doubt elect to minor in statistics, I would be surprised if that route can be accurately described as "popular." In reality, the standard parade of noncalculus statistical methods courses beckons to ecology graduate students at Cornell ("Biometry 601-602-603"), as it does everywhere else.

The comments of AME are confusing and unclear. At times, he misrepresents what I wrote in the paper.

The first two sentences of AME's second paragraph ("Scientists construct convincing explanations . . .") paraphrase the nutshell definition of science I presented. I am glad we agree here.

However, AME goes on to say that "because there is a diversity of reasons that we become scientists, we approach that truth in many ways," citing Feyerabend. This sentence illustrates a postmodern argumentative technique: the use of vague statements with sliding meanings. Just what is he talking about here? Scientist A uses ultraviolet light, scientist B uses radio waves, scientist C uses auspices from chicken sacrifices? Sure, we are scientists for different reasons, but good scientists check most of their baggage at the door of the laboratory, and use practices *designed to eliminate sociology as a possible explanation of the results.*

AME misrepresents my paper by stating that I advocate "straitjacketing ourselves to one method." I have never advocated that science be restricted to one statistical method (is that the straitjacket he means?). Rather, I have raised questions as to whether *Bayesian* statistical methods are firmly rooted in, and motivated by, the scientific method. I welcome developments in statistics that lead to improvements in science (for instance, information criteria for model selection, bootstrapping, generalized linear models, and survival models, among others, are some statistical developments in the past thirty years that have altered the scientific landscape), and I would welcome new developments that could help build a logical and persuasive Bayesian science.

AME's footnoted lesson in the taxonomy of postmodern subgroups is a diversion and does not address the points I raised. AME's definition of "postmodern" is a narrow one by contemporary popular or scholarly usage (Koertge, 1998). It suffices that those humanities scholars who seek to portray science as essentially sociological call themselves postmoderns (Forman, 1997), and I believe most readers of my paper have a clear idea of who and what I was talking about. Also, it is on the surface patently absurd to call Einstein a relativist and Bayesians objectivists without clarification or justification. Muddy waters, indeed.

AME's description of statistics as "a language used to interpret another language (science) to learn about truth and explain error" is vague and uninformative. Language? Is this meant as a metaphor, or more? Certainly statistics and science share some characteristics with languages, but certainly they are not literally languages. In the next sentence, the meaning of "language" slides around: "As a plethora of human languages enriches our social lives, so a multiplicity of statistical languages can enrich our understanding of science and bring us closer to understanding reality." What are we to make of this as an argument for Bayesian statistics? Of course new statistical methods can enrich science, but it does not follow that Bayesian statistics will do so. Bayesian statistics must stand or fall on its merits; science will not admit it to practice simply on an affirmative action/diversity appeal.

AME claims that "ossification of hypothetico-deductivist methods in ecology shares much with the spread of gospel by religious disciples blindly following their master. . . . However, in many (most?) instances it [the hypothetico-deductive falsificationist method] is applied post hoc to ideas conceived inductively and tested with experiments deliberately designed to address the favored, 'alternative' hypothesis." AME leaves these accusations completely unsubstantiated.

AME's claim that I do not consider induction to be science is absolute nonsense, and I challenge any reader to find such a statement in my paper. What, after all, is estimation based on a random sample? Humbug, Dr. Ellison.

AME restates an old Bayesian sound bite: "I only emphasize here that frequentist statistics answer the question 'how probable are my data, given my (null) hypothesis?' . . . whereas Bayesian inference answers the question, 'how probable is my hypothesis (null or alternatives(s)) given my data?'" AME does not engage my criticism of the sound bite, namely, that the Bayesian definition of "probability of a hypothesis" has no operationally consistent and convincing meaning. Moreover, the frequentist approach in AME's statement is imprecisely represented.

AME thinks that Bayesian inference "would allow us to use effectively *unbiased* [italics his] results from previous research." We must assume he means the use of the previous results in the analysis of new data; his statement, though, contains a vague implication that frequentists do not use previous research at all. Frequentists use previous results to design new studies. Frequentists, however, when analyzing *new* data do not presume to decide *which* old results are unbiased enough to be mixed into the inferences; frequentists let a new data set speak for itself.

AME: "Frequentist statistics . . . are not well suited for prediction and inference." Huh? The twentieth century is a monument to frequentist prediction and inference, from manufacturing quality control to vaccine and drug clinical trials to agriculture improvement experiments. We built this city on Neyman and Pearson.

AME misses the point of my "cooked" example involving the copper concentration. To AME, the mining company investigator is "clearly biased."

Who is AME to say who is biased and who is not? If one ignores for a moment the loonier aspects of postmodern "science studies," one can acknowledge a serious current of scholarship (Kitcher, 1998, calls it the "marginalized middle") that indicts many scientists as biased in their individual and collective behavior toward the subject and toward other scientists. Just which investigators, then, should be allowed to use their prior distributions? Frequentists have an easy and reassuring answer: no one. Science's successes are rooted in the *process* of systematically discounting individual and collective biases, and the frequentist methods have been highly useful toward that end. We have yet to learn how Bayesian approaches could enhance the *process*.

AME misrepresents my mining company example as primarily a hypothesis test; a cursory reading will show that I compared frequentist confidence intervals and Bayesian highest probability regions.

AME recommends that "ecologists adopt good statistical practice and use whatever method is appropriate to the question . . . frequentist, exact, likelihood, information-theoretic, or Bayesian." A minor detail: exact and information-theoretic methods are frequentist (i.e., their justification is based on sample space probability). The major problem with such pluralism is our lack of guidance concerning the Bayesian approach. AME does not attempt to inform us just when Bayesian methods are appropriate, and why. Bayesians themselves claim that Bayesian and frequentist methods are logically incompatible, and that frequentist methods are illogical, with the consequence that the entire statistical practice should be a Bayesian one (start with Lindley, 1990). Bayesian research needs to focus on basic questions of scientific method: we need a Bayesian philosophy of science (such as has been constructed by Mayo, 1996, for frequentist statistics) that will be reassuring to practicing scientists on matters of quality, error, reliability, and repeatability. As it stands, Bayesian philosophy smacks of scientific relativism, with different observers finding different results in the same data.

AME's vague suggestions for statistics education ("to know what statistics are, how they are used, how they are interpreted, and why society reveres them") are fine for nonscientists. Unfortunately, such watered-down precalculus courses, based on trendy textbooks that stress data analysis over inferential concepts (that require probability and calculus) are *by themselves* woefully inadequate for scientists. To ecologists-in-training, I warn that AME's suggestions condemn the trainee to a lifetime of insecurity and incompetency in a subject of central importance to the theories and practices of ecology.

To summarize: my two major points were not addressed by either commentator.

First, we do not yet have an adequate description of a Bayesian scientific method. What we have are detailed descriptions of how a Bayesian can make a computer hum with MCMC calculations, with "the probability of a hypothesis" emerging from the millions of generated random numbers. A Bayesian can do it, but a frequentist can't, goes the claim. Unfortunately, the probability of a hypothesis, along with highest probability regions, and all the other entities obtained from posterior distributions, have no meaning except in the sense of the investigator's personal beliefs—precisely what is enjoined in convincing scientific practice. We do not yet know how to map the Bayesian's analyses to the scientist's task of building convincing arguments. Such a map exists for frequentist analyses (see Mayo, 1996, for a rigorous modern philosophical account); indeed, that map has guided workaday science for most of the twentieth century.

Second, there is a curious resemblance between the Bayesians' worldview and the postmodern one in regards to science. To Bayesians and postmoderns alike, beliefs are to be accepted as data, with the strength of beliefs playing the same role as sample size in one's arguments. This resemblance has been noticed only rarely (Dennis, 1996; Bunge, 1996), but is striking enough to cause this observer at least to wonder why Bayesian approaches have not been embraced wholesale by postmoderns (along with New Age shamans, creation scientists, and tobacco company scientists). Perhaps the mathematical difficulty of Bayesian statistics marks it as a tool used by Euro-male techno-hegemonists to exclude cultures with alinear, empirically transgressive consciousnesses (Sokal, 1996). Nonetheless, if the Bayesians and postmoderns could ever live through an awkward first date, I predict it would be a match made in heaven.

In conclusion: data are precious, beliefs are cheap. Before the scientific community adopts Bayesian methods wholesale, would someone, anywhere, please answer the following question: what is the logic in mixing data and beliefs? Why dilute the Rothschild with Boone's Farm?

REFERENCES

Akaike, H. 1973. Information Theory as an Extension of the Maximum Likelihood Principle. In Petrov, B. N., and F. Csaki, eds., *Second International Symposium on Information Theory*. Budapest: Akademiai Kiado.

Akaike, H. 1974. A New Look at the Statistical Model Identification. *IEEE Transactions on Automatic Control AC* 19:716–723.

Albert, J. H. 1996. Bayesian Computation Using Minitab. Belmont, CA: Wadsworth.

Anderson, J. L. 1998. Embracing Uncertainty: The Interface of Bayesian Statistics and Cognitive Psychology. *Cons. Ecol.* [online] 2:2. http://www.consecol.org/vol2/iss1/art2.

Anderson, W. T. 1990. *Reality Isn't What It Used to Be: Theatrical Politics, Ready-to-Wear Religion, Global Myths, Primitive Chic, and Other Wonders of the Post Modern World.* New York: Harper and Row.

Bain, L. J., and M. Engelhardt. 1992. *Introduction to Probability and Mathematical Statistics.* 2nd ed. Belmont, CA: Wadsworth.

Berger, J. O. 1985. *Statistical Decision Theory and Bayesian Analysis.* 2nd ed. New York: Springer.

Berger, J. O., and D. A. Berry. 1988. Statistical Analysis and the Illusion of Objectivity. *Am. Sci.* 76:159–165.

Brown, J. H. 1999. Macroecology: Progress and Prospect. *Oikos* 87:3–14.

Brown, J. H., and B. A. Maurer. 1989. Macroecology: The Division of Food and Space among Species on Continents. *Science* 243:1145–1150.

Bunge, M. 1996. In Praise of Intolerance to Charlatanism in Academia. In Gross, P. R., N. Levitt, and M. W. Lewis, eds., *The Flight from Science and Reason.* New York: New York Academy of Sciences.

Burnham, K. P., and D. R. Anderson. 1998. *Model Selection and Inference: A Practical Information-Theoretic Approach.* New York: Springer.

Cahoone, L. E., ed. 1996. *From Modernism to Postmodernism: An Anthology.* Malden, MA: Blackwell.

Casella, G., and R. L. Berger. 1987. Reconciling Bayesian and Frequentist Evidence in the One-Sided Testing Problem. *J. Am. Stat. Assn* 82:106–111.

Chechile, R. A., and S. Carlisle. 1991. *Environmental Decision Making: A Multidisciplinary Perspective.* New York: Van Nostrand.

Cobb, G. W., and D. S. Moore. 1997. Mathematics, Statistics, and Teaching. *Am. Math. Monthly* 104:801–823.

Connor, E. F., and D. Simberloff. 1979. You Can't Falsify Ecological Hypotheses without Data. *Bull. Ecol. Soc. Am.* 60:154–155.

Connor, E. F., and D. Simberloff. 1986. Competition, Scientific Method, and Null Models in Ecology. *Am. Sci.* 75:155–162.

Cook, P., and L. D. Broemeling. 1995. Bayesian Statistics Using Mathematica. *Am. Stat.* 49:70–76.

Cox, D. R., and D. V. Hinkley. 1974. *Theoretical Statistics.* London: Chapman and Hall.

De Man, P. 1989. *Critical Writings 1953–1978.* Minneapolis: University of Minnesota Press.

Dennis, B. 1996. Discussion: Should Ecologists Become Bayesians? *Ecol. Appl* 6:1095–1103.

Dennis, B., R. A. Desharnais, J. M. Cushing, and R. F. Costantino. 1995. Nonlinear Demographic Dynamics: Mathematical Models, Statistical Methods, and Biological Experiments. *Ecol. Monographs* 65:261–281.

Dennis, B., and M. L. Taper. 1994. Density Dependence in Time Series Observations of Natural Populations: Estimation and Testing. *Ecol. Monographs* 64:205–224.

Derrida, J. 1967. *L'écriture et la différence.* Paris: Editions du Seuil. English trans., *Writing and Difference* (Chicago: University of Chicago Press, 1978).

Dixon, P. M., and K. A. Garrett. 1993. Statistical Issues for Field Experimenters. In Kendall, R. J., and T. E. Lacher, eds., *Wildlife Toxicology and Population Modeling.* Boca Raton: Lewis.

Earman, J., ed. 1983. *Testing Scientific Theories.* Minneapolis: University of Minnesota Press.

Edwards, A. W. F. 1972. *Likelihood: An Account of the Statistical Concept of Likelihood and its Application to Scientific Inference.* Cambridge: Cambridge University Press.

Efron, B. 1986. Why Isn't Everyone a Bayesian? (with discussion). *Am. Stat.* 40:1–11.

Efron, B. 1998. R. A. Fisher in the 21st Century. *Stat. Sci.* 13:95–122.

Ellison, A. M. 1996. An Introduction to Bayesian Inference for Ecological Research and Environmental Decision-Making. *Ecol. Appl.* 6:1036–1046.

Enquist, B. J., J. H. Brown, and G. B. West. 1998. Allometric Scaling of Plant Energetics and Population Density. *Nature* 395:163–165.

Feyerabend, P. 1975. *Against Method.* London: Verso.

Feyerabend, P. 1978. *Science in a Free Society.* London: Verso.

Feyerabend, P. 1987. *Farewell to Reason.* London: Verso.

Fisher, R. A. 1922. On the Mathematical Foundations of Theoretical Statistics. *Phil. Trans. Roy. Soc. London,* ser. A, 222:309–368.

Forman, P. 1997. Assailing the Seasons. *Science* 276:750–752.

Goodman, D. 2004. Taking the Prior Seriously: Bayesian Analysis without Subjective Probability. Chapter 12 in Taper, M. L., and S. R. Lele, eds., *The Nature of Scientific Evidence: Empirical, Statistical, and Philosophical Considerations.* Chicago: University of Chicago Press.

Gopalan, R., and D. A. Berry. 1998. Bayesian Multiple Comparisons Using Dirichlet Process Priors. *J. Am. Stat. Assn* 93:1130–1139.

Gross, P. R., and N. Levitt. 1994. *Higher Superstition: The Academic Left and Its Quarrels with Science.* Baltimore: Johns Hopkins University Press.

Hairston, N. G., Sr. 1989. *Ecological Experiments: Purpose, Design, and Execution.* Cambridge: Cambridge University Press.

Heisenberg, W. 1958. *Physics and Philosophy: The Revolution in Modern Science.* New York: Harper and Row.

Hill, B. M. 1968. Posterior Distribution of Percentiles: Bayes' Theorem for Sampling from a Population. *J. Am. Stat. Assn* 63:677–691.

Howson, C., and P. Urbach. 1993. *Scientific Reasoning: The Bayesian Approach.* 2nd ed. Chicago: Open Court.

Huff, D. 1954. *How to Lie with Statistics.* New York: Norton.

Hurlbert, S. H. 1984. Pseudoreplication and the Design of Ecological Field Experiments. *Ecol. Monographs* 54:187–211.

Jaynes, E. T. 1968. Prior Probabilities: IEEE Transactions on Systems Science and Cybernetics. *SSC* 4:227–241.

Johnson, D. H. 1995. Statistical Sirens: The Allure of Nonparametrics. *Ecology* 76:1998–2000.

Johnson, D. H. 1999. The Insignificance of Statistical Hypothesis Testing. *J. Wildlife Mgmt.* 63:763–772.

Kitcher, P. 1998. A Plea for Science Studies. In Koertge, N., ed., *A House Built on Sand: Exposing Postmodernist Myths about Science.* New York: Oxford University Press.

Koertge, N., ed. 1998. *A House Built on Sand: Exposing Postmodernist Myths about Science.* New York: Oxford University Press.

Lee, P. M. 1989. *Bayesian Statistics: An Introduction.* New York: Oxford University Press.

Lee, P. M. 1997. *Bayesian Statistics: An Introduction.* 2nd ed. New York: Wiley.

Lehmann, E. L. 1983. *Theory of Point Estimation.* New York: Wiley.

Lewis, R. J., and D. A. Berry. 1994. Group Sequential Clinical Trials: A Classical Evaluation of Bayesian Decision-Theoretic Designs. *J. Am. Stat. Assn* 89:1528–1534.

Lindley, D. V. 1990. The 1988 Wald Memorial Lectures: The Present Position in Bayesian Statistics (with discussion). *Stat. Sci.* 5:44–89

Mann, C. C., and M. L. Plummer. 2000. Can Science Rescue Salmon? *Science* 289:716–719.

Marmorek, D. R., ed. 1996. *Plan for Analyzing and Testing Hypotheses (PATH).* Vancouver: ESSA Technologies.

Mayo, D. G. 1996. *Error and the Growth of Experimental Knowledge.* Chicago: University of Chicago Press.

Moore, D. S. 1997. Bayes for Beginners? Some Reasons to Hesitate. *Am. Stat.* 51:254–261.

Mosteller, F. 1988. Broadening the Scope of Statistics and Statistical Education. *Am. Stat.* 42:93–99.

Neyman, J., and E. S. Pearson. 1928. On the Use and the Interpretation of Certain Test Criteria for Purposes of Statistical Inference. *Biometrika* 20:175–240 (part 1), 263–294 (part 2).

Neyman, J., and E. S. Pearson. 1933. On the Problem of the Most Efficient Tests of Statistical Hypotheses. *Phil. Trans. Roy. Soc. London,* ser. A, 231:289–337.

Nicholls, R. J., F. M. J. Hoozemans, and M. Marchand. 1999. Increasing Flood Risk and Wetland Losses Due to Global Sea-Level Rise: Regional and Global Analyses. *Global Envir. Change* 9:S69–S87.

Nolan, D., and T. P. Speed. 1999. Teaching Statistics Theory through Applications. *Am. Stat.* 53:370–375.

Orr, D. W. 1992. *Ecological Literacy: Education and the Transition to a Postmodern World.* Albany: State University of New York Press.

Ott, R. L. 1993. *An Introduction to Statistical Methods and Data Analysis.* 4th ed. Pacific Grove, CA: Brooks/Cole.

Peterman, R. M. 1990. Statistical Power Analysis Can Improve Fisheries Research and Management. *Can. J. Fish. Aquat. Sci.* 47:2–15.

Peters, R. H. 1991. *A Critique for Ecology.* Cambridge: Cambridge University Press.

Platt, J. R. 1964. Strong Inference. *Science* 146:347–353.

Pole, A., M. West, and J. Harrison. 1994. *Applied Bayesian Forecasting and Time Series Analysis.* Chapman and Hall, New York.

Popper, K. R. 1959. *The Logic of Scientific Discovery.* New York: Basic Books.

Postman, N. 1995. *The End of Education.* New York: Vintage.

Postman, N., and C. Weingartner. 1969. *Teaching as a Subversive Activity.* New York: Dell.

Potvin, C., and D. A. Roff. 1993. Distribution-Free and Robust Statistical Methods: Viable Alternatives to Parametric Statistics? *Ecology* 74:1617–1628.

Pratt, J. W. 1965. Bayesian Interpretation of Standard Inference Statements. *J. Roy. Stat. Soc.,* ser. B, 27:169–203.

Press, S. J. 2003. Subjective and Objective Bayesian Statistics: Principles, Models, and Applications. Hoboken, NJ: Wiley.

Reckhow, K. H. 1990. Bayesian Inference in Non-replicated Ecological Studies. *Ecology* 71:2053–2059.

Rice, J. A. 1995. *Mathematical Statistics and Data Analysis.* Belmont, CA: Wadsworth.

Robbins, H. 1964. The Empirical Bayes Approach to Statistical Decision Problems. *Ann. Math. Stat.* 35:1–20.

Robert, C. P., and G. Casella. 1999. *Monte Carlo Statistical Methods.* New York: Springer.

Roberts, R., R. Scheaffer, and A. Watkins. 1999. Advanced Placement Statistics: Past, Present, and Future. *Am. Stat.* 53:307–321.

Rosner, B. 2000. *Fundamentals of Biostatistics.* 5th ed. Pacific Grove, CA: Duxbury.

Royall, R. M. 1997. *Statistical Evidence: A Likelihood Paradigm.* London: Chapman and Hall.

Royall, R. M. 2000. On the Probability of Observing Misleading Statistical Evidence (with discussion). *J. Am. Stat. Assn* 95:760–780.

Saarinen, E., ed. 1980. *Conceptual Issues in Ecology.* Dordrecht: Reidel.

Samaniego, F. J., and D. M. Reneau. 1994. Toward a Reconciliation of the Bayesian and Frequentist Approaches to Point Estimation. *J. Am. Stat. Assn* 89:947–957.

Scheaffer, R. L., W. Mendenhall, and L. Ott. 1996. *Elementary Survey Sampling.* 5th ed. Pacific Grove, CA: Brooks/Cole.

Schrader-Frechette, K. S., and E. D. McCoy. 1993. *Method in Ecology: Strategies for Conservation.* Cambridge: Cambridge University Press.

Simberloff, D. 1980. A Succession of Paradigms in Ecology: Essentialism to Materialism and Probabilism. *Synthese* 43:3–39.

Simberloff, D. 1990. Hypotheses, Errors, and Statistical Assumptions. *Herpetologica* 46:351–357.

Sindermann, C. J. 1982. *Winning the Games Scientists Play.* New York: Plenum.

Smith, A. F. M. 1984. Present Position and Potential Developments: Some Personal Views: Bayesian Statistics. *J. Roy. Stat. Soc.*, ser. A, 147:245–259.

Smith, J. Q. 1988. Decision Analysis: A Bayesian Approach. London: Chapman and Hall.

Smith, S. M. 1995. Distribution-Free and Robust Statistical Methods: Viable Alternatives to Parametric Statistics? *Ecology* 76:1997–1998.

Sokal, A. D. 1996. Transgressing the Boundaries: Toward a Transformative Hermeneutics of Quantum Gravity. *Social Text* 46/47:217–252.

Sokal, A. D., and J. Bricmont. 1998. *Fashionable Nonsense: Postmodern Intellectuals' Abuse of Science*. New York: Picador.

Stewart-Oaten, A. 1995. Rules and Judgments in Statistics: Three Examples. *Ecology* 76:2001–2009.

Strong, D. R. 1980. Null Hypotheses in Ecology. *Synthese* 43:271–285.

Strong, D. R., D. Simberloff, L. G. Abele, and A. B. Thistle, eds. 1984. *Ecological Communities: Conceptual Issues and the Evidence*. Princeton: Princeton University Press.

Stuart, A., and J. K. Ord. 1991. *Kendall's Advanced Theory of Statistics, Volume 2: Classical Inference and Relationships*. 5th ed. London: Griffin.

Student [W. S. Gosset]. 1908. The Probable Error of a Mean. *Biometrika* 6:1–25.

Toft, C. A., and P. J. Shea. 1983. Detecting Community-wide Patterns: Estimating Power Strengthens Statistical Inference. *Am. Nat.* 122:618–625.

Underwood, A. J. 1990. Experiments in Ecology and Management: Their Logics, Functions, and Interpretations. *Australian J. Ecol.* 15:365–389.

Vardeman, S. B. 1994. *Statistics for Engineering Problem Solving*. Boston: PWS.

Wilks, S. S. 1938. The Large-Sample Distribution of the Likelihood Ratio for Testing Composite Hypotheses. *Ann. Math. Stat.* 9:60–62.

Yoccoz, N. G. 1991. Use, Overuse, and Misuse of Significance Tests in Evolutionary Biology and Ecology. *Bull. Ecol. Soc. Am.* 72:106–111.

12 Taking the Prior Seriously: Bayesian Analysis without Subjective Probability

Daniel Goodman

The precise specification of our knowledge is, however, the same as the precise specification of our ignorance. —R. A. Fisher, 1956

ABSTRACT

Environmental decision making is concerned with determining a prudent course of action. This is different in focus from a theoretical search for scientific truth. The search for scientific truth attempts to select among alternative hypotheses about nature, whereas decision making attempts to select an action in light of the uncertainty about all the possible alternatives. Decision theory requires the assignment of probabilities for the different possible states of nature. Bayesian inference provides such probabilities, but at the cost of requiring prior probabilities for the states of nature. Early in Bayesian history, the justifications of the prior distribution were often confused and unsatisfactory. In this century, the justification for prior probabilities has often rested on subjective theories of probability. Subjective probability can lead to internally consistent systems relating belief and action for a single individual; but severe difficulties emerge in trying to extend this model to justify public decisions. Objective probability represents probability as a literal frequency that can be communicated as a matter of fact and that can be verified by independent observers confronting the same information. If, as I show can be done in this paper, prior distributions are developed in this spirit, the frequency interpretation is conferred on the inference as well, so the inference can serve as a shared basis for public decision.

Portions of this research were supported by DOE cooperative agreement DE-FCO794ID13317 to Montana State University.

INTRODUCTION

The statistical power that derives from multiple observations and the use of empirical histograms to represent what we know and what we don't know about underlying phenomena that can be observed only subject to error are of course features of modern statistical practice. But the fundamental importance of these primitive concepts is not always fully appreciated. In this paper, I propose an interpretation of statistical inference for decision making, in which an empirical histogram is the logical starting point and minimal requirement. In a data-rich setting, construction of a convincing histogram is straightforward, and sound conclusions can be drawn without complicated statistics. In a data-poor situation, where the data specific to the case about which we are making the decision are too sparse for a convincing histogram, we need to embed the case in a broader class of cases. The broader class of cases must be broad enough to encompass enough known cases that the information that we have on the aggregate is sufficient for a convincing histogram for the aggregate. Then, to draw conclusions about the case of interest, using the combined information on the aggregate and the data specific to the case, we do need a formal statistical procedure for inference. In this procedure, the case-specific data provide information for refining the aggregate histogram to make an inference specific to the case. This interpretation allows use of Bayes' formula for inference, without recourse to subjective probability.

TRUTH OF HYPOTHESES OR PROBABILITIES FOR DECISIONS

Decision making is concerned with what course of action is prudent to take. This is rather different in focus from a theoretical search for scientific truth, though both may use evidence of a kind that we call scientific. Basically, the search for scientific truth attempts to select one among several alternative hypotheses about the state of nature, whereas decision making attempts to select an action in light of the uncertainty about all the possible alternatives. Application of decision theory requires the assignment of probabilities for the different possible states of nature (Good, 1952). Bayesian inference provides such probabilities, but at the cost of requiring prior probabilities for the states of nature.

Early in the history of Bayesian analysis, the justifications for specification of the prior distribution as an abstract representation of pure uncertainty were confused and unsatisfactory. In the past hundred years, the justification for prior probabilities has often rested on subjective theories of probability. Subjective probability can lead to very nice, internally consistent systems relating belief and action for a single individual; but severe difficulties emerge in trying to extend this model to a justification for public decisions.

In this paper, we will attempt to explain how an objective theory of probability, treating probability as a literal frequency, can be the basis for developing prior distributions for Bayesian analysis, and how the literal frequency interpretation then carries over to the resulting inference. This does not answer the theoretical scientist's question about which hypothesis is true, but it does provide a basis for the decision maker's question about what action to take in light of the probabilities.

In the process of developing this explanation, we will encounter a shift of emphasis in understanding the operation of extracting probability inferences from data. The common attitude is that the likelihood function is the primary conduit for converting information, in the broadest sense, to probabilistic conclusions. In this view, the prior is either an unavoidable nuisance or an avoidable nuisance, but a nuisance in any case. The perspective developed in this paper will assign the prior a more central role, more important to the inference. In this perspective, the validity of statistical inference depends more closely on the fundamental idea of a probability density as the limit of an empirical histogram, and this is connected to the primitive notion that replication is the basis for secure judgment from insecure observations.

TERMINOLOGY: PIECES OF THE PUZZLE

The language of statistical inference carries a lot of baggage. Overinterpretation of the tangential connotations is an obstacle to clarity. The first problem is what we mean by "inference" itself. Here, we will take it to mean a logical process for making defensible statements about the value of a parameter, specific to a particular case in which the value of that parameter has not been observed directly but is assumed to have a unique though unknown value; and some information is available to provide a basis for the resulting statement about the parameter value, though that information is not sufficient to determine the value of the unknown parameter exactly.

There are four kinds of information that bear on a statistical inference. Two of these kinds of information are observables pertaining to the case, and the partitioning of these two will require some conceptual care. One of

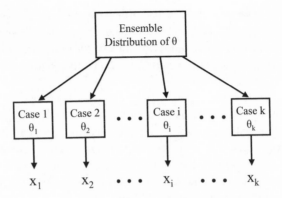

FIGURE 12.1 Cases are sampled from an ensemble, in which the parameter θ literally has a distribution. In each case, θ has an unknown fixed value. Data x specific to each case are obtained by a process that is influenced by the value of θ in that case. Inference for the value of the parameter in case i uses that case's data for the likelihood in Bayes' formula, and the data from the other cases are used to estimate the prior

the kinds of information pertaining to the case is (1) categorizing information, identifying the case as a member of a particular collection of cases that we will call the "ensemble." (The literature also refers to this as a "superpopulation.") The second kind of information pertaining to the case is (2) information, conventionally called "the data," that is a reflection of the unknown value of the parameter of interest in this case but which was not used for categorizing. The remaining two kinds of information used for the inference are (3) distributional information about the ensemble to which the case belongs and (4) knowledge of a stochastic mechanism whereby the parameter of interest in the case gives rise to the observed case-specific data (see figure 12.1).

By "case," we mean an instance where the parameter of interest has a fixed but unknown value. The ensemble to which it belongs is a set of cases that may not share the same value for the parameter of interest but which, as a group, are characterized by a distribution for the values of the parameter of interest. The ensemble membership information and the case-specific data reflecting the value of the parameter are separate and operate independently in the inference process: the assignment of the case to an ensemble based on the categorizing information about the case makes no use of the information contained in "the data." The categorizing information about the case and the distributional information about the ensemble to which the case belongs are conventionally lumped and called "the prior."

Notwithstanding the many connotations of this word, these two sources of information need not have originated prior in time to "the data," they do not intrinsically take precedence over "the data," they are not necessarily obtained by means that are inherently different from the way "data" are measured, nor are they necessarily known by some logically a priori means (though this latter is the origin of the use of this term).

The designation of the case-specific information conventionally called "the data" seems straightforward enough, but such terminology should not lull us into thinking that this partition of the information about the case is the only source of information (or evidence) that goes into the inference.

The knowledge about the nature of the stochastic mechanism underlying the generation of "the data" is conventionally taken for granted, though there is a conventional name, "likelihood function," for representing that knowledge as a function that to a proportionality constant is the probability distribution of "the data" conditional on the value of the unknown parameter.

Bayes' formula combines these four kinds of information to yield distributional information that is nominally "about the case." Some distress arises from this superficial terminology, since a distribution is necessarily about a population, not a case. The resolution of this confusion rests on the recognition that the inference provided by Bayes' formula really is about a population, namely the population of cases that are indistinguishable from the case in question with respect to all the information being used that is thought to bear on the value of the unknown parameter. If these cases are indistinguishable, there is no basis for treating any of them any differently if we encountered them. Hence, a decision for dealing with any one of them, where the outcome depends on the value of the unknown parameter, would naturally be governed by the distribution of the parameter in the population of indistinguishable cases. The conventional Bayesian term to encapsulate the concept of "indistinguishable with respect to the information" is "exchangeable," a concept developed extensively by de Finetti (1937), though he actually used a word that translates to "equivalent." The inference provided by Bayes' formula is conventionally called the "posterior" probability density or distribution. This probability depends on the data in that it is explicitly conditional on the data, so to this extent it has an a posteriori character.

To reduce clutter, in the remainder of this chapter we will mostly dispense with quotation marks around the above terms, but those of the audience, especially the philosophers, who are concerned about such things, are reminded that these terms, and many others associated with theories of statistical inference, are terms of art, the meanings of which are not necessarily what they appear at face value. Indeed, even the term "Bayesian" can encompass approaches with different theories of probability, different theories of origin of the information constituting the prior, different frameworks for assessing the performance of an inference procedure, and even startlingly idiosyncratic meanings for words such as "subjective" and "objective." Note that Berger (2000) classifies as "objective Bayes" inference approaches that rest on subjective probability but adopt a conventionally fixed prior for each model; and that Gelman et al. (1995) object to the label "empirical Bayes" on the grounds that this might be taken to imply that the "hierarchical Bayes" methods that these authors favor in many applications are not "empirical."

NOTATION AND FORMALITIES

As a prelude to the inference, assume that we have identified the case in question as a member of a particular ensemble, within which the parameter of interest, θ, has a known distribution, $p(\theta)$; assume further that we have obtained a measured value x for this case, not included in the information used for assigning the case to the ensemble, and that we know $p(x|\theta)$, the probability of the measured value conditional on the unknown parameter value. The data x and the parameter θ may be vector quantities.

Purely as a formality, application of the rules of the probability calculus yields Bayes' formula,

$$p(\theta|x) = \frac{p(\theta)p(x|\theta)}{p(x)}, \tag{1}$$

where $p(x|\theta)$ is the inference—called the posterior probability of θ or the probability of θ conditional on the data—$p(\theta)$ is the prior for θ, $p(x|\theta)$ is the likelihood of θ (though it is, to a proportionality, a *probability* of x), and $p(x)$ is the unconditional probability of the data or the marginal probability of the data in the ensemble. Because the left-hand quantity of equation (1) is a probability, the integral of which, over the parameter θ, must be 1, and as the denominator of the right-hand side of the equation is not a function of that parameter, we see that this denominator serves only as normalization term. This offers an alternative formulation:

$$p(x) = \int_{-\infty}^{\infty} p(\theta)p(x|\theta)d\theta, \tag{2}$$

which also illustrates the sense in which $p(x)$ is a marginal, or unconditional, distribution.

A COMPOUND SAMPLING PERSPECTIVE

The least mysterious context for interpreting the application of Bayes' formula is compound sampling. Imagine randomly drawing cases from the ensemble where the distribution of the parameter is $p(\theta)$, and then obtaining a measurement on whatever the observable is that constitutes the data. Consider the subset of cases for which, under this compound sampling (repeatedly drawing cases from the ensemble and sampling data from the cases), the data were identically x. In that subset, the distribution of the parameter θ will exactly recapitulate the distribution called the posterior in Bayes' formula, as the term $p(\theta)$ represents the probability of drawing a case with parameter value θ in the first stage of the sampling, the term $p(x|\theta)$ weights for the probability of obtaining exactly x as the data in the second stage of the sampling (conditional on having drawn a case where the parameter value is θ), and the denominator term simply normalizes by the overall probability of obtaining the data x from all cases drawn.

In other words, the probability obtained as the inference in this application of Bayes' formula is a probability that refers to an actual frequency. It is the frequency of values of the parameter exactly equal to θ in the subset of cases that pass the two-stage filter of (1) drawing the parameter from the ensemble where the distribution is $p(\theta)$, and (2) obtaining data from sampling observations from the case according to the likelihood function that yield exactly the value of x.[1]

A CONCRETE EXAMPLE OF COMPOUND SAMPLING

At the Hanford Nuclear Reservation, in Washington State, there are 179 very large buried tanks holding wastes from nuclear weapons production and nuclear fuel reprocessing. Most of the tanks are old, and many are known or suspected leakers, so environmental cleanup plans for the site include removal of the waste material from the tanks and conversion of the high-level

1. Perversely, the literature on empirical Bayes and hierarchical Bayes methods insists on calling the sampling from the ensemble "the second stage" and the sampling of data from the case "the first stage."

fractions to solid forms for long-term storage. These operations will require some knowledge of the composition of the contents of each tank.

Sampling the material in the tanks is difficult and expensive; as a consequence, the number of analyzed samples from most tanks is small. However, some samples have been analyzed from all tanks, so the aggregate number of samples is large.

Some of the material in the tanks is in the form of sludges and precipitates that, because of their heterogeneity, make estimation of the composition of the inventory of a given tank quite uncertain if based on the small number of samples that are available for that tank. But there are enough samples for the entire collection of tanks that the distribution of compositions for the entire collection can reasonably be characterized.

In the vocabulary of this paper, the collection of 179 tanks constitute an *ensemble*. The concentration of the isotope of interest is the *parameter*. In a question about the concentration in a particular tank, that tank constitutes the *case*. The measurements of samples from the tank of interest constitute the *case-specific data*. From the aggregate of measurements on samples from all the tanks we also have a prior that is the *distribution of the parameter in the ensemble*. If we use Bayes' formula to merge these sources of information, the *posterior* is the distribution of values of the parameter of interest in the subset of tanks that would be obtained by *repeatedly sampling tanks from the collection of tanks, sampling data from the tanks, and assigning to the subset those tanks that gave the same data as the data we actually have from the tank of interest*. Whenever we approach a decision about one of these tanks, we are in fact drawing a tank from the ensemble and considering data that are specific to the tank, in conjunction with what we know about the collection of tanks, to make a probabilistic statement that is nominally about *this* tank but is actually about all indistinguishable tanks (which means tanks drawn from this same ensemble and yielding the same data). The decision analysis for remediation operations on these tanks, using this compound sampling perspective, is described in Goodman and Blacker (1998).

THE STATUS OF THE PRIOR

In this fashion, we have arrived at an inference procedure that yields a probability statement about the unknown parameter, where the probability statement is interpretable as an actual frequency appropriate for guiding a decision. It is comforting that the resulting posterior distribution is a real probability in this sense. But we need to understand that in order to obtain a result that has this status, the prior distribution, $p(\theta)$, also has to be a real probability interpretable as a frequency in an actual ensemble (since there has to be a meaning to the idea of drawing cases from this ensemble).

Historically, obtaining a posterior that is this sort of probability, interpretable as a literal frequency, has been the goal of much inference effort, but the commitment to taking the prior seriously has been very slight. Briefly surveying the last two centuries of theory and practice, most Bayesians have made up a prior, either on grounds that the prior represented ignorance or uncertainty for which there was allegedly a *theoretically* correct representation in a particular distribution, or on grounds that probability really is subjective, so it's all made up anyway, and we should be satisfied with priors and posteriors that don't have frequency interpretations.

Most "frequentists" have attempted to develop procedures that omit an explicit prior, at the cost of generating results that are genuine frequencies, but for the *wrong subpopulation*. They have normalized over the sampling space of the data (which of course are fixed in any given case) rather than normalizing over the space of the parameter. Blithely using such frequentist confidence intervals *as if* they were posterior probabilities for the parameter implies a particular prior, namely whatever prior yields that particular posterior with the likelihood function for those data.[2]

One justification sometimes advanced for playing fast and loose with the prior has been the observation that when the data are strong (e.g., when there is a very large amount of data) the details of the prior make little difference to the resulting posterior, motivating imagery of limiting cases where the prior would make no difference. All this is true enough, but it misses the point. When there is that overwhelming amount of data available for a case, there is so much information that the value of the parameter for this case can be deduced with little uncertainty, and any half-reasonable procedure will perform adequately. The actual goal should be to develop an inference procedure that performs correctly when we really need it; that is, when the available case-specific data are limited enough that the uncertainty is substantial. Then the prior *does* matter. Needless to say, most ecological and environmental applications, which is where I earn my living, are characterized by large uncertainty and limited case-specific data.

---

2. Bayesians and frequentists use the same likelihood function, and both agree that the data are the data.

The analyst is free to employ a *hypothetical* prior just to see what posterior results (Bernardo, 1979). But that posterior generally has a correct frequency interpretation in a compound sampling context only if the first stage in the compound sampling consisted of drawing cases from an ensemble where the parameter of interest had a distribution identical to the hypothetical prior. Expecting this correspondence, without adopting a rigorous procedure to ensure it, is to rely on improbable coincidence.

More realistically, if we don't know whether the actual distribution we are drawing the case from corresponds well or poorly (or at all) to the hypothetical prior, we are left not knowing whether the resulting posterior is a good approximation or a bad approximation (or no approximation at all) to the distribution that we *should* be using as a basis for making a decision about this case. What use is that?

The implicit prior underlying the use of a frequentist "confidence interval" as if it were a Bayesian posterior suffers the same unpredictable relation to the actual distribution in the population from which cases are being drawn (Lindley, 1957; Edwards, Lindman, and Savage, 1963; Dempster, 1964; Pratt, 1965; Lindley and Phillips, 1976; Berger and Sellke, 1987). So this procedure constitutes an equally unsatisfactory basis for guiding decisions when the uncertainty about the parameter is substantial.

If we are dealing with an especially tractable compound sampling situation, where the ensemble from which cases are drawn is defined by a *process* that has a known distribution, there is no difficulty. An example of such a process *defining* the prior is the uniform distribution that follows from the process of randomly throwing a marker on a table. This is the first stage of the compound sampling process that Bayes proposed as an illustration in his original paper (1763). Unfortunately, when we *don't* know the process generating the distribution, we are *not* going to arrive at the correct prior distribution just by thinking about it.

BAYES' FORMULA AS AN OPERATION FOR DECOMPOSING THE PRIOR

Because of these difficulties with the prior, notwithstanding our wish to obtain a legitimate posterior, an odd mythology has grown up. This mythology portrays Bayes' formula as a tool for transforming the information captured by the likelihood function into the kind of posterior distribution we are looking for, provided we can come up with any sort of useable prior to sanction the operation. So the focus is on the likelihood and the data and, nowadays, on the numerical strategies for the necessary integration. As a gesture, a prior is obtained by hook or crook—mostly crook—and as little attention as possible is paid to it, as if the prior is statistical inference's dirty secret.

By contrast, the compound sampling perspective suggests a very different emphasis on what Bayes' formula accomplishes: *Bayes' formula is a tool for decomposing the prior distribution, which is a genuine distribution of the values of the unknown parameter in a defined ensemble, into a data-specific component, the posterior, which also is a genuine distribution of values of the parameter, but in a population that is a subset of the ensemble that defined the prior, where the subsetting is with respect to the case-specific data.* In this view, it is clear that the reality of the resulting posterior depends on the reality of the prior that is decomposed.

The following mathematical examples will illustrate the process of decomposing the prior into posteriors, or recomposing the prior as a composite of posteriors.

A Component Specific to the Data

The compound sampling scenario used above as a context for applying Bayes' formula draws cases from the prior and then saves those sampled cases for which specific values of the data are obtained from an observation process. The posterior then is the distribution of the parameter in the new subset, *which is a component of the ensemble that defined the prior.* In this sense, Bayes' formula has decomposed the prior into (1) the posterior distribution which interests us and (2) its complement or residual distribution that describes the distribution of the parameter in all the cases where the data were not x exactly:

$$p(\theta) = \int_{-\infty}^{\infty} p(\theta|y)p(y)dy$$
$$= p(\theta|x)p(x) + \int_{y \neq x} p(\theta|y)p(y)dy. \qquad (3)$$

A Component Specific to a Function of the Data

Imagine randomly drawing cases from the ensemble where the distribution of the parameter is $p(\theta)$, and obtaining a measurement on whatever the observable is that constitutes the data, and then calculating some defined function of the data $f(x)$. Consider the subset of cases for which, under this compound sampling, the function of the data is identically $f(x) = w$. What is the distribution of the parameter θ in that subset?

Define a switch function S that takes the value 1 when its argument is 0, and takes the value 0 otherwise

$$S(y)_{y=0} = 1, \tag{4}$$

$$S(y)_{y \neq 0} = 0. \tag{5}$$

Then, we obtain the distribution of the parameter in the new subpopulation of interest as

$$p(\theta|w) = \frac{p(\theta)\int_{-\infty}^{\infty}p(x|\theta)S(w - f(x))dx}{\int_{-\infty}^{\infty}p(\theta)\int_{-\infty}^{\infty}p(x|\theta)S(w - f(x))dxd\theta}, \tag{6}$$

where the term $p(\theta)$ represents the probability of drawing a case with parameter value θ in the first stage of the sampling, the integral in the numerator is the probability of obtaining exactly w as the value of the function of the data averaged over all possible data weighted by their respective probabilities of being obtained in the second stage of the sampling (conditional on having drawn a case where the parameter value is θ), and the denominator is a brute normalization.

Substituting equation (1) into equation (6), we reexpress the new posterior as

$$p(\theta|w) = \frac{\int_{-\infty}^{\infty}p(\theta|x)p(x)S(w - f(x))dx}{\int_{-\infty}^{\infty}\int_{-\infty}^{\infty}p(\theta|x)p(x)S(w - f(x))dxd\theta}, \tag{7}$$

which shows that the posterior of the parameter, in this new subset defined by a function of the data, can be obtained as a composite of posteriors of the parameter in the subpopulation defined by the data, where each component data-specific posterior is weighted by the probability of the data that give rise to it.

Either way, the probability obtained as the inference in this application of Bayes' formula again is a probability that refers to an actual frequency. It is the frequency of values of the parameter exactly equal to θ in the subset of cases that pass the two-stage condition of (1) drawing the parameter from the ensemble where the distribution is $p(\theta)$ and (2) obtaining data from sampling observations from the case according to the likelihood function that yield exactly the value of the function $f(x)$.

Thus, we see that the decomposition of distributions under Bayes' formula can operate in the reverse direction as well: we can obtain a posterior distribution as a (data-specific) decomposition of the prior, or we can obtain it as a composite of component posterior distributions for components of the data.

One practical application of this maneuver is for validating computer code that solves for the posterior under a particular model. The direct approach of simulating the two-stage sampling process for decomposing the prior into a data-specific component, to compare the computed posterior to the histogram of values of the parameter in the subset of cases in the simulation that pass the two-stage filter, may not be effective with a finite sample of trials if the probability of obtaining exactly x as the data is too small (as it will generally be for continuous distributions). But embedding the code for the posterior in a compositing process, as in equation (7), where the form $f(x) = w$ is defined to allow a range of data values—for example, all instances where the mean of the data falls between a and b—allows the range to be adjusted so that the frequency of cases passing the two-stage filter is large enough for a practical and effective numerical evaluation.

Compositing over All Data in the Ensemble

Imagine randomly drawing cases from the ensemble where the distribution of the parameter is $p(\theta)$, obtaining for each case a measurement on whatever the observable is that constitutes the data, calculating a posterior associated with those data, and cumulating the posteriors as this compound sampling process is repeated. Since this compound sampling does not select on the data, the distribution of θ in the cases that pass the filter is the same as the prior. What does the composite posterior represent?

$$
\begin{aligned}
\int_{-\infty}^{\infty} p(\theta|x)p(x)\,dx &= \int_{-\infty}^{\infty} \left(\frac{p(\theta)p(x|\theta)}{p(x)} \right) p(x)\,dx \\
&= \int_{-\infty}^{\infty} p(\theta)p(x|\theta)\,dx \\
&= p(\theta) \int_{-\infty}^{\infty} p(x|\theta)\,dx \\
&= p(\theta).
\end{aligned}
\tag{8}
$$

Equation (8) indicates that the prior is composed of data-specific posteriors each weighted according to the probability of drawing the data associated with it when sampling cases from the prior. This is interesting enough, though it may not be immediately obvious what this knowledge is good for.

OBTAINING A PRIOR FROM A SAMPLE OF
MEASUREMENTS ON A SAMPLE OF CASES

In a compound sampling situation, where we don't know the process generating the distribution from which we are drawing cases in the first stage
of the sampling, our knowledge of the distribution that constitutes the prior
will have to be empirical. Ideally, we might construct a histogram of values
of the parameter of interest in a probability sample of cases from the population defining the prior. But often enough our knowledge of the sample of
cases from this population will be knowledge of *data* rather than direct
knowledge of the values of the parameter. How do we estimate the distribution of the parameter from a distribution of data?

This is a classic problem for variance components methods, though these
generally have been developed from a frequentist perspective that may seem
a little out of place as a prelude to a Bayesian analysis. Efron (1996) has proposed a maximum likelihood approach to exactly this problem, but this too
is essentially frequentist.

Deely and Lindley (1981) developed a thoroughly Bayesian approach in
which a prior called the "hyperprior" is put on the parameters that describe
the distribution that will be used as the later prior in the case-specific analysis for the parameter of interest in the case of interest. But this hierarchical structure begs the question of where we obtain the hyperprior on the parameters of the prior that is then estimated empirically.

From a practical standpoint, these philosophical concerns should not prevent us from getting adequate estimates of the prior from a reasonably large
sample of data from the defining ensemble. But it would be nice to have a
thoroughly Bayesian formulation with no loose ends, and especially nice if
this formulation lent itself to an easy, practical implementation.

Imagine that we have an indefinitely large probability sample of cases
from the ensemble defining the prior, and that we have limited sample data
from each of these cases. Consider an arbitrary distribution $g(\theta)$ that we use
in place of the prior in a process of forming and then compositing the posteriors from these data. If the composite fails to recapitulate $g(\theta)$, then from
equation (8) we know (with probability 1) that $g(\theta)$ can't be the actual distribution of θ in the population that was sampled, so it would not be correct
to use $g(\theta)$ for the prior. Formally then, substituting a distribution $g(\theta)$ as a
candidate prior, but knowing that we have obtained the probability sample
of cases, which yielded the empirical distribution of data $h(x)$ (which should
approach $p(x)$, the true marginal distribution of data in the prior), from the

ensemble where the distribution of the parameter follows the actual unknown prior $p(\theta)$, we can test for the equality

$$\int_{-\infty}^{\infty} \frac{g(\theta)p(x|\theta)h(x)}{\int_{-\infty}^{\infty} g(\theta)p(x|\theta)d\theta} dx = g(\theta). \qquad (9)$$

If this equality does not hold, then $g(\theta)$ cannot be the correct prior. Thus,
equation (9) constitutes at least part of the definition of what we might
mean by the *true prior*, based on knowledge of the distribution of the data,
$h(x)$, from the population.

   This is a necessary but not sufficient condition, for more than one function g can satisfy the equality, as can be seen by considering functions that
have 0 density over part of their domain. But if, based on our empirical experience of nature, we impose a preference for smoothness among the functions g that satisfy equation (9), this will suffice to identify the prior.

A RECURSION ALGORITHM

Our problem is a variant of the classical mixture problem (Lindsay, 1995);
here both the mixture and the components are unknown, and we are attempting a solution based on a necessary relation between the two.

   We can embed equation (9) in a recursion, where for iteration i

$$g_{i+1}(\theta) = \int_{-\infty}^{\infty} \frac{g_i(\theta)p(x|\theta)h(x)}{\int_{-\infty}^{\infty} g_i(\theta)p(x|\theta)d\theta} dx. \qquad (10)$$

If the initial candidate prior g_0 is too broad to satisfy equation (9), the sequence will converge smoothly to a solution where, at iteration j, equality
does obtain to any arbitrary resolution:

$$|g_{j+1}(\theta) - g_j(\theta)| \leq \varepsilon. \qquad (11)$$

Mathematically (but not in terms of the compound sampling scenario, which
gives meaning to the result), the prior and the likelihood are interchangeable
in equation (1) (Bayarri, DeGroot, and Kadane, 1988); each simply weights
the other.

So over any domain where neither the prior nor the likelihood is 0, equation (1) basically splits the difference between them (in a sort of geometric mean), which is to say that the recursion of equation (10) will converge.

If the mathematical form (or machine precision) allows the nominal g to have 0 density in places, the above algorithm, which does converge to a stable solution from the direction of candidate priors that are too broad, may not converge from the direction of candidate priors that are too narrow. But convergence from the direction of candidate priors that are too broad is rapid, so to be on the safe side it is practical to start with a candidate density that is nearly uniform over the feasible domain.

In practice, actual computer implementation is quite simple. With a finite sample of cases whose parameter values are a probability sample from the true prior, the sampling itself, and the compositing of the resulting nominal posterior densities, accomplishes the outer integration of equation (10), and the integration in the denominator is simply the normalization of each posterior. If a parametric form is specified for the prior, this is *fit* to the composite posterior at the end of each iteration so that the next iteration begins with a prior of this form. Such a parametric approach can impose smoothness, if we think it appropriate. The practical question that remains is how large a sample of cases is needed to obtain a reasonable estimate of the prior by this means.

NUMERICAL EXAMPLE: PARAMETERS OF A NORMAL PRIOR

Consider a process that generates case-specific observables that are normally distributed with a known constant variance and an unknown case-specific mean, μ_c. The unknown case-specific mean is drawn from an ensemble of normally distributed values, and it is the parameters of that normal distribution of values of μ in the ensemble that we wish to estimate for use as a prior in making case-specific inferences. In the numerical example that follows, we carry out that estimation on a sample of k cases drawn from the ensemble, where the case-specific data consist of m observations from each case. The procedure used for the estimation is the recursion of equation (10), where the candidate prior is constrained on each iteration to be a Gaussian, with parameters μ_π and σ_π^2 that are revised on each iteration simply by matching moments. To avoid the known bias in the estimate of a population variance from the sample variance of a finite sample, we compute the variance of the composite, at each iteration, from the mean and variance of each component posterior, where the variance contribution owing to the between-case variance among the means of the case-specific posteriors is normalized by $k - 1$ rather than k, the limited sample size of cases from the ensemble.

Formally, we have

$$x_c \sim N(\mu_c, \sigma), \tag{12}$$

$$\sigma = \text{constant}, \tag{13}$$

$$\mu_c \sim N(\mu_\pi, \sigma_\pi), \tag{14}$$

where the case-specific values of x are the only observables, σ is known, and the objective is to estimate μ_π and σ_π from a sample of m observations of x in each of a sample of k cases (which sample μ_c but do not observe it directly).

To investigate the stability of this process, as a function of the sample size k, we conduct the estimation procedure on 100,000 independent realizations of drawing a sample of k cases, for each value of k considered between 1 and 100, with a sample of $m = 4$ observations in each case. In the example reported here, the true unknown parameters of the distribution for the population defining the prior are $\mu_\pi = 3$ and $\sigma_\pi^2 = 1$. The known constant variance of the process giving rise to the observations that are the data for each sampled case is $\sigma^2 = 0.75$. The initial parameter values specifying a candidate prior that is intentionally too broad were $\mu_\pi = 1.5$ and $\sigma_\pi^2 = 6$. The convergence criterion was that both μ_π and σ_π^2 change by less than 0.001 on successive iterations. In all 10,000,000 trials the convergence was rapid.

Figure 12.2 shows the distribution of estimates of μ_π for values of k between 1 and 100, and figure 12.3 shows the distribution of estimates of σ_π^2. As expected from a process dependent on sampling a distribution, the mean of that distribution, μ_π, was estimated without bias and with a precision that improved rapidly with increasing sample size (k); and the variance of that distribution, σ_π^2, was estimated with a bias that diminished rapidly with increasing sample size and with a precision that improved slowly with increasing sample size. The same general behavior would have been observed with a variance components estimation procedure.

For many purposes, the accuracy and precision of the estimates obtained with a sample of 20 cases from the population defining the prior might be

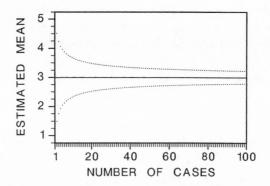

FIGURE 12.2 The distribution of estimates of the mean of the prior as estimated, by the recursion of equation (10), from data from samples of different numbers of cases, in a simulation experiment. The solid line gives the mean of 100,000 realizations at each number of
cases (the true value was 3), and the dotted lines locate the 2.5% tails. The mean is estimated
with no bias and with precision that improves rapidly with the number of cases. Details of
the simulation are given in the text.

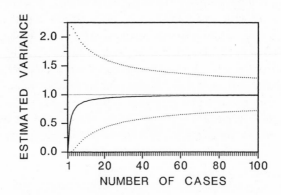

FIGURE 12.3 The distribution of estimates of the variance of the prior as estimated by the recursion of equation (10) from data from samples of different numbers of cases, in a simulation experiment. The heavy solid line gives the mean of the estimates of the variance from
100,000 realizations at each number of cases, and the dotted lines locate the 2.5% tails. The
thin solid line shows the true value (1). The variance is estimated with a bias that diminishes
rapidly with increasing number of cases, and the precision of the estimates improves slowly
with the number of cases. Details of the simulation are given in the text.

deemed acceptable (for this amount of data, with this level of variability, per case), and the performance at a sample size of 100 cases might be deemed good. Of course, the real criterion of acceptability of an estimated prior is not how it *looks,* but rather how it performs in a given decision-theoretic application (with a given utility function, a given amount of case-specific data, and a given spectrum of actions to choose among). Some applications will be more sensitive than others to small discrepancies in the prior.

STATISTICAL POWER FROM THE PRIOR:
MORE TERMINOLOGY AND HISTORY

The compound sampling perspective developed in this paper requires that the distribution used as the prior represent the distribution of the parameter of interest in an actual population that we call the ensemble. When we don't know the mechanism generating this distribution, the only way we can know the distribution is by empirical information obtained from sampling that ensemble. This latter is at the heart of the approach now generally called empirical Bayes (Robbins, 1995), where the prior is estimated from the marginal distribution of the data from the ensemble.

The compound sampling perspective predates the use of the term "empirical Bayes." Von Mises (1942) clearly describes a compound sampling situation, allowing him a frequentist definition of probability while using Bayesian techniques. Fisher, Corbet, and Williams (1943) set up a compound sampling framework and estimated a parameter from the marginal distribution of the data in an ensemble, but avoided Bayesian terminology. Other expositions of a compound sampling context include Gini (1911) (described in Forcina, 1982), Barnard (1954), Neyman (1962), Pratt (1963), and Rubin (1984). It is enlightening to find Fisher (1956, 13–18) describing such an application in which he was perfectly comfortable with the use of Bayesian inference and using language of "compound event," "recognizable subset," and "superpopulation."

The notion of *somehow* drawing statistical power from data on a population that includes more cases (with different values for the parameter) than the case of interest was treated explicitly in an ideologically impure frequentist setting by James and Stein (1961). An empirical Bayes explanation of James-Stein estimation was given by Efron and Morris (1973).

The literature that called itself empirical Bayes at first was not very Bayesian, looking for point estimates of the parameters of the prior and judging the performance of the procedure by coverage criteria that are conditional on the parameter, not the data. One thoroughly Bayesian description of the empirical Bayes setup (Berry and Christiansen, 1979; Deely and Lindley, 1981) is to imagine a "vague hyperprior" for the prior, use Bayes' formula to infer a distribution for parameters of the prior from the data on the population, and use this prior, in Bayes' formula, to infer a posterior distribution for the case of interest from the data on this case. The hierarchical form of this approach gives the label "hierarchical Bayes" (also called "full Bayes"). While the hierarchical Bayes literature (Carlin and Louis, 1996) is clear that compound sampling underlies the relation of the case of interest to the ensemble that supplies data for inference on the prior, this literature is not clear on whether a similar compounding relates that ensemble (which they often call a "population") to a "hyperpopulation" associated with the hyperprior or whether we need to be concerned about the relation between the conventional vague hyperprior and an actual physical distribution for relevant parameters in the hyperpopulation.

As empirical Bayes methods evolved, authors became dissatisfied with the simple empirical assessments of the prior and looked for ways, albeit non-Bayesian, to allow for uncertainty in the parameters of the prior (Laird and Louis, 1987; Carlin and Gelfand, 1991; Efron, 1996). This literature in fact views hierarchical Bayes as the underlying, but generally impractical, framework and searches for less Bayesian, but hopefully more practical, methods for approximating the prior. The less than full embrace of all the Bayesian aspects of hierarchical Bayes may be reflected in some peculiar language, such as calling the hyperprior a "structural distribution" or the phrase "borrowing strength" from other cases, which leads one to wonder exactly what "strength" is and to worry when we might have to give it back.

Hierarchical Bayes approaches do, in general, envision an ensemble, one step back from the case of interest that supplies the data that go into the likelihood function. It is then just a matter of interpretation whether the hierarchical Bayes approach does or does not contemplate the case of interest as an instance sampled from that ensemble. And it is a matter of commitment whether the specifications for the distribution of the parameter in the larger population in the ensemble are obtained from a Bayesian or frequentist method; and if one is a Bayesian, whether a prior somewhere further back in the hierarchy is made up. Here, I have shown the extent to which the analysis can be Bayesian in its entirety, without a need to make up anything.

The approach developed here, of composing the prior from a sample of cases from the ensemble, is not a panacea. The core identity, equation (9), establishes a necessary condition, but it is not sufficient. In order for the recursion algorithm, equation (10), to progress, we had to impose an assumption of smoothness for the prior, which is an extraneous constraint. Smoothness may accord well with our experience of nature, but that in itself is an empirical assertion. In the simple numerical example, smoothness was enforced with a parametric form. Finally, the approach requires that the sample of cases be a probability sample from the population, and the core identity was for the limiting situation of a large sample of cases.

In practice, if the sample of cases from the ensemble is too small, then the implicit histogram upon which the prior is based will be too uncertain. One logically consistent solution, then, would be to treat this ensemble as a hierarchical layer in a fully Bayes treatment, but to define a hyperensemble, which is more inclusive and therefore offers more opportunities for finding useful data, from which the hyperprior would be calculated according to the empirical Bayes treatment we have developed here. Thus, no prior, or hyperprior, or hyperhyperprior, need be made up; the daisy chain uses fully Bayesian inference for each layer that is data-poor to an extent requiring the statistical machinery for propagating uncertainty and ends in a layer where there are enough cases with data that a histogram formed of case-specific posteriors is an adequate estimate of the distribution for the parameter(s) in that level.

The approach of composing the prior, because of its Bayesian purity, may aid understanding of the limitations of induction, and therefore inference, rather like the way the Deely and Lindley (1981) hierarchical formulation answers the theoretical question of whether the data from the target case should be included with the data from the other cases in developing the inference for the prior. The answer from the Deely and Lindley derivation clearly is that the data from the target case should not be included at that stage. The empirical Bayes literature continued to wrestle with the question (Efron and Morris, 1972; Laird and Louis, 1987; Carlin and Gelfand, 1991). The answer that emerges from this chapter is that if the number of cases in a given level is small enough that inclusion or exclusion of one of them makes a difference, this is a sure sign that this level should be treated as a hierarchical level with the target case excluded, and at least one more level must be sought to provide an ensemble with enough cases that an empirical Bayes calculation of the unconditional distribution of the parameter will suffice.

Ultimately, all inference providing a probability of the parameter conditional on the data must be supported by a large sample somewhere in the chain, but not necessarily at the level of the case of interest.

POSTSCRIPT

It is amusing to conjecture where Bayes himself stood in the interpretation
of the prior as a literal physical distribution. The record is clouded, because
the one publication defining Bayesian inference that is attributed to Bayes
(and indeed the first description of any legitimate procedure for statistical
inference on an unknown parameter) was actually submitted for publication by an amateur mathematician friend of Thomas Bayes, after Bayes'
death, and no earlier manuscript is known (Stigler, 1982). In the main text
of Bayes' paper, the example for which the method is derived is a compound
sampling scheme, where the distribution of the parameter in the population
defining the prior is determined by a known process with a physical frequency interpretation. The philosophical speculation in Bayes' paper about
the distribution to describe knowing "absolutely nothing" does not appear
in the body of the text, but rather in a separate section (sort of an appendix)
called a scholium, from which we might wonder whether Bayes, or his
friend, had second thoughts on this score.

12.1   Commentary

Nozer D. Singpurwalla

The title of this chapter prompts several questions. What does one mean by
a Bayesian analysis? Are there Bayesians who admit to *not* taking their priors seriously? Are all Bayesians necessarily subjectivistic? The answers to
questions 2 and 3 are, I should think, in the negative. An answer to question
1 is that a Bayesian analysis is one whose premise is solely the calculus of
probability; much has been written on this topic, the most recent contribution being that by Lindley (2000). What then is the essence of Goodman's
paper?

I was able to identify two points; others may see more. The first is that
the prior on a parameter, say θ, should have a frequency-based interpretation. The second is that Bayes' formula is a mechanism for decomposing the
prior into a data-specific component (which is the posterior) and its complement. I will first comment on the former.

A frequency-based interpretation of probability has been embraced by
many, despite the fact that its underlying logic has been the subject of criticism. This interpretation is made operational by conceptualizing a collec-

tive, and probability of an event is interpreted in terms of encountering the event in the collective. When a collective cannot be conceptualized, as is sometimes the case, then probability is not the appropriate means for quantifying uncertainty. Thus, requiring that the prior have a frequency-based interpretation—"taking the prior seriously"—is de facto a limitation. There are many Bayesians who do not subscribe to subjectivity (but mostly on practical grounds) and who interpret the prior objectively. To them a prior does not have to reflect true personal beliefs. Rather, the prior is a device that produces a posterior, and it is the posterior on which they are willing to place their bets.

Regarding the second point, a subjectivist sees Bayes' law as "the mathematics of changing one's mind," i.e., a prescription for how one's bets should change in the light of new evidence. As mentioned above, an objective Bayesian sees Bayes' law as a device for producing a means to bet in the light of the data. Goodman views Bayes' law as a way to decompose the prior into a data specific component and its complement; see equation (3). In the absence of a subjectivistic interpretation of the probability, I have trouble seeing the operational value of this decomposition. Perhaps the author can provide clarification.

There are some other issues not worthy of a serious objection, such as distinguishing between taking a prudent action and "searching for scientific truth," and the claim that inference is a logical process of making defensible statements about the value of a parameter. I see the need for searching for scientific truth as a prelude to an action, and inference as a way to quantifying uncertainty about an *observable*.

12.2	Rejoinder

Daniel Goodman

> *During the last few decades, the Anglo-Saxon countries have seen a good deal of progress in the practical application of probability theory to statistical problems in biology and industry. This development started with the misunderstanding of one of the classical formulas of probability calculus (Bayes' rule): there followed a period that was characterized by the erroneous practice of drawing statistical conclusions from short sequences of observations. —R. von Mises (1957)*

What constitutes a "good" inference procedure? I am crass enough to believe that the question can be resolved by considering the performance of the inference procedure in a betting game. The test is to use the candidate inference procedure to infer the probabilities of outcomes and then to place bets on the outcomes, betting against other players who have access to the same information about the game but are using other inference procedures. In this tournament, a good inference procedure demonstrably will, in the long run, accumulate the largest winnings.

> The probability that an event has happened is the same as the
> probability I have to guess right if I guess it has happened.
> —T. Bayes (1763)

Aside from inference, the winning betting strategy is obvious: place wagers where the price of the wager is less than the product of the probability of the specified outcome and the payoff in the event of that outcome. Viewed from the perspective of the stochastic process on which bets are placed, the probability of each outcome is unquestionably a frequency. Therefore, if we use an inferred probability as the basis for betting, we must at least pretend that this too has a frequency interpretation. Furthermore, if we want to achieve the best possible long-run performance, the numerical value of the inferred probability must match that of the actual long-run physical frequency of outcome of the stochastic process as closely as possible.

Bayesian inference provides a means for putting a probability on outcomes, but, depending on the kind of Bayesian inference, the probability may or may not have a frequency interpretation, and the numerical value of that probability may or may not correspond, in some definable way, to the actual physical probability of the outcome. A player who espouses a subjective Bayes theory of probability will have the satisfaction of knowing that he can place his bets so as to be consistent with his beliefs. A player who employs an inference procedure that delivers probabilities that correspond to the probabilities of the game will have the satisfaction of placing bets so as to come out ahead in the game in the long run. I know whose shoes I would prefer to be in. And if the player is an environmental regulator making decisions on my behalf, I know what theory I want that regulator to use.

> It has been said, that the principle involved in the above and
> in similar applications is that of the equal distribution of our
> knowledge, or rather our ignorance—the assigning to different
> states of things of which we know nothing, and upon the very

ground that we know nothing, equal degrees of probability. I apprehend, however, that this is an arbitrary method of procedure.
—G. Boole (1854)

Almost everything that is unsatisfactory in the long history of Bayesian inference has to do with approaches to the prior. The likelihood function is philosophically unproblematic—it relies on data, and, when normalized over the data space, it is a probability that has an uncontroversial physical frequency interpretation.

It was desperation over the prior that forced a retreat to subjective probability or frequentist hypothesis testing and confidence intervals.

The cost of this retreat was considerable to both camps. The frequentists retreated to real probabilities in the wrong space. The inward-turning character of this part of the discipline is revealed in their choice to evaluate "performance" in the data space, even though the actual performance of a betting procedure plays out in the parameter space. This led some to the ironic observation that frequentist confidence intervals may provide a better assessment of the probabilities before the data have been taken into account than afterward (Dempster, 1964).

The Bayesian retreat was to probabilities that weren't frequencies and weren't "out there" in the world at all.

The prior matters when the data are weak. Practically, the prior affects the posterior, so the open-ended option of a subjective prior creates a loose end that is at odds with the desire that the inferred probability correspond to the actual probability. In other words, the prior has to be right. Philosophically, a subjective prior merged with an objective likelihood produces a posterior that is at least partly subjective. The philosophical dodge of declaring all probability to be subjective doesn't really solve the problem when we want an inferred probability to correspond to a real physical frequency that we are betting on. The weakness of the extreme subjectivist position is perhaps exposed just as a matter of consistency—if it is all subjective, why even bother with data and the likelihood?

If the population sampled has itself been selected by a random procedure with known prior probabilities, it seems to be generally agreed that inference should be made using Bayes' theorem.
—D. R. Cox (1958)

That the difficulties really originated with the conceptualization of the prior is perhaps illustrated by the striking instances in which figures with a lot invested in the frequentist school still accepted a Bayesian solution as the correct solution when an unambiguous prior was available (Fisher, 1956; Neyman, 1962).

> *The only relevant thing is uncertainty—the extent of our own knowledge and ignorance. The actual fact of whether or not the events considered are in some sense determined, or known by other people, and so on, is of no consequence.* —B. de Finetti (1974)

Among the writers who contributed to the subjective theory were many great and subtle thinkers, so the actual position could not have been simple-minded. And there are real insights to be salvaged from this literature. One is that we can shift the meaning of "subjective" from the realm of personal belief to refer instead to the possibility that the individual may have access to more, or less, information even when that information itself is objective in the sense that other observers would interpret it in the same way if they had access to it. This serves as a point of reference to develop deeper arguments about the consequences of limited information, even when all that information is shared, and in that sense objective. Practically, probability arises when our information is limited to knowledge of frequency of occurrence.

> *Do we mean unknown through lack of skill in arguing from given evidence, or unknown through lack of evidence. The first alone is admissible, for new evidence would give us a new probability, not a fuller knowledge of the old one . . . We ought to say that in the case of some arguments a relation of probability does not exist, and not that it is unknown.* —J. M. Keynes (1921)

Outside the set of situations where a known physical mechanism is responsible for the prior, the frustration in attempting to develop a useable prior lies in the recognition that in the context just of the given case, where the data are sparse enough that they are not convincing in themselves, the prior really is unknown.

When speaking of probability, the universe of cases from which the frequency of occurrences is of interest is ours to choose. When the universe of cases is going to be employed as an ensemble, as the basis for an empirical prior, the choice is guided by two considerations. (1) So as not to neglect information that we actually have, the chosen universe of cases should be narrow enough that our actual knowledge of the relevant characteristics of cases in this universe is limited to frequency of occurrence. (2) The universe should be inclusive enough that our knowledge of the relevant frequencies of occurrence is convincing in its own right.

> *And it appears from this Solution that where the Number of Tri-*
> *als is very great, the Deviation must be inconsiderable: Which*
> *shews that we may hope to determine the Proportions, and by*
> *degrees, the whole Nature, of unknown Causes, by a sufficient*
> *Observation of their Effects.* —D. Hartley (1749)

Imagine that the "case" is a round of betting on the outcome, heads or tails, of spinning a given coin on its edge, and that the data for this case are the results of observed outcomes of past spins of this same coin. The outcome of spinning a coin is sensitive to very small imperfections in the geometry of the coin and its edge, so a given coin is prone to a nontrivial bias in the probability of how it will land, unlike the outcome of flipping a coin, where the probability in practice is essentially 1/2 (Diaconis and Ylvisaker, 1985). Thus, a player placing bets on the outcome of a spun coin by assuming the probability is 1/2 would, in the long run, be outperformed by a player who based his bets on a better knowledge of the bias of that particular coin. (More esoterically, knowledge that this bias exists in general, but with no knowledge whether the bias for a particular coin is for heads or tails, creates an exploitable small margin in favor of double heads or double tails, over that expected of a theoretical fair coin, in a game where the same unexamined coin is spun twice and bets are placed on whether the outcome is a double.) The information to draw upon in forming a better assessment of the particular coin, for purposes of bets on the outcome of the next spin of that coin, would be the record of observed outcomes of past spins of that coin. If the number of observed past spins is not large, the assessment would have to be statistical, and if the inference is Bayesian, as it should be, that analysis will require a prior distribution for the binomial parameter governing the outcome of spinning the coin. A prior that is "subjective" in the whimsical sense will be problematic.

Assigning a uniform prior to the parameter in its range of (0, 1), on the grounds of ignorance, is nonsensical, since the bias of individual coins, while noticeable, is not that large. Assigning a sharp prior at the value of .5 for the parameter, on the grounds that this is the way a perfect coin would behave, defeats the point of the inference, which is to quantify the imperfection in that particular coin. Just by thinking about the problem, we can understand that the correct prior is some sort of compromise between these two extremes of a flat distribution and a delta function at .5, but the details of the correct compromise do not emerge just by thinking.

It turns out that sampling a number of coins will reveal a bimodal distribution to their biases, indicating a tendency toward noticeable biases with either sign (Diaconis and Ylvisaker, 1985). This is an empirical fact, and correct characterization of the prior for inference on spinning a coin will be an empirical undertaking. The sample of coins that is used to form a prior will be "subjective" in the sophisticated sense of being comprised of that sample and not another and being limited to the information that is available.

Dr. Singpurwalla asks, Why would we want to decompose an objective prior? We can imagine many ways of restricting the ensemble of "other" coins for the prior so as to improve its specificity to the case at hand. We might want the coins to be of the same denomination, the same year, and from the same mint, and exhibiting the same degree of handling wear. All these characteristics can be determined readily from inspection of the coin. Yet we may not be in a position to stratify so narrowly for a particular analysis if this would result in too small a sample size of cases with available data for a reliable empirical prior. Furthermore, we can imagine that it would help to form the empirical prior from a sample of coins struck from the same die, and perhaps from the same portion of the run from that die, as the die itself wore. And we might want to make very precise dimensional measurements of the coins. All this would make sense, and we would form the prior that way if we had the information. But if we didn't have the information, we couldn't use it, and we would adopt a broader categorization of "other" coins, for which we did have enough information, on a large enough sample, to form the prior.

These different ensembles would result in different priors, which in turn would result in different posteriors when used on inference from the same data. Interpreted as probabilities, these different posteriors are different probabilities—not just different numbers—for each defines the frequency from compound sampling of a different universe. The most useful, the one with the greatest predictive power, will be the one from the narrowest universe that still offers a large enough sample of comparative information. Analysis with a prior formed from a narrower universe will fall prey to sampling error in the prior. Analysis from a broader universe will deliver a posterior that is true, in the sense of correctly representing the probability of outcome of compound sampling from that universe. But such an analysis, if used in the betting game, will not perform as well as one based on a more specific universe for the prior, because it neglects the available information that could be extracted by stratifying within the sample forming the prior.

Demystifying probability, and taking the witchcraft out of the prior, frees philosophical inquiry to pursue the really deep question about the foundations of statistical inference, namely, the role of the law of large numbers.

REFERENCES

Barnard, G. A. 1954. Sampling Inspection and Statistical Decisions. *J. Roy. Stat. Soc,* ser. B, 16:151–174

Bayarri, M. J., M. H. DeGroot, and J. H. Kadane. 1988. What Is the Likelihood Function? *Statistical Decision Theory and Related Topics IV,* vol. 1. New York: Springer.

Bayes, T. 1763. An Essay Towards Solving a Problem in the Doctrine of Chances. *Phil. Trans. Roy. Soc. London* 53:370–418. (Reproduced with slight editing and with updating of the mathematical notation in Barnard, G. A. 1958. Studies in the History of Probability and Statistics. IX. Thomas Bayes' essay towards solving a problem in the doctrine of chances. *Biometrika* 45:293–315.)

Berger, J. O. 2000. Bayesian Analysis: A Look at Today and Thoughts of Tomorrow. *J. Am. Stat. Assn* 95:1269–1276.

Berger, J. O., and T. Sellke. 1987. Testing a Point Null Hypothesis: The Irreconcilability of p Values and Evidence (with discussion). *J. Am. Stat. Assn* 82:112–122.

Bernardo, J. M. 1979. Reference Posterior Distributions for Bayesian Inference (with discussion). *J. Roy. Stat. Soc.* ser. B. 41:113–147.

Berry, D. A., and R. Christiansen. 1979. Empirical Bayes Estimation of a Binomial Parameter via Mixtures of Dirichlet Processes. *Ann. Stat.* 7:558–568.

Boole, G. 1854. *The Laws of Thought.* London: Macmillan. Reprinted, New York: Dover, 1958.

Carlin, H. P., and A. E. Gelfand. 1991. A Sample Reuse Method for Accurate Parametric Empirical Hayes Confidence Intervals. *J. Roy. Stat. Soc.* ser. B, 53:189–200.

Carlin, H. P., and T. A. Louis. 1996. *Bayes and Empirical Bayes Methods for Data Analysis.* New York: Chapman and Hall.

Cox, D. R. 1958. Some Problems Connected with Statistical Inference. *Ann. Math. Stat.* 29:357–372.

Deely, J. J., and D. V. Lindley. 1981. Bayes Empirical Bayes. *J. Am. Stat. Assn* 76:833–841.

de Finetti, B. 1937. La prevision: ses lois logiques, ses sources subjectives. *Ann. Inst. Henri Poincaré* 7:1–68. English trans., "Foresight: Its Logical Laws, Its Subjective Sources." In Kyburg, H. E., Jr., and H. E. Smokler, eds., *Studies in Subjective Probability.* Huntington, NY: Krieger, 1962.

de Finetti, B. 1974. *Theory of Probability.* Chichester: Wiley.

Dempster, A. P. 1964. On the Difficulties Inherent in Fisher's Fiducial Argument. *J. Am. Stat. Assn* 59:56–66.

Diaconis, P., and D. Ylvisaker. 1985. Quantifying Prior Opinion. In Bernardo, J. M., M. H. DeGroot, D. V. Lindley, and A. F. M. Smith, eds. *Bayesian Statistics 2.* Amsterdam: North-Holland.

Edwards, W., H. Lindman, and L. Savage. 1963. Bayesian Statistical Inference for Psychological Research. *Psych. Rev.* 70:193–242.

Efron, B. 1996. Empirical Bayes Methods for Combining Likelihoods (with discussion). *J. Am. Stat. Assn* 91:538–565.

Efron, B., and C. Morris. 1972. Limiting the Risk of Bayes and Empirical Bayes Estimators. Part 2: The Empirical Bayes Case. *J. Am. Stat. Assn* 67:130–139.

Efron, B., and C. Morris. 1973. Stein's Estimation Rule and Its Competitors: An Empirical Bayes Approach. *J. Am. Stat. Assn* 68:117.

Fisher, R. A. 1956. *Statistical Methods and Scientific Inference.* New York: Hafner.

Fisher, R. A., A. S. Corbet, and C. H. Williams. 1943. The Relation between the Number of Species and the Number of Individuals in a Random Sample of an Animal Population. *J. Animal Ecol.* 12:42–58.

Forcina, A. 1982. Gini's Contributions to the Theory of Inference. *Internat. Stat. Rev.* 50:65–70.

Gelman, A., J. B. Carlin, H. S. Stern, and D. B. Rubin. 1995. *Bayesian Data Analysis.* London: Chapman and Hall.

Gini, C. 1911. Considerazioni Sulla Probabilita a Posteriori e Applicazioni al Rapporto dei Sessi Nelle Nascit Umane. *Metron* 15:133–172.

Good, I. J. 1952. Rational Decisions. *J. Roy. Stat. Soc.,* ser. B. 14:107–114.

Goodman, D., and S. Blacker. 1998. Site Cleanup: An Integrated Approach for Project Optimization to Minimize Cost and Control Risk. In *Encyclopedia of Environmental Analysis and Remediation.* New York: Wiley.

Hartley, D. 1749. *Observations on Man, His Frame, His Duty, and His Expectations.* London: Richardson. Reprint, Gainesville, FL: Scholars' Facsimiles and Reprints, 1966.

James, W., and C. Stein. 1961. Estimation with Quadratic Loss. In *Proceedings of the Fourth Berkeley Symposium on Mathematical Statistics and Probability.* Berkeley: University of California Press.

Keynes, J. M. 1921. *A Treatise on Probability.* London: Macmillan.

Laird, N. M. and T. A. Louis. 1987. Empirical Bayes Confidence Intervals Based on Bootstrap Samples (with discussion). *J. Am. Stat. Assn* 82:739–757.

Lindley, D. V. 1957. A Statistical Paradox. *Biometrika* 44:187–192.

Lindley, D. V. 2000. The Philosophy of Statistics. *Statistician,* vol. 49, part 3, 293–337.

Lindley, D. V., and L. D. Phillips. 1976. Inference for a Bernoulli Process (a Bayesian view). *Am. Stat.* 30:112–119.

Lindsay, B. G. 1995. *Mixture Models: Theory, Geometry, and Applications.* Hayward, CA: Institute of Mathematical Statistics; Arlington, VA: American Statistical Association.

Neyman, J. 1962. Two Breakthroughs in the Theory of Statistical Decision Making. *Rev. Int. Stat. Inst.* 30:112–127.

Pratt, J. W. 1963. Shorter Confidence Intervals for the Mean of a Normal Distribution with Known Variance. *Ann. Math. Stat.* 34:574–586.

Pratt, W. 1965. Bayesian Interpretation of Standard Inference Statements. *J. Roy. Stat. Soc.,* ser. B, 27:169–203.

Robbins, H. 1955. An Empirical Bayes Approach to Statistics. In In *Proceedings of the Third Berkeley Symposium on Mathematical Statistics and Probability.* Berkeley: University of California Press.

Rubin, D. B. 1984. Bayesianly Justifiable and Relevant Frequency Calculations for the Applied Statistician. *Ann. Stat.* 12:1151–1172.

Stigler, S. M. 1982. Thomas Bayes' Bayesian Inference. *J. Roy. Stat. Soc.,* ser. A, 145:250–258.

von Mises, R. 1942. On the Correct Use of Bayes' Formula. *Ann. Math. Stat.* 13:156–165.

von Mises, R. 1957. *Probability, Statistics, and Truth.* London: Allen and Unwin.

13   Elicit Data, Not Prior: On Using Expert Opinion in Ecological Studies

Subhash R. Lele

ABSTRACT

Ecological studies are often hampered by insufficient data on the quantity of interest. For example, because of cost and time limitations, hard data on attributes, such as pollutant contamination or presence or absence of species, are difficult to come by. Limited data usually lead to a relatively flat likelihood surface that is not very informative. One way out of this problem is to augment the available data by incorporating other possible sources of information. A wealth of information in the form of "soft" data, such as expert opinion about whether pollutant concentration exceeds a certain threshold (Journel, 1986; Kulkarni, 1984), may be available. The purpose of this paper is to propose a mechanism to incorporate such soft information and expert opinion in the process of inference. A commonly used approach for incorporating expert opinion in statistical inferences is via the Bayesian paradigm. In this chapter, I discuss various difficulties associated with the Bayesian approach. I introduce the idea of eliciting data instead of priors and discuss its practicality. Subsequently, a general hierarchical model setup for combining elicited data and the observed data is introduced. The effectiveness of this method is illustrated for presence-absence data using simulations.

BAYESIAN APPROACH TO USING EXPERT OPINION

One of the standard approaches for incorporating expert opinion is the subjective Bayesian paradigm (Steffey, 1992; DeGroot, 1988; Genest and Zidek, 1986; von Mises, 1942). In this approach, the basic idea is to obtain prior beliefs from the scientists regarding the possible values of the parameters of the statistical model. These prior beliefs are quantified in terms of a prob-

410

ability distribution, called the "prior distribution." This distribution is then modified in the light of the data to obtain the "posterior distribution." The posterior distribution signifies the changed beliefs in the light of the data. The posterior distribution may also be used for decision making and prediction of the unobserved values.

For the most part, scientists are more interested in summarizing and assessing evidence in the data than reporting personal beliefs. It is clear that the Bayesian approach is inappropriate for such a task (Royall, 1997). The posterior probability distribution, even if obtained as a consequence of a noninformative prior, cannot be used to compare evidence for one hypothesis over the other hypothesis. For example, Lele (2004 [chapter 7 of this book]) discusses general conditions for functions that may quantify strength of evidence. One of the conditions involves parameterization invariance. The posterior distribution does not satisfy parameterization invariance and hence may be considered invalid for summarizing evidence.

Many practicing scientists are reluctant to use the Bayesian paradigm for a variety of other reasons (Royall, 1997; Mayo, 1996; Dennis, 1996; Efron, 1986). It is the subjectivity inherent in specifying the prior that is most troubling to the scientists. For example, Efron (1986) observes that "subjectivism . . . has failed to make much of a dent in scientific statistical practice" because "strict objectivity is one of the crucial factors separating scientific thinking from wishful thinking." A related problem is that the prior provided by the expert may be intentionally misleading. For example, in environmental sampling for pollution investigation, an industry expert may want the sampling effort to be concentrated on the top of a hill in the study area where there is little chance of finding any contamination. An environmental activist, on the other hand, may want it to exclusively concentrate at the bottom of the hill, where a high degree of contamination may be expected. Dennis (1996; 2004 [chapter 11, this volume]) discusses other instances in ecological applications where subjective priors elicited from intentionally misleading experts could have serious public policy implications. He argues that "the Bayesian philosophy of science is scientific relativism" where "truth is subjective" (Dennis, 1996). This "scientific relativism" is unacceptable to many scientists. Moreover, Royall (1997) and Journel (1986) have demonstrated that the standard Bayesian practice of expressing lack of prior information as a uniform distribution can also be patently misleading.

Even if we set aside the philosophical objections to the Bayesian formulation, the quantification of prior belief in the form of a probability distribution on the parameter space is a complex problem (Garthwaite and Dickey, 1992). West (1988) observes that "it is often (or rather, always) difficult to elicit a full distribution with which an expert is totally comfortable." Yang and Berger (1997), who are strong proponents of the Bayesian paradigm, discuss the difficulties in the elicitation of prior distributions. They claim: "Frequently, elicitation of subjective prior distributions is impossible, because of time or cost limitations, or resistance or lack of training of clients."

- Subjective elicitation can easily result in poor prior distributions, because of systematic elicitation bias and the fact that elicitation typically yields only a few features of the prior, with the rest of the prior (e.g., its functional form) being chosen in a convenient, but possibly inappropriate, way.
- In high-dimensional problems, the best one can typically hope for is to develop subjective priors for the "important" parameters, with the unimportant or nuisance parameters being given noninformative priors.

There are other problems as well. The elicited priors are usually not invariant to reparameterization. For example, suppose in the population dynamics modeling, one elicits a prior from an expert for the density dependence and the intrinsic growth rate parameters. The same expert when asked to provide a prior on the parameters corresponding to carrying capacity and intrinsic growth rate, which are simply functions of the density dependence and intrinsic growth rate parameters, in general will not provide a prior that is compatible with the prior for the first parameterization. Consequently, the same data with the same expert and the same model will provide different inferences! This noninvariance to reparameterization of the elicited prior distribution is disturbing.

Why is it so difficult to elicit a prior distribution? It is important to recognize that the concept of a prior probability distribution on the parameters of a statistical model is a *statistical* construct that is hard for most scientists to visualize. This makes it difficult for a scientist to express his opinions in a precise, clear, and consistent fashion. Thus, it is important that a mechanism be devised that will enable scientists to express their opinions in a practical fashion.

The next question regarding the Bayesian approach to the incorporation of the expert opinion in ecological studies is, Does it buy anything substantial over and above a likelihood-based approach?

An area where the Bayesian approach has found relevance is in the area of prediction (Geisser, 1993). Consider a situation where one may want to know the amount of contamination at the unsampled locations or one may want to know if a particular species is present at a particular location or one may want to know the probability that a particular species will go extinct in the next hundred years. A typical, not necessarily Bayesian, approach to address the prediction problem is via a statistical model.

Let us consider a simple situation. Suppose Y denotes the amount of contaminant at a given location. We model the possible values of Y using a probability density function $f(y, \theta)$ or the corresponding distribution function $F(y, \theta)$. In the prediction problem, instead of providing a single value, one usually provides a range of values, namely the prediction interval. Suppose we wish to obtain the prediction interval for a new observation. If the true parameter value is θ_0, then the prediction interval is given by $(F^{-1}(.05, \theta_0),$ $F^{-1}(.95, \theta_0))$ where F^{-1} is the inverse-distribution function. This prediction interval says that with probability .90 the new value will be somewhere in this interval.

In practical applications, however, the true parameter value is unknown. Let $Y_1, Y_2, \ldots, Y_n$ be independent identically distributed random variables from the probability density function $f(y, \theta)$ where the parameter θ is unknown. Based on these observations one can estimate the value of the true parameter. This estimator is denoted by $\hat{\theta}$. One may then behave as if the estimated value is the true value and use $(F^{-1}(.05, \hat{\theta}), F^{-1}(.95, \hat{\theta}))$ to obtain a 90% prediction interval. This approach is known as a "plug-in" approach. It is also well known (Zimmerman and Cressie, 1991) that such a prediction interval is too optimistic; that is, the actual coverage is smaller than the nominal coverage.

The Bayesian answer to this problem is both simple and elegant. In the Bayesian formulation the parameter is itself a random variable. Suppose the prior distribution on the parameter θ is $\pi(\theta)$. Then the joint distribution of (Y, θ) can be easily derived as $f(Y, \theta) = f(y|\theta)\pi(\theta)$. To obtain prediction intervals for a new observation in the presence of no data, one can use the marginal distribution of Y, namely $\int f(y|\theta)\pi(\theta)d\theta$. If data are available, one uses the distribution $f(y|y_1, y_2, \ldots, y_n) = \int f(y|\theta)\pi(\theta|y_1, y_2, \ldots, y_n)d\theta$ to obtain prediction intervals. Here $\pi(\theta|y_1, y_2, \ldots, y_n)$ denotes the posterior distribution.

Even if one does not subscribe to the Bayesian paradigm fully, one can be a "reluctant Bayesian" (Diggle, Moyeed, and Tawn, 1998) and look at the above solution as a practical way out of the less than nominal coverage of the plug-in approach. If the data are sparse the Bayesian approach generally provides wider prediction intervals than the plug-in approach and tends to have better coverage properties. Thus, one can look at the Bayesian approach as providing a fudge factor that makes the prediction intervals wider than the plug-in approach, in the process making them conservative. This justification for using Bayesian inference is put forward in kriging and other applications of spatial statistics (Diggle et al., 1998). Unfortunately, the Bayesian fudge factor depends on the choice of the prior distribution that is necessarily subjective.[1] An intentionally misleading expert can potentially have a disastrous effect on the prediction problem.

Let us look at this Bayesian solution when we have large amount of data. In such a situation, the posterior distribution is highly concentrated around the maximum likelihood estimator, and hence the Bayesian solution corresponds very closely to the plug-in approach. Thus, in the case where we have enough data, the likelihood is very peaked, and we practically "know" the true parameters, the Bayesian approach and the plug-in approach provide similar inferences. On the other hand, if the data are sparse, the Bayesian inferences are driven more by the choice of the prior than by the information in the data. The key to the problem of prediction is not the accommodation of a Bayesian fudge factor but the increase in the information about the value of the unknown parameter by utilizing all possible sources, including expert opinion.

A typical argument advocated by practicing Bayesians against the sensitivity to the choice of the prior is that one should always conduct sensitivity analysis, checking the sensitivity of the inferences to the choice of the prior (Yang and Berger, 1997). Unfortunately, this seems to raise the problem that if the posterior is not sensitive (probably because the data have substantial information), the Bayesian paradigm does not buy anything substantial over and above what the likelihood provides; on the other hand, if the posterior is sensitive, the subjectivity of the prior makes its use suspicious in the scientific context. Sensitivity of the posterior to the choice of the prior only says that the information content in the data is small. Not only could we have learned that the data contain little information by looking at the likelihood function but also quantified it by, for example, examining the curvature of the likelihood function.

The problems in using the Bayesian approach to incorporate expert opinion can be summarized as follows.

Philosophically, scientists, in general, have difficulty accepting the subjectivity inherent in the Bayesian paradigm. We, as scientists, almost always want to assess evidence in the data and not report somebody's personal beliefs about the model or quantities derived from the model. Moreover, the posterior distribution is not a valid evidence function because of its lack of invariance to reparameterization.

---

1. Sometime the use of "objective" priors is advocated. However, there does not seem to be a unique definition of what constitutes an objective prior.

Operationally, even the proponents of the Bayesian paradigm agree that elicitation of priors is a difficult if not an impossible task. Even in the hands of the expert, elicitation is more an art than science; one can only imagine the problems faced by an applied statistician or a quantitative ecologist working in the Yellowstone National Park trying to do this.

Pragmatically, even if one can elicit a prior from an expert, one is immediately faced with the issue of whose expert. Should we take the ecologist working for the Sierra Club as an expert, or should we take the ecologist working for the mining industry as an expert? Unfortunately, Bayesian inferences are extremely sensitive to such a choice.

Without any doubt, the Bayesian paradigm is one of the most elegant developments in modern statistical theory. As Lindley (1972) correctly puts it, "Bayesian statistics is an axiomatic branch of mathematics (like Euclidean geometry) and must stand or fall as applied mathematics by the relevance of those axioms to the real world." Unfortunately, in regards to the scientific questions posed in ecological studies, the Bayesian approach for incorporation of expert opinion seems neither practical nor relevant.

Shall we then conclude that expert opinion cannot be utilized in a sensible way in ecological analysis? I think the situation is not bleak. There is a way to use expert opinion in ecological data analysis while avoiding many of the shortcomings of the Bayesian approach. This chapter describes such an approach. A detailed study of the statistical properties of the suggested approach is reported in Lele and Das (2000). An ecological application is under way.

ELICIT DATA, NOT PRIOR

Let us start with the operational aspect of quantifying the expert opinion in a fashion useable in statistical analysis. If the elicitation of priors is so difficult for the experts in the field, what is more practical? What kind of information can we expect to obtain from an expert with comparative ease?

In my experience working with scientists, I have seen that it is natural for an expert to think in terms of the process or the observables under study. A sensible approach, then, is to ask the expert to provide guess values for observable data itself. Consider some practical situations. Suppose we want to predict and map the presence or absence of a particular species in the southwestern U.S. Surveying every bit of land would be prohibitively costly. However, we may have substantial information on the habitat characteristics at all the locations of interest. For example, we may have satellite images of the area, soil maps, digital elevation maps, climate data, vegetation cover types, and other relevant ecological data. Given this data, it is not an impossible task for an expert to give his/her guess on the presence or absence of a given species at a given location. The same is true of pollutant contamination surveys; experts can use geomorphological features, soil type, etc., to guess whether contamination will be above or below a given threshold. This approach is quite operational and not just the musing of a theoretical statistician. For example, two practicing geologists, Kulkarni (1984) and Journel (1986), suggest the possibility of using of such expert *guesses* to improve kriging predictions in geological studies. Furthermore, McCoy, Sutton, and Mushinsky (1999) discuss successful application of "educated guesses" in predicting optimal habitats for a sand skink, *Neoseps reynoldsi.* Clearly, it is operationally easier to obtain the guess values of the observable data than a prior distribution on the parameters of a statistical model.

Given my discussion of the difficulties in the Bayesian incorporation of expert opinion posed by misleading experts, it is imperative that we ask, What if the expert providing such guesses is intentionally misleading? Can we still use the "information" provided by such an expert? Would the subjectivity in the expert opinion affect the inferences in adverse fashion the same way it affects the Bayesian approach?

Let us consider this in the environmental sampling situation discussed earlier. Suppose the expert is trying to mislead us by guessing large amounts of contamination where there is very little, or vice-versa. As soon as we have some real, hard observations, it will be clear to the analyst that this particular expert's opinion is negatively correlated with the reality. As an extreme case, let us suppose the expert opinion and the reality turn out to be perfectly negatively correlated. Then, to predict the amount of contamination at an unsampled location all that we have to do is flip the sign of the expert opinion at the locations that we have not sampled and we will have perfect prediction based on a perfectly misleading expert. The point is that any correlation, negative or positive, is information. Even a so-called misleading expert provides useful information, and that information should be used. This *calibration of an expert* is the key step in fruitfully using the elicited data in a statistical analysis. This is also the step that distinguishes the approach suggested in this paper from the standard Bayesian approach. Although the expert guesses themselves are subjective, because they are calibrated against the real data, the inherent subjectivity is reduced.

The third question is, Can we use expert opinion in the form of elicited data to augment the evidential content of the hard, observed data? How do we weigh the hard data and the soft data when quantifying evidence?

Toward this goal, we apply a hierarchical model as described below. This approach is in spirit very much the same as that suggested in Efron (1996) for combining several data sets. The main difference is that in the cases Efron considers, the data sets are independent of each other whereas in our case the elicited data and the real data are correlated to each other.

A HIERARCHICAL MODEL FOR THE ELICITED DATA

In this section, we propose a hierarchical modeling approach that provides a framework for the augmentation of observed data with data elicited from experts. We couch our discussion in terms of the finite population setup, which is natural for many ecological studies. Suppose there are N locations indexed by $1, 2, \ldots, N$, in a study area D. Let the true values of the quantity of interest at these locations be denoted by $\underline{Y} = (Y_1, Y_2, \ldots, Y_N)$. The Ys may be continuous or discrete random variables. For example, we may be interested in the amount of pollutant contamination (continuous values), level of pollutant contamination—low, medium, or high (ordinal values)—or the presence or absence of particular species (binary values). Suppose we have only one expert. We ask the expert to provide his/her best guess for the values of $\underline{Y}$. Let the expert guess values, the elicited data, be denoted by $\underline{E} = (E_1, E_2, \ldots, E_N)$. These may, in turn, be continuous or discrete. For instance, if the expert is prepared to guess exact values for the amount of pollutant contamination, then we have continuous expert data. On the other hand, it may be much easier for an expert to provide ordinal (high, medium, or low contamination) or binary values (high or low) over the whole study area, rather than continuous data. We assume the following relationship between the truth, $\underline{Y}$, and expert guess about the truth, $\underline{E}$:

1. The vector $\underline{Y}$ is a realization from a distribution $f(y, \theta)$ indexed by unknown parameters θ.
2. Given $\underline{Y} = y$, expert opinion $\underline{E}$ has a distribution $g(e \mid y; \eta)$ where η denotes unknown parameter or parameters that characterize the dependence between $\underline{Y}$ and $\underline{E}$.

The parameters η play a central role in the inferential process. They capture the degree of dependence between the truth and the expert guess, thus providing a mechanism to evaluate the knowledge and sincerity of an expert or calibrate the expert. Information from a good expert would closely mimic the reality, and this should be reflected in a high positive value of η. Conversely, a high negative value for η would indicate misleading expert opinion. If η is very small, then expert opinion is purely random, bearing no relation to reality. Thus, η may be thought of as an "honesty or credibility parameter" that tells us whether the expert is believable. Let us denote the sampled values at n locations by $\underline{Y}_s = (Y_1, Y_2, \ldots, Y_n)$, and the unknown values at unsurveyed locations by $\underline{Y}_{ns} = (Y_{n+1}, \ldots, Y_N)$, $n < N$. The sampled observations, along with the expert guesses, enable us to estimate the credibility parameter η. This estimate helps in suitably calibrating any information provided by an expert, so that proper inferences can be made.

Notice that the model presented here is very general. We do not assume any particular distributional structure for either $\underline{Y}$ or $\underline{E}$. This enables different kinds of expert information to be combined with real data. For instance, while the actual response may be continuous, the elicited data could be either continuous or discrete.

In many ecological studies, the question centers on the presence or absence of a species and its relationship with the habitat characteristics. The responses $\underline{Y}$ are thus binary instead of continuous. In the following, I discuss the use of elicited data in such a situation. A detailed study involving continuous data, dependence, and other complications is reported in Lele and Das (2000).

THE BINARY-BINARY MIXTURE MODEL

The following probability model is adopted to combine binary expert opinion with an underlying process that is also binary:

Let Y_i be Bernoulli random variables with $P(Y_i = 1 | \underline{x}_i) = p_i$, where

$$\log\left(\frac{p_i}{1 - p_i}\right) = \underline{x}_i \beta,$$

$i = 1, 2, \ldots, N$, the $\underline{x}_i$s are covariates for location i, and $\underline{\beta}$ is the vector of regression coefficients.

Given $Y_i = y_i$, E_i are independent Bernoulli random variables with $P(E_i = 1 | y_i) = \tilde{p}_i$, where

$$\log\left(\frac{\tilde{p}_i}{1 - \tilde{p}_i}\right) = \eta y_i,$$

$i = 1, \ldots, N$, where η characterizes the dependence between $\underline{Y}$ and $\underline{E}$.

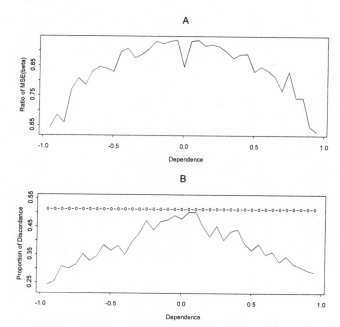

FIGURE 13.1 Simulation results for binary-binary mixture. Subfigure A plots the ratio of observed $MSE(\hat{\beta})$ with elicited data to $MSE(\tilde{\beta})$, calculated without data. Subfigure B plots the observed proportion of discordance in predicted values for the two scenarios, solid line with elicited data and dotted line with no elicited data. Estimation error and proportion of discordance in predicted values for elicited data scenario gets smaller as dependence between real and expert data differs from 0 in either direction.

One may also introduce an intercept in this model to incorporate systematic bias.

Maximum likelihood estimation of the parameters in this model can be based on the likelihood

$$L(\beta, \eta \mid \underline{Y_s}, \underline{E}) = \exp\left(\eta \sum_{i=1}^{n} Y_i E_i + \beta \sum_{i=1}^{n} x_i Y_i \right) \prod_{i \in s} \frac{1}{(1 + e^{\eta Y_i})(1 + e^{\beta x_i})}$$
$$\times \prod_{i \notin s} \left\{ \frac{1}{2}\left(\frac{1}{1 + e^{\beta x_i}} \right) + \left(\frac{e^{\eta}}{1 + e^{\eta}} \right)^{E_i} \left(\frac{1}{1 + e^{\eta}} \right)^{1 - E_i} \left(\frac{e^{\beta x_i}}{1 + e^{\beta x_i}} \right) \right\}.$$

To explore the effect of binary expert data on the estimation of parameters, we conducted a small simulation study. We fixed $\beta = 2$, $N = 100$, and $n = 5$. Figure 13.1A plots the observed average MSEs for β (both with and without elicited data), based on 1,000 simulations, over the range of η. Estimates utilizing elicited data (both good and misleading) are closer to the truth than those based solely on observed data, unless $\eta \approx 0$ and the elicited data is random. Thus, when both the underlying response and elicited data are in a binary form, a combination of the two significantly improves parameter estimates.

Next, we study the predictive performance based on $P(Y_i|E_i)$. If $P(Y_i = 1|E_i) \geq P(Y_i = 0|E_i)$, we take the predicted value to be 1; otherwise it is 0. For the ordinary situation where there is no elicited data, these predictions are based on $P(Y_i)$. Figure 13.1B plots the proportion of discordance in predicted values for these two situations. Again, expert opinion in the form of elicited data greatly improves predictions, as long as the expert is informative. Predictions for a random expert are not appreciably different from those based solely on the observed data. These results indicate that augmentation of observed data with data elicited from experts is advantageous even when the elicited data is in binary form and the underlying process is also binary in nature.

Why are the expert data helping inference substantially? As I mentioned previously, the key to better inference and prediction is to increase the information about the parameter of interest. The elicited data, if associated with the truth, whether negatively or positively, provide information over and above what the observed hard data do. In Lele and Das (2000), this result, quantified in terms of Fisher information, is proved for the normal-normal mixture situation. Even with the addition of an extra parameter η, the Fisher information about the parameter of interest β is larger in the elicited data/observed data combination than in just the observed data. This additional information is reflected in better prediction intervals and better MSEs. By eliciting data from the expert, one can effectively add information about the parameter of interest. This presumably is what expert opinion should do. The more information one has, the more precise are the inferences. I have illustrated how this can be achieved effectively by eliciting data.

BAYESIAN WOLF UNDER THE LIKELIHOOD SKIN?

I have received three kinds of comments on the above proposal. I list them along with my responses.

1. The proposed method is as subjective as the Bayesian method. The predictions obtained from the experts are certainly subjective.

What makes this method more objective than the Bayesian method is that one can actually check/validate the model that relates the real data with the expert opinion. For example, one can plot the expert opinion predictions versus the real data at those locations and see if the postulated model, $g(e|y; \eta)$, is reasonable or not. If this model is inappropriate one can change the model and use a more appropriate model.

Such model checking on the prior distribution is fundamentally unacceptable in the Bayesian paradigm. A prior is chosen before any data are observed, and it remains whatever it was throughout the analysis. A prior is not data based, whereas the model that relates the real data the predictions based on expert opinion is data based in the same way as a regression model that relates covariates with the responses.

2. This is a statistically wrong method: Suppose we collect lots of expert data without collecting any real data, can we estimate the underlying parameters with very high accuracy? This seems very counterintuitive and one wonders if there is a flaw in the method.

There are two possible answers.

First, in most situations, if only expert opinion is available, some of the parameters are not identifiable. In this situation, we cannot even estimate all the parameters, let alone estimate them well! The MSE cannot go to 0 by simply collecting a large amount of expert data. To give a simple example, consider the model $Y_i \sim N(\mu, \sigma^2)$ and $E_i|Y_i \sim N(\eta Y_i, \tau^2)$. Suppose we obtain only the expert guesses in this situation. It is easy to see that the marginal distribution of $E_i \sim N(\eta\mu, \eta^2\sigma^2 + \tau^2)$. The parameters η, μ are nonidentifiable even if τ^2 is assumed known.

Second, suppose the model is such that, based on the marginal distribution of E_is, the parameters are in fact identifiable. In such a situation, indeed if the expert opinion is truly informative—that is, the association between the expert data and the real data is nonzero—the MSE will go to 0. Recall that the MSE of an estimator goes to 0 when the *information* about that parameter goes to infinity. If the expert can truly provide us with usable information about the real phenomenon, the information about the parameter is indeed going to infinity and MSE should go to 0 in such a case. It does not matter if the information comes from an expert opinion or real data. There is nothing wrong statistically.

However, there is an important caveat in the above discussion. The MSE will converge to 0 *only if* the full model specification, $f(y, \theta)$ and $g(e|y; \eta)$, is correct. If only the expert opinion is available and no real data are collected, model checking is impossible. The above procedure will then become solely model dependent without any objective checking of the model assumptions. Thus, if the model is off the mark, inferences will be erroneous (very similar to what happens in the Bayesian approach when the prior is off the mark). *The possibility of model validation and calibration of the expert opinion is what makes the procedure suggested here objective* and hence distinguishes itself from the Bayesian approach. Obviously, I do not recommend collecting only expert data, in which case inference would blindly rely on the correctness of the model specification.

(3) There has to be a prior corresponding to the expert guesses. Why not conduct full Bayesian analysis using this prior?

Aside from the fact that the Bayesian approach is inappropriate for quantifying and comparing evidence, there are two answers.

First of all, there is no reason that there *should* exist a prior that leads to the observed expert predictions. It is quite likely that, implicitly or explicitly, the expert does not think in a Bayesian fashion. A mathematical answer is that this is a mixture problem. There is no guarantee such a prior (a unique mixing distribution) would always exist. And even if it exists, in practice it is almost an impossible problem to actually compute such a prior using the data at hand. The rate at which such an estimator converges to the truth is excruciatingly slow (Lindsay, 1995).

An issue could be raised about the role of expert opinion in model specification and the subjectivity associated with use of any specific model. Is specification of a model for the data at par with the specification of the prior? It clearly is not, for the simple reason that a specific model for the data is potentially refutable whereas the prior is not. We accept priors on belief whereas the models for the data are supposed to be tested and validated. See Dennis (1996; 2004 [chapter 11, this volume]) for a clear discussion of this issue.

DISCUSSION

In many environmental studies, expert opinion can be used to augment the frequently sparse observed data. To incorporate expert opinion, we present an approach that is based on elicited data. There are some key advantages to this approach:

Scientists generally think in terms of the data generation process. Hence, eliciting data is, in our experience, more natural than eliciting a prior distribution on the parameters of a statistical model.

Both an intentionally misleading expert and a genuinely honest expert provide information about the underlying process. Elicited priors from an intentionally misleading expert can badly influence inferences in the Bayesian paradigm, whereas in my approach elicited data from such an expert proves as useful as information from an honest expert. It is the calibration part of my scheme that separates it from the Bayesian paradigm.

A crucial issue, of course, is whether it is possible to elicit data in practice. I refer the reader again to Journel (1986), Kulkarni (1984), and McCoy, Sutton, and Mushinsky (1999). I myself have been able to elicit data from experts regarding presence/absence of mammal species in Montana. A detailed ecological analysis of these data is in Dover (2002) and Dover, Lele, and Allen (2003).

To conclude, I have illustrated, both theoretically and using simulations, that incorporating expert opinion via elicited data substantially improves estimation, prediction, and design aspects of statistical inference for spatial data. The incorporation of model selection procedures should reduce the sensitivity of this approach to model misspecification. More research is needed in this direction.

13.1   Commentary

R. Cary Tuckfield

INTRODUCTION

The problem that Lele addresses concerns the purveyance of ecological science; I'll explain what this means. It is not a matter of choosing up sides, nor does he advocate such. But choosing sides is tacitly demanded and generates most of the heat and rhetoric in this historical contest over methods. "Are you a Bayesian or a frequentist? Which?" Submission to the demand brings little scientific insight and much misspent energy in defending your position. It derives from the argument over objectivity versus subjectivity, typically portrayed as a fight over real science versus pseudoscience. As arguments go, it is often contentious and in my experience devolves into an unengaging tug-of-war over equally justifiable preferences. This is a problem to be sure, but it is not the problem Lele is seeking to address, not directly anyway.

Much effort has been expended in the literature to ensure that ecologists understand the issues around the proper use of statistical methods such as ensuring sufficient real replication (Hurlbert, 1984) or addressing false negative as well as false positive error rates (Parkhurst, 1990). Statistical methods are a means of assessing the evidence in the data, the "likeliness" or "propensity" for some hypothesis to be more reliable than another in explaining a natural phenomenon. From this perspective, using several different methods and comparing results would seem to be preferable to a single best-method restriction. It's that likeliness though that most ecologists want to discover and publish, not the methods. I believe scientists in general are excited to find a way to know what the world is really like. New methods in cosmology, for instance, are only interesting to the extent that they allow us to see something about the universe we couldn't see before, but it's what we couldn't see that's ultimately and profoundly exciting. The method is only the means, not the aim, unless you intend to publish in *JASA*.

Hypothesis testing methods can be as novel or clever as the ecologist is determined. The bothersome thing about most statistical hypotheses, however, is that they are cloaked in the guise of parameter space. So technically, the scope of inference is about the most likely properties of the probability distribution that generated the data. Such formal hypotheses have to be translated back into science space, into statements about the hypotheses we are seeking evidence for in the first place. Lele admits that "scientists are more interested in summarizing and assessing evidence in the data [for their stated hypotheses] than reporting their personal beliefs." Ah, yes, the evidence. It's the evidence that convinces the jury of our peers, be they scientists or otherwise. Lee (1989), for example, laments his introduction to traditional statistics not because of any insufficient mathematical prowess but because of the seemingly torturous logic invoked to say something permissible about the evidence. Classical frequentist statistical tests arrive at statements of uncertainty about the data, given the hypothesis, while Bayesian statistical tests arrive at statements of uncertainty about the hypothesis, given the data. What's curious here is that these latter kinds of less popular and nonclassical statements are just the sort we would rather make.

> I found it difficult to follow the logic by which inferences were arrived at from the data. It sounded as if the statement that a null hypothesis was rejected at the 5% level meant that there was only a 5% chance of that hypothesis being true, and yet the books warned me that this was not a permissible interpretation. . . . It appeared that the books I looked at were not answering the questions that would *naturally* occur to a beginner, and that instead they answered some rather recondite questions that no-one was likely to want to ask. (Lee, 1989, vii)

What was exposed as beginners' naïveté in Lee's Statistics 101 class is precisely the kind of inquiry to which he thinks (and I do too) scientists are naturally inclined—it's a human thing.

Even though choosing up sides over methods is not a fruitful exercise, Bayesian thinking seems to have something to offer since it is more like the thinking we naturally do. Humans are an endpoint of universal reality with considerable reasoning ability, and it is neither satisfying nor tenable to presume that natural selection has disposed us to ask questions we can't answer. Why is it naïve to presume the opposite: that we in fact can answer the very questions we are naturally inclined to ask? What if we were to begin with this adventurous assumption? Where might it lead us? I think that it leads to the kind of inquiry Lele is suggesting. What he is trying to do may be perceived by some as a lure to the dark-side by combining traditional objective and therefore palatable methodology with always intriguing, arguably useful, but usually unpalatable subjective methodology. It is apology at worst, but nonetheless a curious blending of ideas that gives it a centrist character. At best, Lele's method is a fresh look at portraying empirical evidence for the kind of hypotheses we're naturally inclined to ask and in the end may provide more insight about scientists than science.

Consider, for example, the gist of his ingenuity in combining real data $Y = [Y_1, Y_2, \ldots, Y_N]$ with the elicited expert guess data $E = [E_1, E_2, \ldots, E_N]$ for the same N corresponding sample locations. This avoids that uncomfortable feeling of dangling one's professional credibility over a snake pit by valuing subjective opinion alone and allows for the calibration of the expert instead. Both the real and the guess data can be used together to provide some statement of evidence concerning the hypothesis of interest. It's the expert calibration feature of this thinking that Lele says is the key to reinstating a respectable measure of objectivity to this Bayesian approach and allaying the concerns of the skeptical traditionalists. While I agree that the elicited data model is an advance over guessing at a prior, the acid test is whether or not the National Academy of Science would be willing to make recommendations to the president on this basis. Would they risk their reputations by combining real data with guess data—Congress, maybe, attorneys, always, but NAS? Aye, there's the rub.

The real and elicited data model suffers from one other minor problem, and that's the concept of the expert. In the estimation problem, Lele proposes that negatively correlated expert opinion elicited from a misleading expert is as informative as positively correlated expert opinion. As long as the credibility parameter (η) is not near 0, then the expert opinion is informative. This may be, but I have difficulty referring to someone as an expert whom you can nearly always count on to give you, in quantity or quality, a misleading response. Bias is one thing; sabotage is another. Even completely uninformed nonexperts provide reliable data some of the time, you just wouldn't know when that was. I am equally uncertain about the identity of a "random expert" consulted in the prediction problem. Is this strictly an uninformative consultant or an educated person who is devoted to reading the *Scientific American?* I believe uninformed experts are not experts. My point is not to quibble over English semantics, but to clarify that reliable data elicitation occurs when $\eta > 0$, substantially.

Finally, the more fascinating estimation problem is the one for which the elicited data are the only data. Suppose we want an interval estimate of the mean Sr^{90} radioactivity (pCi/kg) in *Microtus* bone tissue sampled near the Chernobyl reactor facility in Ukraine. The sampling costs to answer such a question could be very high and are measured in the incidences of specific carcinomas as well as in rubles. So we seek a safer method for estimating this quantity. Once a set of k experts (e.g., Ukrainian mammalogists) has been calibrated, we could elicit their expectation on this random variable. That is, let $[X_1, X_2, \ldots, X_k]$ be an elicited sample from k calibrated experts wherein each expert is asked to guess at the average Sr^{90} activity among *Microtus* rodents in the immediate vicinity of Chernobyl. Now, let $[X_{1i}^*, X_{2i}^*, \ldots, X_{ki}^*]$ be the ith bootstrapped (Efron and Tibshirani, 1993) random sample with replacement where $i = 1, 2, \ldots, N$. Then a bootstrapped interval estimate by the "plug-in" (Efron and Tibshirani, 1993) method for N random samples is given by

$$P\{\overline{X}^* - 1.96\hat{\sigma}_{SE}^* \leq \mu \leq \overline{X}^* + 1.96\hat{\sigma}_{SE}^*\} \cong 95\%$$

where $\overline{X}^* =$ bootstrap sample mean

$$= \sum_{i=1}^{N} \left(\sum_{j=1}^{k} X_{ij}^*/k \right) / N$$

$$= \sum_{i=1}^{N} \overline{X}_i^*/N.$$

$$\hat{\sigma}_{SE}^* = \left\{ \sum_{i=1}^{N} (\overline{X}_i^* - \overline{X}^*)^2/(N-1) \right\}^{1/2}.$$

This quasi-Lele approach using elicited data only could provide useful information in the ecological or human health risk assessment processes under prohibitive sampling conditions.

PURVEYORS AND STATISTICIANS

Several years ago, I attended the annual meetings of the American Statistical Association and made an effort to attend the president's address. Dr. J. Stuart Hunter delivered that address and he presented the observation that "statisticians are the purveyors of the scientific method." I thought this statement was awfully brash, if not arrogant. But after much reflection I see wisdom in it, however self-serving it may sound.

To purvey means to sell, provide, supply, or deal in. Consulting statisticians though are less like salespersons and more like antique dealers of data. Data require careful scrutiny and a thorough explanation of how they were obtained, the materials and/or circumstances under which they were produced. In the end the "authenticity" or verity of the hypothesis is a matter of interpreting the data. Many collect these relics with the hope of a real find. But when they think they've found something they take it to the dealer to be sure. "What do you think, Dr. Carter?" says the ecology graduate student to the statistics professor. "Is it significant? . . . at the .05 level, I mean."

So how well are statisticians or statistical ecologists currently purveying ecological science? How can we discover more promising advances in statistical thinking that will answer the questions we are naturally inclined to ask? The answers depend on the purveyors' understanding of what ecologists want to know. Here's my view. An ecologist wants to know how likely her hypothesis is given the samples she collected and the data she obtained. That's the long and short of it. Presenting the weight of evidence, insightfully interpreted from the data, makes for a successful scientific career.

The purveyees of ecological science will be better served by the purveyors when the latter acknowledge what the former are seeking. This requires an alteration in the kind of thinking we have traditionally fostered and prescribed. We should stand back from methods dominated by parameter space long enough to explore an approach based more on the weight of evidence, one that results in statements about the verity of the hypothesis, given the data. For starters, we should consider the widely acknowledged but largely unused methods of Tukey (1977). These have an unabashed empirical orientation, are much less mathematically rigorous, require fewer initial conditions or assumptions, and allow the scientist (not only the ecologist) more freedom to interpret his or her data. Two of Tukey's (1977) methods have gained fairly widespread acceptance, viz., the box and whisker plot and the stem and leaf diagram—the former more than the latter. These methods are what purveyors should be purveying as initial recommendations. But what is always wanted thereafter is a formal probability statement regarding the hypothesis. It is less clear to me now than when I was a graduate student why such statements are essential. Is it because probability statements are more reliable than data displays? Is it because they are more succinct and take up less journal space than graphics? The latter is probably true, but I doubt the former. In fact, a reliance on the former promotes a view of the .05 level of significance as a dividing line between truth and falsehood, a sort of action limit that permits the purveyee to publish, given that he or she does not exceed that critical value. Truth is what we seek, tested explanations are what we get. The only certainty is in knowing that the two will never be one and the same. That is von Mises's legacy—in the long-run, of course (von Mises, 1957). The demand for probability statements by the scientific community in order to reduce the potential for subjective and erroneous interpretations of the data have crafted a parameter-space world for hypothesis testing. It's mathematically rigorous and an admirable linguistic translation, but we have sacrificed our license to interpret a hypothesis directly, given the data.

I think Lele has got something here; it should be explored.

13.2 Commentary

Lance A. Waller

Lele provides an insightful discussion involving hierarchical approaches for incorporating expert opinion in statistical analyses and outlines an elegant approach for combining information from observed data with information from data elicited from subject-matter experts. Lele emphasizes differences between his approach and traditional Bayesian inference, often criticizing the latter for philosophical and practical shortcomings. Without overplaying the role of a Bayesian apologist, I briefly contrast Lele's approach with Bayesian analyses in several key aspects.

I applaud the use of a hierarchical approach for combining data, as such formulations clearly illustrate the conceptual structure underlying statistical approaches for combining data of different types, sources, or formats. In the present case, the hierarchical structure clarifies some differences between Lele's and traditional Bayesian approaches.

Lele uses the probability model $g(E|Y)$ to calibrate the elicited data, E, with the observed data, Y. Lele contrasts this with the Bayesian notion of a prior distribution of parameters defined independently of the data. One of the strengths of Lele's elicited data approach is the ability to calibrate the expert's opinion with the observed data; indeed, this calibration is at the root of the approach: The stronger the correlation (positive or negative) between the elicited data and the observed data, the greater benefit for including the expert opinion (as one essentially gains more data informing on the same underlying parameters). In Lele's approach, one assesses elicited opinion (in the form of the elicited data) in the light of the observed data, then uses the observed and elicited data (properly calibrated through $g(E|Y)$) to provide inference regarding model parameters. Lele is correct in saying there is no way to calibrate priors to the data, but the Bayesian would respond that s/he has no interest in doing so—the prior reflects belief about the value of an unknown quantity before including the impact of the data. This is not to say that the Bayesian is not interested in "uncalibrated" prior information; e.g., Kadane et al., (1980) describe an approach to calibrate elicited prior information within the expert's responses to avoid self-contradictory information. Here, the analyst calibrates elicited information within the responses of a single expert, rather than with the set of observed data.

Lele's comments regarding the difficulty in eliciting prior distributions (statistical concepts) from subject-matter experts (typically nonstatisticians) are true. I agree that most scientists will be better positioned to describe their current state of knowledge through hypothetical data or observable quantities than through guesses as to parameter values for distributions (outside of means and standard deviations). However, this alone is not reason to abandon a Bayesian approach. Gelman et al. (1995, 120–122) discuss the general idea of generating prior distributions from historical data, which extends directly to defining priors based on an expert's elicited data (provided the expert is not aware of the observed study data). Also, Chaloner et al. (1992) provide a detailed example incorporating physicians' prior opinions regarding population proportions expected to respond to various treatments in a clinical trial of HIV treatments. Naturally, such approaches require some statistical assumptions as to the parametric form of the prior distribution, but these strike me as no less subjective than the assumptions required to define $g(E|Y)$ in any particular implementation of Lele's approach (e.g., the logistic assumption in the example). In another example of eliciting observable quantities, Kadane et al. (1980) describe a prior elicitation approach based on eliciting quantiles of the predictive distribution of the dependent variable conditional on values of the independent variables in a regression setting. Such an approach is quite similar to Lele's approach in concept, in that one elicits properties of the outcome data (here, quantiles rather than actual data values) given hypothetical sets of values of covariates. In the presence/absence and soil contamination examples, Lele mentions the use of remote sensing habitat and geomorphological covariates, respectively. In these cases, it is not clear to me whether such covariate effects appear in $g(E|Y, \eta)$, $f(Y, \theta)$, or both. If the analyst provides observed covariate values (denoted x) to the expert and elicits data based on x (resulting in, say, $g(E|Y, x, \eta)$), conditioning appears to mean different things for Y and x. In the case of x, the expert's elicited data is conditional on the observed values. In the case of Y, conditioning serves primarily to calibrate the elicited data with the observed data (as I doubt the analyst would provide the observed data, Y, to the expert prior to data elicitation).

Lele notes elicited data strongly correlated (positively or negatively) to observed data improve inference with respect to decreased mean square errors for estimation and proportion of discordance for prediction, which argues for including elicited data from either intentionally misleading or genuinely honest experts. While I agree that these situations add information, Lele's approach does not obviate all opportunities for subjective manipulation of results by disingenuous experts. Consider the following situation: A selectively deceptive expert provides accurate elicited contamination levels for all locations except for one where contamination values are particularly contentious. In this location, our (biased) expert deliberately over- or underestimates the value they suspect. If there are many locations overall, we obtain high correlation between the expert and the observed data resulting in an undue influence of the expert's selectively deceptive value and biased predictions at a location of primary interest. In general, prior elicitation involves global parameters, or global observable quantities, arguably making it more robust to the selective manipulation described above. Also, I suspect such a selectively deceptive expert would be more common than the expert who intentionally misrepresents all observations.

Finally, in contrast to Lele's experience, I generally find scientists open to the fundamental Bayesian philosophy of research updating previous knowledge. I expect most scientists with whom I collaborate would agree with the following loose paraphrase of the scientific method: what we know now (posterior) is a combination of what we knew before collecting data (prior) and what we observe in the present data (likelihood). While issues remain as to the degree one agrees with technical and other philosophical aspects of Bayesian statistics, many analytic issues lend themselves nicely to Bayesian analysis (see Breslow, 1990, for examples in biostatistics, and Omlin and Reichert, 1999, for relevant discussion in an ecology setting). In addition, developments in computational methods in the last two decades (e.g., the development of Markov chain Monte Carlo algorithms) allow application of Bayesian methods and models (particularly hierarchical models) in a wide variety of settings overcoming previous computational and analytic challenges (see Gelman et al., 1995, Carlin and Louis, 1996, and Gilks, Richardson, and Spiegelhalter, 1996, for many examples).

13.3   Rejoinder

Subhash R. Lele

I thank Drs. Tuckfield and Waller for kindly commenting on my paper. Given the controversial and unorthodox nature of the ideas in my paper, I expected much firework, but both commentators have avoided the dogmatic outlook that many Bayesian statisticians tend to take, and have presented insightful comments.

I am glad that Tuckfield agrees with me that scientists try to quantify the strength of evidence, or as he puts it, the likeliness of one hypothesis over another. I also agree with him that the frequentist approach leads to substantial confusion as well as difficulty in its interpretation. That is really the reason why this book was conceived! However, I do not think that the Bayesian approach is getting to the quantification of evidence either, as implied by Tuckfield's commentary. One cannot quantify strength of evidence as a statement of the probability of a hypothesis being true (Lele, 2004 [chapter 7, this volume]). To do so would require specification of a prior distribution, which is where the rub lies.

I agree with Waller that scientists like the idea of modifying their prior knowledge in the light of the present knowledge. However, this can be achieved simply by combining the two likelihoods rather than the prior-posterior paradigm that involves subjectivity and beliefs. It is this part of the Bayesian paradigm that is disturbing to many scientists.

One of the major problems with the frequentist approach is that it cannot explicitly incorporate expert opinions in the analysis. The Bayesian approach has an explicit mechanism for allowing incorporation of the expert opinion. I do think there is substantial information that experts possess and incorporation of this knowledge should strengthen the observed data. The difference between the Bayesian and this evidential approach lies in *how* expert opinion is incorporated in the analysis. Covariates or auxiliary variables in regression analysis can be looked upon as sort of "expert opinions." In my formulation "expert guesses" are almost being used as covariates.

I was thrilled to find out that I could very easily obtain expert guesses in a few different situations. One situation related to the presence or absence of shrews at certain locations in Montana (Dover, Lele, and Allen, 2003; Dover, 2002). The other involved the abundance of Lyme disease–carrying ticks on an army base in Maryland. Whenever I have talked about this idea with the scientists, I have received an appreciative nod from them. The main advantage is that neither the elicitor nor the elicitee needs to be particularly experienced in the exercise.

As Tuckfield notes, turning scientific statements into statistical models is an art and needs a lot of care. See Lewin-Koh, Taper, and Lele (2004 [chapter 1, this volume]) for some comments on this subject. This difficulty again points to the elicitation of data rather than elicitation of prior distributions.

I agree with Tuckfield that exploratory data analysis is important, but I do not think that it can be an end in itself. One has to be able to predict or forecast the future and exploratory data analysis, in general, does not allow that.

I am somewhat at a loss with regard to Tuckfield's Chernobyl example. I explicitly state that one cannot use expert guesses alone for a proper quantification of the strength of evidence. It lacks the calibration step so important in my methodology. However, they may be quite useful in designing sampling strategies (Lele and Das, 2000). I think perhaps that is what Tuckfield's comments relate to.

Whom to call an expert? I feel that anybody who provides an opinion is an "expert." The η parameter provides the reliability of the expert. Sometimes, experts are useful; sometimes, they are not. It is characterized through the parameters in the calibration model.

Both Waller and Tuckfield think of an expert who may be "selectively misleading." Calibration procedures could certainly adapt to this as well. We could check for "outliers" in the relationship between the expert and the reality. These outliers indicate selective misleading by an expert. Model validation and model selection are an essential part of the evidential paradigm. One can then probably down-weigh the possible outliers in the expert opinion. I would like to reemphasize here that the outliers are data based and not belief based.

Eliciting prior beliefs through eliciting data is not new. Waller points out its precedence in Kadane et al (1980). The main problem with what Kadane et al. do is that they construct a prior based on these elicited data *without any calibration with the observed data*. They only make sure that the expert is not self-contradicting. However, once such a prior is constructed, it is accepted as is. The observed data, even if apparently contradictory to this prior, does not allow us to question the so-called expert! Goodman (2004 [chapter 12, this volume]) also discusses construction of priors using the existing data.

In the formulation suggested in this paper, I have introduced nuisance parameters corresponding to the parameters in the calibration model. Thus, in order for this formulation to be useful, the cost of estimation of these parameters should be less than the benefit attained by the inclusion of the expert guesses in the data. In Dover, Lele, and Allen (2003) this question is dealt with in detail. In general, because the relationship between the expert opinion and the observed data tends to be a mathematically simpler function than the relationship between the covariates and the observations, smaller samples generally suffice to calibrate the expert opinion and use it to improve upon the quality of statistical inference. The same sample size may not permit the effective estimation of the relationship between the covariates and the observations (without the expert guesses).

Scientists seem to have a love-hate relationship with Bayesian inference. They like the idea of using expert opinion and "soft data"; however, they balk at the subjectivity it introduces. L. J. Savage admonished R. A. Fisher for trying to make the "Bayesian omelet" without breaking the Bayesian egg; here, I think I have utilized expert opinion to season an evidential omelet that is made without any Bayesian egg.

REFERENCES

Breslow, N. 1990. Biostatistics and Bayes (with discussion). *Stat. Sci.* 5:269–298.
Carlin, H. P., and T. A. Louis. 1996. *Bayes and Empirical Bayes Methods for Data Analysis*. London: Chapman and Hall.
Chaloner, K., T. Church, T. A. Louis, and J. P. Matts. 1992. Graphical Elicitation of a Prior Distribution for a Clinical Trial." *Statistician* 42:341–353.
Das, A. 1998. Modeling and Robust Inference in Spatial Studies. Unpublished Ph.D. dissertation, Department of Biostatistics, Johns Hopkins University School of Hygiene and Public Health.
DeGroot, M. H. 1988. A Bayesian View of Assessing Uncertainty and Comparing Expert Opinion. *J. Stat. Planning Inference* 20:295–306.
Dennis, B. 1996. Discussion: Should Ecologists Become Bayesians? *Ecol. Appl.* 6:1095–1103.

Dennis, B. 2004. Statistics and the Scientific Method in Ecology. Chapter 11 in Taper, M. L., and S. R. Lele, eds., *The Nature of Scientific Evidence: Statistical, Philosophical, and Empirical Considerations.* Chicago: University of Chicago Press.

Diggle, P. J., R. A. Moyeed, and J. A. Tawn. 1998. Model-Based Geostatistics (with discussion). *Appl. Stat.* 47:299–350.

Dover, D. C. 2002. Elicited Data and the Incorporation of Expert Opinion. Master's project, University of Alberta.

Dover, D. C., S. R. Lele, and K. L. Allen. 2003. Using Expert Opinion to Improve Predictive Accuracy of Species Occurrence: Application to Masked Shrews (*Sorex cinereus*) in Montana. Unpublished manuscript.

Efron, B. 1986. Why Isn't Everyone a Bayesian? (with discussion). *Am. Stat.* 40, 1–11.

Efron, B. 1996. Empirical Bayes Methods for Combining Likelihoods (with discussion). *J. Am. Stat. Assn* 91:538–565.

Efron, B., and R. J. Tibshirani. 1993. *An Introduction to the Bootstrap.* New York: Chapman and Hall.

Garthwaite, P. H., and J. M. Dickey. 1992. Elicitation of Prior Distributions for Variable Selection Problems in Regression. *Ann. Stat.* 20:1697–1719.

Geisser, S. 1993. *Predictive Inference: An Introduction.* London: Chapman and Hall.

Gelman, A., J. B. Carlin, H. S. Stern, and D. B. Rubin. 1995. *Bayesian Data Analysis.* London: Chapman and Hall.

Genest, C., and J. V. Zidek. 1986. Combining Probability Distributions: A Critique and an Annotated Bibliography. *Stat. Sci.* 1:114–135.

Gilks, W. R., S. Richardson, and D. J. Spiegelhalter. 1996. *Markov Chain Monte Carlo in Practice.* Boca Raton: Chapman and Hall.

Goodman, D. 2004. Taking the Prior Seriously: Bayesian Analysis without Subjective Probability. Chapter 12 in Taper, M. L., and S. R. Lele, eds., *The Nature of Scientific Evidence: Empirical, Statistical, and Philosophical Considerations.* Chicago: University of Chicago Press.

Hurlbert, S. H. 1984. Pseudoreplication and the Design of Ecological Field Experiments. *Ecol. Monographs* 54:187–211.

Journel, A. G. 1986. Constrained Interpolation and Qualitative Information: The Soft Kriging Approach. *Math. Geol.* 18(3): 269–286.

Kadane, J. B., J. M. Dickey, R. L. Winkler, W. S. Smith, and S. C. Peters. 1980. Interactive Elicitation of Opinion for a Normal Linear Model. *J. Am. Stat. Assn* 75: 845–854.

Kulkarni, R. B. 1984. Bayesian Kriging in Geotechnical Problems. In Verly, G., ed. *Geostatistics for Natural Resources Characterization Part 2.* Boston: Reidel.

Lee, P. M. 1989. *Bayesian Statistics: An Introduction.* New York: Oxford University Press.

Lele, S. R. 2004. Evidence Functions and the Optimality of the Law of Likelihood. Chapter 7 in Taper, M. L., and S. R. Lele, eds., *The Nature of Scientific Evidence: Statistical, Philosophical, and Empirical Considerations.* Chicago: University of Chicago Press.

Lele, S. R., and A. Das. 2000. Elicited Data and Incorporation of Expert Opinion in Spatial Studies. *Math. Geol.* 32:465–487.

Lewin-Koh, N, M. L. Taper, and S. R. Lele. 2004. A Brief Tour of Statistical Concepts. Chapter 1 in Taper, M. L., and S. R. Lele, eds., *The Nature of Scientific Evidence: Empirical, Statistical, and Philosophical Considerations.* Chicago: University of Chicago Press.

Lindley, D. V. 1972. *Bayesian Statistics: A Review.* Philadelphia: Society for Industrial and Applied Mathematics.

Lindsay, B. G. 1995. *Mixture Models: Theory, Geometry, and Applications.* Hayward, CA: Institute of Mathematical Statistics; Arlington, VA: American Statistical Association.

Mayo, D. G. 1996. *Error and the Growth of Experimental Knowledge.* Chicago: University of Chicago Press.

McCoy, E. D., P. E. Sutton, and H. R. Mushinsky. 1999. The Role of Guesswork in Conserving the Threatened Sand Skink. *Cons. Biol.* 13(1): 190–194.

Omlin, M., and P. Reichert. 1999. A Comparison of Techniques for the Estimation of Model Prediction Uncertainty. *Ecol. Modeling* 115:45–59.

Parkhurst, D. F. 1990. Statistical Hypothesis Tests and Statistical Power in Pure and Applied Science. In von Fursteberg, G. M., ed., *Acting under Uncertainty: Multidisciplinary Conceptions.* Boston: Kluwer.

Royall, R. M. 1997. *Statistical Evidence: A Likelihood Paradigm.* London: Chapman and Hall.

Steffey, D. 1992. Hierarchical Bayesian Modeling with Elicited Prior Information. *Comm. Stat.,* ser. A, 21:799–821.

Tukey, J. W. 1977. *Exploratory Data Analysis.* Reading, MA: Addison-Wesley.

von Mises, R. 1942. On the Correct Use of Bayes' Formula. *Ann. Math. Stat.* 13: 156–165.

von Mises, R. 1957. *Probability, Statistics,and Truth.* 2nd ed. Mineola, NY: Dover.

West, M. 1988. Modelling Expert Opinion. In Bernard, J. M., M. H. DeGroot, D. V. Lindley, and A. F. M. Smith, eds., *Bayesian Statistics 3.* New York: Oxford University Press.

Yang, R., and Berger, J. 1997. A Catalogue of Noninformative Priors. Institute of Statistics and Decision Science, Duke University, discussion papers # 97-42.

Zimmerman, D., and N. A. C. Cressie. 1992. Mean Squared Prediction Error in the Spatial Linear Model with Estimated Covariance Parameters. *Ann. Inst. Stat. Math.* 44:27–43.

PART 5

MODELS, REALITIES, AND EVIDENCE

Mark L. Taper and Subhash R. Lele

OVERVIEW

An oft-quoted remark about statistical models (commonly attributed to G. E. P. Box) is that all models are wrong. We prefer to say that all models are approximations, some more adequate than others. This part deals with the question of model adequacy and model selection. Because evidence corresponds to comparison of the estimates of distances of models from the "true" but unknown data-generating process, model adequacy and model selection clearly are central to the study of evidence.

Bruce Lindsay (chapter 14) takes on the problem of model adequacy. Echoing Bozdogan (1987), he argues that there are two components of errors in any statistical analysis. One component is due to model misspecification; that is, the working model is different from the true data-generating process. The other component is due to the uncertainty in parameter estimation. Most of statistical theory focuses on the second component. One also needs to consider the first component when making statistical inferences. Lindsay develops ideas on the impact of model misspecification on statistical inference and ways to quantify this effect. It is well known that there is a difference between statistical significance and scientific significance. Statistical significance asks whether an effect is reliably detectable; scientific significance asks whether a real effect of this magnitude would substantially affect a scientific conclusion or a management decision. Lindsay proposes to compare confidence intervals on model misspecification error with external knowledge of the scientific relevance of prediction variability to address the issue of scientific significance. This naturally leads to the point that the idea of divergence between the truth and the working model is "scientific problem dependent." There is no single measure that will be suitable for answering this question. He then discusses several divergence measures that might be useful for various scientific purposes. This paper lays out the basics of his philosophy of model adequacy.

Mark Taper (chapter 15) discusses the Akaike information criterion and its variants, such as Schwarz's information criterion. The AIC penalizes the likelihood for the complexity of the model, represented as the number of parameters in the model, whereas the SIC (also known as the BIC) takes into account the sample size as well. Taper argues that one should also take into account the complexity of the model set, the set of all models that are being compared. He shows how a data-based penalty can be developed to take into account the working model complexity, model set complexity, and sample size. His method is reminiscent of inspection of scree plot for determining the number of informative principle components.

While lauding Lindsay's approach, Sir D. R. Cox points out some of the potential pitfalls to be wary of when exploring such new territory. One of his most important cautions is that model fit to data is only one criterion for model selection; often model fit to theory is of equal importance. The discussions of Taper's chapter by Hamparsum Bozdogan and by Isabella Verdinelli and Larry Wasserman are exciting. Bozdogan points out that model complexity is inadequately represented by model order and offers an alternative approach based on Fisher information to penalizing for complexity. Verdinelli and Wasserman provide a discussion of recent developments in the statistical literature on both testing and Bayesian-based model selection procedures. They also present an example indicating that Taper's approach does not work as well for models with sparse effects as it does for models with tapered effects. This underscores the contention made in both chapters of this section that evidence procedures should be problem specific.

REFERENCE

Bozdogan, H. 1987. Model Selection and Akaike's Information Criterion (AIC): The General Theory and Its Analytical Extensions. *Psychometrika* 52:345–370.

14   Statistical Distances as Loss Functions in Assessing Model Adequacy

Bruce G. Lindsay

An argument is made for reformulating the way model-based statistical inference is carried out. In the new formulation, we never "believe" in the model; that is, we do not treat the model as "truth." It is instead an approximation to truth. Rather than testing for model fit, an integral part of the proposed statistical analysis is to assess the degree to which the model provides adequate answers to the statistical questions being posed. One method for doing so is to create a single overall measure of inadequacy that evaluates the degree of departure between the model and truth. Small values of this measure should indicate high adequacy; large values should represent breakdown of the model, in the sense that the model provides answers to important statistical questions that are far from the truth. Interval estimation of this measure can then be the basis for determining the limitations of the model-based inference. We analyze several familiar distance measures in terms of their possible use as inadequacy measures. Along the way it is noted that good parameter estimation in this setting logically means choosing the parameters so that the corresponding model elements are the most adequate.

ASSESSING MODEL UNCERTAINTY

The fundamental job of a statistician is to incorporate uncertainty into the assessment of scientific data. This is normally achieved by the creation of a probability-based framework that mimics the data generation process and appears reasonable in light of the particular nature of the data collected.

For example, based on a sampling scheme that generated the data, we might make the *basic modeling assumption* that "the data $Y_1, Y_2, \ldots, Y_n$ are independent and identically distributed from an unknown distribution τ." In doing so, we mean that data generated from such a probability mechanism mimics closely the properties of the data generated from the actual scientific experiment. We then might make a secondary modeling assumption, such as saying that τ belongs to some restricted family of probability distributions $\mathcal{M}$, like the normal distribution family.

This essay is focused strictly on secondary modeling errors. That is, we will take the basic modeling assumption as correct and use it as a tool to assess the error we make when we say that τ is in $\mathcal{M}$ when it is not. It is certainly possible to propose a basic modeling assumption that is seriously flawed, as when "independent and identically distributed" is assumed but not plausible. We will suppose that the basic modeling assumption has sufficient subject matter or sampling-based credibility to be believable. The interpretation of our methods if it is false will be discussed in the conclusion.

The first section of this chapter is devoted to setting up a framework for assessing the errors made in building a model. Along the way we develop the idea of a distance function as a tool for measuring model adequacy. It summarizes the overall error in model building by a single number. Admittedly, such a single-number summary could wildly oversimplify a complex problem and so should be just a portion of an adequacy assessment. However, I think it is an informative starting point.

The sections that follow are more technical and address the question of how one should choose a distance function. In the second section, I examine carefully a number of well-known statistical distance measures that can be used on discrete data. The goal is to identify which of those distance measures might be most widely useful for assessing the adequacy of a model. This requires a detailed look at the mathematical properties of those distances.

The third section moves to the case of continuous data and discusses the new issues that arise in these data. Here the picture is much more complex, as the cost of assessing the model adequacy is increased many times, particularly in multivariate data.

One resolution to the difficulty of constructing distances in continuous problems is through the construction of a quadratic distance function, which turns an infinite-dimensional problem into a more realistic finite-dimensional one. This methodology is discussed in the fourth section.

THE LOSS FUNCTION

Our goal then is to measure the cost in uncertainty due to specification of a restricted statistical model $\mathcal{M}$ when it is likely to be incorrect. We follow a very traditional course when we propose as a starting point the creation of a meaningful loss (or cost) function that will describe the magnitude of the error that we make in using a model element instead of τ.

There is an elegant statistical theory about the selection of methods based on their ability to minimize statistical risk. In a parametric model, with parameter θ, we start by defining a loss function $L(\theta_0, \theta)$ that describes the magnitude of the mistake made when θ_0 is the true value but θ is used. A common loss function for estimation is $L(\theta_0, \theta) = \|\theta_0 - \theta\|^2$, so called *squared error* loss. Then, for a given estimator $T(x)$ based on data x, we can interpret $L(\theta_0, T(x))$ as the loss incurred when $T(x)$ is used as an estimator of θ but θ_0 is correct. Finally, the risk for the estimator is the expected loss under the true distribution, namely $R(\theta_0, T) = E_{\theta_0}L(\theta_0, T(X))$. If we label L as a cost, then R is the mean price.

The reason for setting up a decision-based theory in this problem is that it enables a simple description of the tradeoffs involved in model building. On the whole, it is an oversimplification, but it is a starting point in a problem that is otherwise complex almost beyond description.

In the model-fitting problem, we will need to define a loss function $\delta(\tau, g)$ that describes the loss incurred when τ is the true distribution, but g is used for statistical inference. We use the symbol delta, for "distance," to indicate that such a loss function tells us how far apart (in a statistical sense) the two distributions are, just as squared error is also the squared Euclidean distance between parameter values. Note, however, that there is no reason that the loss function should be symmetric in its arguments, as we are giving different roles to τ and g.

If τ is in the model, say equal to m_{θ_0}, and g is in the model, say m_θ, then the distance $\delta(\tau, g)$ induces a loss function L on the parameter space via $L_\delta(\theta_0, \theta) = {}_{def}\delta(m_{\theta_0}, m_\theta)$. Hence, if τ is in the model, then the losses are strictly parametric losses.

However, if τ is not in $\mathcal{M}$, the overall cost can be split into two parts. We start by defining the *best model element* m_{θ_τ} as that element of $\mathcal{M}$ closest to τ in the given distances. That is, $\delta(\tau, m_{\theta_\tau}) = \inf_\theta \delta(\tau, m_\theta)$. In this case, then the loss made with a given estimator $\hat{\theta}$ can be broken into two parts,

$$\delta(\tau, m_{\hat{\theta}}) = \delta(\tau, m_{\theta_\tau}) + [\delta(\tau, m_{\hat{\theta}}) - \delta(\tau, m_{\theta_\tau})]. \tag{1}$$

The first term after the equality sign is an unavoidable error that arises from using the model $\mathcal{M}$, which we call the *model misspecification cost*, and write as $\delta(\tau, \mathcal{M})$. The nonnegative term in brackets represents the error made due to the point estimation, which we call the *parameter estimation cost*. How we balance these two costs depends on our basic modeling goals.

First, one might wish to use a particular parametric model and assess the additional model misspecification cost that has been incurred from using the model. That is, we might wish to do postdata inference on the model misspecification to see if the approximation of the model to the true distribution is "adequate," relative to some standard. We will call this *model adequacy checking*. This will be the focus of this essay.

However, if one is *model hunting*, searching through a collection of models $\mathcal{M}_\alpha$ for α an index of the models, then one would logically pursue a model building strategy that minimizes the overall risk. The first term on the right-hand side of (1) is now random, corresponding to randomness in the selection of the model. In this case, the overall risk is a sum of the expected model misspecification cost and the expected parameter estimation cost. We call this *risk-based model selection*.

The primary objective of this essay is to address the following question: How can one design the loss function δ to be scientifically and statistically meaningful? That is, at the end of the day, we would like to attach a very specific scientific meaning to the numerical values of the loss, so that $\delta = 5$, for example, has an explicit interpretation in terms of our statistical goals.

The remainder of this section provides our views on the role of statistical modeling in scientific work so as to clarify what we find important in a loss function. In the second section we will look at some important discrete distance measures so as to assess them as loss functions.

The construction of a loss function for continuous models creates a new set of problematic issues. These issues are discussed in the third section.

In the fourth section, we will describe a general framework for the construction of a loss function that assesses the truth of a finite list of model fit and parameter definition statements, thus providing one avenue around the problems with construction of a continuous model loss function.

THE EXISTENCE OF TRUTH

The scientific process starts with a set of *scientific questions* that are the focus of the investigation. The questions are often vaguely stated (asking for a description of the relationship among several variables), but they can be quite precise (asking if drug A provides better headache relief than drug B by some fixed, precisely defined notion of better). A study or experiment is then designed to provide data that should enable the investigator to address the scientific questions. Ideally, the design is constructed with a full understanding of statistical principles of design. The study is then conducted and the data collected.

For the analysis stage we need a statistical model. Our basic assumption is that the experimental data Y is the realization of a random process that has probability distribution τ, where τ stands for "true distribution" and which we will sometimes call "truth." Of course, the objective reality of "truth" is questionable because the basic modeling assumption can never be exactly true.

We might illustrate by a coin tossing experiment. In the abstract, we can imagine a coin having a probability p of having heads that would be the proportion attained by infinitely many independent and identically distributed coin tosses. However, given a real coin, we cannot hope to create a systematic way of independently and identically flipping the coin—indeed, if we could flip it identically, we would get the same outcome on every toss!

To take a subtler example, we might suppose that with true random sampling from a population, we have a well-defined truth as determined by the population and the sampling mechanism. However, if we use random numbers from a computer to generate the sample, we must realize that computers generate numbers that are only pseudorandom and pseudoindependent. Of course, we test random number generators so that the distinctions can be presumed to be very subtle, but there is still a gap between the basic assumption and the actual experiment. That is, truth here, as expressed by the random sampling model, is still no better than an approximation to the actual experiment.

That said, we believe that real experiments generate data that are very similar to data from idealized models, and so we proceed to operate as if the basic assumption is true, leaving deeper questions aside. That is, we presume that there is an objective truth τ that generates the data and which we would know if we could collect an infinite amount of data.

MODEL-BASED INFERENCE

After the basic assumption, we create, based on selection from a catalog of models, and possibly incorporating scientific information, a tentative family of probability measures $\mathcal{M}$ that we will call the *model*. The model is narrower than the class of all possible distributions in the sampling framework. The model is often described by a set of "modeling assumptions," such as the assumption that the data are a random sample from a normal distribution. We assume for now that the model distributions can be represented by density functions, although this feature is not essential.

The elements of the model will be indexed by a parameter θ, so that $\mathcal{M} = m_\theta$. The individual densities, written as m_θ or m, will be called the *model elements*. Note that our point of view will sometimes force us to think of $\mathcal{M}$ not in terms of its parameters, but in terms of the densities m that it contains, in which case we can suppress the parameter from the notation.

Why do we build a model? A key aspect to the construction of the model is that the *scientific questions* of interest can be turned into *statistical questions* phrased in terms of the values of the parameter θ. However, the issues here are larger, as the model $\mathcal{M}$ not only provides parameters, it provides meaning to them. For example, if we know the mean and variance of a model element of the normal model family, we can *completely* describe the chances of any future event. It is this additional interpretation given by the model that contrasts with model-free nonparametric inferences. We return to this point later.

In an ideal world, the statistician or scientist would then use a method appropriate to the selected model to analyze the data and report a statistical analysis of the original scientific questions. However, it is *critical* to this statistical approach that the truth τ be in the model $\mathcal{M}$. If not, the meaning of the parameter is lost. That is, if $\tau \in \mathcal{M}$, so that $\tau = m_{\theta_0}$ for some θ_0, we are in luck—the parameter is defined. However, if $\tau \notin \mathcal{M}$, it is not even clear what we mean by inference on the now nonexistent true θ. (We will give meaning to the parameter θ for arbitrary τ by use of the distance function.)

One standard solution to this dilemma is to examine the truth of the model, asking whether $\tau \in \mathcal{M}$ through tests or diagnostics, but then to return to assuming $\tau \in \mathcal{M}$ to carry out the inference on θ.

MODELING WITHOUT BELIEF

It is a tenet of this chapter that it is reasonable to carry out a model-based scientific inquiry without being forced to "believe in the model" $\mathcal{M}$. That is, it is possible to carry out inference from the point of view that $\tau \notin \mathcal{M}$, provided that we know that the quality of the approximation to τ is sufficiently good that the model can be used to address the original scientific questions with reasonably small error from the true answers. It is the role of the loss function $\delta(\tau, m)$ to measure the quality of the approximation.

To give a simple concrete example, suppose that the data are bivariate discrete counts collected by random sampling from a population and put into a two-way contingency table. Suppose that the model $\mathcal{M}$ is the standard model for the independence of the two factors, whereas τ represents the true distribution of the two factors in the population.

One might ask, as is done in hypothesis testing, whether these factors are truly independent. The focus here is instead on the ways in which we can say that the model $\mathcal{M}$ of independence is not quite true, but that it is a very good descriptor of the population at hand. For example, is it good enough to make accurate predictions about future observations? How accurate? We want the statistical distance measure δ to answer these questions.

This view of the model differs fundamentally from goodness-of-fit hypothesis testing. We do not decide yes or no, the model is correct or not; we instead ask how well the model describes the population. Inferential tools for this approach were developed in Rudas, Clogg, and Lindsay (1994). However, that paper focused on the statistical meaning of one particular loss function, called *pi-star*. Here, we lay out a more extensive philosophical argument, as well as catalogue a set of possible statistical distance measures to better understand how they might serve the goal of evaluating model adequacy.

How can we consistently operate from the point of view that $\tau \notin \mathcal{M}$, a state of disbelief in the model? To create a framework for statistical analysis, we propose adopting an *assumption-free point of view* that regards τ as completely unknown. When we select between models, we should do so based on the quality of the approximation as measured by the model's ability to provide answers to important scientific questions very nearly equal to the true answers. This is what the loss function δ should measure. If we never assume the model is correct, then it is a necessary part of the statistical analysis that we include an assessment of the adequacy of the approximation for the statistical purposes at hand.

WHY USE MODELS?

It is possible to conduct statistical analyses without using models at all by using nonparametric methods. We here address the question of why one would use a model if it is only approximately correct. In days of hand calculation, one might have said that modeling leads to tractable analyses, but there is more to it than that.

In past or present, the approximate validity of the model gives *greater scientific meaning* to the parameters we estimate. To illustrate, suppose we are investigating the relationship between two variables (X, Y). We could do a model-free analysis by examining the correlation coefficient ρ, a "nonparametric" parameter. However, it is clear that the correlation coefficient has limited interpretative value in a banana-shaped data cluster or data that have multiple clusters of points. That is, the correlation parameter seems most meaningful if the data pattern is approximately elliptical, as then the numerical value of ρ contains precise shape information.

Whether it is because nature is kind, or modelers are clever, it is often true that a simple model works pretty well. Suppose we are considering a multiway ANOVA model. Even though we would find it hard to believe that the additive model is really *exactly* true, we might think that it is sufficiently close to truth that using it will provide some advantages through parsimony, not only in describing the truth, but fitting the cell means. While the truth could be that all the treatment effects are higher-order interactions, we would be surprised to find many experiments with that structure.

In other contexts, such a nonparametric regression modeling or density estimation, there seems to be a basic belief by modelers that nature will be smooth, and so a model building strategy should start with the smooth and work towards the rough.

Which goodness-of-fit features of the model add important additional meaning to the parameters? This is a difficult question, and is not answerable without knowing how the model will be used. Additionally, this question seems to be rarely addressed in the goodness-of-fit literature, where the focus has been on determining whether $\tau \in \mathcal{M}$. Our purpose here is to initiate a discussion of this issue.

THE MODEL FIT QUESTIONS

It is clear that the meaning of "best-fitting model" should depend on the "model fit questions" to which we would like it to give accurate answers. One way to give a precise meaning to the phrase "model fit question" is through the use of statistical functionals. A statistical functional $h(q)$ assigns a number to every distribution q. For example, the mean functional assigns to each distribution q the expected value of a random variable from q. The median functional assigns to each distribution its median value. A model fit question regarding the difference between τ and m would then be "is $|h(\tau) - h(m)| \leq B$," where B is some fixed bound. Thus, we can ask a question like "how far apart are the medians of the true and the model distribution; is it as big as B?"

More generally, we might create a list of functionals $h(m)$ of the model that we wish to be close to the same functionals $h(\tau)$ of the truth if the model is to be deemed adequate. That is, we would like to know if $|h(\tau) - h(m)| \leq B$, where B possibly depends on h, for every h. Each functional generates a question: is $h(m)$ sufficiently close to $h(\tau)$? We will call these the list of model fit questions. The actual value of $|h(\tau) - h(m)|$ will be the *model misspecification error* in functional h.

To illustrate this idea, suppose we wish the model to give accurate prediction statements about a forthcoming observation X_{n+1}. For example, we might wish the model-based probability for A to be close to the true probability for several sets A. In this case, $h(q) = \mathrm{Pr}_q\{X_{n+1} \in A\}$, where the subscript q indicates the distribution under which the calculation was made.

The model misspecification error in functional h plays a role in the overall statistical error in estimation. If we estimate the probability of A using the model with a parameter estimator $\hat{\theta}$, then the *total error*

$$\mathrm{Pr}_\tau\{X_{n+1} \in A\} - \mathrm{Pr}_{\hat{\theta}}\{X_{n+1} \in A\}$$

can be decomposed into the sum of the model misspecification error

$$\mathrm{Pr}_\tau\{X_{n+1} \in A\} - \mathrm{Pr}_{\theta_\tau}\{X_{n+1} \in A\},$$

where θ_τ is the parameter value corresponding to the density m_{θ_τ} that best fits τ, and the within model *parameter estimation error*

$$\mathrm{Pr}_{\theta_\tau}\{X_{n+1} \in A\} - \mathrm{Pr}_{\hat{\theta}}\{X_{n+1} \in A\}.$$

The second kind of error is always accounted for in a model-based statistical analysis. However, if we do not insist that the model contain τ, then we cannot eliminate the possibility of model misspecification error. The best we can do is to mitigate it by careful selection of the model. At the same time, if we also measure it, it gives us a way of describing the potential limitations of the model.

Now the hardest part of our problem is to generate the list of model fit questions in such a way that it would strengthen our answers to the list of *scientific questions*. The scientific list might be short, such as asking what the true mean of variable X is, assuming that X is normally distributed. The model fit questions could include a global question: is the normal model correct in every particular? However, this is probably more than we need to feel that the model is adequate.

For example, we might feel that making normal theory inference on the mean will not be satisfactory if the data are heavily skewed, or if the data histogram appears bimodal, or if the distribution is heavy tailed—cases in which we might well expect considerable difficulty in interpreting the mean. That is, use of the mean is more satisfactory if there is also some degree of symmetry and unimodality to the distribution, and the distribution is relatively light tailed. These questions are similar to, but weaker than, asking if the density is exactly normal.

Suppose our scientific questions lead us to desire a model that widely describes the population, as might be the case for the model of statistical independence in a two-way table, mentioned earlier. We might then specify that the model fit questions include the evaluation of the prediction probabilities for a new observation from the population, $P_\tau\{X_{n+1} \in A\}$, for a finite or infinite list of sets A. If so, saying that independence is nearly true would mean that these probabilities are close to those generated by the independence model.

Or, if one is fitting a regression model, the questions might be those whose answers describe the regression function, as well as questions about the fit of the function to the conditional mean, or might go further in determining prediction probabilities about the conditional mean.

The list of model fit questions might be long enough to describe completely the density τ, in the sense that if $\tau \notin \mathcal{M}$, it would eventually be detected by one of the questions. However, we might wish the model only to provide additional descriptive power for a few key characteristics of τ that are of central interest.

FROM QUESTIONS TO DISTANCE

We now summarize the overall plan for constructing a meaningful loss function δ for inference. All the measures we consider will be nonnegative; small values of δ will indicate that the model is adequate, while large values will indicate model failure.

1. Construct a nonnegative distance-type measure between probability distributions represented by densities τ and m, say $\delta(\tau, m)$, that measures the adequacy of the model density m as a substitute for truth τ. It is important that it have *meaningful statistical interpretation* as a measure of the risk of using m in place of τ. One approach is to set it

equal to the worst-case model misspecification error that would be made in a list of statistical functionals.

2. Then set the distance from the model family $\mathcal{M}$ to the truth τ by finding the best possible fit of the model to the truth:

$$\delta(\tau, \mathcal{M}) = \min_{m \in \mathcal{M}} \delta(\tau, m).$$

This corresponds to choosing a model density that minimizes the model misspecification risk associated with using the model.

3. Correspondingly we assume that there is a best-fitting model density m_τ in $\mathcal{M}$, best in the sense that it is closest to the truth:

$$m_\tau = \arg \min_{m \in \mathcal{M}} \delta(\tau, m).$$

Since the elements of $\mathcal{M}$ are indexed by θ, there is a corresponding parameter value θ_τ, so that m_τ is an abbreviation for m_{θ_τ}. That is, we have now assigned a parameter value to every true distribution.

FROM DISTANCE TO PARAMETER DEFINITION

Adopting this framework leads to the following important observation. In this model-free setting, a goal for estimation should be to determine this "best model," m_τ or equivalently the "best parameter" θ_τ, as it minimizes the model misspecification cost. To each $\tau \notin \mathcal{M}$ we have assigned a value, namely θ_τ, to the parameter θ even when the model is incorrect. It corresponds to the model that best approximates the truth. Of course, if $\tau \in \mathcal{M}$ then m_τ is τ and θ_τ is just the true θ, so we have not changed the nature of the model-based inference, merely widened its meaning.

This is a different way of thinking. We will fit models based on their quality as approximations. If their quality is high enough, then we will use the model that best approximates the truth as our inferential model.

The above setup has led us to a change in emphasis for the *model-based statistical inference*. In the usual model-based framework, we focus on attributes such as consistency and efficiency for estimators of θ, but these are evaluated strictly within the model. Consistency for θ_τ is a deeper problem, as we wish the estimated model to be consistent for estimating the best approximating model m_τ or equivalently, $\hat{\theta}$ should be consistent for the best parameter θ_τ, for *all* distributions τ. This desideratum now takes precedence

over the usual estimation goal of statistical efficiency when the model is assumed to be exactly true. Fortunately, there do exist some distance measures
that lead to this type of model-based efficiency, as will be described later.

In a coming section, we will consider common measures such as the chi-
squared and the Kullback-Leibler divergences and evaluate them as possible
adequacy measures. That is, we address the question, "what does this measure tell me about the adequacy of the model in approximating the truth?"
First, we elaborate a bit more on how we might carry out inference in this
framework.

INFERENCE

Once we have established a reasonable loss function for assessing model adequacy, one needs to develop strategies for minimizing the expected loss. I
next argue that a clear definition of model adequacy can also lead to an *inference function* that can be used as a single integrated framework for addressing both scientific and model-fit statistical questions.

A natural starting point is to create estimators based on minimizing
$\delta(d, m_\theta)$ over θ, where d is a nonparametric estimator of τ. We will then call
$\delta(d, m_\theta)$ the *δ-inference function*. Such an estimator will be denoted $\hat{\theta}_\delta$ the
corresponding density estimator will be $\hat{m}_\delta$. This method will generally pro-
vide consistent estimators of θ_τ as well as measures of the statistical error in
estimating θ_τ.

In our *secondary statistical analysis,* we can then treat $\delta(\tau, \mathcal{M})$ as an unknown "parameter" in the statistical sense, with a point estimator $\delta(d, \hat{m}_\delta)$.
Better yet, we would construct a nonparametric confidence interval for its
values, say $L \leq \delta \leq U$. For a statistically meaningful distance this interval
gives us a sense for the range of errors that might have occurred in our
model-based statistical statements. The lower end would indicate the best
possible situation, the upper end would tell us about the worst case. Further
diagnostics could then indicate which questions the model might answer
most poorly.

The above description is familiar to anyone who has done contingency
table analysis. In these models, we can use the likelihood method, which is
based on Kullback-Leibler distance, to estimate parameters, compare nested
models, and evaluate overall fit against the "nonparametric" alternative. The
distinctions between this proposal and likelihood-based contingency table
analysis are:

(1) We argue that goodness of fit should *not* be treated as a testing prob-

lem, but rather an estimation problem, in which one estimates by confidence interval the distance between model and truth; i.e., the model misspecification cost $\delta(\tau, \mathcal{M})$.

(2) We think that one should select the inference function to meet a variety of desirable features when the model is *not* correct. That is, when creating a loss function, one takes the view that τ is correct and that $\mathcal{M}$ is in error. The distance function corresponding to likelihood methods has a questionable value in this regard.

CONFIDENCE INTERVALS VERSUS TESTING

We contrast these suggestions with *goodness-of-fit testing,* wherein one treats $\tau \in \mathcal{M}$ as a null hypothesis. In this case, we treat $\tau \in \mathcal{M}$ as the truth unless the evidence from the data is overwhelmingly against it. Unfortunately, goodness-of-fit testing for model assessment places us in an accept/reject game that is not scientifically realistic. We know in advance that the model is most likely to be false; it is more valuable to know whether the difference is *scientifically important.* That is, are the answers to our statistical questions provided by the model sufficiently accurate for our scientific purposes?

We might compare goodness-of-fit testing to hypothesis tests about parameters, such as $H: \theta = \theta_0$ versus $A: \theta \neq \theta_0$. When one performs the latter test, it is fully apparent that point null hypothesis is artificial—it is almost certainly never exactly true, and so we will surely reject it for a large enough sample. However, a small difference from the null may not be a *scientifically* significant failure even if it causes a *statistically* significant outcome.

We might imagine two medical treatments whose differential effect on survival is sufficiently small as to be dwarfed by other issues, such as side effects, quality of life, and economic expense. One way to clarify the issues is an informal use of confidence intervals, which can show the degree to which the plausible range of the parameter contains scientifically significant values of the parameter.

Suppose, for example, that we would consider the interval $I = [\theta_0 - \delta, \theta_0 + \delta]$ to be the values of θ not *scientifically* distinct from θ_0. We might propose using a confidence interval in an *informal* decision making fashion as follows:

- If the confidence interval C for θ is contained in I, we should "accept" the extended null hypothesis $H_0^* : \theta \in I$; there is little evidence of scientific significance.

- If the confidence interval does not overlap I, we can be confident that the result is scientifically significant, and we should "reject" the extended H_0^*.
- And if I and C have a nonempty intersection, the experimental results are *ambiguous* about scientific importance of the true state of nature. Only a larger sample size could resolve the ambiguity. (Of course, if we adopt scientific conservatism, we would "fail to reject" in this ambiguous case.)

Similarly, suppose we were to set a standard for model misspecification cost, saying $\delta(\tau, m_\tau) \geq k$ means that the model is inadequate. Then the confidence intervals $[L, U]$ for model adequacy δ would indicate whether

- the model was *clearly adequate,* with $L < U < k$,
- *clearly inadequate,* with $k < L < U$,
- or *ambiguously adequate,* with $k \in [L, U]$.

In the first case, we need to build a better model, and in the second, we stay pat. But if the result is ambiguous, there is not much we can do: one should be cautious about the use of the model's predictions, even though it is not possible to reject the model's adequacy. In this last case, we would certainly also want to know, through diagnostics, which questions the model might answer poorly. Of course, sample size is highly relevant here, as in small samples one will always have a large upper limit U and so never find the model adequate unless one chooses a distance δ that does not ask too many questions.

SOME OBSERVATIONS

If we adopt the point of view that the truth τ can almost never safely be presumed to be in the model family $\mathcal{M}$, but that $\mathcal{M}$ might be a very useful simplifying description, one has to rethink some of the standard statistical story line.

(1) One must be cautious about the meaning of the new parameter θ_τ. When we carry out a parametric analysis, it is on the parameters θ_τ corresponding to the best model m_τ, not on the θ in the model. The θ in the model might be extendable to model-free interpretations in other ways. For example, as applied with a normal model, inference on μ_τ will be inference on

the *mean of the best-fitting normal distribution,* not the mean of the true distribution. What we mean by best fitting is strictly a function of the distance measure, and it can differ dramatically from the mean. It is the secondary analysis that warns us about the limitations of the parametric analysis. For example, it might tell us that the normal model does not fit well, so the parameter μ_τ might not be a good summary of the population characteristics (e.g., if the population is bimodal).

(2) Given that we do not "believe" in the model, we should carry out the parametric error analysis using unknown τ as the underlying distribution, not m_τ. We would, for example, construct confidence intervals for the parameters in the best model in a nonparametric way, without assuming the best model is true.

(3) Additionally, in order to obtain the widest validity, the probability analysis should avoid making assumptions about τ. It is conventional in nonparametric analysis to replace model type assumptions with assumptions about the smoothness properties of τ. We view these as just wider models, ones that are still suspect, but much more difficult to evaluate than the usual parametric ones. We will gain strength in summarizing the data through the simplifying nature of the model structure itself. In our setting, simple models will tend to be adequate for small data sets, in the sense that there will not be enough statistical power to reject their adequacy. The wide confidence interval for the model adequacy will, however, alert us to the dangers of using the model's predictions. Larger data sets might force us to choose richer models, as the inadequacy of the simple smooth models will become apparent. This description of model building is in close accord with the "method of sieves," an approach to statistical model building in which the complexity of the model used is increasing in the sample size. However, that theory does not generally allow one to minimize complexity by accepting an adequate but false model.

(4) On the other hand, we should not ignore the characteristics our methods display when the model is actually correct; that is, when $\tau \in \mathcal{M}$. The model is often chosen because we have some reason to believe that it should approximate the truth well, and so our concerns about the optimality properties of procedures should focus on neighborhoods of the model. In particular, we would like our estimators to be asymptotically normal, with high statistical efficiency in this setting, as in some sense we have assigned a high prior to the model being nearly correct. We must therefore grade our statistical distance measures on whether the inference they generate is similar to classical model-based inference when the model is correct.

(5) An intriguing and difficult issue that we must consider when constructing statistical distance measures is that it is possible to ask the data "too many statistical questions" about the truth. That is, it is *impossible* to answer infinitely many questions equally well with only finitely many pieces of data. One should think carefully about what to ask, and one should have a way of deciding if too much was asked. This point will be discussed further later.

(6) We might contrast the risk-based approach with the standard robustness theory. In this literature there is a similar denial of the model's truth, but it seems, at least to this author, that the focus has been on a narrow set of statistical questions regarding the centers of symmetric distributions. The goodness-of-fit question is then ignored. In some sense, it appears that secondary analysis of model fit has been treated as unnecessary once one has used a robust statistical procedure. What we advocate here turns this approach around. We start with goodness of fit by identifying a measure that assesses whether the model element m approximates the truth τ adequately. It is this distance measure that will be robust or not. Although we do not digress here into robustness questions, Lindsay (1994) extensively investigated the robustness properties of a number of important distances. If the distance is robust, the estimators that result will generally display classical robustness properties in location-scale problems, but will also have a goodness-of-fit interpretation.

(7) We have not yet commented on selection from a class of models, as opposed to model evaluation, and the problem is of sufficient importance to offer some thoughts on how the present theory could be extended. A natural strategy for minimizing risk is to build successively complex models until $\delta(d, m_{\hat{\theta}})$, the empirical distance, is sufficiently small. However, it is likely that one can reduce this empirical distance to 0 with a sufficiently large model. Unfortunately, this procedure naïvely "overfits" the data, ignoring the fact that the data is noisy. Suppose that one adopts instead a model selection strategy that chooses a model that satisfies $\delta(d, m_{\hat{\theta}}) = E[\delta(d, \tau)]$. The quantity $E[\delta(d, \tau)]$ is the *inherent statistical noise* in the problem, as it measures the average distance between the data and the truth. It can be shown that this strategy spreads the risk evenly between model misspecification and parameter estimation costs. If δ is a metric, then it also offers the greatest opportunities for risk reduction, in the sense that there is a lower bound for risk that is 0 for this strategy and increases monotonically for any strategy whose mean empirical distance departs from the inherent statistical noise.

DISCRETE DISTANCE MEASURES

We now introduce some classical measures of discrepancy between truth and model and discuss their merits as measures of model adequacy. Our purpose here is to evaluate them from our new point of view. We note that these *statistical distances* will not generally be metrics, which would require the distance to be symmetric in its arguments and to satisfy a triangle inequality. However, they will be nonnegative and have 0 value when the truth is in the model. We have used the word "statistical" to emphasize that they should carry statistical meaning. In essence, these measures provide us with a list of statistical questions to ask our data. The point of this section is to ask what those questions might be.

One problem with the distances in this section is that they force us to ask a specific set of statistical questions; in a later section we introduce a distance that allows the user to create his or her personal list of important questions.

In the material that follows, a number of technical assertions are made without proof for the sake of space and smoothness of flow.

THE DISCRETE FRAMEWORK

We will start within the discrete distribution framework so as to provide the clearest possible focus for our calculations and their interpretations, while avoiding certain mathematical difficulties. Of course, we can approximate all continuous data problems by discretizing the data and the model. I will discuss this point further below.

We start with a discrete sample space X, which we label as $t = 1, 2, \ldots,$ T. On this sample space we have a true probability density $\tau(t)$, as well as a family of densities $\{m_\theta(t)\} = \mathcal{M}$ that we will call the model.[1] If we have n i.i.d. observations $X_1, \ldots, X_n$ from $\tau(.)$, then we can record the data as counts $n(t) = $ "the number of times $x_i = t$ occurred" or as sample proportions $d(t)^\circ = n(t)/n$.

We will call any functional $\delta(f, g)$ of two densities f and g a *statistical distance measure* if it is nonnegative, taking on the value 0 if f and g generate the same probability measure. We will allow it to be 0 even when $f \neq g$,

1. Some authors use the term *probability mass function* instead of density.

however, if there is no reason to distinguish between f and g based on our statistical goals.

In practice, our best nonparametric representative of $\tau(t)$ is $d(t)$, the data density. In a discrete sample space the sample proportions $d(t)$ converge to $\tau(t)$ as the sample size increases. An integrated statistical inference would involve minimizing the inference function $\delta(d, m_\theta)$ over θ to give an estimator $\hat\theta$ of θ_τ. If δ is "sufficiently smooth," then we can expect $\hat\theta \to \theta_\tau$ because $d(t) \doteq \tau(t)$ almost surely. Further, we expect $\delta(d, \hat m) \to \delta(\tau, m_\tau)$ as n becomes large.

TOTAL VARIATION DISTANCE

We start with a simple measure of statistical difference in densities that has major implications for prediction probabilities. Let

$$V(f, g) = \frac{1}{2} \sum |f(t) - g(t)|$$

be the *total variation distance* between f and g. Looking at its form, there is no obvious interpretation as a measure of the risk in answering a list of statistical questions.

However, we can create an alternative representation of total variation as

$$V(f, g) = \sup_A \{|\mathrm{Pr}_f(A) - \mathrm{Pr}_g(A)| : A \subset \chi\}.$$

Here, we have used the symbol Pr_f to represent the probability measure corresponding to density f. One set A where the supremum occurs is $A_{\max} = \{t : f(t) > g(t)\}$; note that we could add to this set any points from the set $\{t : f(t) = g(t)\}$ without altering the value of $|\mathrm{Pr}_f(A) - \mathrm{Pr}_g(A)|$.

Therefore $V(\tau, m)$ bounds the maximum possible absolute error over *all* prediction probability statements that we could make using m when τ is correct.

We collect some observations on the total variation distance V as a measure of model misspecification costs:

1. The model misspecification measure $V(\tau, \mathcal{M})$ has a "minimax" expression

$$V(\tau, \mathcal{M}) = \inf_{m \in \mathcal{M}} \sup_{A}\{|\mathrm{Pr}_\tau(A) - \mathrm{Pr}_m(A)| : A \subset \chi\}.$$

This indicates the sense in which the measure assesses the overall risk of using m instead of τ, then chooses the m that minimizes that risk.

2. When describing a population, it is very natural to describe it via the proportion of individuals in various subgroups. Having $V(\tau, \mathcal{M})$ small would ensure uniform accuracy for all such descriptions. On the other hand, we also often like to describe a population by the *means* of various variables, or means of subgroups. Having this measure small does not imply that means are close on the scale of standard deviation, a point we return to later.

3. By asking that the model be accurate for all sets A, we are asking many questions simultaneously. There will be a price to pay for asking so many questions, in the sense that it will be more difficult to make inference on $V(\tau, \mathcal{M})$ than if we had used just a subfamily of sets of particular interest. Reducing the number of sets of interest would narrow the confidence interval for V, and hence for small samples the goodness-of-fit intervals would be less likely to be ambiguous. Indeed, it is doubtful that in practice one would want the model to be accurate over all sets; one might prefer to have accuracy just for tail sets, or sets of a certain minimum size, for example.

4. On the other hand, V does not seem to provide us with the sharpest possible answers to the questions we ask. That is, suppose we are interested in predicting probability for set B where $\tau(B)$ is small. Our measure gives a guaranteed upper bound on the error made in answering that question with the model

$$|\mathrm{Pr}_\tau(B) - \mathrm{Pr}_{\theta_\tau}(B)| \le V(\tau, \mathcal{M}).$$

But we know that we could carry out inference on this error using the sample estimator n_B/n, where n_B is the number of observations in set B, with standard error $\sqrt{\tau(B)(1 - \tau(B))}$. That is, the statistical error in answering this question gets small as $\tau(B)$ decreases to 0, whereas our upper bound $V(\tau, \mathcal{M})$ is independent of the size of $\tau(B)$. That is, this distance will not provide the best possible upper bounds to the errors we might make on sets of small probability. As we shall see, chi-squared measures do improve this bound. Of course, if these "small" sets do not interest us as much, this is not important.

5. Using $V(d, m_\theta)$ as an inference function yields estimators of θ that have the disturbing feature of not generating smooth asymptotically normal estimators when the model $\mathcal{M}$ is true (Xi, 1996). This feature is related to the "pathologies" of variation distance described by Donoho and Liu (1988b).

6. As an aside, we note that V has a second statistical interpretation. Suppose we wish to test the simple hypotheses $\tau = f$ versus $\tau = g$. If, instead of fixing size we choose a test based on maximizing $\beta - \alpha$, the difference in the two probabilities of error, then the optimal test has rejection region A_{max}, and $\beta - \alpha = V(f, g)$.

THE MIXTURE INDEX OF FIT

Rudas, Clogg, and Lindsay (1994) proposed a measure closely related to total variation distance V. It was motivated by the techniques of mixture/latent-class modeling. The idea is as follows: Suppose that we can decompose the population τ into two subgroups, with proportions $(1 - \pi)$ and π. The first subgroup is a population where the model holds exactly; that is, where the variable X follows a density m in $\mathcal{M}$. The second subgroup is an error component in which X has an arbitrary distribution e. We can think of e as being the distribution in a population of "outliers" that was mixed together with the pure population. Then the overall distribution of X in the population is a mixture of the two components:

$$\tau(t) = (1 - \pi)m(t) + \pi e(t).$$

Now for any given τ and model m there is ambiguity in this decomposition in that there are multiple representations of τ in this form. To take an extreme case, if τ is in $\mathcal{M}$ we could use $m = \tau$ and $e = \tau$, together with *any* value of π.

We can resolve the ambiguity by defining $\pi^* = \pi^*(\tau, m)$ to be the smallest π for which such a representation exists. That is, choose the representation that gives the greatest possible mass to m. The fact that e must be a density, with nonnegative entries, enables us to provide an explicit formula for π^* in the discrete case. That is, it can be shown that if we let

$$\pi^*(\tau, m) = \sup_t \left\{ 1 - \frac{\tau(t)}{m(t)} \right\} \tag{1}$$

be our measure of distance between two densities τ and m, then $\pi^*(\tau, \mathcal{M})$ $= \inf_{m \in \mathcal{M}} \{\pi^*(\tau, m)\}$. If $\tau \in \mathcal{M}$ then $\pi^* = 0$.

For our purposes, the value of this decomposition is simple. In terms of prediction, it is clear that every prediction from the model applies to the pure component of the population, which is $1 - \pi^*$ of the full population. However, it does not tell us the magnitude of the error in the second group of the population.

Rudas, Clogg, and Lindsay (1994) provided an extensive discussion of the computation and asymptotic statistical inference for this measure. It is this paper that laid out in detail the style of inference we describe here, using the particular measure of fit π^*.

We make the following observations on this measure:

(1) The statistical interpretations made with this measure are attractive, as *any* statement made with the model applies to at least $1 - \pi^*$ of the population involved.

(2) However, unlike V, it measures error from the model only in a "one-sided" way. For example, if there exists t_0 such that $\tau(t_0) = 0$, but no model element satisfies $m(t_0) = 0$ then π^* equals its maximum value, 1. This would seem to indicate a complete lack of fit, *regardless* of how well other cells might be fit, or how close to 0 we can make $m(t_0)$. A similar problem arises in data analysis, as $d(t_0)$ will possibly be 0 by sampling error.

(3) When we look at (1), there does not appear to be a sense in which the measure π^* measures a natural risk we would incur when we use the model instead of τ.

(4) Once again, we find that the method does not generate asymptotically normal estimators when the model is actually true (Xi, 1996).

CHI-SQUARED DISTANCES

If we employ chi-squared distances, we will find that the aforementioned lack of sensitivity of V and π^* to our potential statistical knowledge of an individual question can be corrected. At the same time, however, the problem of asking too many questions becomes more severe, as we are raising the bar on what we require of the model.

We start with a class of chi-squared measures of distance between two densities τ and m defined by

$$\chi_\alpha^2(\tau, m) = \sum \frac{[\tau(t) - m(t)]^2}{\alpha(t)}.$$

The function $\alpha(t)$, assumed to be a density, provides a scaling factor. If we let $\alpha(t)$ equal the second argument, m, then we get *Pearson's chi-squared measure:*

$$P^2(\tau, m) = \sum \frac{[\tau(t) - m(t)]^2}{m(t)}.$$

If we employ instead $\tau(t)$ for $\alpha(t)$, we get *Neyman's modified chi squared:*

$$N^2(\tau, m) = \sum \frac{[\tau(t) - m(t)]^2}{\tau(t)}.$$

Finally, a chi-squared measure of particular interest to us is the *symmetric chi-squared:*

$$S^2(\tau, m) = \sum \frac{[\tau(t) - m(t)]^2}{[m(t) + \tau(t)]/2}.$$

It corresponds to using for $\alpha(t)$ a 50-50 blend of the two densities, $[m(t) + \tau(t)]/2$.

Note that if we used the data $d(t)$ for $\tau(t)$ we would need to multiply these distance measures by n in order to obtain the usual chi-squared test statistics.

CHI-SQUARED RISK ASSESSMENT

Like total variation distance V, the chi-squared distances intrinsically measure the risk present in a natural set of statistical inferences. Consider the following problem.

One method for testing whether model element m was equal to τ would be to take a function $g(t)$ on the sample space (i.e., a random variable) and compare the means generated by the data d and the model m using a squared *z-type* statistic:

$$z^2_\alpha(g) = n \left[\frac{\frac{1}{n}\sum g(x_i) - E_m[g(x)]}{\sigma_\alpha(g)} \right]^2. \tag{2}$$

Here, the denominator is the variance of $g(X)$ under a selected density $\alpha(t)$. We could use $\alpha(t) = m(t)$, the model-based estimator, which gives a z-statistic. If we used $\alpha(t) = d(t)$, the denominator is the sample variance, and so we have a t-statistic. As a compromise we might use $\alpha(t) = 0.5m(t) + 0.5d(t)$.

One could extend this test by considering the supremum of the z-squared statistics over a class of functions g that was considered statistically important. This would generate a bound on the scaled mean deviations over this class. This could be technically difficult to compute for some classes of functions. However, there is a simplification if we consider all possible functions g, because the chi-squared distance between d and m equals n^{-1} times the largest z-test one could construct over this class: $\chi_\alpha^2(d, m) = \sup_g n^{-1} z_\alpha^2(g)$.

We note the strong resemblance of this representation with the representation of an F statistic as the supremum over a collection of t^2 tests for a collection of contrasts, which in turn yields the Scheffé method for simultaneous inference on all contrasts.

Moreover, there is an explicit representation of the function $g(t)$ that gives the supremum:

$$g_{max}(t) = \frac{d(t) - m(t)}{\alpha(t)}.$$

One usually thinks of standardizing the residuals $d(t) - m(t)$ by their standard deviation to assess normality and find outliers. This gives a different perspective: these residuals, standardized by something more like variance, give an estimator of the most powerful function g one could use to test the difference between τ and m using a z-statistic. For example, if $g_{max}(t)$ has a quadratic shape in t, then the model and data differ most strongly in the second moment.

THE CHOICE OF $\alpha(t)$

If we think about the choice of the denominator $\alpha(t)$ from the point of view of testing the null hypothesis of $\tau = m$, one might feel that using the model-based variances from $m(t)$ would give the greatest power (just as z-tests are more powerful than t). This leads us to the Pearson chi-squared statistic, with $\alpha(t) = m(t)$.

On the other hand, the corresponding difference between truth and model is

$$\chi_\alpha^2(\tau, m) = \sup_g \left[\frac{E_\tau[g(X)] - E_m[g(X)]}{\delta_\alpha(g)} \right]^2.$$

If we think about this as a distance between truth and model, we are comparing the difference in means of arbitrary functions, standardized by our ability to detect differences. Among other features, knowing that this is small would enable us to know that for any subgroup of the population, the mean of any variable would be approximated well by the model *on the scale of standard deviation under* α.

Viewed from the perspective that τ is not m, and hence that the proper variance for g is based on τ, using a variance based on m no longer gives a sharp measure of the potential accuracy available for determining the mean of $g(X)$. Since d is our stand-in for τ, it suggests using the sample variances, $\alpha(t) = d(t)$, to standardize the t-tests. That is, it leads to Neyman's chi-squared measure N^2.

On the other hand, in a data-analytic sense, there is instability inherent in using the sample variance to estimate variances over a large class of functions g. We believe that S^2, the symmetric chi-squared measure, provides a nice compromise by averaging the model and sample variance estimates. Moreover, it gives estimators that are highly robust, whereas Pearson's P^2 gives highly nonrobust estimators (worse than likelihood, see Lindsay, 1994), and N^2, while robust, is unstable for small data sets.

We conclude this discussion with some further observations.

(1) Notice that we could include among our test functions g all possible indicator functions $I_A(x)$ for sets A, so we have

$$\sup_A \left[\frac{\tau(A) - m(A)}{\sigma_\alpha(I_A(X))} \right]^2 \leq \chi_\alpha^2(\tau, m).$$

That is, in this way we obtain bounds on prediction errors for probabilities that account for our potential accuracy in assessing them.

(2) Chi-squared measures on discrete sample spaces yield asymptotically normal and fully efficient estimators when $\tau \in \mathcal{M}$. As noted by Lindsay (1994), the robustness properties depend quite strongly on the choice of $\alpha(t)$, as do the second-order efficiency properties under the model.

(3) We have raised the level of difficulty for the distance by insisting that

it be accurate for the means of all functions on the sample space, at the level of accuracy of standard deviation. We can expect attaining accuracy under these measures to be yet more difficult than under V.

(4) As a note of mathematical interest, we can calculate the effect of sample size on the chi-squared measures. That is, suppose that f_n and g_n are two densities for the data $X_1, \ldots, X_n$, where X_i has density f and g respectively. Then the Pearson distance between f_n and g_n is

$$P^2(f_n, g_n) = [1 + P^2(f, g)]^n - 1.$$

This establishes a bound on the difference in means of functions of n independent variables, $h(x_1, x_2, \ldots, x_n)$.

HELLINGER DISTANCE

It would seem that an essay on statistical distances must include Hellinger distance, as it has wide use in theoretical analyses, and so would seem to measure something important. Indeed, it is very closely related to the symmetric chi-squared distance, although this is not immediately obvious.

The formula for squared Hellinger distance is

$$H^2(f, g) = \sum [\sqrt{f(t)} - \sqrt{g(t)}]^2.$$

We can more readily see its relationship to chi-squared distances by rewriting it as

$$H^2(f, g) = \sum \frac{[f(t) - g(t)]^2}{[\sqrt{f(t)} + \sqrt{g(t)}]^2}.$$

A little manipulation of the proceeding relationship, using

$$f + g \leq [\sqrt{f(t)} + \sqrt{g(t)}]^2 \leq 2(f + g),$$

gives a very strong near equivalence between Hellinger distance and the symmetric chi-squared (Le Cam, 1986):

$$\frac{1}{4} S^2(f, g) \leq H^2(f, g) \leq \frac{1}{2} S^2(f, g).$$

However, we know of no other direct statistical interpretation of H^2 in terms of model adequacy beyond this relationship with the chi-squared measure, and so we prefer S^2 over H^2 for use in statistical inference.

Some further notes:

(1) Another reason to prefer S^2 over H^2 as a measure of adequacy arises from the properties of H^2 and S^2 as inference functions in point estimation. Although both are equally robust to outlying observations, the measure H^2 does not behave as well for sampling zeroes.

(2) On the other hand, H^2 is better than S^2 for some theoretical calculations. If, for example, we wish to calculate the effect of sample size on the measure, then, using the notation from the previous section on chi-squared measures, we can find a simple relationship

$$H^2(f_n, g_n) = 2 - 2\left(1 - \frac{H^2(f, g)}{2}\right)^n.$$

LIKELIHOOD

In the discrete context, the method of maximum likelihood can be derived as a minimum distance estimator based on a version of the Kullback-Leibler divergence measure. That is, let the likelihood distance be

$$L^2(f, g) = \sum f(t) \ln \frac{f(t)}{g(t)}.$$

Minimizing the inference function $L^2(d, m_\theta)$ over θ results in the maximum likelihood estimators. This measure can be shown to be nonnegative using Jensen's inequality. Alternatively, we can write

$$L^2(f, g) = \sum g(t)\left[\frac{f(t)}{g(t)} \ln \frac{f(t)}{g(t)} - \frac{f(t)}{g(t)} + 1\right].$$

In this representation, the argument in the brackets, being of the form $r \ln r - r + 1$, for $r = \dfrac{f(t)}{g(t)}$, is nonnegative.

We can use this last relationship to give some idea of the relationship between the chi-squared measures and the likelihood measure. If we write P^2 in a similar fashion, we get

$$P(f, g) = \sum g(t)\left[\frac{f(t)}{g(t)} - 1\right]^2.$$

From the functional relationship $r \ln r + r - 1 \leq (r - 1)^2$, we have the distance relationship $L^2 \leq P^2$. However, it is also clear from the right tails of the functions that there is no way to bound L^2 below by a multiple of P^2, and so the measures are not equivalent in the same way that Hellinger and symmetric chi-squared were. In particular, knowing that L^2 is small is no guarantee that all the Pearson z-statistics are uniformly small.

On the other hand, we can show by the same mechanism that $\chi_\alpha^2 \leq 2L^2$, where $\alpha(t) = [f(t) + 2g(t)]/3$. Moreover, this blended chi-squared distance is equal to twice the likelihood distance "up to second order," in the sense that it generates second-order equivalent estimators when the model is true. It is therefore true that small L^2 implies small z-statistics with blended variance estimators. However, the reverse is not true because the right tail in r for χ_α^2 is of magnitude r, as opposed to $r \ln r$ for likelihood.

Although this gives us some feeling for the statistical interpretation of the likelihood measure, we are still unsure about its meaning as a measure of model misspecification. What would a confidence interval for the true likelihood distance tell us in a direct sense about our accuracy in a set of statistical questions? This puzzles us, because on every other count likelihood has been among the great miracles of statistical methods.

In addition to having no clear meaning as a measure of risk, our general impression is that likelihood, like Pearson's chi-squared, is "too optimal" for the model, and so does not work so well for inference when τ is not in $\mathcal{M}$. For example, for any given truth, the best-fitting normal model based on the likelihood distance simply matches the mean and the variance of the normal to the true mean and variance. This fit ignores many other features of the truth that might provide a more meaningful description. Of course, the entire robustness literature grew out of this inadequacy.

OUR PREFERENCES

Although our investigation is far from exhaustive, it seems to us that the family of chi-squared measures has a simple and attractive interpretation as a loss function. Indeed, it gets at the very problem of making sure the model and the truth have the same structure for means, as adjusted for standard deviation.

On the other hand, these measures may be too stringent for our needs. In such a case, we might prefer the total variation distance V. It seems closest to a robust measure, in that if the two probability measures differ only on a set of small probability, such as a few outliers, then the distance must be small.

Outliers can influence chi-squared measures more, because they are based on means. For example, the empirical Pearson's chi-squared distance can be made dramatically larger by increasing the amount of data in a cell with small model probability $m_\theta(t)$; in fact, if there is data in a cell with $m_\theta(t) = 0$, the distance is infinite. We note that data occur in such a cell with probability 0 under the model, and thus the model is not possibly true; nevertheless, we still might wish to say that m_θ is a good approximation to the data based on the other cells.

The symmetric chi-squared, by averaging data and model in the denominator, avoids such extreme behavior. It also has two other attractive features. First, it generates a metric on the space of probability densities. This guarantees a certain metric robustness of the estimator, as per Donoho and Liu (1988a).

Second, it is closely linked to the total variation distance via the relationship

$$V^2(\tau, m) \le S^2(\tau, m)/4 \le V(\tau, m).$$

That is, the two distances generate equivalent topologies on the space of distributions. (There is a V-ball inside every S^2-ball and vice-versa.) Thus, if we think of the distance V as being robust, not strongly affected by the location of a few outliers, these bounds show that this property carries over in some degree to the symmetric chi-squared.

In fact, the investigations of Lindsay (1994) show that the symmetric chi-squared distance generates a highly efficient and robust method of estimation. There is a slight amount of second-order information lost at the model; in fact, using $a(t) = [\tau(t) + 2m(t)]/3$ yields second-order efficiency in estimation.

For these reasons, we recommend the symmetric chi-squared distance as a good overall measure of statistical distance.

THE CONTINUOUS CASE

We next turn to more model misspecification measures that one might use for continuous data. We consider how we might construct a distance function that discriminates between all probability measures on $\mathfrak{R}$ or $\mathfrak{R}^d$. That is, we write $\delta(\tau, G)$ now to represent the distance between two probability measures τ and G, whether discrete or continuous. The goal of this section is to introduce some of the key issues in their construction without giving an in depth analysis of particular measures.

DISCRETIZATION ROBUSTNESS

The first issue regards the robustness of the distance measure to the difference between continuous and discrete measurements. The discrete measures we have considered have natural continuous analogues when there exist absolutely continuous densities for the measures. For example, it would seem natural to extend the V distance to $V(F, G) = \int |f(x) - g(x)| dx$. This measure would still bound the maximum difference $|F(A) - G(A)|$ in set probabilities for F and G, but now over all measurable sets A.

As natural as this measure might seem, it will result, in effect, in saying that every discrete distribution is maximally different from every continuous distribution, as every discrete distribution F gives full probability to its support, which is of continuous G probability 0. They would thus have V distance 1, the maximum possible.

This is undesirable for two reasons. First, actual continuous data should have infinite numerical accuracy and could no more be put into a computer than could the value of π. Thus, if our data is in a computer, it must in a sense be discrete data through the necessary limitations on numerical accuracy. That is, a reasonable distance measure should follow the rule that a discretized version of a continuous distribution becomes closer to the continuous distribution as the discretization gets finer. Otherwise, for example, we could *never* say that a data distribution was close to being normal.

A second reason for wanting discretization robustness is that we will want to use the empirical distribution $\hat{F}$ to estimate the true distribution τ, but without this robustness, there is no hope of the discrete empirical distribution being close to any model point.

THE PROBLEM OF TOO MANY QUESTIONS

One of the largest problems we face when constructing a *sensitive distance*, one for which $\delta(\tau, G) > 0$ whenever $\tau \neq G$, is that we must carry out a delicate balancing act between sensitivity and statistical noise, which we now discuss.

We will use the chi-squared distance to illustrate the problem of "too many questions." To make this concrete, we consider the partitioning of a continuous univariate variable X so as to create a discrete variable Y, with $\{Y = j\} \Leftrightarrow X \in (a_{j-1}, a_j)$, for some ordered set $-\infty = a_0 < a_1 < \ldots < a_T = +\infty$.

In this setting, we can think of the chi-squared distance corresponding to Y as equaling the supremum of a class of squared z tests based on functions $g(x)$ of variable X, where we restrict the functions $g(x)$ to be constant on the individual intervals. If we further partition the x-axis, say by splitting each previous interval into two pieces, creating $2T - 1$ intervals, then we have strictly enlarged the class of functions we are considering in the supremum, essentially allowing them greater local variation, and so the chi-squared distance is bigger.

What are the statistical implications of this refinement in partition? First, if we are indeed interested in a statistical question that cannot be asked without the finer partition, then we need to use it. However, we will pay a price in the following sense. When we increase the partition, the variability in the chi-squared statistic will increase. However, for all questions already determined completely, or nearly so, from the coarser partition, such as $\mathrm{Pr}_\tau\{X \in (a_{j-1}, a_j]\}$, we do not gain information—but we add the noise from the chi-squared variable.

To illustrate the issue simply, suppose we have T independent normal variables, $Z_1, \ldots, Z_T$. Suppose that they satisfy $E[Z_j] = \theta_j$, and $Var[Z_j] = 1$, but the model specifies that all means are 0. Now consider the problem of detecting the true distribution, which has a single $\theta_j \neq 0$, from the model with all θs equal to 0. Suppose we use the chi-squared statistic $Z_1^2 + Z_2^2 + \ldots + Z_T^2$. As T increases, the power that one would have to detect this departure would decrease to the size, as the degrees of freedom are increasing while the noncentrality parameter is constant.

Kallenberg, Oosterhoff, and Schriever (1985) showed that this simple heuristic applies to the partitioned chi-square. Suppose we are testing a simple null hypothesis against a local alternative density using a partitioned chi-squared test. Which gives greater asymptotic power, using a fixed finite number of cells or letting the number of cells become infinite as the sample size increases?

In fact, the authors show that for many alternatives one obtains better power from the chi-squared test if one uses a *finite* number of cells; otherwise, if more and more cells are used, the power will shrink down to the size. In particular, this will occur if the Pearson's chi-squared distance between null and alternative densities is finite.

To put this into our context, increasing the number of cells widens the sensitivity to model departures in new "directions," because we gain some power in them, but it also increases statistical noise and so decreases power in every existing direction.

There are a number of ways to address this problem, but they all seem to involve a loss of statistical information. That is, we cannot seem to ask all model fit questions with optimal accuracy. Two immediate solutions come to mind. First, limit your investigation to a finite list of questions. This is addressed in a later section. The second approach is covered next.

ANSWERING INFINITELY MANY QUESTIONS

One might fairly ask if there are natural ways to answer infinitely many statistical questions with a finite sample. The answer is yes, if we allow the possibility that we can prioritize the questions.

Returning to our simple normal means example, suppose we could prioritize the components of our mean vector $\theta_1, \theta_2, \ldots$, so that we would have greatest power in detecting changes in θ_1, next greatest for θ_2, and so forth. To do so, we might consider a test statistic of the form $Z_1^2 + \lambda_2 Z_2^2 + \lambda_3 Z_3^2 + \ldots$, where $\lambda_2, \lambda_3, \ldots$, is a decreasing sequence. If the sequence decreases fast enough, then the statistic has a bounded mean and variance as the number of variables increase. Further, it is clear that the test has some power against all alternatives, but the best power is for alternatives in the first few coordinates.

There are a number of classical goodness-of-fit tests that create exactly such a balance. For example, the Cramer–von Mises test statistic has an asymptotic distribution equal to an infinite linear combination of squared normals, where the weights are $\lambda_j = 1/j^2$ (Durbin, 1973).

KERNEL-SMOOTHED DENSITY MEASURES

The problem with many classical measures is that the balance between sensitivity and statistical noise is fixed, whereas one might wish to have a flexible class of distances that allows for adjusting the sensitivity/noise tradeoff. We briefly indicate one such class.

The following method of distance construction was introduced in Basu and Lindsay (1994). We start with a kernel function $k_h(x, y)$; to be concrete, let k be the normal density at x when the mean is y and the variance h^2.

Given any distribution F for X, discrete or continuous, the variable $\xi = X + hZ$, where Z is standard normal, has the continuous density

$$f^*(y) = \int k(x, y)\,dF(x).$$

In particular, if we carry out this operation with the empirical distribution function $\hat{F}$, we get

$$d^*(y) = \int k(x, y)\,d\hat{F}(x).$$

This kernel-smoothed data is often treated as a density estimator. We could construct model adequacy methods based on comparing $d^*(y)$ with our continuous model elements. However, we go one step further and carry out an identical smoothing operation on the model elements, creating

$$m_\theta^*(y) = \int k(x, y)\,d\mathcal{M}_\theta(x).$$

In a similar way, we have a kernel-smoothed true density, $\tau^*(y)$.

We may now create measures of discrepancy based on the kernel-smoothed densities, such as the *kernel-smoothed likelihood measure:*

$$L^*(d, m_\theta) = \int d^*(y) \ln \frac{d^*(y)}{m_\theta^*(y)}\,dy,$$

and the *kernel-smoothed chi-squared measures:*

$$\chi_\alpha^2(d, m_\theta) = \int \frac{[d^*(y) - m_\theta^*(y)]^2}{\alpha^*(y)}\,dy.$$

Here, once again, $\alpha^*(y)$ could be constructed from d, m_θ, or a 50-50 mixture of them, yielding kernel-smoothed versions of Pearson's chi-squared, Neyman's chi-squared, and the symmetric chi-squared. In this notation, we have suppressed the dependence on the kernel used and the value of the smoothing parameter h used, although they clearly have some relevance.

The advantage gained through smoothing both data and model distributions is that these measures automatically yield valid distances between any two distributions and so give Fisher consistent estimators, *without regard* to the smoothing operation. This would not be true if we smoothed only the data to obtain a density estimator. In fact, if we use a kernel that preserves identifiability of the distributions, we will very generally get asymptotically consistent estimators for the parameters with the bandwidth *h fixed*. Moreover, we can fine-tune the sensitivity/noise tradeoff via the manipulation of the bandwidth parameter *h*.

We gain further insights if we consider the way in which such a measure would assess the adequacy of a model. We return to the minimax idea of the chi-squared to notice that we again have a measure of the worst-case z^2 difference

$$X_\alpha^2(\tau, m) = \sup_g\{z^2(g)\},$$

where

$$z_\alpha(g) = \frac{E_{\tau^*}[g(Y)] - E_{m^*}[g(Y)]}{\sqrt{Var_{\alpha^*}(g(Y))}}$$

is the discrepancy in the means of τ^* and model element m^*, on the scale of standard deviation under α^*.

But how do we interpret these differences for the original unsmoothed data? We note that by reversing orders of integration, we can write

$$E_{\tau^*}[g^*(Y)] = \int g(y) \int k(x, y)d\tau(x)dy = \int g^*(x)d\tau(x) = E_\tau[g^*(X)].$$

Here, $g^*(x) = \int g(x)k(x, y)dy$ is a kernel-smoothed version of the variable $g(x)$. Thus, the numerator of $z_\alpha(g)$ is the difference in means for the kernel-smoothed function $g^*(x)$, where the means are now calculated *without* the kernel smoothing.

Looking next at the denominator of $z_\alpha(g)$, elementary manipulations reveal that

$$Var_\alpha(g^*(X)) \le Var_{\alpha^*}(g(Y)),$$

so that the denominator can be thought of as a penalized version of the standard deviation of $g^*(X)$. The difference between the two sides of the last display depends on the smoothness of the original function g.

To summarize, the kernel-smoothed chi-squared adequacy measure evaluates the worst-case difference in means between the truth and the model, for a class of smoothed functions, where the differences are standardized by a *penalized* standard deviation. It is through this operation of smoothing and penalizing that we obtain a measure that can address an infinite list of questions.

OTHER ISSUES

There are many other issues in the construction of distances in continuous spaces. One that should be mentioned is the problem of invariance.

First, in a continuous space, the behavior of the distance measures under transformations of the data is relevant. If we take a monotone transformation of the observed variable X and use the corresponding model distribution, does the distance stay the same?

In the univariate case, one can use the probability integral transform of a continuous distribution to create such an invariant class of distances. The Cramer–von Mises distance is of this type (Durbin, 1973). The kernel-smoothing distances mentioned above are space dependent in the sense that transformations of observed variable X do change the distance.

Invariance seems desirable from an inferential point of view, but difficult to achieve without forcing one of the distributions to be continuous and appealing to the probability integral transform for a common scale.

In multivariate continuous space, the problem of transformation invariance is even more difficult, as there is no longer a natural probability integral transformation to bring data and model onto a common scale. For this reason, there is a paucity of important multivariate distance functions.

At this point we are not ready to make any strong recommendations. We can only argue that the problem of constructing meaningful distance functions for continuous models provides a large arena for interesting research. It will require carefully thinking about what we would like to identify as model misspecification costs.

QUADRATIC DISTANCE FUNCTIONS

We conclude our investigation of statistical distance with a generalization of the partitioned chi-squared statistical distance. It will allow us to generate a user-selected set of questions to assess. The idea goes back to Ferguson (1958), who called it the *minimum chi-squared method,* and to Hansen

(1982), who called it the *generalized method of moments*. We think both names are correct, but also somewhat misleading in the statistical context. The method belongs in the genre of estimating functions, as it provides a very general way to combine multiple estimating functions to yield a single inference. We will here identify it with the technique used to combine the estimating functions and call it the *quadratic distance method*.

In this context, the model need be described not directly by density m_θ, but rather by a set of unbiased estimating functions, which will become our list of statistical questions. As will be seen, this method is not restricted to discrete models.

The Model Scores

We start with K functions $s_k(x, \theta)$, $k = 1, \ldots, K$, of the data and the parameter θ, written as a vector $s(x, \theta)$. This will be called the *score vector*. A distribution m will be in our model, and be assigned parameter value θ, if there exists some θ such that the following *model fit equations* are satisfied:

$$E_m[s(x, \theta)] = 0.$$

Note that we usually use at least as many equations as parameters, $K \geq \dim(\theta)$, so that there is some hope that there is just one θ that satisfies the equation for a fixed m. In other words, we wish the parameter to be *identifiable*.

Sometimes the vector of scores s is created by assuming a *parametric model* $\mathcal{M} = \{m_\theta\}$ and then creating a set of scores such that $E_\theta[s(x, \theta)] = 0$. One such set would be the likelihood scores for the model. Another set, corresponding to the classical method of moments, are the moment scores,

$$s_p(x, \theta) = \overline{x}^p - E_\theta[\overline{X}^p].$$

As another example, given an underlying model m_θ, we could use as score functions the indicator functions for a partition of the x-axis minus their expectation under a particular parametric model m_θ:

$$s_k(x, \theta) = I(x \in (a_{k-1}, a_k]) - \mathcal{M}_\theta((a_{k-1}, a_k]).$$

The Semiparametric Model

However, even if the model started as a parametric one, there is a natural extension to a *semiparametric model* as follows. First define the set of all measures that satisfy the model fit equations for a fixed θ:

$$\mathcal{M}_\theta = \{F : \int \mathbf{s}(x, \theta)dF(x) = 0\}.$$

We will say that any element of $\mathcal{M}_\theta$ has parameter value θ. We then let $\mathcal{M}_{sp}$, the semiparametric version of the model, be all measures that satisfy the score mean equations for some θ:

$$\mathcal{M}_{sp} = \bigcup \mathcal{M}_\theta.$$

The semiparametric model could be as large as the class of all distributions, which would occur if for every F there existed some solution θ_F to the model fit equations:

$$\int \mathbf{s}(x, \theta)dF(x) = 0.$$

This might occur if the dimension of θ were equal to $\dim(\mathbf{s})$. However, if there are more equations than unknowns, $\dim(\theta) < \dim(\mathbf{s})$ we would expect there to be distributions F for which there was *no* solution to the score equations, and so have no value of the parameter assigned to them. These distributions would display lack of fit.

DEFINING THE DISTANCE

Given this or any other set of functions, we can define a vector of discrepancies γ between the model and truth to be

$$E_\tau[\mathbf{s}(x, \theta)] = \gamma(\theta).$$

Recall that the model fit equations say that $\mathbf{s}$ has mean zero under θ, so γ equals the difference in means between τ and any element of $\mathcal{M}_\theta$.

If $\gamma(\theta) = 0$, then τ is in the semiparametric version $\mathcal{M}_\theta$ of the model, and has parameter value θ, whether or not it was not in the original parametric model m_θ. That is, we cannot, based on $\mathbf{s}$ alone, distinguish between m_θ and any other measure with the same expectations for $\mathbf{s}$.

If γ is not 0, we have a measurable distinction between the models with parameter value θ and truth. In the partitioned chi-squared example, $\gamma(\theta)$ measures the difference in probability for each cell between τ and the model. In an i.i.d. sampling situation, $\gamma(\theta)$ will be estimated by $n^{-1}\sum s(x_i, \theta)$.

If we wish to use a measure that standardizes these discrepancies based on our ability to estimate them, then we need to use the variance-covariance matrix of the scores. An overall discrepancy over all functions s_k, based on our best ability to determine simultaneously the values of these functions, is given by the *quadratic distance:*

$$Q^2(\tau, m_\theta) = \gamma(\theta)^t[Var_\alpha(s(x, \theta))]^{-1}\gamma(\theta).$$

If the variance is not invertible, we replace it with the Moore-Penrose generalized inverse. Once again, the function $\alpha(x)$ represents the distribution that we use to evaluate the variance.

THE INFERENCE FUNCTION

Our sample space might be continuous, and so the cell proportions $d(t)$ are no longer defined. However, we can readily create an inference function from $Q^2(\tau, m)$ by replacing the expectations under τ in the definition of Q^2 with expectations under the empirical distribution function. We will use $Q^2(d, m)$ to denote this object, thinking now of d and m as the empirical and model probability measures respectively. That is, we have the *quadratic inference function:*

$$Q^2(d, m) = \left[\sum s(\theta, x_i)\right]^t V_\alpha^{-1}\left[\sum s(\theta, x_i)\right],$$

where V is the α-covariance for the scores.

We could again use for α the model element m_θ or τ or a blend of m and τ. The original arguments for using a blend of the two still hold, but now there exists an additional advantage to using τ for α. If done this way, we do not need to specify *any* moment properties of the model other than the original model fit equations, as we will be using a sample covariance estimator, not a model-based one, when we construct the inference function.

The standard partitioned chi-squared statistic is an example of a quadratic inference function. It is easily checked that if we use as scores s the partitioning indicators described above, then the quadratic inference function Q^2 equals the usual partitioned chi-squared distance, with denominators determined by α. But for other scores, we have generalized the chi-squared distance function to allow other types of statistical questions. For example, we could also use a list of moment functions, $s_k(x, \theta) = x^k - E_\theta[X^k]$. In this manner, we would be led to a "generalized method of moments" in the more conventional meaning of this expression.

The measure Q^2 has an alternative interpretation that relates to the maximum cost over a list of statistical questions. In fact, we can describe it as the supremum over all the z^2 test statistics based on using linear combinations $\Sigma \, b_k s_k(x, \theta)$ of the score functions as test statistics. In the case of the partitioning indicators, this yields the original chi-squared interpretation regarding all z-tests based on functions $g(x)$ that are constant on the intervals. If we use the moment functions, then it becomes all z-tests based on polynomial functions of degree no greater than K. Just as for the chi-squared measure, we can then construct an estimator of the worst-fitting function.

OTHER INFERENTIAL PROPERTIES

We can summarize some properties of the quadratic inference function found in Qu (1998) as follows:

(1) As in the chi-squared measure, the quadratic inference function yields highly efficient procedures. In fact, it can be shown that the estimators correspond to the estimators based on the best linear combinations of the scores **s** that were used.

(2) This means that if we do have a particular model in mind, such as the normal, and we include the likelihood scores in our set **s**, then the estimators will be fully efficient when the model is true. Thus, if one were using a partitioned chi-squared statistic to test fit to a parametric model, one could add the likelihood scores to the indicators and obtain full efficiency in the parameter estimators—otherwise the discretization causes a loss of efficiency.

(3) As an inference function, $Q^2(d, m_\theta)$ behaves much like a likelihood function. In particular, (i) $Q^2(d, m_\theta) - Q^2(d, m_{\hat\theta})$ is asymptotically chi-squared in distribution, regardless of the truth of the semiparametric model, with degrees of freedom equal to the dimension of θ; and (ii) $Q^2(d, m_{\hat\theta})$, the point estimator of the adequacy $Q^2(\tau, m_\theta)$, is asymptotically chi-squared as a test statistic for testing whether the truth is in the model. This can be thought of as a test of whether there exists a value of θ that satisfies the model fit equations.

COMBINING PRIMARY AND SECONDARY QUESTIONS

We illustrate how one might use a quadratic inference function to create a semiparametric model that simultaneously defines a parameter and asks a model fit equation. In the process, the estimation will retain high efficiency.

Suppose we are interested in the "center of location" of a distribution, but not particularly in predictive probabilities. It might be close to normally distributed or it might have heavier tails. It might be skewed, which would affect our interpretation of the answer.

We could construct a quadratic inference function based on the mean score $s_1 = x - \theta$ and the median score $s_2 = I(x > \theta) - 1/2$. The resulting semiparametric model specifies only that the mean and the median are the same. We have used the likelihood scores from the normal and double exponential model. As a consequence, the estimator of θ based on minimizing Q^2 will, asymptotically, be fully efficient if the true distribution is *either* normal or double exponential, and otherwise will be the most efficient linear combination of the two scores. Moreover, $Q^2(d, m_{\hat{\theta}})$ is a test statistic for testing the model hypothesis that the mean and the median are identical. As a point estimator, it estimates the greatest squared and standardized difference in means between τ and $\mathcal{M}$ using linear combinations of s_1 and s_2.

The author and Ph.D. student Chanseok Park have investigated this and similar methods numerically and theoretically. One interesting finding is that in the mean/median problem, the resulting minimum quadratic inference estimator has a 25% breakdown point, putting it intermediate to the mean (0%) and the median (50%). One-sided outliers cause asymmetry and so are detectable by the model adequacy measure.

DISCUSSION

I have presented an analysis of some of the many important issues involved in building a loss function for model misspecification analysis and model selection. The subject is in fact enormous. There are many issues, some simple to solve and others complex.

Along the way in this presentation, we have offered some potential ways to think through the problem from the foundations. We anticipate that with these basic principles one could arrive at reasonable methods for the selection of a distance measure for use as a loss function for model misspecification.

It would be immodest for me to suggest that such measures should be used widely in reporting scientific results. That is a question for the scientific community to decide. I can only give my personal opinion; namely, I would find statistical analyses more compelling if there were a handful of generally recognized measures of model adequacy that one could report along with the P-values and standard errors and other measures of within-model errors. I don't think that these analyses have to be decisions on model adequacy based on formal rules; they can be of the "let the reader decide if this fit is inadequate" type. As an analogy, although a P-value can be used for formal decision making, it is much more used as a guide to the strength of the conclusions.

Nonparametric methods have taken on great favor in the statistical community. I have found many of them unsatisfying because, unlike parametric models, they do not offer a complete description of the data generation mechanism. This complete description is necessary if one wishes to use it for prediction. I would hope that the methods suggested here could bridge the gap between parametric and nonparametric analyses.

To tell the truth, much of the motivation for this treatise is that I find it very difficult to say, "Assume the data is normally distributed," but I would have no problem saying, "Assume the data is approximately normally distributed." Making the second statement in full honesty requires the development of a language for approximation. I hope the ideas in this paper are a starting point to becoming more honest.

Since the initial draft of this manuscript I have been working, along with coauthor Marianthi Markatou, on turning the bare bones presented here into a book-length document. Therein one will find a more detailed analysis of candidate distances and more methodology that would enable one to use them for model adequacy analysis.

14.1   Commentary

D. R. Cox

The very word "model" implies simplification and idealization. It is, in any case, inconceivable that, say, a complex social phenomenon could be totally described by some set of mathematical formulae (or by a computer representation thereof). This applies far beyond the role of models in formal statistical analysis.

Indeed, it may only be theoretical physicists at the most fundamental level who believe their models to be, in some sense, really true and, greatly though one may be in awe of that line of work, some note of skepticism may be in order even there.

Bruce Lindsay has launched a characteristically original and lucid direct assault on the issue that we use models knowing perfectly well that they are, at best, imperfect. Before addressing the central question—"How successful has he been?"—one or two general comments are in order.

First, the suggestion that nonparametric methods can be regarded as assumption-free is usually misleading, in that they involve very strong independence assumptions that are often much more critical than assumptions of distributional form.

Next, it is not really the case that the safe use of means is restricted to symmetrical distributions. For any variable that is what physicists call extensive, i.e., physically additive, the mean determines a long-run property of the system, whatever the shape of the distribution. Yield of product and cost of a unit of output are examples. The highly robust interpretations of mean and variance in such situations are central.

Traditional tests of goodness of fit are probably best regarded as addressing the question of whether the direction of departure from the model is reasonably firmly established. There is then the implication that, by and large, it may be best to keep to the model, unless one knows the direction in which to amend it. Now with modest amounts of data, it may well be that such a direction is not clearly established and the goodness-of-fit test is a satisfactory component of analysis. It is then not so much a question of whether one believes in the model, but of whether one knows how to amend it fruitfully. Now with very large amounts of data, it is highly likely that formal tests will be very significant. Then the question is different: is the departure big enough to matter? This is getting close to the spirit of Bruce Lindsay's approach, suggesting that his is more relevant to large than to modest sets of data. There is an important note of caution about large sets of data, however; they frequently have internal structure that makes assessment of error based on conventional assumptions of independence quite misleading.

It would be a severe limitation indeed, if parameters were meaningful only under the model for which they are notionally defined. It is, in fact, important that those parameters that are the primary basis for interpretation should retain a clear meaning, at least under modest perturbations of the initial model. When the parameters are generalized moments—for example, regression coefficients—this is clear. The least squares estimate of the slope of a fitted line is an estimate of the slope of the best-fitting line in the numerical-analytical sense of least squares; the more nonlinear the relation, the less relevant is the least-squares slope. Again, it can be regarded as an estimate of the average gradient over some region. More generally, a maximum likelihood estimate of a parameter in an incorrect model estimates the parameter in the model best fitting the "true" model in a reasonable sense (Cox, 1961).

While I appreciate the clarity of exposition and thought gained via decision-theoretic formulations, I think one has to be very cautious about the specification of objectives, especially in contexts where the gaining of understanding is a goal. In making this point, I am not wishing to emphasize a largely false antithesis between science and technology. Cox and Wermuth (1996, 18–19) listed seven requirements of a model of which fitting the data was only one and by no means necessarily the most important. Others include the ability to link with underlying theory or previous work and the interpretability of the parameters of primary interest. A model that links strongly with underlying theory and fits over most, but not all, the range of explanatory variables of interest may be preferable to a wholly empirical model that fits better, provided, of course, that discrepancies with the first model are reported.

Indeed, I think my main reservation about the paper is its treatment of distance between fitted and observed distribution as, in some sense, the overriding criterion for assessing models, especially in the discrete case. While it is hard to proceed very differently without some subject-matter-specific assessment of those aspects of importance, it seems essential to introduce such aspects.

While I have sounded above some notes of caution about certain aspects of the paper, let me conclude by stressing that it represents a major innovative piece of work, and that I greatly welcome it and look forward very much to seeing future developments.

14.2   Commentary

Stephen P. Ellner

The first and most important core idea of Lindsay's paper is that *all models are wrong, but some models are useful* (a motto that was posted on a colleague's door until his retirement). We never really believe our models (if we're sane), but as Lindsay observes that doesn't keep us from publishing an ANOVA when its scientific implications are interesting. Ecological modelers (the modeling community to which I belong) are constantly building models that are far simpler than the real systems, and constantly answering biologists who decide that our model is invalid because some favorite biological detail has been left out. I am, therefore, fully sympathetic to Lindsay's arguments that statistical theory and practice should begin from the accurate premise that models are flawed, rather than the pretence that the fitted model is perfect.

The problem is figuring out which models are useful and which ones are just plain wrong. Lindsay's second core idea is to fit models and evaluate their adequacy using a loss function whose value will dictate whether we should publish that ANOVA or instead go back and produce a better model. In principle that is hard to argue with, especially if the loss function meets Lindsay's goal of providing precise information about the risk of scientifically meaningful errors that would be entailed by accepting the model.

However, more than half of the paper is devoted to evaluating candidates for a general-purpose loss function, and I am less sympathetic to that aspect of Lindsay's prescription. If all models are wrong, it necessarily follows (as Lindsay explicitly acknowledges) that a model's adequacy or inadequacy is entirely dependent on what it's being used for. A quest for a general-purpose loss function seems somewhat inconsistent with this principle. For example, the "list of scientific questions" underlying a particular loss function may or may not be appropriate for the particular scientific issues at hand. Even the quadratic loss function (a.k.a. method of moments) imposes specific forms on the list of questions that can be used to interrogate a model and on how they are combined into a scalar measure.

I would like to suggest that Lindsay's goals would be better realized by using a highly case-specific, computer-intensive approach that would involve simulating the entire process from data collection and modeling through to the process of drawing conclusions from the model. In the context of Lindsay's main example, univariate probability distributions, one could resample the data or sample from a nonparametric estimate of the distribution, and (for example) determine the distribution of errors when the population median is estimated by the mean of a fitted normal. The programming is essentially a bootstrap, but the focus at the end is not on model parameters but rather on the risk of reaching scientifically unsound conclusions. The essential feature of this procedure, which maps it onto Lindsay's proposal, is that the data-generating model is deliberately far more complex and realistic ("nonparametric") than the fitted model.

Ecological modelers have used this kind of approach for some time. In general (Caswell, 1988) a model that has *theoretical* validity (accurate representation of mechanistic structure) may be too simple or too complex to optimize *operational* validity (accurate numerical predictions of system dynamics and response to manipulations). Ludwig and Walters (1985) give a practical example in fisheries management where a theoretically valid model is overly complex. Intrinsic limitations on the available data create a situation where management is more successful, in terms of meeting quantitative targets, if the population model omits known aspects of fish biology (age and size structure) in order to reduce the number of fitted parameters. Conversely, my student John Fieberg and I recently found that more detailed models than those typically used can be advantageous for estimating the rate of population growth or decline in a threatened animal species (Fieberg and Ellner, 2001). The more complex models include the effect of an environmental covariate on birth and death rates, but use long-term environmental monitoring data to parameterize a finite-state Markov chain model for the covariate.

In both cases, an appropriate loss function with "an explicit interpretation in terms of our statistical goals" was constructed by automating the scientific inference step and estimating the risk of serious errors by simulation. However, these studies also illustrate the problem of choosing a baseline model to serve as the stand-in or best approximation for the truth. Ludwig and Walters (1985) used a fish population model with full detail on age and size structure. Fieberg and I used an avian population model (based on Beissinger, 1995) with continuously varying water level as the environmental covariate, while our fitted model reduced the water variable to three discrete states (drought, postdrought, high). Lindsay assumes that a nonparametric estimator is available. Sometimes it is; univariate density estimation is one example. But sometimes, nonparametric estimation is harder (multivariate smoothing) and often it would be a venture into unexplored territory (nonlinear state-space models, as in the ecological examples above). If we must have a nonparametric, truly assumption-free model as the benchmark to evaluate our parametric models, then the problem of identifying a trustworthy parametric model has been "reduced" to the thornier problem of identifying a trustworthy nonparametric model.

Despite these practical concerns, it is a step forward to have the right problem under discussion: finding a trustworthy model, rather than finding a correct model. Lindsay's reconceptualization of model evaluation is a step towards reconciling the actual decisions facing modelers with the theory that we use as our guide.

14.3   Rejoinder

Bruce G. Lindsay

Let me thank Sir D. R. Cox for his thoughtful commentary on my treatise. As always, he adds considerable insight into the problem of building models in the face of their imperfection. His wariness about accepting any single framework for model assessment is certainly merited. My defense would be that my intention was to see whether one could create a formal framework, not necessarily perfect, that would allow one to use a model while not believing that it was true, and then see what the implications of this might be.

Do we really need to build standardized frameworks for model assessment? I think that having a few benchmarks that most users could understand would be valuable, but I am not claiming that I have the ideal formulation for this. One problem is the vast variation in statistical sophistication among the users of statistics. Another is that statistical models play a wide variety of roles in scientific work.

A number of Dr. Cox's points are worth reviewing. First, I certainly agree with his point that the presence of an independence assumption in the nonparametric framework limits its application. I would also say, however, that other nonparametric frameworks, such as stationarity plus ergodicity, could certainly replace it. It is mandatory, however, that one assume that there is some larger truth and that one should be capable of measuring the model's departure from it in a statistically meaningful way, through a consistent measure of distance or loss. In general, what is needed is a model plus some directions of departure from the model that are viewed as central to the interpretation of the results and whose magnitude can be statistically determined.

More generally, I think it is better to do something than to say that nothing can be done. Certainly, many times the nonparametric assumptions are untenable. Indeed, if we can agree that the essential power of statistics in parametric models arises from doing *something* even though it is not quite right, then my defense is that I am just taking this to the next level by doing it in a larger model that is not quite right.

I agree as well with the point that the mean of a process can be meaningful even if the distribution is not symmetrical about the mean. However, a counterargument might be that if the distribution is bimodal, then the process has important substructure that could be worth considering. In this case, the data might have more to say about the construction of a reasonable alternative model, one that would enable the investigator to learn more about the distribution of the yield of product or cost of a unit while still learning about the mean of this distribution. The subject of much of my research work has been on mixture models that allow one to investigate such structures.

I might also add that many statisticians feel that robustness to outliers is an important attribute for a statistical procedure. Certainly, means and variances are robust in interpretation, but some would say that they are too much influenced by a few observations and so are in some way unrepresentative.

Dr. Cox correctly pointed out that sample size would be an important factor in the analyses I have discussed. In small samples, one does not have the luxury of investigating too many questions of fit while still building a model sufficiently rich to describe the scientific phenomena. In a small study, one inherently gains less information about all aspects of the problem, including parameter values and model failure. I am afraid that the most that one can say in a small sample is that the model-based statements *could* have substantial model misspecification error in addition to parameter estimation error. Perhaps it is better to say that up front than not at all.

On the other hand, if there is sample size to support it, one might be able to recognize that a parsimonious model is a good fit even if there is sufficient statistical power to reject its correctness. If, as Dr. Cox suggested, lack of independence does make the error assessment implausible, then, as in many statistical analyses using observational data, it should be taken with a grain of salt. In such uncontrolled situations one inevitably deals in hypothetical truths, such as "if the data had been independently sampled from the population." If there is a more realistic model that allows dependence it should be used for parametric and nonparametric versions of the problem. One might hope, at least, that the orders of magnitude are correct, in which case truly important model breakdown might be found.

Dr. Cox and I agreed on the need to have parameters that are meaningful when the model is perturbed. As he indicated, the maximum likelihood estimator does indeed estimate something; I believe that he also implied that it usually estimates a reasonable thing. I am not convinced of that, and we disagree on this because we disagree on what it means to be a small perturbation. I think that in many applications, and for the kinds of interpretations users would like to have, a more robust distance should be used to measure perturbation.

I do think maximum likelihood is our most sensitive tool for model-based statistical inference when the model is correct. Problems arise only when the model is incorrect, in which case the parameters can change dramatically with seemingly small changes in the distribution. One might ask then if there is not some way to use likelihood as the inference tool while still retaining a robust measure of distance. I am currently working on a hybrid technology in which one enlarges the model slightly to include nearby distributions (using a robust distance) and then uses maximum likelihood to do inference in this extended model. Doing this robustifies the model interpretation while using the sensitivity of the likelihood for parameter estimation and testing.

Dr. Cox expressed appreciation for the clarity of exposition and thought gained from a decision-theoretic formulation but was concerned about the specification of objectives that such a formulation requires. To be honest, I created a formal decision-theoretic framework for model adequacy exactly because I could not find any other way to see the forest for the trees, not because I thought such rules would dictate our behavior. In the end, I think that the clarity of thought one can gain from such a formulation is valuable for understanding quite independently of its use to create a formal decision making rule. To make a parallel, in the theory of parameter estimation a loss function approach leads to an appreciation of the tradeoff between bias and variance, with mean squared error being just one way to form a single objective. In model adequacy, there is a tradeoff between model misspecification costs and parameter estimation costs.

Finally, let me address Dr. Cox's main reservation. If I seemed, through my content, to hold that the data-fitting aspect of model building overrode subject matter considerations, then it was a failure of communication. I certainly do not believe that model adequacy *should* be the primary consideration in model building. Indeed, to take his example, I would certainly not take a wholly empirical model over a theory-based model on the grounds of better fit; indeed I would rather say that if there was lack of fit, its scientific meaning should be assessed. One of my main points was that we should be openly willing to tolerate some error in a simple model rather than insisting on having one that is totally correct, or at least not provably incorrect. If I had taken more care to add that the simple model I had in mind should be based on subject matter theory, then I think Dr. Cox and I would have been in perfect agreement.

I am happy to see that Stephen Ellner and I are in agreement with the statement "all models are wrong but some models are useful." And I think he agreed with me that it is important to face this fact and try to make some sense of it.

In addressing the details of my proposals, he indicated that I was probably mistaken in searching for one measure of distance that would fit all needs. My response is that single-number summaries do carry some information, but clearly not all. A good part of the art of statistics is in finding useful and well-recognized single-number summaries such as the P-value. In my chapter, I examined several possible distance measures for their meaningful interpretations because much of the use of statistics is "canned," and so I thought that there might be merit in determining whether a standard distance carried the potential of wide applicability, much as one uses R^2 in regression.

One aspect of the problem not discussed in detail in my chapter is the role that I would envision for secondary diagnostics. Once the model is found inadequate, then the first question is "How did it fail?" and the second is "How important is this failure?" A number of statistical distances discussed can be used to create diagnostic plots. Thus, one might use the standard distance measure to learn of potential failures and then individualize the assessment to fit the scientific context at the secondary level.

That is, if there were such a standard measure, I believe it should act as an alarm signal that, when sounded, not only means that one should more thoroughly investigate the failures of the model but also itself becomes a tool for that investigation. On the other hand, if the alarm is not sounded one might be reassured, and save some work. I am not sure if such a multitool is possible, but I thought it worth considering.

I applaud Dr. Ellner's considerably more sophisticated bootstrap simulation approach to assessing model adequacy. I consider this a tailor-made distance assessment, something that I would strongly encourage when study time and scientific importance make it feasible. The main point of my essay is exactly that whenever possible one should think carefully about what the model is supposed to be providing and from this find a way to measure its failings. I am glad to hear that this is part of the scientific culture in ecology. If it were more universal, I would not have felt the need to write this article.

In agreement with Dr. Cox, Dr. Ellner pointed out the difficulty of creating a larger nonparametric framework in which to insert the parametric model. Of course, I concur. Just the same, as long as we create some larger framework that includes the smaller, we can at least learn about some directions of model failure.

In conclusion, I considered the work of this paper to be my humble first try at making a useful theory for model adequacy. I am working on a book with Marianthi Markatou of Columbia University that is an attempt to flesh out the ideas of this paper, as well as to gain further understanding of the statistical implications of using general-purpose distances.

REFERENCES

Basu, A., and B. G. Lindsay. 1994. Minimum Disparity Estimation for Continuous Models: Efficiency, Distributions, and Robustness. *Ann. Ins. Stat. Math.* 46:683–705.

Beissinger, S. R. 1995. Modeling Extinction in Periodic Environments: Everglades Water Levels and Snail Kite Population Viability. *Ecol. Appl.* 5:618–631.

Caswell, H. 1988. Theory and Models in Ecology: A Different Perspective. *Ecol. Modeling* 43:33–44.

Cox, D. R. 1961. Tests of Separate Families of Hypotheses. In *Proceedings of the Fourth Berkeley Symposium on Mathematical Statistics and Probability.* Berkeley: University of California Press.

Cox, D. R., and N. Wermuth. 1996. *Multivariate Dependencies.* London: Chapman and Hall.

Donoho, D. L., and R. C. Liu. 1988a. The "Automatic" Robustness of Minimum Distance Functionals. *Ann. Stat.* 16:552–586.

Donoho, D. L., and R. C. Liu. 1988b. Pathologies of Some Minimum Distance Estimators. *Ann. Stat.* 16:587–608.

Durbin, J. 1973. *Distribution Theory for Tests Based on the Sample Distribution Function.* Philadelphia: Society for Industrial and Applied Mathematics.

Ferguson, T. S. 1958. A Method of Generating Best Asymptotically Normal Estimates with Application to the Estimation of Bacterial Densities. *Ann. Math. Stat.* 29: 1046–1062.

Fieberg, J., and S. P. Ellner. 2001. Stochastic Matrix Models for Conservation and Management: A Comparative Review of Methods. *Ecology Letters* 4(3): 244–266.

Hansen, L. P. 1982. Large Sample Properties of Generalized Method of Moments Estimators. *Econometrica* 50:1029–1054.

Kallenberg, W. C. M., J. Oosterhoff, and B. F. Schriever. 1985. The Number of Classes in Chi-Squared Goodness of Fit Tests. *J. Am. Stat. Assn* 80:959–968.

Le Cam, L. M. 1986. *Asymptotic Methods in Statistical Theory.* New York: Springer.

Lindsay, B. G. 1994. Efficiency versus Robustness: The Case for Minimum Hellinger Distance and Related Methods. *Ann. Stat.* 22:1081–1114.

Lindsay, B. G., and M. Markatou. Forthcoming. *Statistical Distances: A Global Framework for Inference.* New York: Springer.

Ludwig, D., and C. J. Walters. 1985. Are Age-Structured Models Appropriate for Catch-Effort Data? *Can. J. Fish. Aquat. Sci.* 42:1066–1072.

Qu, P. 1998. Adaptive Generalized Estimating Equations. Thesis in statistics, Pennsylvania State University.

Rudas, T., C. C. Clogg, and B. G. Lindsay. 1994. A New Index of Fit Based on Mixture Methods for the Analysis of Contingency Tables. *J. Roy. Stat. Soc.,* ser. B, 56: 623–629.

Xi, L. 1996. Measuring Goodness-of-fit in the Analysis of Contingency Tables with Mixture-Based Indices: Algorithms, Asymptotics, and Inference. Thesis in Statistics, Pennsylvania State University.

15 Model Identification from Many Candidates

Mark L. Taper

ABSTRACT

Model identification is a necessary component of modern science. Model misspecification is a major, if not the dominant, source of error in the quantification of most scientific evidence. Hypothesis tests have become the de facto standard for evidence in the bulk of scientific work. Consequently, because hypothesis tests require a single null and a single alternative hypothesis there has been a very strong tendency to restrict the number of models considered in an analysis to two. I discuss the information criteria approach to model identification. The information criteria approach can be thought of as an extension of the likelihood ratio approach to the case of multiple alternatives. However, it has been claimed that information criteria are "confused" by too many alternative models and that selection should occur among a limited set of models. I demonstrate that the information criteria approach can be extended to large sets of models. There is a tradeoff between in the amount of model detail that can be accurately captured and the number of models that can be considered. This tradeoff can be incorporated in modifications of the parameter penalty term.

HYPOTHESES, MODELS, AND SCIENCE

The hypothesis concept plays an important role in science. The classic scientific method (Popper, 1959) continually reiterates a cycle of hypothesis

This paper was first presented in August 1998 at a symposium titled "Scientific Evidence" hosted by the Ecological Statistics section of the Ecological Society of America. I thank Subhash Lele, Jay Rotella, Prasanta Bandyopadhyay, and Brad Shepard for commenting on drafts of this manuscript.

488

formulation, hypothesis testing, and hypothesis refinement. A hypothesis is a concrete description of the way the universe might be or operate. The statement that the world is flat is interesting as a scientific hypothesis only if it can be coupled with some alternative, which may be nothing more complex than that the world is not flat. A hypothesis is a construct of the human mind that organizes our understanding of the universe. To quantitatively assess a scientific hypothesis, it must be translated into a model. This paper refers to structural models and to specific models. Structural models are mathematical formulations of a hypothesis, including parameterization for error and noise. Specific models are structural models with specific parameter values (see Mayo, 2004 [chapter 4, this volume], and Miller and Frost, 2004 [chapter 8, this volume]).

For example, consider the scientific hypothesis that a population exhibits density-dependent growth. A structural model representing this hypothesis might be the Ricker dynamics equation for change in population size: $\ln\left(\frac{N_{t+1}}{N_t}\right) = a + bN_t + \varepsilon_t$, where N_t is population size at time t, a is the intrinsic or maximum growth rate, b is an intraspecific competition parameter providing density dependence to the model, and ε_t is a time-dependent stochastic noise term. A specific model for this hypothesis and structural model might be $\ln\left(\frac{N_{t+1}}{N_t}\right) = 0.75 - .003N_t + \varepsilon_t$, where ε_t is normally distributed with mean 0 and variance 0.01.

THE NATURE OF EVIDENCE AND MODEL IDENTIFICATION

As human beings, we can fruitfully think about the world only in terms of constructs or models. The alternative of viewing the world in full detail is frighteningly paralytic if not impossible (Borges, 1967). As scientists, we are interested in the validity of our constructs. How do we know that our models are appropriate? This is an important question. Model misspecification is a major, if not the dominant, source of error in the quantification of most scientific analysis (Chatfield, 1995). As scientists, we are trained to seek evidence supporting or refuting our ideas. But what is evidence? We can think of evidence as information contained in data about the validity of our models. Unfortunately, when we ask "Is this model true?" we have begun a failed quest. The answer is always no! No model represents the real world exactly. It is better to ask, "Does this model do a better job for me than that one?" What you can do is compare models by asking questions, such as whether this hypothesis or model is more similar to underlying truth than that model. Evidence compares models, not only to one another, but also to an

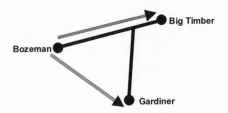

FIGURE 15.1 Geographic distance versus road distance. Solid lines represent roads while arrows represent geographic distance.

unknown truth. In other words, we should consider evidence to be some measure for the models under consideration of the relative distance to the truth. Further, whatever evidence is, it should be measured on a continuous scale.

How can one choose an evidence measure? Unfortunately, it is unlikely that there is a single best evidence measure. Evidence measures should be adapted to the problems at hand. Consider the schematic map in figure 15.1 of towns near my home. Is Bozeman closer to Big Timber than it is to Gardiner? The answer depends on whether you are flying or driving. Insisting on "distance" always being geographic distance is not as useful as a more thoughtful consideration of the problem.

Royall (1997; 2004 [chapter 5, this volume]) proposes the likelihood ratio as the fundamental measure of the evidence for one model over another (see also Forster and Sober, 2004 [chapter 6, this volume]; Edwards, 1992). The likelihood ratio is expressed as $P(X = x|A)/P(X = x|B)$. Here, $P(X = x|A)$ and $P(X = x|B)$ represent the probability of a data set X being equal to the observed data set x assuming models A and B respectively. Instead of assuming the likelihood ratio as a primitive postulate, Subhash Lele develops (2004 [chapter 7, this volume]) quantitative requirements for distance measures[1] to be effective evidence measures. It is easy to see by expanding equation 1 that the likelihood ratio represents a relative distance measure and is thus a special case of Lele's evidence functions. Here, $P(X = x|T)$ is the true probability of data set x occurring.

$$\frac{P(X = x|A)}{P(X = x|B)} = \left(\frac{P(X = x|A)}{P(X = x|T)} \right) \bigg/ \left(\frac{P(X = x|B)}{P(X = x|T)} \right) \tag{1}$$

1. Technically, these are discrepancy measures, as they are not required to be symmetric.

As Lele demonstrates, the likelihood ratio is in some sense an optimal evidence measure for comparing simple single-parameter models. However, interpretation of the likelihood ratio as evidence in more complex situations can become problematic. For example, with nested linear models, likelihood increases monotonically with the number of parameters. Consequently, the probability of misleading evidence can be made arbitrarily close to 1 by adding spurious covariates. This indicates that the number of parameters in the models being compared needs to be taken into consideration in an effective evidence measure (see Forster and Sober, 2004 [chapter 6, this volume]).

One measure that compares an approximating distribution (model) to a "true distribution" is the Kullback-Leibler distance. It is the expected log-likelihood ratio of the approximating model to the "true model."

$$KL = \int f(x) \cdot \log\left(\frac{f(x)}{g(x)}\right) dx, \tag{2}$$

where $f(x)$ is the true probability of x and $g(x)$ is the probability of x under the model.

The KL distance itself cannot be an evidence measure because it draws no information from data. In addition, it requires that "truth" be known. Fortunately, the KL distance can be estimated from data (up to an unknown constant) by various information criteria (Forster and Sober, 2004 [chapter 6, this volume]). Burnham and Anderson (1998) provide a very readable introduction to the field. There are a variety of approximating forms; the AIC (Akaike, 1973), the AIC_c (Hurvich and Tsai, 1989), the SIC (Schwarz, 1978), the CAIC (Bozdogan, 1987), AIC_{hq} (Hannan and Quinn, 1979), and others (see table 15.1). All take the form of constant $- 2 \ln(L) + f(k, n)$, where L is the likelihood of a model given data, k is the number of parameters in the model, and n is the number of observations.

A single information criterion (IC) value is meaningless because the unknown constant cannot be evaluated. Nevertheless, information criteria are very useful for comparing models. A model is selected as the best model in a suite of models if it has the minimum value of the information criterion being utilized. The magnitudes of differences in the information criterion values, ΔIC, are used as measures of the strength of evidence for one model over another. We are relieved of the need to evaluate the unknown constant because the differencing of IC values eliminates it.

TABLE 15.1 Common information criteria. Minimum total discrepancy forms are designed to minimize prediction error, while order consistent forms attempt to correctly identify the underlying order of the data generating model.

Class	Name	Formula
minimum total discrepancy forms	AIC	$-2\ln(L) + 2k$
	AICc	$-2\ln(L) + 2k\left(\dfrac{n}{n-k-1}\right)$
order consistent forms	SIC	$-2\ln(L) + \ln(n)k$
	CAIC	$-2\ln(L) + (\ln(n) + 1)k$
	IC_{hq}	$-2\ln(L) + c\ln(\ln(n))k, c > 1$

A ΔIC measure is related to the likelihood ratio measure supported by Royall (2004 [chapter 5, this volume]) by the fact that ΔIC are penalized log-likelihood ratios ($\Delta IC = -2\ln(L_1/L_2) + \Delta f(k, n)$) (see also Forster and Sober, 2004 [chapter 6]) with the penalty depending on both the order of the models and (except for the AIC) the number of observations. The ΔIC between two models reduces to minus twice the log-likelihood ratio when the number of model parameters is the same.

One can get a feel for the interpretation of ΔIC values by translating them to equivalent likelihood ratios and P-values. For this exercise, I am comparing models with a one-parameter difference, and using a two-sided test, as might happen in the construction of a confidence interval. A ΔAIC of 2 is equivalent to a likelihood ratio of 7.4 and a realized P-value of .091, and a ΔAIC of 3 is equivalent to a likelihood ratio of 12.2 and a realized P-value of .051.

COMPONENTS OF ERROR IN THE APPROXIMATION OF TRUTH

Linhart and Zucchini (1986), Bozdogan (1987), Forster (2000), and others have described two kinds of discrepancies in approximating truth with models. These are (1) the discrepancy due to approximation and (2) the discrepancy due to estimation. Even if estimation were not a problem, there is a limit as to how close a particular approximating model can get to the truth. This limitation yields discrepancies of the first type. Discrepancies of the second type are a result of errors in parameter estimation that cause the discrepancy to be greater than the minimum possible for a given model. The total discrepancy between truth and an identified and estimated model is of course due to both kinds of error.

Much of the controversy surrounding information criteria–based model identification stems from the relative importance placed on different kinds of error. Using simulations, Hooten (1995) has investigated the behavior of a suite of information criteria under a variety of model identification performance criteria. He found that no criterion is superior for all purposes.

Information criteria can be divided into two broad classes. Minimum total discrepancy (MTD) criteria such as the AIC and the AIC_c explicitly try to minimize the total distance between truth and an identified and fitted model. Given this intent, it is not surprising that model identification using AIC has been shown to be asymptotically equivalent to model identification using jackknife cross-validation (Stone, 1977). On the other hand, the SIC, AIC_{hq}, CAIC, and other Order-Consistent (OC) criteria seek to estimate the "true model order." These criteria focus on minimizing the discrepancy due to approximation.

Practice, simulation, and theory all indicate that the AIC "overfits" (Bhansali and Downham, 1977; Bozdogan, 1987; Hooten, 1995). Of course, there are objections to the use of the OC criteria as well. These have been reviewed forcefully by Burnham and Anderson (1998). Some of these objections are: (1) truth has infinite order, and thus there is no "true order" to identify; (2) in the same vein, if truth has infinite order, then overfitting is impossible; (3) the OC IC do not estimate the K-L distance; and (4) the OC IC underfit. Burnham and Anderson (1998) have also objected to the use of the AIC rather than the AIC_c to represent the MTD criteria, as the AIC_c includes an important sample size correction.

These criticisms are not as damning to the use of OC criteria as they might at first appear, because they can be countered. (1) Truth may have an infinite order, but one's set of approximating models does not. There is in the set a model that could (ignoring estimation error) minimize the K-L distance between the approximating model and truth. (2) Overfitting is possible and potentially deleterious. I will discuss this at greater length shortly. (3) The OC IC do estimate the K-L distance, but there is a sense in which these estimates are not optimal. (4) Whether a selected model is considered underfit or overfit depends in part on the intended use of the model.

What is overfitting/underfitting? This is not a trivial question. It is not fruitful to define an underfitted model as one with fewer parameters than truth. If truth has infinite order, then all models are underfit. One could consider a model overfit if it contained more parameters than needed to minimize some criterion. This terminology may be useful if one is committed to a particular criterion but only confuses the issue when comparing criteria. In this paper, I do not describe a model as being overfit; instead, I enumer-

ate both the numbers of "overfitting errors" and "underfitting errors" that occur in the model. Overfitting error occurs if a variable not in truth is included in the approximation. Underfitting error occurs if a variable in truth
is left out of the approximation. Both errors may occur in a single model.

MODELING GOALS AND SELECTION CRITERIA

Cox (1990) distinguishes three broad categories for models based on the
purpose for which models are used. Lehmann (1990) creates a similar categorization. Cox classifies models as (1) empirical models, which are designed
primarily for predictions; (2) mechanistic models, whose purpose is to explain observation in terms of processes; or (3) descriptive models, which are
used to discover patterns that may suggest the application of other models,
either empirical or mechanistic.

All of these can appear in either structural or specific forms. For empirical models, minimizing prediction errors is the goal, and so either the AIC
or the AIC_c is the appropriate selection criterion. For mechanistic and descriptive models, OC criteria are probably more appropriate than MTD criteria. For example, in Zeng et al. (1998) our interest was in identifying structural models for population growth in natural time series, and the SIC was
our selection criterion. On the other hand, in Kramer et al. (2001), a predictive model of forest blowdown was desired, and we utilized the AIC in
model building.

However, in modern science models may often serve multiple purposes,
so the appropriate technique may not be clear. A researcher planning to use
a model for multiple purposes may choose to select the model using a criterion appropriate for the primary purpose or may reselect models for each
use with the appropriate criterion.

FEW- VS. MANY-CANDIDATE MODELS

Another central tension in the model identification approach is between use
of a large or small set of candidate models (Burnham and Anderson, 1998).
Several arguments militate for using a small set of models: (1) the proportion of model misidentifications increases as the number of potential models increases (Hooten, 1995); and (2) in large candidate sets, there may not
be a model with an intermediate number of parameters for which the AIC
is minimized (Tong, 1990).

There are two ways that IC selection may be confused by many models. The first may be termed "selection indifference." If there are a number of models in the candidate set that are similar to the best structural model (the model that would be closest to truth if parameters were known), then sampling variability may lead to the selection of a model other than the best model. Further, if there are only subtle differences in the fits of candidate models, the best model may not be selected because of the parameter penalty. The selection of any model other than the true best structural model is generally considered a mistake when quantifying model identification procedures. However, these mistakes are not very pernicious because the selected models are good approximations (Hooten, 1995).

The second type of confusion may result from what is termed "model selection bias" (Zucchini, 2000). If a researcher applies a large number of models to a given data set, then by chance one may fit well without necessarily describing the underlying generative processes, or having strong predictive power (low prediction error sum of squares, or "PRESS," values). This is a general problem in inference, and applies to hypothesis testing–based inference, as well as model identification–based inference. The prohibition against the post hoc testing of observed patterns in data and Bonferroni adjustments of critical values for multiple comparisons are frequentist devices that have been developed to circumvent this problem in the hypothesis testing framework. Burnham and Anderson (1998) also warn against the perils of "data dredging" within the model identification context.

Some arguments for the use of many models are equally compelling as arguments for the use of compact set of models. For model selection by information criteria to work well, one needs to have a "good model" in the candidate set. The use of large number of models increases the chances that at least one of the candidates will be close to truth. In a sense, using a large number of models allows the data to speak.

One's attitude about the size of the candidate model set is influenced by how one feels the information criterion approach extends classical frequentist statistics. As discussed earlier, ΔIC can be related to hypothesis tests, and thus the information criteria approach can be thought of as an extension of the hypothesis test to the case of multiple alternative models. On the other hand, information criteria have been considered the "likelihood of the model" (Burnham and Anderson, 1998). Akaike (1973) felt that the AIC was an extension of the maximum likelihood principle, while years earlier Fisher (1936) speculated that eventually the functional form of truth might be derivable from the data.

If a researcher conceives of IC model identification as an extension of hypothesis testing to more than a single null and alternative model, then a small set of candidate models will probably seem appropriate. On the other hand, researchers who view the information criterion value as the "likelihood of a model" will naturally gravitate to large sets of models. Maximum likelihood estimates for parameters are found not by comparing likelihoods for a few selected parameter values but by comparing likelihoods for all possible parameter values. This analogy indicates all potential models should be investigated.

Whatever one's view, large candidate model sets arise quite reasonably and frequently in ecological problems. The problem of very large candidate sets was recently brought to my attention when I was asked to aid in the analysis of a project designed to identify the influence of abiotic and biotic factors on abundance of stream-resident westslope cutthroat trout *Oncorhyncus clarki lewisi* in Montana streams (Shepard and Taper, 1998). The study estimated cutthroat and brook trout (*Salveinus fontinalis*) densities at seventy-one sites for three years. At all sites, physical covariates such as elevation, slope, aspect, latitude, longitude, stream size, and bed type were measured. Also measured were several anthropic factors such as the levels of grazing, mining, logging, and road use in the drainage. Dimension reduction with principal components analysis reduced thirty-eight variables to eight factors. Nevertheless, after including interactions and quadratic terms, both of which are reasonable given the ecology of these fish, we were left with seventy-two predictor variables, which could be combined into a staggeringly large number of credible candidate models.

Two alternative approaches seem reasonable in dealing with large numbers of alternative models. The first is to increase the threshold for discerning between models. Subhash Lele is currently investigating how the probability of misleading evidence for a given threshold depends on the number of alternative models. A difficulty with this approach is that the researcher is likely to end up with a very large number of alternative models that are statistically indistinguishable. A second tactic is to increase the penalty imposed for complexity as a function of the number of alternative models. Below, I demonstrate the utility of this approach with a method that determines the appropriate parameter penalty empirically from the data.

Once the complexity penalty depends on the number of models in the candidate set, then the tension between small and large candidate model sets resolves itself into a tradeoff. An increased parameter penalty clearly reduces the complexity of selected models. If a researcher considers only a few models, the penalty for complexity need not be so severe and models of greater complexity can be identified. Why should a researcher give up the ability to identify complexity? If one has considerable knowledge of the system, complex models can be selected from a small candidate set. On the other hand, a researcher uncertain of the fundamental workings of the system can avoid missing important effects by considering large candidate model sets. Thus, a researcher using only a few models retains the ability to detect model fine structure, while a researcher using many models reduces the probability of missing important effects.

A CLARIFYING EXAMPLE

My intention is to present an example complex enough to circumvent the criticisms directed at previous simulation tests of IC-based model selection, yet simple enough to be readily understandable. Some of the criticisms of simulation tests of IC model identification are (Burnham and Anderson, 1998) (1) that the models considered are too simple to represent the real complexities of nature; (2) that the true model is included in the candidate model set; and (3) that nature has tapering effects, but most simulation models do not.

With these criticisms in mind, I simulate a multiple regression problem not unlike the real data-analytic problems described above. The data for this exercise were constructed as

$$Y = \mu + \sum_{i=1}^{10} a_i \cdot X_i + \sum_{j=1}^{1000} b_j \cdot Z_j, \tag{3}$$

where Y is a response vector of 3000 observations. The parameters a_1 and b_1 are set to 0.7 and 0.5 respectively. Subsequent parameters are given by the recursions; $a_{i+1} = a_i \cdot 0.7$ and $b_{j+1} = b_j \cdot 0.5$. The X_i are 10 known covariate vectors. The Z_j are 1000 covariate vectors treated as unknown in the analysis. Also included in the analysis are 30 W vectors of spurious covariates that have nothing to do with the construction of the data. Thus, data construction was deterministic and not stochastic, truth has near infinite order, there are tapering effects, and full truth is not a candidate because most covariates are unknown.

The unknown portion was constructed deterministically for heuristic purposes. The sum of the 1000 independent Z vectors is indistinguishable from a normal random vector. The ignorance of the observer may reduce a complex reality to a simple set of models. Thus, the early simple simulation studies disdained by Burnham and Anderson may be interesting after all.

TABLE 15.2 Results with training set of 1500 observations.

Selected by:	Order	Underfit error	Overfit error	PR^2 (%)	Variable list
Optimum	12	0	0	77	$X1$–$X10$
AIC	21	0	9	76	$X1$–$X10$, $9W$
AICc	17	0	5	76	$X1$–$X10$, $5W$
SIC	11	1	0	76	$X1$–$X9$
CAIC	11	1	0	76	$X1$–$X9$

The data were analyzed using all subsets regression of order less than or equal to 20 covariates plus an intercept term. Model selection and parameter estimation were undertaken on data subsets of 1500, 150, and 50 observations. Information criteria were compared on the basis of underfit and overfit errors and on prediction accuracy of the selected model. Predictions were made for the values of all remaining observations using the covariates included in each model. The accuracy of predictions is reported as the prediction R^2 (PR^2 = variance of predicted values divided by total variance of the validation set). The performance of the AIC, AIC$_c$, SIC, CAIC, and SIC(x) were compared.

SIC(x) is an ad hoc criterion created for this exercise, with adjustable parameter penalty falling into Bozdogan's class of OC criteria (Bozdogan, 1987). I define the SIC(x) as:

$$\text{SIC}(x) = -2\ln(L) + (\ln(n) + x)k. \tag{4}$$

SIC(0) is identical to SIC and SIC(1) is identical to CAIC. The SIC(x) was constructed to add flexibility in the complexity penalty to the commonly used SIC and CAIC. In retrospect, the IC$_{hq}$ of Hannan and Quinn (1979) could have been used as effectively because one can always choose a c that equates the IC$_{hq}$ with the SIC(x).

This example has vast data-dredging potential, with 1.141×10^{14} candidate models being considered. Unfortunately, as indicated above, one can't dismiss candidate sets this large as unrealistic.

The results are summarized in tables 15.2 through 15.4. With large amounts of data (table 15.2), all criteria investigated identified models with essentially identical prediction accuracy. Each of the minimum total discrepancy criteria included all true covariates and therefore made no underfit errors. However, use of the MTD criteria did lead to a number of overfitting

TABLE 15.3 Results with training set of 150 observations.

Selected by:	Order	Underfit errors	Overfit errors	PR^2 (%)	Variable list
Optimum	12	0	0	77	$X1–X10$
AIC	>22	2	12	67	$X1–X7, X10, 12W$
AICc	16	2	6	74	$X1–X7, X10, 6W$
SIC	10	3	1	74	$X1–X7, 1W$
CAIC	9	3	0	74	$X1–X7$
SIC(2)	9	3	0	74	$X1–X7$

TABLE 15.4 Results with training set of 50 observations.

Selected by:	Order	Underfit error	Overfit error	PR^2 (%)	Variable list
Optimum	12	0	0	77	$X1–X10$
AIC	>22	2	12	17	$X1–X8, 12W$
AICc	>22	2	12	17	$X1–X8, 12W$

Intervening rows have been omitted.

SIC(3)	>22	2	12	17	$X1–X8, 12W$
SIC(4)	5	7	0	67	$X1–X3$

errors with the AIC making 9 overfit errors and the AIC$_c$ making 5 overfit errors. In contrast, the OC SIC and CAIC both included 9 true covariates and no spurious covariates in their selected model, leading to one underfit error and no overfit errors.

When the training set is reduced to 150 observations (table 15.3), all selected models have a reduced accuracy. The models selected by the AIC have somewhat lower prediction accuracy than do models selected by OC criteria. However, the AIC$_c$, with its small sample size correction, selects a model with a prediction accuracy just as good as the models selected by the OC criterion. Both the AIC and the AIC$_c$ make large numbers of overfit errors while the OC criteria make few, if any. The AIC selects a model of the maximum order allowed.

The most striking results can be seen in the small-sample case (table 15.4). Here, both the MTD and the standard OC criteria lead to massively overfit models with very low predictive ability until the adjustable parameter penalty is substantially raised. Both the AIC and the AIC$_c$ selected models with the maximum order allowed. With the SIC(4) selected model there are no overfitting errors, and predictive ability is only modestly lower than in the medium and large data size cases despite a large number of underfitting errors.

There may be a substantial cost in the use of small model sets. If an important covariate is missed there can be a dramatic drop in predictive as well as explanatory power. This can have a larger impact than the effect of underfitting errors forced by large suites of candidate models. In this example if covariate X_1 is eliminated from consideration, no identification technique can identify a model with predictive power greater than 35%.

INFORMATION CRITERION VALUE CURVES

Determining parameter penalty appropriate for use with large candidate model sets is not a fully solved problem. Nonetheless, it seems that information criterion value curves such as figure 15.2 can be used to diagnose the degree of parameter penalty needed to locate the real information in the data. Operationally, one can increase the parameter penalty until one finds a well-defined minimum. Furthermore, all of the curves, whether or not they have a minimum, seem to carry information about the appropriate model order in the shape of the curves. All curves have a noticeable change in slope at about the same order. This process is similar to determining the

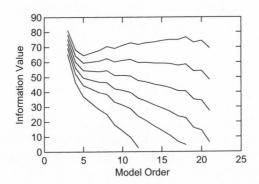

FIGURE 15.2 Graphically determining parameter penalty. IC curves for small sample size ($n = 50$). The parameter penalty increases from the lowest curve to the highest. The curves represent, from bottom to top, the AIC, SIC, CAIC, SIC(2), SIC(3), and SIC(4) information criteria.

number of factors to retain in a principal components analysis through the inspection of screeplots (Jolliffe, 1986).

CONCLUSIONS

In this chapter, I support the definition given by Subhash Lele (2004 [chapter 7, this volume]) of evidence as a data-based relative distance measure of models to truth. Because there are a many categories of models and many uses for models, it seems reasonable to use a variety of distance measures for choosing among models in different contexts. The minimum total discrepancy criteria and the order consistent criteria are designed to answer subtly different questions, and both have domains where their use is appropriate.

Model selection bias has restricted the effective use of large candidate model sets. I have proposed an ad hoc procedure that adjusts the parameter penalty in a fashion designed to determine from the data what level of model complexity is supportable by the data given the number of alternatives in the candidate model set. This device gives researchers another tool and another tradeoff about which investigators can make reasoned decisions. The few- versus many-models choice seems less stark now. By adjusting the parameter penalty, the researcher can reasonably use large candidate model sets and avoid missing important effects, but with a cost in that the degree of model detail identifiable has been reduced.

15.1	Commentary

Isabella Verdinelli and Larry Wasserman

Professor Taper raises the interesting question of how to choose among many models. He also suggests the possibility of using the data to choose the form of the penalty function in penalty-based methods. These are indeed important and timely problems. We would like to draw attention to some recent results in the statistical literature. We give special attention to three developments: adaptive penalties based on metric entropy (Barron, Birgé, and Massart, 1999), adaptive minimax estimation of sparse vectors (Abramovich et al., 2000), and Bayesian methods. We conclude with an example in which we examine Professor Taper's suggestion for choosing a data-based penalty.

ADAPTIVE PENALTIES

The question that Professor Taper raises is how we choose a model when there are many candidate models. An extreme version of this problem occurs when there are infinitely many models. Many statistical methods are versions of this problem. For example, consider fitting a polynomial of order p to a data set. Now p can take any value in $\{1, 2, \ldots \}$, so in choosing the order of the polynomial we are really choosing a model from an infinite list of potential models.

Suppose we have i.i.d. data $Y_1, Y_2, \ldots, Y_n$ from an unknown probability density function $f_o(y)$. Consider an infinite list of models $M_1, M_2, \ldots$, where each model consists of a set of density functions indexed by a parameter:

$$M_j = \{f_{\theta_j}; \theta_j \in \Theta_j\}.$$

Let $\hat{\theta}_j$ be the maximum likelihood estimator of θ_j in model M_j. If we choose model M_j, then we can estimate f_o with $\hat{f}_j \equiv f_{\hat{\theta}_j}$. As Professor Taper notes, one way to select a model is to use a penalized likelihood (or penalized least squares in the regression case). Let $L(\theta_j) = \prod_{i=1}^{n} f_{\theta_j}(Y_i)$ be the likelihood function for the jth model and let $\hat{\ell}_j = \log L(\hat{\theta}_j)$. Let $\hat{j}$ maximize

$$\gamma_j = \hat{\ell}_j - r_j, \tag{1}$$

where r_j is the penalty term. (If there is no maximizer, it suffices to choose a model that is sufficiently to the supremum of γ_j over j. For simplicity, we assume that the supremum does occur at some value $\hat{j}$.) Let $\hat{f} \equiv \hat{f}_{\hat{j}}$ be the corresponding estimate of f_o. A reasonable goal is to choose the penalties to make $\hat{f}$ as close to f_o as possible. Barron, Birgé, and Massart (1999) have proved the following result. For an "appropriate choice" of r_j, there is a constant $C > 0$, such that

$$E(d^2(f_0, \hat{f})) \le C \inf_j \left\{ \inf_{g \in M_j} D(f_0, g) + \frac{r_j}{n} \right\}, \tag{2}$$

where $d(f, g) = \{ \int (\sqrt{f(y)} - \sqrt{g(y)})^2 dy \}^{1/2}$ is the Hellinger distance between f and g and $D(f, g) = \int f(y) \log f(y)/g(y) dy$ is the Kullback-Leibler divergence. The result requires some explanation. The "appropriate choice" of r_j depends on the "metric entropy" of the model M_j that is a measure of the complexity of the model. Essentially, it is the dimension of the model. The $E(\cdot)$ on the left-hand side of (2) means expectation (or average), so the result bounds the average distance of the estimator from the truth. The infimum on the right-hand side of (2) occurs at the j that best balances the approximation error of the model (the first term) and the penalty (the second term). This is like the usual bias and variance tradeoff. This is only an upper bound, but it can be shown that, in many cases, this bound is close to the optimal risk; it cannot be improved. This means that penalized likelihood, with appropriate penalties, is an optimal approach. What do these penalties r_j look like? This depends on the models under consideration, and they have been worked out in special cases by Barron, Birgé, and Massart (1999) and others. A full discussion involves some technicalities that we cannot go into here. But generally, the penalties tend to be of one of two forms. In one case, the penalty is of the form $r_j = cd_j$, where d_j is the dimension of the parameter space Θ_j and c is a carefully chosen constant. This is similar to AIC. If there are many models with the same dimension (think of model selection in regression), then r_j will typically also involve a log n term and will be more similar to SIC. Hence, AIC and SIC are unified from this perspective. These theoretical results are, in some cases, not yet practical, but they do give important qualitative insight. In particular, we see that the choice of penalty depends on the richness of the class of models.

SPARSE MODELS

Some of the most exciting breakthroughs in model selection have appeared in a series of papers by Iain Johnstone and David Donoho and their colleagues. Let us briefly mention here some results from Abramovich, Benjamini, Donoho, and Johnstone (2000), from now on referred to as ABDJ. They study the following stylized model selection problem. Let $Y_i \sim N(\theta_i, 1)$ for $i = 1, \ldots, n$ where n is large. Let $\theta = (\theta_1, \ldots, \theta_n)$ be the vector of parameters. We will consider submodels where some of the θ_is are set to zero. Note that the number of parameters is the same as the number of data points. Thus, as the sample size increases, so does the number of parameters and the number of models. This is another way of formally capturing the "many models" problem. ABDJ were inspired to study this problem because it has a deep connection with nonparametric function estimation.

Because there are so many parameters, we have no hope of estimating θ

well unless the vector is sparse. By sparse, we mean that many of the θ_is are small. This is like the regression problem in Professor Taper's chapter, where we want to estimate a regression model with many potential regression coefficients. We cannot do well if all the regression coefficients are important. We assume instead that some might be big, but many are small. The problem is to find the big (important) coefficients.

As ABDJ note, there are many ways of measuring sparseness. The simplest is to assume that $\|\theta\|_0$ is small where $\|\theta\|_0$ is the number of elements of the vector θ that are nonzero. Another way to measure the sparseness of θ is with the ℓ_p norm defined by

$$\|\theta\|_p = \left\{ \sum_{i=1}^{n} |\theta_i|^p \right\}^{1/p}.$$

When p is a small positive number, this norm provides a measure of sparseness: if $\|\theta\|_p$ is small, then θ must be sparse.

Let $\hat{\theta} = (\hat{\theta}_1, \ldots, \hat{\theta}_n)$ be an estimate of θ. Because we expect θ to be sparse, it makes sense to set $\hat{\theta}_i = 0$ for some i's. But we must decide which ones to set to 0. This, of course, is just model selection. A model M in this context is a subset of $\{1, \ldots, n\}$, and selecting a particular model M will be interpreted as using an estimator $\hat{\theta}$ whose elements are defined by

$$\hat{\theta}_i = \begin{cases} Y_i & if\, i \in M \\ 0 & if\, i \notin M \end{cases}. \tag{3}$$

The question is how to select the model M. We could try any of the various penalty methods mentioned in Taper. ABDJ show, instead, that a method based on false discovery rates (FDR) due to Benjamini and Hochberg (1995)—originally designed for multiple testing problems—can be adapted for this model selection problem. The procedure works as follows. Consider testing the hypothesis H_{0_i} that $\theta_i = 0$. The P-value for that test is $P_i = P_r(|Z| > |Y_i|) = 2(1 - \Phi(|Y_i|))$ where Z has a standard normal distribution and Φ is the standard normal cumulative distribution function. Let $P_{(1)} \leq P_{(2)} \leq \cdots \leq P_{(n)}$ be the ordered P-values and let j be the largest integer such that $P_{(j)} < \alpha j/n$. Take $M = \{i; P_{(i)} \leq P_{(j)}\}$. This defines our model selection procedure. As noted above, this may also be thought of as a method for producing an estimate of θ via (3), which we denote by $\hat{\theta}(\alpha)$. We can measure how well $\hat{\theta}(\alpha)$ estimates θ by its minimax risk, defined by

$$\sup_{\theta \in \Theta_n} E_\theta \|\theta - \hat{\theta}(\alpha)\|_r^r = \sup_{\theta \in \Theta_n} E_\theta \sum_{i=1}^{n} |\theta_i - \hat{\theta}_i(\alpha)|^r,$$

where Θ_n is the parameter space for θ and r defines the type of loss function. For example, $r = 2$ gives the usual squared error loss.

The parameter α has the following interpretation: every $i \in M$ may be viewed as a rejection of the null hypothesis that $\theta_i = 0$; the expected fraction of rejections that are false (i.e., for which $\theta_i = 0$) is bounded above by α. The fraction of false rejections is called the false discovery rate (FDR).

ABDJ prove the following theorem that unifies hypothesis testing (which FDR was originally designed for) and estimation. Let $\Theta_n = \{\theta; n^{-1}\|\theta\|_p^p \leq a_n^p\}$, where $\log^5 n/n \leq a_n \leq n^{-\delta}$ for $\delta > 0$. Let $\hat{\theta}(\alpha_n)$ be the estimate based on the above procedure using level α_n where $\alpha_n \to 0$. Then, for any $0 \leq p < r \leq 2$, they proved that

$$\sup_{\theta \in \Theta_n} E_\theta \|\theta - \hat{\theta}(\alpha_n)\|_r^r \sim \inf_{\hat{\theta}} \sup_{\theta \in \Theta_n} E_\theta \|\theta - \hat{\theta}\|_r^r. \tag{4}$$

Here, $b_n \sim c_n$ means that $b_n/c_n \to 1$ as $n \to \infty$ and the infimum on the right-hand side is over all possible estimators.

The result has the following interpretation. The right-hand side is the minimax risk that represents the best risk (expected loss). Any estimator must have a risk at least this big. The theorem says that the FDR procedure attains (asymptotically) this optimal risk. Moreover, it does so simultaneously for various values of p (which measures the sparseness) and r (which defines the loss function). This is a remarkable adaptivity result, and it suggests that the FDR procedure might be ideal for large-model selection procedures. ABDJ conjecture that the FDR procedure is similar to the penalty-based methods using a penalty of the form $k \log(k/n)$, where k is the number of parameters in the model. This suggests (though it has not yet been proved) that such a penalty might be optimal in a wide variety of model selection problems.

BAYESIAN METHODS

Bayesian methods for model selection have also received much attention lately. Reviews include Kass and Raftery (1995) and Wasserman (2000). In the Bayesian framework, one treats the model index j as a parameter. One then places a prior on j and for each j, one places a prior on the parameter θ_j. Thus, the prior is of the form $\pi(j)\pi(\theta_j|j)$. It is then possible (sometimes with much computational burden) to find the model with highest posterior probability. One can also produce predictions that are obtained by averaging over the possible models with respect to their posterior probabilities.

Bayesian methods are attractive for their conceptual simplicity. However, it is important to understand the frequentist properties of these Bayesian methods. George and Foster (2000) have shown that a particular Bayesian model selection method has very good frequentist properties. ABDJ note that the George-Foster method seems to be related to the FDR method. Barron, Schervish, and Wasserman (1999), Ghosal, Ghosh, and van der Vaart (2000), Huang (2000), Shen and Wasserman (2000), and Zhao (1993, 1999) have shown that, if one places priors carefully over an infinite list of models, then the resulting posterior distribution has optimal frequentist performance.

The advantage of these Bayesian methods is that, in principle, the methods are very general. The disadvantage is that to ensure that the posterior has good frequentist properties—which strikes us as essential in complex problems—requires very carefully chosen priors. Indeed, the sharpest results so far appear to be those in Huang (2000) and the choice of priors used there was extremely delicate.

CHOOSING PENALTIES: AN EXAMPLE

Professor Taper suggests that one should plot the model selection scores for a variety of choices of the penalty. This is wise advice, especially given that most of the available theoretical results are large-sample results leaving the data analyst, who has only finitely many data points, with some reasonable doubt about optimality theorems. Professor Taper hints at choosing the penalty such that the resulting model score γ_j definitively chooses a best model. This suggestion intrigued us, so we conducted a very small simulation study.

We generated $Y_i \sim N(\theta_i, 1)$, $i = 1, 2, \ldots, n$, with $n = 1,000$, and we used the estimator $\hat{\theta}$ proposed in (3). We chose the model M by penalized likelihood, and we then plotted the model selection scores for the family of penalty functions

$$\gamma_k = -2\ln(L) + 2k\left(\frac{\ln(n)}{2}\right)^{\alpha}. \tag{5}$$

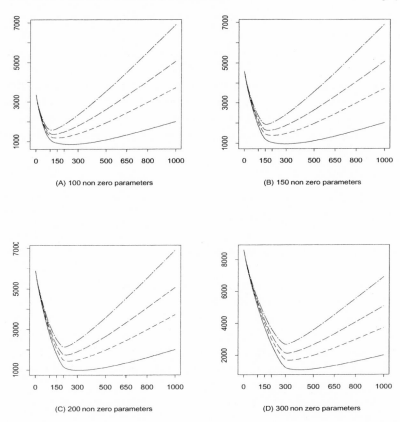

FIGURE 15.1.1 $\alpha = 1$ (or AIC criterion), 0.75, 0.5, 0 (or SIC criterion) from top to bottom in each plot.

Recalling that $\alpha = 0$ in (5) produces the SIC(0) score and $\alpha = 1$ gives that AIC score, we chose four values of α: respectively 0, 0.5, 0.75, and 1. This is a slight variation of the criteria SIC(x) suggested by Professor Taper, which would correspond to changing (5) to the form

$$\gamma_k(x) = -2\ln(L) + 2k\left(\frac{\ln(n)}{2} + x\right)^\alpha.$$

Results from our simulation are shown in the four plots of Figure 15.1.1, which correspond to different levels of "sparseness" of the model's parameters. The sparseness of the model is governed by the number of parameters that are set to 0 in the simulation. We fixed k (= 100, 150, 200, 300) of the θ_is to 5 while the remaining $(1,000 - k)$ θ_is were kept equal to 0.

All the model selection scores we considered consistently show low values in the neighborhood of the correct model order. However, note that if we choose α by finding the curve that sharply delineates a model, it seems that one is always led to $\alpha = 1$, and hence one would use SIC. Of course, this is only an example and without further investigation, we cannot make any definitive statement. But the example suggests that α may have to be chosen by some other criterion.

15.2   Commentary

Hamparsum Bozdogan

INTRODUCTION, AND A NEW INFORMATION MEASURE OF COMPLEXITY (ICOMP) CRITERION

Mark Taper raises several important issues and concerns. I wish to isolate these issues and briefly introduce what I call *new generation information criteria,* which I believe address these important issues data-adaptively under correct and misspecified models.

Taper's important and repeated theme throughout this chapter can be summarized as follows:

1. Tradeoff can be incorporated in modifications of the parameter penalty term that are sensitive to the size of the candidate model set. Taper proposes a model identification procedure that empirically determines an appropriate parameter penalty considering both the data and the candidate model set.
2. "Model misspecification is a major, if not the dominant, source of error in the quantification of most scientific analysis (Chatfield, 1995)."
3. "Why should a researcher give up the ability to identify complexity?"

This research was partially supported by the Scholarly Research Grant Program (SRGP) Award at the University of Tennessee during January 2000–August 2001. I extend my gratitude to Professor Mark Taper for inviting and encouraging me to make a commentary to his chapter and make a contribution to this important manuscript.

4. "Determining parameter penalty appropriate for use with large candidate model sets is not fully solved problem."
5. "By adjusting the parameter penalty, the researcher can reasonably use large candidate model sets and avoid missing important effects, but with a cost in that the degree of model detail identifiable has been reduced."

These important issues bring us to the point of "overall" model complexity. In general statistical modeling and model evaluation problems, the concept of model complexity plays an important role. At the philosophical level, complexity involves notions such as connectivity patterns and the interactions of model components. Without a measure of "overall" model complexity, prediction of model behavior and assessing model quality is difficult. This requires detailed statistical analysis and computation to choose the best-fitting model among a portfolio of competing models for a given finite sample. As I have argued with Akaike on many occasions during my visit to the Institute of Statistical Mathematics (ISM) in Tokyo, Japan, as a Visiting Research Fellow during 1988, in AIC and AIC-type criteria, counting and penalizing the number of parameters in a model is necessary but by no means sufficient. Model complexity in statistics depends intrinsically on many factors other than the model dimension, such as the several forms of parameter redundancy, parameter stability, random error structure of the model, and the linearity and nonlinearity of the parameters of the model, to mention a few (see, e.g., Bozdogan, 2000, 64).

The purpose of this short note and commentary is to affirm that we need to pay attention to the important issues listed above, and to develop and present information-theoretic ideas of a measure of overall model complexity in statistical estimation to help provide new approaches relevant to statistical inference.

As is well known, based on Akaike's (1973) original AIC, many model selection procedures that take the form of a penalized likelihood (a negative log likelihood plus a penalty term) have been proposed (Sclove, 1987). For example, for AIC, this form is given by

$$AIC(k) = -2 \log L(\hat{\boldsymbol{\theta}}_k) + 2m(k), \tag{1}$$

where $\hat{\boldsymbol{\theta}}_k$ is the maximum likelihood estimator (MLE) of the parameter vector $\hat{\boldsymbol{\theta}}_k$ for model M_k, and $L(\hat{\boldsymbol{\theta}}_k)$ is the maximized likelihood function, and $m(k)$ is the number of free parameters in M_k. According to Akaike (1987,

319), the accuracy of parameter estimates is measured by the expected log likelihood of the fitted model, where the expectation is taken with respect to the true data generating process. AIC is asymptotically unbiased estimator of minus twice this expected log likelihood.

Motivated by considerations similar to those in AIC, Rissanen's (1976) *final estimation criterion* (FEC) as well as the analytical extensions of AIC in Bozdogan (1987), Bozdogan (1988, 1990, 1994, 2000) introduced a new generation model selection criterion called ICOMP. "I" stands for "information," and "COMP" stands for "complexity." ICOMP criterion is based on the complexity of an element or set of random vectors via a generalization of van Emden's (1971) entropic covariance complexity index. Using an information-theoretic interpretation, ICOMP views complexity as the discrimination information of the joint distribution of the parameter estimates against the product of their marginal distributions. Discrimination information is equal to 0 if the distributions are identical and is positive otherwise (van Emden, 1971, 25). The most general form of ICOMP is based on the estimated *inverse Fisher information matrix* (IFIM) of the model. For a general multivariate linear or nonlinear model defined by statistical model = signal + noise, the ICOMP criterion for model M_k is defined by

$$\text{ICOMP(IFIM)} = -2 \log L(\hat{\boldsymbol{\theta}}_k) + 2C_1(\hat{F}^{-1}(\hat{\boldsymbol{\theta}}_k)), \tag{2}$$

where $\hat{F}^{-1}(\hat{\boldsymbol{\theta}}_k)$ is the estimated IFIM under model M_k, and

$$C_1(\hat{F}^{-1}) = \frac{s}{2} \log\left[\frac{tr(\hat{F}^{-1})}{s}\right] - \frac{1}{2} \log |\hat{F}^{-1}| \tag{3}$$

is the maximal covariance complexity of $\hat{F}^{-1}(\hat{\boldsymbol{\theta}}_k)$, and $s = \dim(\hat{F}^{-1}) = \text{rank}(\hat{F}^{-1})$. The term $C_1(\hat{F}^{-1})$ in (3) is an upper bound to van Emden's (1971) covariance complexity index, measuring both inequality among the variances and the contribution of the covariances in $\hat{F}^{-1}(\hat{\boldsymbol{\theta}}_k)$. Large values of C_1 indicate a high interaction between the parameter estimates. C_1 takes into account the fact that as we increase the number of free parameters in a model, the accuracy of the parameter estimates decreases. As preferred according to the principle of parsimony, ICOMP chooses simpler models that provide more accurate and efficient parameter estimates over more complex, overspecified models.

The first component of ICOMP in (2) measures the lack of fit of the model, and the second component measures the complexity of the estimated IFIM, which gives a scalar measure of the celebrated Cramér-Rao lower bound matrix, which takes into account the accuracy of the estimated parameters and implicitly adjusts for the number of free parameters included in the model (see, e.g., Cramér, 1946, and Rao, 1945, 1947, 1948). According to ICOMP, the best model among a set of competitors is the one that minimizes ICOMP.

Under well-known regularity conditions, the MLE $\hat{\theta}$ is asymptotically normally distributed with covariance matrix $\hat{F}^{-1}$. In this case, $C_1(\hat{F}^{-1})$ measures the Kullback-Leibler (KL) (1951) information divergence against independence of the parameter estimates (Kullback, 1968; Harris, 1978; Theil and Fiebig, 1984) and ICOMP can be viewed as an approximation to the sum of two KL discrepancies. Similar to AIC, the first measures the KL divergence between the true data generating process and the fitted model. Beyond AIC, the second measures the KL divergence against independence of the parameter estimates.

With ICOMP, complexity is viewed not as the number of parameters in the model, but as the *degree of interdependence* (i.e., the correlational structure among the parameter estimates). By defining complexity in this way, ICOMP provides a more judicious penalty term than AIC, Rissanen's (1978, 1986) minimum description length (MDL), Schwarz's (1978) SBC (or), and Bozdogan's (1987) consistent AIC (CAIC). Lack of parsimony and profusion of complexity are automatically adjusted by $C_1(\hat{F}^{-1})$ across the competing models in a portfolio. This measures the quantum of information contained in a particular likelihood function (see Ghosh, 1988, 34). Therefore, ICOMP has the attractive feature of implicitly adjusting for the number of parameters and for the sample size, and thus controlling the risks of both insufficiently paramaterized and overparameterized models.

Similar to AIC, ICOMP can be made invariant with respect to scaling and orthonormal transformations by using the correlational form of IFIM given by

$$\text{ICOMP(IFIM)}_R = -2 \log L(\hat{\theta}_k) + 2C_1(\hat{F}_R^{-1}(\hat{\theta}_k)), \tag{4}$$

where

$$\hat{F}_R^{-1}(\hat{\theta}) = D_{\hat{F}^{-1}}^{-1/2} \hat{F}^{-1} D_{\hat{F}^{-1}}^{-1/2}, \tag{5}$$

and $D = \text{Diag(IFIM)}$.

In this way, ICOMP becomes invariant to one-to-one transformations of the parameter estimates and more. In the literature, several authors such as McQuarie and Tsai (1998, 367) and Burnham and Anderson (1998, 69), without reviewing the impact and the applications of ICOMP to many complex modeling problems, have erroneously interpreted the contribution of this novel approach over AIC and AIC-type criteria.

ICOMP FOR MISSPECIFIED MODELS

"Model misspecification is a major, if not the dominant, source of error in the quantification of most scientific analysis (Chatfield, 1995)." In this section, we generalize ICOMP above to the case of a misspecified model and develop ICOMP under misspecification. To our knowledge, in the literature, there do not appear to be any existing criteria that handle misspecification, except the Takeuchi's (1976) information criterion (TIC), or AIC_T. Our approach will provide new tools for researchers to determine for themselves whether or not the probability model is misspecified as we fit and evaluate them. This is very important in practice, but is often ignored. Furthermore, we will be able to do model evaluation in the presence of skewness and/or kurtosis, especially when we carry out subset selection of best predictors, detection of outliers, and so on in linear or nonlinear regression settings. Our approach will eliminate the serious errors made in inference when the usual standard techniques are used when there is skewness and/or kurtosis present in the data and normality is always assumed.

Following White (1982), we give a simple test of information matrix equivalence to check the misspecification of a model. We construct a consistent estimator of the covariance matrix $Cov(\theta_k^*)$ given by

$$\widehat{Cov}(\hat{\theta})_{Misspec} = \hat{F}^{-1}\hat{R}\hat{F}^{-1}. \tag{6}$$

When a model is misspecified, then the two forms of the Fisher information matrices, that is, the inner- $(\hat{F}^{-1})$ and outer-product $(\hat{R})$ forms, are not the same. Thus, White's (1982) covariance matrix in (6) would impose greater complexity than the inverse Fisher information matrix (IFIM).

For example, in a regression model, there are a number of ways a researcher can misspecify the model. Some of these are discussed in Godfrey (1988, 100). The most common misspecification occurs:

1. when the functional form of the model is not correctly specified
2. when there are "near-linear" dependencies among the predictor vari-

ables, which is known as the problem of multicollinearity in regression models

3. when there are high skewness and kurtosis in the variables, which causes nonnormality of the random error or disturbances

4. when there is autocorrelation and heteroscedasticity

Therefore, under misspecification, ICOMP is defined by

$$\text{ICOMP(IFIM)}_{Misspec} = -2 \log L(\hat{\boldsymbol{\theta}}) + 2C_1(\widehat{\text{Cov}}(\hat{\boldsymbol{\theta}})_{Misspec}). \tag{7}$$

When we assume that the true distribution does not belong to the specified parametric family of pdfs, that is, if the parameter vector $\boldsymbol{\theta}$ of the distribution is unknown and is estimated by the maximum likelihood method, then it is no longer true that the average of the maximized log likelihood converges to the expected value of the parameterized log likelihood. That is,

$$\frac{1}{n} \log L(\mathbf{x}|\hat{\boldsymbol{\theta}}) = \frac{1}{n} \sum_{i=1}^{n} \log f(x_i|\hat{\boldsymbol{\theta}}) \nrightarrow E_x[\log f(\mathbf{X}|\hat{\boldsymbol{\theta}})]. \tag{8}$$

In this case, the bias, b, between the average of the maximized log likelihood and the expected maximized log likelihood is given by

$$b = E_G\left[\frac{1}{n}\sum_{i=1}^{n} \log f(x_i|\hat{\boldsymbol{\theta}}) - \int_R \log f(x|\hat{\boldsymbol{\theta}})dG(x)\right]$$

$$= \frac{1}{n} tr(F^{-1}R) + O(n^{-1}). \tag{9}$$

We note that $tr(\hat{F}^{-1}\hat{R}^{-1})$ is the well-known Lagrange-multiplier test statistic. See, for example, Takeuchi (1976), Hosking (1980), and Shibata (1989). Thus, we have the generalized Akaike's (1973) information criterion (GAIC) defined by

$$\text{GAIC} = -2 \log L(\hat{\boldsymbol{\theta}}) + 2tr(\hat{F}^{-1}\hat{R}^{-1}). \tag{10}$$

GAIC is the same as Takeuchi's (1976) information criterion, discussed above. For more details, see Konishi and Kitagawa (1996) and Konishi (1999).

The basic idea underlying the information matrix equivalence test (IMET) is that it relies on the equality between the two forms of the Fisher information matrices. These are useful to check the misspecification of a model. So if $\hat{F} = \hat{R}$, then the bias reduces to

$$b = \frac{1}{n}k + O(n^{-2}) \tag{11}$$

which gives AIC as a special case. In this respect, when the true model that is entertained is not in the model set considered, which is often the case in practice, AIC will confuse identification of the best-fitting model and it will not penalize the presence of skewness and kurtosis.

ICOMP AS A BAYESIAN CRITERION IN MAXIMIZING A POSTERIOR EXPECTED UTILITY

From Bozdogan and Haughton (1998), we define ICOMP as a Bayesian criterion close to maximizing a posterior expected utility given by

$$\text{ICOMP(IFIM)}_B = -2 \log L(\hat{\boldsymbol{\theta}}_M) + k + 2C_1(\hat{F}^{-1}(\hat{\boldsymbol{\theta}}_M)). \tag{12}$$

OTHER NEW GENERATION MODEL SELECTION CRITERIA

More recently, Bozdogan and Ueno (2000) extended Bozdogan's (1987) CAICF both in Bayesian and in frequentist frameworks. Their approach is based on a different estimation of the expectation of the quadratic variation in AIC. The result obtained includes both Akaike's approach and that of Schwarz (1978) and Rissanen (1978) as special cases. Further, it includes the additional term $tr(\hat{F}^{-1}\hat{R}^{-1})$. It generalizes to the case of model selection under misspecification. This new criterion is given by

$$\begin{aligned} \text{CAICF}_{Gen} &\doteq -2 \log L(\hat{\boldsymbol{\theta}}) + k \log(n) + \log|\hat{F}(\hat{\boldsymbol{\theta}})| \\ &\quad + 2\, tr\{\hat{F}^{-1}(\hat{\boldsymbol{\theta}})\hat{R}(\hat{\boldsymbol{\theta}})\}. \end{aligned} \tag{13}$$

Further, we give an approximation to (13) that corrects the bias for small as well as large sample sizes:

$$\text{CAICF}_c \doteq -2 \log L(\hat{\boldsymbol{\theta}}) + k \log(n) + \log|\hat{F}(\hat{\boldsymbol{\theta}})| + 2\left(\frac{nk}{n-k-2}\right). \tag{14}$$

Indeed as $n \to \infty$, the term $2n/(n - k - 2)$ goes to 2, and CAICF$_c$ reduces to CAICF; see Bozdogan (1987).

The Bayesian model selection (BMS) criterion of Bozdogan and Ueno (2000) is given by

$$\begin{aligned} \text{BMS} = &-2 \log L(\hat{\boldsymbol{\theta}}) - 2 \log \pi(\hat{\boldsymbol{\theta}}) + k \log(n) - k \log(2\pi) \\ &+ \log |\hat{F}(\hat{\boldsymbol{\theta}})| + tr\{\hat{F}^{-1}(\hat{\boldsymbol{\theta}})\hat{R}(\hat{\boldsymbol{\theta}})\}. \end{aligned} \tag{15}$$

Dropping the constant term $-k \log(2\pi)$ in (15), because it will not affect the comparisons of the models, BMS further simplifies to

$$\begin{aligned} \text{BMS} = &-2 \log L(\hat{\boldsymbol{\theta}}) - 2 \log \pi(\hat{\boldsymbol{\theta}}) + k \log(n) + \log |\hat{F}(\hat{\boldsymbol{\theta}})| \\ &+ tr\{\hat{F}^{-1}(\hat{\boldsymbol{\theta}})\hat{R}(\hat{\boldsymbol{\theta}})\}. \end{aligned} \tag{16}$$

For a constant prior $\pi(\hat{\boldsymbol{\theta}})$, we have

$$\text{BMS} = -2 \log L(\hat{\boldsymbol{\theta}}) + k \log(n) + \log |\hat{F}(\hat{\boldsymbol{\theta}})| + tr\{F^{-1}(\hat{\boldsymbol{\theta}})\hat{R}(\hat{\boldsymbol{\theta}})\}. \tag{17}$$

The minimum message length (MML) criterion of Wallace and Freeman (1987) is defined by

$$\text{MML87} = -2 \log L(\hat{\boldsymbol{\theta}}) + k + \log |\hat{F}(\hat{\boldsymbol{\theta}})| - 2 \log \pi(\hat{\boldsymbol{\theta}}) + k \log(\kappa_k). \tag{18}$$

In (18), $\pi(\hat{\boldsymbol{\theta}})$ is the estimated prior distribution and κ_k is the quantification constant (Baxter, 1996).

OBSERVATIONS

The increase of $|\hat{F}(\hat{\boldsymbol{\theta}})|$ with the complexity of the models is not so obvious, since it depends on the type of models that are being entertained and fitted. When the observed data are i.i.d., we can write

$$|\hat{F}(\hat{\boldsymbol{\theta}}_k)| = O(n^{k/2}). \tag{19}$$

If we take the first two terms of CAIFC$_{\text{Gen}}$ or CAIFC$_c$, we can deduce the naive MDL of Rissanen (1978) and the SBC of Schwarz (1978):

$$\text{MDL/SBC} = -2 \log L(\hat{\boldsymbol{\theta}}_k) + k \log(n). \tag{20}$$

A MONTE CARLO SIMULATION EXAMPLE

We carried out a subset selection of variables in multiple regression analysis based on the following simulation protocol. Let

$$x_1 = 10 + \varepsilon_1,$$
$$x_2 = 10 + 0.3\varepsilon_1 + \alpha\varepsilon_2, \text{ where } \alpha = \sqrt{1 - 0.3^2} = \sqrt{0.91} = 0.9539$$
$$x_3 = 10 + 0.3\varepsilon + 0.5604\alpha\varepsilon_2 + 0.8282\alpha\varepsilon_3,$$
$$x_4 = -8 + x_1 + 0.5x_2 + 0.3x_3 + 0.5\varepsilon_4,$$
$$x_5 = -5 + 0.5x_1 + x_2 + 0.5\varepsilon_5,$$

where ε_i are independent and identically distributed, according to $N(0, \sigma^2)$, for $i = 1, 2, \ldots, n$ observations, and also $\varepsilon_1, \varepsilon_2, \varepsilon_3, \varepsilon_4, \varepsilon_5, \approx N(0, \sigma^2 = 1)$. The response variable is generated from the true model:

$$y_{(nx1)} = [1, \mathbf{X}_{or}]_{(nx4)}\boldsymbol{\beta}_{(4x1)} + \boldsymbol{\varepsilon}_{(nx1)},$$

with

$$\mathbf{X}_{or} = [x_1, x_2, x_3],$$

where

$$\boldsymbol{\beta}_{4x1} = \begin{bmatrix} -8 \\ 1 \\ 0.5 \\ 0.3 \end{bmatrix}$$

$$\varepsilon_{(100\times1)} \sim N(0, \sigma^2 I), \sigma^2 = 1.$$

Note that the generation of y does not contain variables x_4 and x_5, but x_4 and x_5 are implicitly related to the other variables. The multicollinearity is controlled by α, and indeed the generated model is misspecified. As the sample size gets larger, the correlation matrix of the predictors depicts more of the relationship among the variables, and thus makes it difficult for the model selection criteria to pick the true model that generated the data. Under this setup, we carried out subset selection of variables using y, and $X = [x_1, x_2, x_3, x_4, x_5]$. So there are $2^5 = 32$ possible subset models including the constant term (constant model), and $2^5 - 1 = 31$ possible subset no-constant models. In this case, we expect to pick $X = [x_1, x_2, x_3]$ subset variable model as the best-fitting model over the saturated full model and other subset models, and study the relative performance of the information criteria. All the computations are carried out using a new cutting-edge open architecture and self-learning computational toolboxes using informational model selection procedures in MATLAB developed by this author. These modules are available from the author. We now summarize our results in carrying out the subset selection of variables to determine the best predictors in 100 replications of the simulation experiment for different sample sizes.

Frequency of choosing the best subset regression model in 100 replications of the Monte Carlo experiment for $n = 75$: ICOMP(IFIM) chooses the subset $\{0, 1, 2, 3, -, -\}$ 82 times; CAIC$_{Gen}$, 70 times; BMS, 82 times; and AIC, 69 times. This indicates the confusion of AIC in misspecified and highly collinear models. For AIC, overfitting and underfitting are not well balanced. On the other hand, ICOMP balances the overfitting and underfitting better than CAIC$_{Gen}$ and BMS in this experiment.

Frequency of choosing the best subset regression model in 100 replications of the Monte Carlo experiment for $n = 100$: ICOMP(IFIM) chooses the subset $\{0, 1, 2, 3, -, -\}$ 91 times; CAIC$_{Gen}$, 87 times; BMS, 92 times; and AIC, 73 times. This too indicates the confusion of AIC in misspecified and highly collinear models. In this case, AIC overfits the model more than any other criterion. Again, ICOMP balances the overfitting and underfitting better than CAIC$_{Gen}$ and BMS.

Frequency of choosing the best subset regression model in 100 replications of the Monte Carlo experiment for $n = 200$: ICOMP(IFIM) chooses the subset $\{0, 1, 2, 3, -, -\}$ 97 times; CAIC$_{Gen}$, 100 times; BMS, 99 times; and AIC, 74 times. Once again AIC is confused in misspecified and highly collinear models. On the other hand, overfitting and underfitting disappear for ICOMP, CAIC$_{Gen}$, and BMS as the sample size gets larger, which is the case (see, e.g., Bozdogan, 1987, on consistent criteria for more details).

We note that AIC's performance has not improved beyond 70% compared to ICOMP(IFIM), CAIC$_{Gen}$, and BMS in moderate to large samples despite the fact that the empirical power of AIC is about 84.27%. This is, again, due to the fact that counting and penalizing the number of parameters may be a necessary but not a sufficient condition to capture the entire model complexity. By measuring the degree of interaction between the parameter estimates via their measure of complexity of the covariance matrix, we are now able to determine not only which regression estimates are degraded by the presence of collinearity, but also which are not adversely affected. It is because of these problems that the tradeoff between the interaction of the parameter estimates of a model and the interaction of the residuals is very important in both linear and nonlinear models. Therefore, it is important to take parameter interdependencies and accuracies into account in the model evaluation criterion. ICOMP(IFIM), $CAIC_{Gen}$, and BMS do these for us data-adaptively. For more on this within the multiple regression context, we refer the reader to Bozdogan and Haughton (1998), where the asymptotic consistency properties of the class of ICOMP criteria are studied in detail, both when one of the models considered is the true model and when none of the models considered is the true model.

CONCLUSION

In this short note and commentary, we introduced several new generation model selection criteria that take into account misspecification of the model, data-adaptively modify the parameter penalty, determine the complexity of the model, and balance overfitting and underfitting risks more judiciously. ICOMP considers situations both with correlated and with uncorrelated model residuals by including dependence, and both linear and non-linear model parameters. As it is formulated, ICOMP is very general in the sense that it can be easily extended to other distributional assumptions on the model structures. It requires minimal assumptions and measures the strength of the structure of a model. It represents relations such as similarity, dissimilarity, and higher-order correlations within the model. If the variables show neither similarity nor dissimilarity, then the complexity becomes 0 and one should not use ICOMP. The difference between ICOMP, AIC, and SBC/MDL is that with ICOMP we have the advantage of working with both biased and unbiased estimates of the parameters and measure the complexities of their covariances to study the robustness properties of different methods of parameter estimates. AIC and AIC-type criteria are based on MLEs, which often are biased, and they do not take into account the concept of parameter redundancy, accuracy, and the parameter interdependencies in the model fitting and selection process. In the literature (see, e.g., Li, Lewandowsky, and DeBrunner, 1996), a measure of a model's total sensitivity to all of its parameters is often defined in terms of the trace of FIM, and, in some cases, it is defined by the determinant of IFIM, called a generalized variance. Using such measures alone as performance measures has serious disadvantages to which one should pay attention. In concluding, we note that our Monte Carlo simulation of subset selection of variables in multiple regression analysis under collinearity of variables clearly demonstrates the superiority of ICOMP-class criteria to AIC in model selection, prediction, and perturbation studies. We believe the set of potentially fruitful applications of information-theoretic model selection criteria are vast in scientific investigations and in empirical work to discover the nature of scientific evidence (see, e.g., Bearse and Bozdogan, 1998). We hope that future research will continue to explore these avenues.

Finally, both robust and misspecification versions or forms of ICOMP have been developed by this author, and the results of this will be reported elsewhere.

15.3   Rejoinder

Mark L. Taper

I thank Drs. Verdinelli, Wasserman, and Bozdogan for their interesting and useful commentaries on my chapter.

Drs. Verdinelli and Wasserman present an example showing the breakdown of the SIC(x) in the analysis of model sets with sparse effects of uniform magnitude. As mentioned in my chapter, the SIC(x) was explicitly designed for models with tapered effects, but it is very useful to have the range of the SIC(x)'s utility clarified.

They have also brought us current on related articles in the statistical literature published since my chapter was written in 1998. Another important recent article not mentioned either in the commentary or in my chapter is Bai, Rao, and Wu (1999).

Bai, Rao, and Wu propose a class of general information criteria (GIC) that consider the data and all compared models in the construction of the penalty term. The SIC(x) that I proposed would fall into their class, with an additional restriction on the size of x relative to the data size.

Bai, Rao, and Wu compare the efficacy of a member of their class and find it superior to information criteria not taking the model set into consideration. I am gratified by this work. I do not consider the SIC(x) a definitive but only an ad hoc solution, whose primary purpose is to demonstrate the problem. I welcome further development by mathematical statisticians.

Although the effectiveness of the GIC and the SIC(x) have not been compared yet, I can point out one computational difference. The GIC requires the evaluation of all regressions in the model set, whereas the SIC(x), because it can take advantage of the "leaps and bounds" algorithm, does not. This may allow the SIC$(x)_x$ to be employed in the analysis of larger model sets than the GIC.

I appreciate Dr. Bozdogan's recognition of the importance of the questions that I have raised. Further, I agree with him that a model's order (number of parameters) is only a crude index of its complexity. The information-based penalties that Bozdogan proposes should give better corrections for the parameter estimation component of the total discrepancy. Further, because sample size correction is implicit rather than explicit, ICOMP should be superior for the identification of hierarchical models and other cases where observations are not independent.

Unfortunately, the ICOMP does not consider the magnitude of the model set. However, I see no reason why a hybrid criterion might not be utilized. One could define an ICOMP(x) as:

$$-2 \log(L(\hat{\theta}_k)) + 2 \exp(x) C_1(\hat{F}^{-1}(\hat{\theta}_k)).$$

As in the SIC(x), the x in the above would be increased until a clear minimum was observed.

Dr. Bozdogan extends the ICOMP criterion in a fashion that acknowledges model misspecification. This is an exciting development, in keeping with the belief expressed in this chapter and in chapters 7 (Lele), 14 (Lindsay), and 16 (Taper and Lele), that models are always misspecified. I am gratified that my paper may have stimulated a portion of this work.

The bias correction term implicit in the ICOMP and explicit in the $\text{CAICF}_{\text{gen}}$ and the CAIF_c should prove to be very useful. The structurally similar correction in the AIC_c certainly has been a major contribution. However, I have one caveat to raise about these criteria based on the empirical information matrix. The information number requires considerable data to estimate accurately. I have found that the CAICF that Dr. Bozdogan proposed in his classic 1987 paper is quite unstable if the number of observations is low, or even moderate. Even with bias corrections, the behavior of these forms may deteriorate rapidly with small sample size.

REFERENCES

Abramovich, F., H. Benjamini, D. Donoho, and I. Johnstone. 2000. Adapting to Unknown Sparsity by Controlling the False Discovery Rate. Technical report, Statistics Department, Stanford University.

Akaike, H. 1973. Information Theory as an Extension of the Maximum Likelihood Principle. In Petrov, B. N., and F. Csaki, eds., *Second International Symposium on Information Theory*. Budapest: Akademiai Kiado.

Akaike, H. 1987. Factor analysis and AIC. *Psychometrika* 52:317–332.

Bai, Z. D., C. R. Rao, and Y. Wu. 1999. Model Selection with Data-Oriented Penalty. *J. Stat. Planning Inference* 77:103–117.

Barron, A., L. Birgé, and P. Massart. 1999. Risk Bounds for Model Selection via Penalization. *Probability Theory Related Fields,* 113:301–413.

Barron, A., M. Schervish, and L. Wasserman. 1999. The Consistency of Posterior Distributions in Nonparametric Problems. *Ann. Stat.* 27:536–561.

Baxter, R. A. 1996. Minimum Message Length Inference: Theory and Applications. Unpublished doctoral dissertation, Department of Computer Science, Monash University, Clayton, Victoria, Australia.

Bearse, P. M., and H. Bozdogan. 1998. Subset Selection in Vector Autoregressive (VAR) Models Using the Genetic Algorithm with Informational Complexity as the Fitness Function. *Systems Analysis, Modeling, Simulation (SAMS)* 31:61–91.

Benjamini, Y., and Y. Hochberg. 1995. Controlling the False Discovery Rate: A Practical and Powerful Approach to Multiple Testing. *J. Roy. Stat. Soc.,* ser. B, 57: 289–300.

Bhansali, R. J., and N. Y. Downham. 1977. Some Properties of the Order of an Autoregressive Model Selected by a Generalization of Akaike's FPE Criterion. *Biometrika* 64:547–551.

Borges, J. L. 1967. Funes, the Memorious. In *A Personal Anthology.* Secaucus, NJ: Castle Books.

Bozdogan, H. 1987. Model Selection and Akaike's Information Criterion (AIC): The General Theory and Its Analytical Extensions. *Psychometrika* 52:345–370.

Bozdogan, H. 1988. ICOMP: A New Model-Selection Criterion. In Bock, H. H., ed., *Classification and Related Methods of Data Analysis.* Amsterdam: Elsevier.

Bozdogan, H. 1990. On the Information-Based Measure of Covariance Complexity and Its Application to the Evaluation of Multivariate Linear Models. *Comm. Stat. Theor. Methods* 19:221–278.

Bozdogan, H. 1994. Mixture-Model Cluster Analysis Using Model Selection Criteria and a New Informational Measure of Complexity. In Bozdogan, H., ed., *Multivariate Statistical Modeling,* vol. 2. Dordrecht: Kluwer.

Bozdogan, H. 2000. Akaike's Information Criterion and Recent Developments in Information Complexity. *J. Math. Psych.* 44:62–91.

Bozdogan, H., and D. M. A. Haughton. 1998. Informational Complexity Criteria for Regression Models. *Computational Statistics and Data Analysis* 28:51–76.

Bozdogan, H., and M. Ueno. 2000. A Unified Approach to Information-Theoretic and Bayesian Model Selection Criteria. Invited paper presented at the sixth world meeting of the International Society for Bayesian Analysis (ISBA), May 28–June 1, 2000, Hersonissos-Heraklion, Crete.

Burnham, K. P., and D. R. Anderson. 1998. *Model Selection and Inference: A Practical Information-Theoretic Approach.* Berlin: Springer.

Chatfield, C. 1995. Model Uncertainty, Data Mining, and Statistical Inference (with discussion). *J. Roy. Stat. Soc.,* ser. A, 158:419–466.

Cox, D. R. 1990. Role of Models in Statistical Analysis. *Stat. Sci.* 5:169–174.

Cramér, H. 1946. *Mathematical Methods of Statistics.* Princeton: Princeton University Press.

Edwards, A. W. F. 1992. *Likelihood* (expanded ed.). Baltimore: Johns Hopkins University Press.

Fisher, R. A. 1936. Uncertain inference. *Proc. Am. Acad. Arts Sci.* 71:245–258.

Forster, M. R. 2000. Key Concepts in Model Selection: Performance and Generalizability. *J. Math. Psych.* 44:205–231.

Forster, M. R., and E. Sober. 2004. Why Likelihood? Chapter 6 in Taper, M. L., and S. R. Lele, eds. *The Nature of Scientific Evidence: Statistical, Philosophical, and Empirical Considerations.* Chicago: University of Chicago Press.

George, E. I., and D. P. Foster. 2000. Calibration and Empirical Bayes Variable Selection. *Biometrika* 87:731–747.

Ghosal, S., J. K. Ghosh, and A. van der Vaart. 2000. Rates of Convergence of Posteriors. *Ann. Stat.* 28:500–531.

Ghosh, J. K., ed. 1988. *Statistical Information and Likelihood: A Collection of Critical Essays.* New York: Springer.

Godfrey, L. G. 1988. *Misspecification Tests in Econometrics.* Cambridge: Cambridge University Press.

Hannan, E. J., and B. G. Quinn. 1979. The Determination of the Order of an Autoregression. *J. Roy. Stat. Soc.,* ser. B, 41:190–195.

Harris, C. J. 1978. An Information Theoretic Approach to Estimation. In Gregson, M. J., ed., *Recent Theoretical Developments in Control.* London: Academic Press.

Hooten, M. M. 1995. Distinguishing Forms of Statistical Density Dependence and Independence in Animal Time Series Data Using Information Criteria. Ph.D. dissertation, Montana State University-Bozeman.

Hosking, J. R. M. 1980. Lagrange-Multiplier Tests of Time-Series Models. *J. Roy. Stat. Soc.,* ser. B, 42:170–181.

Huang, T. 2000. Convergence Rates for Posterior Distributions and Adaptive Estimation. Technical report, Department of Statistics, Carnegie Mellon University.

Hurvich, C. M., and C. L. Tsai. 1989. Regression and Time Series Model Selection in Small Samples. *Biometrika* 76:297–307.

Jolliffe, I. T. 1986. *Principal Component Analysis.* Berlin: Springer.

Kass, R. E., and A. E. Raftery. 1995. Bayes Factors. *J. Am. Stat. Assn* 90:773–795.

Konishi, S. 1999. Statistical Model Evaluation and Information Criteria. In S. Ghosh, ed., *Multivariate Analysis, Design of Experiments, and Survey Sampling.* New York: Dekker.

Konishi, S., and G. Kitagawa. 1996. Generalized Information Criteria in Model-Selection. *Biometrika* 83:875–890.

Kramer, M. G., A. Hansen, E. Kissinger, and M. L. Taper. 2001. Abiotic Controls on Windthrow and Forest Dynamics in a Coastal Temperate Rainforest, Kuiu Island, Southeast Alaska. *Ecology* 82:2749–2768.

Kullback, S. 1968. *Information Theory and Statistics.* New York: Dover.

Kullback, S., and R. A. Leibler. 1951. On Information and Sufficiency. *Ann. Math. Stat.* 22:79–86.

Lehmann, E. L. 1990. Model Specification: The Views of Fisher and Neyman, and Later Developments. *Stat. Sci.* 5:160–168

Lele, S. R. 2004. Evidence Functions and the Optimality of the Law of Likelihood. Chapter 7 in Taper, M. L., and S. R. Lele, eds., *The Nature of Scientific Evidence: Statistical, Philosophical, and Empirical Considerations.* Chicago: University of Chicago Press.

Li, S. C., S. Lewandowsky, and V. E. DeBrunner. 1996. Using Parameter Sensitivity and Interdependence to Predict Model Scope and Falsifiability. *J. Experimental Psych.* 125:360–369.

Lindsay, B. G. 2004. Statistical Distances as Loss Functions in Assessing Model Adequacy. Chapter 14 in Taper, M. L., and S. R. Lele, eds., *The Nature of Scientific Evidence: Statistical, Philosophical, and Empirical Considerations.* Chicago: University of Chicago Press.

Linhart, H., and W. Zucchini. 1986. *Model Selection.* New York: Wiley.

Mayo, D. G. 2004. An Error-Statistical Philosophy of Evidence. Chapter 4 In Taper, M. L., and S. R. Lele, eds., *The Nature of Scientific Evidence: Statistical, Philosophical, and Empirical Considerations.* Chicago: University of Chicago Press.

McQuarie, A. D. R., and C. L. Tsai. 1998. *Regression and Time Series Model Selection.* Singapore: World Scientific.

Miller, J. A., and T. M. Frost. 2004. Whole Ecosystem Experiments: Replication and Arguing from Error. Chapter 8 in Taper, M. L., and S. R. Lele, eds. *The Nature of Scientific Evidence: Statistical, Philosophical, and Empirical Considerations.* Chicago: University of Chicago Press.

Popper, K. R. 1959. *The Logic of Scientific Discovery.* London: Hutchinson.

Rao, C. R. 1945. Information and Accuracy Attainable in the Estimation of Statistical Parameters. *Bull. Calcutta Math Soc.* 37:81.

Rao, C. R. 1947. Minimum Variance and the Estimation of Several Parameters. *Proc. Cam. Phil. Soc.,* 43:280.

Rao, C. R. 1948. Sufficient Statistics and Minimum Variance Estimates. *Proc. Cam. Phil. Soc.* 45:213.

Rissanen, J. 1976. Minimax Entropy Estimation of Models for Vector Processes. In Mehra, R. K., and D. G. Lainiotis, eds., *System Identification.* New York: Academic Press.

Rissanen, J. 1978. Modeling by Shortest Data Description. *Automatica* 14:465–471.

Rissanen, J. 1986. Stochastic Complexity and Modeling. *Ann. Stat.* 14:1080–1100.

Royall, R. M. 1997. *Statistical Evidence: A Likelihood Paradigm.* London: Chapman and Hall.

Royall, R. M. 2004. The Likelihood Paradigm for Statistical Evidence. Chapter 5 in Taper, M. L., and S. R. Lele, eds., *The Nature of Scientific Evidence: Empirical, Statistical, and Philosophical Considerations.* Chicago: University of Chicago Press.

Schwarz, G. 1978. Estimating the Dimension of a Model. *Ann. Stat.* 6:461–464.

Sclove, S. L. 1987. Application of Model-Selection Criteria to Some Problems in Multivariate Analysis. *Psychometrika* 52:333–343.

Shen, X., and L. Wasserman. 2000. Rates of Convergence of Posterior Distributions. Technical report 678, Statistics Department, Carnegie Mellon University.

Shepard, B. B., and M. L. Taper. 1998. Influence of Physical Habitat Characteristics, Land Management Impacts, and Non-native Brook Trout *Salveinus fontinalis* on the Density of Stream-Resident Westslope Cutthroat Trout *Oncorhyncus clarki lewisi* in Montana Streams. In Influence of Abiotic and Biotic Factors on Abundance of Stream-Resident Westslope Cutthroat Trout *Oncorhyncus clarki lewisi* in Montana Streams. Final Report to USDA, Forest Service, Rocky Mountain Research Station on Contract INT-93845-RJVA.

Shibata, R. 1989. Statistical Aspects of Model Selection. In Willems, J. C., ed., *From Data to Modeling.* Berlin: Springer.

Stone, M. 1977. An Asymptotic Equivalence of Choice of Model by Cross-Validation and Akaike's Criterion. *J. Roy. Stat. Soc.,* ser. B, 39:44–47.

Takeuchi, K. 1976. Distribution of Information Statistics and a Criterion of Model Fitting. *Suri-Kagaku* (Mathematical Sciences) 153:12–18 (In Japanese).

Taper, M. L., and S. R. Lele. 2004. The Nature of Scientific Evidence: A Forward-Looking Synthesis. Chapter 16 in Taper, M. L., and S. R. Lele, eds., *The Nature of Scientific Evidence: Statistical, Philosophical, and Empirical Considerations.* Chicago: University of Chicago Press.

Theil, H., and D. G. Fiebig. 1984. *Exploiting Continuity: Maximum Entropy Estimation of Continuous Distributions.* Cambridge, MA: Ballinger.

Tong, H. 1990. *Non-linear Time Series: A Dynamical System Approach.* Oxford: Oxford University Press.

Van Emden, M. H. 1971. *An Analysis of Complexity.* Amsterdam: Mathematisch Centrum.

Wallace, C. S., and P. R. Freeman. 1987. Estimation and Inference by Compact Coding (with discussion). *J. Roy. Stat. Soc.,* ser. B, 49:240–265.

Wasserman, L. 2000. Bayesian Model Selection and Model Averaging. *J. Math. Psych.* 44:92–107.

White, H. 1982. Maximum Likelihood Estimation of Misspecified Models. *Econometrica* 50:1–26.

Zeng, Z., R. M. Nowierski, M. L. Taper, B. Dennis, and W. P. Kemp. 1998. Complex Population Dynamics in the Real World: Modeling the Influence of Time Varying Parameters and Time Lags. *Ecology* 79:2193–2209.

Zhao, L. 1993. Frequentist and Bayesian Aspects of Some Nonparametric Estimation. Ph.D. thesis, Cornell University.

Zhao, L. 1999. A Hierarchical Bayesian Approach in Nonparametric Function Estimation. Technical report, University of Pennsylvania.

Zucchini, W. 2000. An Introduction to Model Selection. *J. Math. Psych.* 44:41–61.

PART 6   CONCLUSION

16

The Nature of Scientific Evidence:
A Forward-Looking Synthesis

Mark L. Taper and Subhash R. Lele

ABSTRACT

This chapter presents a synthesis of various topics and ideas discussed in the book. As such it is a synopsis; however, at the same time it reflects our understanding and is affected by our biases. As scientists, we are interested in learning about the workings of nature. One can gain such an understanding in both inductive and deductive manner. A method that has proved extremely successful in the history of science is to take ideas about how nature works, whether obtained deductively or inductively, and translate them into quantitative statements. These statements, then, can be compared with the realizations of the processes under study. The major focus of this book is on the foundations of *how* this comparison should be carried out. This is the core issue in the quantification of evidence. The main two schools of statistical thought, frequentist and Bayesian, do not address the question of evidence explicitly. However, in practice, scientists inappropriately use the results of statistical techniques to quantify evidence. Various approaches to quantifying evidence have been discussed by a number of authors. In this chapter, we summarize, compare, and contrast these various approaches, and we compare them to frequentist and Bayesian ideas. A key concept learned from this discussion is that evidence is necessarily *comparative*. One needs to specify two hypotheses to compare, and the data may support one hypothesis better than the other. There cannot be such a thing as quantification of support for a single hypothesis. Evidence, even when correctly interpreted, can be misleading. At times, it can also be ambiguous. The control

We thank Prasanta Bandyopadhyay, Chris Jerde, and Bernard Taper for detailed comments on an earlier version of this paper.

of these errors is an essential part of statistical inference. The contributors to this book have different thoughts about what these errors are and how to measure and control them. Models are only approximations of reality. How good are these approximations? Ideas on model adequacy and model selection in the context of quantifying evidence are discussed. The role and scope of the use of expert opinion and data from related studies for quantifying evidence are explored. Some of the authors point out that it is not necessary that all evidence be "statistical evidence." The concept of "consilience" of different forms of evidences is emphasized. Of course, many of these musings have to be checked against the reality of life. Some of the contributors discuss the role of mechanistic models in avoiding some of the pitfalls of inappropriate designs in ecology. Replication is usually highly desirable but in many ecological experiments difficult to obtain. How can one quantify evidence obtained from unreplicated data? Nuisance parameters, composite hypotheses, and outliers are realities of nature. We summarize the discussion of how one can quantify evidence in such situations. Finally, we raise what we think are important unresolved issues, such as using evidence to make decisions without resorting to some sort of subjective probabilities. Although the evidential approach has strong affinities with Bayesian and classical frequentist analysis, it is distinct from both; thus, it may have much to offer science.

THE ROLE OF EVIDENCE IN SCIENCE

Scientists strive to understand the operation of natural processes, creating and continually refining realistic and useful descriptions of nature. But which of many descriptions or models most closely approximate reality? To help in this evaluation, scientists collect data, both experimental and observational. The objective and quantitative interpretation of data as evidence for one model or hypothesis over another is a fundamental part of the scientific process.

Many working scientists feel that the existing schools of statistical inference do not meet all their needs.

> The real problem is that neither classical nor Bayesian methods are able to provide the kind of answers clinicians want. That classical methods are flawed is undeniable—I wish I had an alternative. Perverting classical methods to give answers that sound like what people want (e.g., using the P value as the probability that the null hypothesis is true) is certainly not the answer. However, clinicians

should be aware that Bayesian methods, no matter how beguiling their form, are not the answer either. (Browne, 1995, 873)

Richard Royall puts it more baldly:

Statistics today is in a conceptual and theoretical mess. The discipline is divided into two rival camps, the frequentists and the Bayesians, and neither camp offers the tools that science needs for objectively representing and interpreting statistical data as evidence (Royall, 2004 [chapter 5, this volume])

To some extent, it is safe to say, scientists have had to twist their thinking to fit their analyses into existing paradigms. We believe that a reexamination and extension of statistical theory is needed in order to reflect the needs of practicing scientists.

We believe that at the heart of scientific investigation is the determination of what constitutes *evidence* for a model or a hypothesis over other models or hypotheses. Our goal for this volume was to demonstrate the need for a statistical viewpoint that focuses on quantifying evidence, and through a dialog among practicing scientists, statisticians, and philosophers, to outline how such a new statistical approach might be constructed.

To discuss these issues, and to explore new directions, this book brought together an eminent but disparate catalog of authors including ecologists (both theoretical and empirical), statisticians, and philosophers. Although the scientists represented in this volume are ecologists, we do not believe the book is of interest only to scientists of this discipline. It should speak to all sciences where uncertainty plays an important role, to all statisticians interested in scientific applications, and to all philosophers of science interested in the formalities of the role of evidence in science.

This book is not designed to be synoptic. We do not pretend to have compiled a comprehensive review of the foundations of statistical and scientific inference. Nor has it been our intention to resolve the frequentist/Bayesian debate. Many volumes have addressed this issue. Rather than encouraging our contributors simply to review the existing literature, we asked them to think originally about evidence—a topic insufficiently treated until now. Our wish is to stimulate thought and research in a new field. We, the editors, have developed strong opinions as to profitable directions for this field. However, we aggressively sought contributors whose expertise and opinions differ from ours.

In this final chapter, we attempt a synthesis of what we have learned. However, this chapter is not a consensus report. There are differences of opinion among our contributors. We have striven to represent the positions of our contributors clearly and fairly; nonetheless, this chapter represents our understanding of the volume, its internal connections, and most importantly its implications.

"MODELS" AS REPRESENTATIONS OF REALITY

The most basic theme of this book is that models act as descriptions or representations of reality. Chapters[1] by Maurer (2), Scheiner (3), Mayo (4), Taper and Lele (9), and Taper (15) deal explicitly with this conception. Maurer, referring to Giere (1999), describes science as

> a program of establishing models of reality that then are compared to the real world by scientific investigators to determine their utility. The utility of a model lies in its ability to produce accurate conceptual statements about specific natural systems.

Maurer argues that such descriptions are both possible and useful.

A scientific hypothesis is a statement of how the world is or operates. To be scientifically probed, a hypothesis must be translated into a specific quantitative model. As discussed by Lindsay (chapter 14) and Taper (chapter 15) it is unlikely any model will ever be "true." However, hypotheses can be true by being general. The hypothesis "the world is not flat" is true, but it is also very general—it is a composite of an infinite series of models such as the world is round, the world is ovoid, etc. Scientific hypotheses rarely correspond to a single model clearly and cleanly. Scientific hypotheses are better represented by composite statistical hypotheses. In practice, however, single models are chosen to capture the "important features" of the scientific hypotheses. Translation of a scientific hypothesis into a statistical model is one of the most perilous steps in scientific inference. The strength of Lindsay's model adequacy approach is that it continually forces us to challenge this step.

Most of this book deals with the problems of investigating statistical models of finite complexity. It is important to recall that this is only a single step in the progression of scientific reasoning. Modeling, even statistical modeling, is not a monolithic enterprise. As discussed by Maurer (chapter 2; see also Taper, chapter 15), statistical models can be constructed for a num-

---

1. In this synthesis, chapter designations refer to this volume.

ber of purposes. Some of these can be categorized as exploration, explanation, and extrapolation.

An important use of models is in the recognition of patterns in nature. This is often referred to as exploratory data analysis. Employing such techniques as principal component analysis, factor analysis, classification, and regression trees as well as a host of graphical display techniques, the goal of this class of analysis is to discover unknown patterns in nature. A subsequent modeling purpose may be to demonstrate that the patterns discovered are in some sense real. As an extension of this process, modeling and analysis projects may be undertaken whose goal is to demonstrate the patterns discovered are general.

Another purpose for much statistical modeling is extrapolation. Here the goal is not just to describe the internal structure of an existing data set, but also of data unobserved. A vast amount of scientific work falls into this category, in fields as diverse as economic forecasting to predictions of species occurrence for conservation purposes. Finally, models are frequently used in science to encapsulate an explanation of the processes occurring in nature. An explanatory model is a model that depicts the causal relationships that exist among natural events.

These goals have been described as separate. To some extent, the distinctions are semantic. All of these goals are facets of the representation of the real and complex world by models. They are demarcation points in a continuum, with much overlap in the categories. A precise enough description of pattern can often be used for prediction. An explanatory model may be useful for prediction as well. But, just because the distinctions are fuzzy, it should not be thought that these goals are synonymous. Although Hempel (1965a) argues for the structural identity of explanation and prediction, Scriven (1959), Ruben (1990), Curd and Cover (1998), and others present examples showing that some predictions may not involve explanations at all, and that an important explanatory model may have large predictive uncertainty.

How do scientists move through the levels of this hierarchy, how does a descriptive pattern suggest a predictive procedure to the scientist, and how does a predictive pattern suggest a causal explanation? The process of hypothesis generation and the link from pattern recognition to causal explanation is only weakly addressed in this book. But then, the subject of this book is evidence for models, not generation of models.

Along with a diversity of *goals* for modeling, there are also several *approaches* to modeling. Two dominant tactics are descriptive and mechanistic models. A descriptive model is designed only to produce a pattern that accurately mimics reality. A mechanistic model, on the other hand, is designed not only to produce an accurate pattern, but also to do so in a fashion mimicking the processes by which real-world observations are generated. As discussed by Taper and Lele (chapter 9), mechanistic models have great utility as tools both for understanding the natural world and for making causal statistical inference.

As with the goals discussed above, these approaches to modeling are not as distinct as first described. A descriptive model may make up a component of a model designed to be a mechanistic model of a higher-level phenomenon. In other cases, a mechanistic model may be the easiest way to generate a descriptive pattern for a complex phenomenon.

An essential element in representing reality with models, whether statistical or dynamic, is the incorporation of adequate model structure. In both Brown et al. (chapter 10) and Taper and Lele (chapter 9), we find illuminating discussion of how models with inadequate structure can be not only uninformative, but sometimes importantly misleading. Modelers should be concerned with two types of difficulties involving model structure. On the one hand, models may fail to adequately capture real-world constraints; on the other hand, constraints implicit in model structure may have unexpected inferential consequences

Determining adequate model structure is an elusive component of modeling. It requires deep knowledge of the natural system, a subtle understanding of the behavior of models, a clear statement of the questions to be explored, and a command of statistical methods by which data can be queried by models. Rarely do all these qualities reside in a single scientist. Collaboration is generally the most effective way to probe nature with models.

HOW DO DATA INFORM US ABOUT THE RELATIONSHIP OF MODELS AND REALITY?

The job of a scientist is to find a "better" description than is currently being used. Science is a cyclic process of model (re)construction and model (re)evaluation. Models are (re)evaluated by comparing their predictions with observations or data.

As Taper and Lele (chapter 9) mention, Whewell (1858) laid out a hierarchy for the evidential import of prediction: At the first, or lowest, level there is the fit of observations for data in hand. Second is the prediction of new observations of a kind similar to the data in hand. Third is the prediction of new observations of a different kind. And fourth is the increase in the consilience of science.

The first two levels, with exemplars in maximum likelihood techniq
and in cross-validation, are the foundation of modern statistical pract
Powerful scientific arguments are sometimes made based on level three.
The fourth level is rarely explicitly evoked formally in scientific arguments.
However, as Scheiner (chapter 3) points out, consilience is frequently used
informally. If several disparate lines of research imply the same conclusion,
scientists and people in general feel more confident than if presented with
only a single account. In environmental policy and decision analysis, this is
sometimes referred to as the weight of evidence, but perhaps a better term
would be the consilience of evidence.

Consilience plays an important part in the aesthetics of science as well as
in the inspiration for discovery. In the scientific community, there has been
a considerable resurgence of interest in consilience, as evidenced by Maurer
(chapter 2) and Scheiner (chapter 3) and books such as Pickett, Kolasa, and
Jones (1994) and Wilson (1998). Unfortunately, there has been little technical work on methods for combining evidence. This is an important direction
for future exploration.

Popper (1935, 1959) made a clear statement about what is meant by a
"valid scientific hypothesis." His main contention was that a hypothesis is
valid if it is, at least, potentially falsifiable. Thus, such hypotheses should
make predictions about observable quantities that can be compared with
actual observations. If the predictions and observations match each other
closely, we have stronger corroboration for the hypothesis; if they do not
match, the hypothesis is falsified and hence should be modified. An interesting implication is that one can never prove a hypothesis; one can only refute a hypothesis. An important extension and communication to the scientific community of Popper's ideas came in Platt (1964). Platt championed
the idea of multiple hypotheses. He emphasized that good scientists design
experiments in such a fashion that some of this multitude of hypotheses can
be refuted. The influence these ideas have had on the practice of science can
hardly be overstated.

However, by 1963, Popper seems to have concluded that one can support
a hypothesis in terms of its "verisimilitude" (Popper, 1963). The concept of
verisimilitude or "truthlikeness" recognizes that a hypothesis can have elements of truth and falseness simultaneously. Verisimilitude is very much a
conceptual discrepancy between the hypothesis and the truth. In this framework, one does not so much seek to refute hypotheses as to replace them
with hypotheses of greater verisimilitude. In many ways, verisimilitude is
similar to the model adequacy discussed in Lindsay (chapter 14). Comparison of the verisimilitude of alternative hypotheses is very similar to the con-

struction of an evidence function (Lele, chapter 7). We return to a discussion of verisimilitude at the close of this chapter.

It is quite clear to most scientists that hypotheses are usually neither completely refuted nor accepted with full confidence. This might be because there are infinitely many possible explanations for a particular observed phenomenon and the hypothesis under consideration is only one of many such explanations (see Bandyopadhyay and Bennett, chapter 2.1). Such alternative explanations may depend on the existence of unobserved, or confounding factors. Another factor that makes refutation or acceptance with certainty impossible is stochasticity in the observations. Strictly speaking, all observations are possible but some are more probable than others. Statisticians and scientists have tried to address the issue of quantifying support for a hypothesis in the presence of stochasticity.

Models need to be ratified by observations. Royall (1997) succinctly puts the issues of relating data to models in terms of three questions:

1. Given these data, what do we do?
2. Given these data, how do we change our beliefs about the reality of nature?
3. Given these data, how do we quantify the strength of evidence for one explanation over the alternative explanation?

Clearly, from these questions there is a distinction between data and evidence. Data are the direct quantification of observations. Data are considered by scientists to be, for the most part, largely objective. For example, a scientist's feelings about gold are not very likely to influence any record of observation of the melting point of the metal. Subjectivity, if it does enter, will only affect interpolation and rounding at the limits of precision of the instrument.

Of course, there is a continuum to the objectivity of data. For example, a scientist surveying ranchers regarding the condition of their range may receive reports such as "poor," "good," or "excellent." Such data has tremendous potential for subjectivity. Scientists are suspicious of data such as this, using it only with caution.

Evidence is a summarization of data in the light of a model or models. In traditional frequentist statistical approaches, the strength of evidence is explicitly quantified as the probability of making an error, and an error is defined either as rejecting a true statistical model or accepting a false model.

Fisherian P-value tests, which are the usual tests of significance, set up a null hypothesis and reject this hypothesis if under the null model there is only a small probability of generating a set of observations as discrepant or more discrepant than the realized data. Thus P-value tests consider only a single or unary hypothesis. The logic behind P-value tests is that an event of small probability under the null hypothesis indicates that an alternative explanation is needed. The P-value test is meaningless if interpreted literally. As discussed above, no model of finite order can be true. Because we already knew that the null model isn't true, if the null hypothesis is rejected, the scientist has not really learned anything. On the other hand, if the scientist fails to reject the null hypothesis, then all one can really say under strict interpretation is that sufficient data have not been collected.

The Neyman-Pearson approach requires an explicit a priori specification of both a null hypothesis and an alternative hypothesis. It then provides a rule for deciding between these two hypotheses based on the data (Lewin-Koh, Taper, and Lele, chapter 1; Dennis, chapter 11). The rules for such decision making are created so that on an average the decisions are correct a high percentage of times. The logic is that if the rule is good on an average, it is likely to be correct for the particular situation we have at hand.

The Neyman-Pearson approach tests whether the data are more likely under an alternative hypothesis or under the null hypothesis. Thus, Neyman-Pearson tests seem explicitly comparative in nature. However, the procedure described above is not as comparative as it appears. It is true that a pair of hypotheses is used to create a test statistic of maximum power. However, a realized P-value is computed from the null hypothesis; and a decision is made, by comparing this value to the a priori size of the test. The decision is actually made based only on the expected behavior of the data under the null hypothesis. Thus, the test described is truly neither unary nor binary.

Lewin-Koh, Taper, and Lele (chapter 1) and Dennis (chapter 11) discuss the mechanics of formulating P-value tests as well as Neyman-Pearson tests. Unlike the Fisherian test, the Neyman-Pearson test does not yield a continuous measure of evidence. The result from a Neyman-Pearson test is a black-and-white decision; accept the null hypothesis or accept the alternative. The dichotomous decision of Neyman-Pearson tests creates the distressing circumstance that minor changes in the data can make major changes in the inference by pushing the test statistic over a knife-edge boundary.

Both Fisherian and Neyman-Pearson tests involve integration over hypothetical sample spaces in order to calculate the probabilities of error. This probability of error does not really attach to the specific case of the data in hand; instead, it describes the average performance of the experiment and analysis protocol would have under infinite repetition. Implicit in the use of

Chapter 16

these tests is the assumption that good average performance implies good specific performance.

Although the majority of working scientists continue to use frequentist statistics in the analysis of their research, the approach is currently not as fashionable in philosophical and academic statistical circles as is the Bayesian analysis. Bayesian approaches address the question of belief in various hypotheses. Although mathematically correct and intuitively attractive, this does not seem to be the question that scientists are interested in. Scientists, as is evident in the work of Popper and Platt, are interested in the question of quantifying the strength of evidence. This book is a collection of thoughts on the theme of evidence, what it really constitutes, what we do with such quantification, and so on. Bayesian approaches do not seem to address this question explicitly.

The error-statistical approach advocated by Mayo (1996; chapter 4) strongly rejects Bayesian statistics and attempts to create a coherent philosophical framework for scientific inference that is fundamentally frequentist. For error statisticians, the strength of evidence for a hypothesis is the reliability of the test procedures that the hypothesis has passed. A test is reliable if it is unlikely the alternative hypothesis will be accepted as true when the null is true.

Mayo's statistical approach is very similar to the Neyman-Pearson test, with a subtle but important difference. Null and alternative models are formulated and a test statistic constructed that will maximize the power of the test. However, under Mayo's approach the test is not considered dichotomous, but contoured. The severity of the test passed is roughly 1 minus the realized P-value. As the null hypothesis is changed, a severity curve emerges. This procedure should, if accepted, alleviate the sinking feeling a scientist gets when the data provide a test with an unpublishable realized P-value of .06. In effect, what Mayo (chapter 4) suggests is an extension of the already general practice of reporting realized P-values instead of a priori test size. One should note that a very crucial shift has occurred from the interpretation of classic Neyman-Pearson tests; both realized P-values and severity are postdata quantities.

The goal of the error-statistical approach is the understanding and control of error. Given the defining role of the probability of error, it is obvious that replication has an important position in the error statistician's epistemology. Replication is useful in understanding the probability of error. But it is control of error that is important, not the replication per se. Error is controlled by the statistical test, but perhaps just as importantly, error is controlled by the experimental design and the intent of the experimenter.

Miller and Frost (chapter 8) discuss procedural control of error where replication is impossible.

Consideration of how the data were collected and what decisions were made during the analysis are critical to the error statistician because they influence the calculation of the probabilities of making an error and therefore the strength of evidence. Consideration of data collection and design procedures distinguishes Mayo's error statistics from Bayesian analysis, but does not sharply demarcate it from the evidential approach advocated by the authors of this chapter. As demonstrated by Lele (chapter 7.3), evidence functions can depend on stopping rules and other forms of investigator intent. Lindsay (chapter 14) introduces us to the concept of model adequacy. The discrepancy of a model from truth, as represented by an empirical distribution function, is estimated for some distance measure designed to answer specific scientific questions. A model is declared adequate for a researcher's purposes if the estimated discrepancy is less than some arbitrary but meaningful level. Lindsay's ideas bear a strong relationship to P-values (see Lewin-Koh, Taper, and Lele, chapter 1), but with important differences. P-values measure the unlikeliness of data under a specific model, while model adequacy measures the distance between the empirical distribution function and a model-based distribution function. These seem very similar, but differ in their focus. P-values are focused or centered on an unreal and untrue null hypothesis while model adequacy is centered on reality as revealed by the empirical distribution. The model adequacy approach shares several key elements with the evidence function approach. First, both steadfastly reject the "true model" formalism. Second, both attempt to construct empirical distances from models to "truth." And third, both consider these distance measures problem specific, with the specific distance function designed to help a researcher answer specific questions. Thus, there are numerous points of philosophical agreement between model adequacy and statistical evidence. Nevertheless, there is a difference in the approaches in that model adequacy is not explicitly comparative.

Goodness-of-fit tests have effectively complimented Neyman-Pearson tests in science for years. Model adequacy may have a similar complementary relationship with statistical evidence. Evidence functions may indicate that one model is more truthlike than another without indicating whether either model is reasonably close to truth. Model adequacy can serve this function. Lindsay (chapter 14) proposes that model adequacy not be thought of in a framework of testing of hypotheses as is common in the use of goodness of fit. Instead, he advocates that model adequacy be approached as an estimation problem.

Suppose we have formulated several statistical models that are all pre-sumed to approximate reality. How do we relate observed data to such mod-els and come up with quantitative measures of the support that can be found in the data for each model relative to the others? Chapters in this book, primarily Royall (chapter 5), Forster and Sober (chapter 6), Lele (chap-ter 7), and Taper (chapter 15), discuss attempts at providing such explicit measures.

Royall (1997; 2000; chapter 5) elaborates on the law of the likelihood (Hacking, 1965) and proposes the likelihood ratio as the measure of the strength of evidence for one model over another. An important aspect of the likelihood ratio as a measure of the strength of evidence is that it empha-sizes that evidence is a comparative measure. One needs two hypotheses in order to quantify evidence in the data.

Another aspect of the use of the likelihood ratio as an evidence measure is that only the models and the actual data are involved. This is quite differ-ent from the classical frequentist and error-statistical approaches, where the strength of evidence is the probability of making an error, calculated over all possible configurations of potential data.

While there may be difficulties in using the probability of error as a di-rect measure of evidence (Casella and Berger, 1990; Cox and Hinkley, 1974; Cox, 1977; Johnson, 1999), the probability of error is an important concept. Any measure of evidence, even properly interpreted distance-based mea-sures, can be misleading in finite data sets (Royall, chapter 5). Thus, error probabilities can be an important adjunct to evidence in the inferential pro-cess. Royall (1997; 2000; chapter 5) introduces the concepts of "misleading evidence" and "weak evidence." He shows how probabilities could be at-tached to these concepts, thus addressing the need of the scientists for con-trolling different kinds of errors. These concepts, although similar to the type I and type II errors of the Neyman-Pearson approach, are fundamen-tally different. Probabilities of misleading and weak evidence converge to 0 as the information goes to infinity. On the other hand, in the Neyman-Pearson setting, the probability of type I error is held constant throughout.

Lele's work on evidence attempts to extend Royall's groundbreaking be-ginning. Although Lele's thinking is inspired by Royall's, there are distinc-tions. For instance, although Royall recognizes that the model set need not contain a true hypothesis (Royall, 1997, 6), much of his theoretical develop-ment assumes that one of the hypotheses is the true hypothesis. However, since there is no such thing as the correct or true model under which the data were generated (except in simulation experiments), Lele (chapter 7) asks what happens to the use of the likelihood ratio if the true model is different from either of the models under consideration. He shows that the use of the likelihood ratio corresponds to the choice of the model that is closest to the true model in the Kullback-Leibler divergence.

If the likelihood ratio is, in fact, a comparison of Kullback-Leibler divergences, what about the use of other divergence measures such as the Hellinger distance for quantifying the strength of evidence. Lele (chapter 7) provides a list of regularity conditions that need to be satisfied by a good measure of the strength of evidence. He shows that this class is nonempty and that the Kullback-Leibler divergence is an optimal evidence function within this class, in the sense that probability of strong evidence for the true hypothesis converges to 1 fastest. However, a further study of this class reveals there are evidence functions that are as good as the Kullback-Leibler divergence but are also robust against outliers. This constitutes a second distinction between Royall and Lele. To Royall, statistical evidence is measured solely and axiomatically by the likelihood ratio. To Lele, the likelihood ratio is but one of a broader class of evidence functions.

Lele also shows that one can define evidence in the data when the models are specified only by the mean and the variance functions, similar to the use of the quasi-likelihood and estimating function methods in estimation (Godambe, 1991; Heyde, 1997; McCullogh and Nelder, 1989). The broad flexibility in the definition of evidence functions opens up a possibility of extending the ideas of evidence to the thorny problems of comparing composite hypotheses, model selection, and nuisance parameters. One of the interesting features coming out of the use of evidence functions is that the likelihood principle (Berger and Wolpert, 1984), also emphasized by Forster and Sober (chapter 6), Boik (chapter 6.2), and Royall (chapter 5), is not a defining part of the evidential paradigm. The stopping rule and other design aspects of the study appear explicitly in the quantification of the strength of evidence, thus addressing the concerns of Mayo (chapter 4), Dennis (chapter 11), and other scientists.

The central idea that evidence is a comparative concept occurs repeatedly through the contributions of Royall (chapter 5), Lele (chapter 7), and Taper (chapter 15). However, Lindsay (chapter 14) seems to suggest the possibility of quantifying model adequacy not in the context of competing models but for a single model. Royall constructs a logical trichotomy for evidence, strong evidence in favor of the correct hypothesis, strong evidence in favor of an incorrect hypothesis or misleading evidence, and weak evidence where we cannot make a strong evidential statement. Similarly, Lindsay's paper describes the trichotomy of the model being adequate, not adequate, or of indeterminable adequacy.

Although we have distinguished model adequacy from evidence because model adequacy does not explicitly compare models, there are interesting comparative aspects implicit in model adequacy. One may think of adequacy as a bound on all potential model comparisons. As an example, consider a distance measure δ for assessing model adequacy defined as the ratio of the likelihood under the empirical distribution function to the likelihood under the model. If a model has a distance of δ_1, it is a simple algebraic exercise to show that for any superior model ($\delta_2 < \delta_1$) the likelihood ratio of model 2 to model 1 must be less than δ_1.[2] If a model is considered "adequate" whenever $\delta \leq 8$, then any possible better model can only have "weak" evidence (Royall, chapter 5) in support of it over the "adequate" model.

Thus, there is a unification of the evidential and model adequacy approaches under the common foundation of estimated discrepancies of models from "truth." It will be of interest to study in detail the relationship between the evidential paradigm and the model adequacy paradigm.

One of the main attractions of the Bayesian approach is its explicit use of other kinds of information such as expert opinion and/or past data in statistical inference. However, several statisticians and scientists (for good summaries, see Efron, 1986; Dennis, 1996 and chapter 11 of this volume) have made a strong case against the use of beliefs in science and public decision making. On the other hand, it seems a waste of information if such soft data are not utilized, especially in sciences such as ecology that are starving for data. Goodman (chapter 12) shows how related data could be used to build a data-based prior distribution and then utilize it for decision making. Lele (chapter 13) discusses several problems with eliciting prior distributions from the subject experts and suggests the use of elicited data for incorporation of expert opinion in the evidential setup. The main feature of this latter approach is that it allows explicit evaluation and calibration of the expert opinion, thus mitigating the subjectivity inherent in the classical Bayesian approach.

UNRESOLVED ISSUES IN THE EVIDENTIAL PARADIGM

There are several issues in the development of evidential paradigm for statistical analysis of scientific data needing substantial further thought. In the following, we discuss some of these unresolved issues and our thoughts re-

2. See Taper, chapter 15, eq. 1.

garding their possible resolution. This is not an exhaustive list of the issues, nor do we claim that our suggestions are necessarily the right way to the solutions.

Multiple Hypotheses

One of the thorniest problems in the evidential paradigm is making evidential statements in the presence of multiple hypotheses (Taper, chapter 15; Nelson, chapter 7.2). Taper (chapter 15) discusses model selection from very large model sets, and proposes an ad hoc procedure to remedy model selection bias in models with tapered effects. His solution is to increase the parameter penalty in an SIC-like information criterion until a clear minimum is found. As effective as this technique appears, it currently has little theoretical justification. We present some possible approaches that may help to lay a more solid theoretical foundation for this class of problem.

We start with the concept of identifiability of the hypotheses. The concept of identifiability of parameters is well known in statistical literature (Lehmann, 1983). One can extend this concept to the evidential paradigm.

Consider a collection of several hypotheses. We consider these hypotheses or explanations for a scientific phenomenon to be identifiable if the collection of discrepancies from the truth to these explanations consists of distinct values. Only such a collection should be considered scientifically valid.

Definition of identifiability: Let M_0 and M_1 be two competing models. Let T denote the true distribution. Then we say that the models are identifiable if and only if one of the following inequalities holds: $d(T, M_0) > d(T, M_1)$ or $d(T, M_0) < d(T, M_1)$. If strict equality holds, then we say the models are nonidentifiable under the given discrepancy structure. Notice this definition of identifiability expands the concept so that identifiability is not just a property of a set of models but also involves the discrepancy measure used for quantifying the strength of evidence.

Identifiability of models is critical for scientific process. If models are not identifiable, no evidence can be found to distinguish them and inference is not possible. Although this may seem obvious, it is not a trivial point. There are at least two examples in the scientific literature where nonidentifiable models are commonly used. Factor analysis, which is commonly used both in psychology and in ecology, is based on nonidentifiable models. Morphometrics, the quantitative study of forms, many times employs superimposition and deformation methods (Dryden and Mardia, 1998). These methods also base their inference on nonidentifiable models as shown in Lele and Richtsmeier (2000). The concept of identifiability is, unfortunately, ignored when statistical techniques are taught in many courses designed for work-

ing scientists. The above definition of identifiability can easily be extended
to the case with multiple hypotheses. The concept of identifiability is related to a unique ordering of the hypotheses/models. Thus, in the case of
multiple models, identifiability demands that there should be a unique ordering of the hypotheses in terms of their discrepancy from the truth, with
no ties allowed.

Assuming the models are identifiable, there are still issues about the possibility of distinguishing these models using the data at hand. Estimability
problems might arise because of the wrong observational studies or design
of experiments. A *definition of estimability* could be "P_T (correct estimate of
the ordering of the models under consideration) goes to 1 as the amount of
information (the sample size) goes to infinity." In most situations, this corresponds to the possibility of estimating the discrepancies consistently.

Given the above definitions of identifiability and estimability, we are
now in a position to define misleading evidence in the presence of multiple
hypotheses. Misleading evidence corresponds to obtaining incorrect ordering of the hypotheses. The probability of incorrect ordering is thus the probability of misleading evidence. Operationally, one might also consider the
probability of partially correct orderings, where the first k models are correctly ordered.

Composite Hypotheses, Nuisance Parameters, and Model Selection

The issue of quantifying strength of evidence in the presence of nuisance
parameters is a difficult issue. This problem also closely relates to the quantification of evidence when one is comparing composite hypotheses. Royall
(2000) suggests the use of profile likelihood for evidential inference in the
presence of nuisance parameters and composite hypotheses. Lele (chapter 7.3) suggests a solution to the problem of which profile likelihood is a
particular case.

So far we have dealt with the problems when the number of nuisance parameters is the same in the competing models. An interesting generalization
will be when the number of nuisance parameters is not the same between
models. For example, we may be interested in the effect of one risk factor.
One model might consist only of the intercept (in addition to the covariate
of interest) and the other model might consist of a confounding factor in
addition to the covariate of interest. This is closely related to the model selection issues. The profile likelihood ratio in this situation does not belong
to the class of evidence functions. Without a bias correction, the profile likelihood ratio violates the Lele's condition R1 (chapter 7). Such a bias correction seems to lead to AIC and the related methods.

Evidence and Prediction

Estimation, in the traditional sense, implies obtaining information about fixed quantities such as parameters, e.g., coefficients in the regression model; prediction traditionally refers to obtaining information about "random quantities" such as the response at a new covariate value in regression. Likelihood approaches have been suggested (Hinkley, 1979; Butler, 1986) for prediction, but they seem unsatisfactory. How does one quantify evidence about a future observation? This brings us to yet another distinction between evidence as conceived of by Royall on the one hand and the authors of this chapter on the other. Royall believes that Bayesian methods are most appropriate for prediction, and hence for decision. We are not convinced that evidential approaches will not be profitable here. Certainly this is an important topic for research.

SALIENT AND CONSILIENT FEATURES OF THE EVIDENTIAL PARADIGM

The evidential paradigm has a number of features promoting clarity of thought on a number of troublesome statistical problems. If not illusory, such clarity may be more important to furthering good science than any technical advances.

Model-Centered Inference

Model misspecification is a major cause of error in statistical analysis (Chatfield, 1995). The evidential perspective allows and encourages evaluation of multiple models. With a richer set of models under consideration than allowed by traditional frequentist statistics, the probability of model misspecification is reduced. The rapid acceptance of model identification based on information criteria demonstrates the great scientific utility of a rich model set (Burnham and Anderson, 1998). It can be easily shown that differences of information criteria are evidence functions. Thus, the evidential paradigm can draw support from the success of model identification. At the same time, the evidential paradigm can provide a more formal framework for the interpretation and further development of information criteria.

Scope of Inference

Evidential analysis makes no claims about the truth or falsehood of any of the models considered, not even just semantically. Instead, an evidential report makes a simple statement that the data at hand support model A more than they support model B, and by how much. The scope of inference is clearly and intrinsically delimited.

Indisputably, all statistical inference must be subjective because it is based on the "contingent" knowledge and understanding of the statistician (e.g., Hempel, 1965b). We discuss the distinction between evidence and inference below, but it suffices to say here that statistical evidence, as we have defined it, partitions off and exposes an objective component of statistical inference.

Multiple Comparisons

One of most vexing difficulties for the frequentist paradigm has been the problem of multiple comparisons. If an analysis tests more than one pair of hypotheses, problems with the interpretation of P-values as the strength of evidence emerge. If multiple tests are made, the chance of a type I error being made somewhere in the entire set of comparisons is greater than the nominal size α of each test. The frequentist solution is Bonferroni-type adjustments, which take the number of comparisons made into consideration and reduce the size of the individual tests so the probability of a type I error in the entire analysis is at the desired level. Thus, the strength of evidence for individual tests depends on how many other tests are being made. But which comparisons should be considered? Any reasonably complex data set admits a vast number of potential tests. Do we adjust for all possible tests, or do we adjust only for the ones the researcher intended to make a priori? Do we ever know which comparisons a researcher intended to make? Does even the researcher really know about them?

The problem of multiple comparisons is real, and taking an evidential approach does not make it disappear entirely. Nonetheless, a certain clarity again emerges from the evidential analysis. The strength of evidence in any individual comparison does not change with the number of models considered. What do change, however, are the probabilities of misleading and weak evidence. Because of their flexibility, evidence functions can be adapted to compensate for the model selection bias caused by large numbers of candidate models (Taper, chapter 15).

A Priori versus Post Hoc Comparisons

As with the problem of multiple comparisons, it seems odd that the evidence for a model should depend on when the model was formulated. Scheiner (chapter 3) strongly states, "That a theory is erected after the data are in hand does not weaken the strength of the agreement between data and theory." Nevertheless, the principle of prior prediction is deeply established in the philosophy of science (e.g., Hempel, 1966). A framework for resolving this tension can be found in the above discussion of multiple comparisons. The evidence for a model relative to any other does not depend on when it was formulated. The difference is in the probability of misleading evidence. A post hoc model is essentially only one of a vast set of models cryptically in the comparison—everything the researcher has consciously or subconsciously considered since viewing the data. We cannot yet calculate the impact of these unexpressed models on the probability of misleading evidence. Nevertheless, we can at least understand the distinction between a priori and post hoc analysis, and why post hoc analysis is weaker inferentially than a priori analysis.

The evidential approach is alone among the paradigms explored in this book in having its measure of evidence invariant to intent, belief, and time of hypothesis formulation. The evidence is the evidence. Both belief and error probabilities have been separated from evidence. This is not to say that belief and error probabilities are unimportant in making inferences, but only that belief, error probabilities, and evidence can be most effectively used for inference if they are not conflated.

Causation

Causal inference is another field that benefits from the explicit scope of inference of evidential analysis. Until recently, causal analysis has been almost entirely abdicated by classical statistics except in the case of direct experimental manipulation (Stone, 1993; Greenland, Robins, and Pearl, 1999). The extension of causal analysis to observational studies has not been warmly received by the statistical community in general (Freedman, 1999). Assumptions regarding the absence of confounding variables must be made to make causal inferences in observational studies. Of course, the same is true even in randomized experiments (Stone, 1993; Greenland, Robins, and Pearl, 1999). Furthermore, a treatment may affect an experimental outcome by means other than the putative mechanism. While there are grounds for caution in inferring causal mechanisms from observational studies, there is in fact no fundamental difference between the assumptions needed in observational and experimental studies. Snow's 1855 demonstration of the causal impact of drinking water pollution on cholera in London is strikingly convincing despite the absence of any experimental manipulation.

In an evidential analysis, causal models are treated equally with other models. Causation cannot be "proved," but the evidence in a data set for a given causal model relative to each of a suite of specific causal and noncausal models can be quantified. Lurking variables still lurk, but the scope of inference is clear. Thus, interesting causal inferences can be cleanly made even in observational studies (e.g., Taper and Gogan, 2002).

STATISTICAL EVIDENCE IN THE WEB OF SCIENTIFIC ARGUMENT

We have clarified the distinction between data and evidence. It will be useful to clarify the distinction between evidence and inference. The third edition of the *American Heritage Dictionary of the English Language* defines inference as "the act of reasoning from evidence," while evidence is "a thing or things helpful in forming a conclusion."

A statistical inference may consider more than the evidence in a particular data set in deciding between or among models. Other factors, such as expert opinion/knowledge, related data, data collection and reduction procedures, experimental design, sampling design, and observational procedures, all may legitimately influence a statistical inference. These considerations may also affect the statistician's assessment of the probabilities of weak and misleading evidence, which of course should be part of a complete statistical inference.

As discussed by Scheiner (chapter 3), Mayo (chapter 4), and Dennis (chapter 11), a scientific inference is not made in isolation. Every scientific conclusion is embedded in a web of dependent theories. Statistical interferences validate many of the nodes of this complex web of scientific thought. Once a node (e.g., a hypothesis, statement of fact, or description of pattern) has been validated with statistical inference, it becomes relatively safe to use that node as the basis for scientific argument at a higher level. In effect, statistical inference becomes scientific evidence.

Statistical inference traditionally has been local and very concrete (at least as conceived of by scientists). In this view, statistical inference is confined to the estimation of the parameters of distributions, means, variances, and confidence intervals. It may be a mistake to so strongly compartmentalize, separating scientific inference from statistical inference and scientific thinking from statistical thinking.

Many of the attributes of thought supposedly distinguishing scientific thinking can be found in or be found emerging in statistical thought. In recent decades, statistics has accepted the challenge of causal analysis. Horizontal and vertical linkages can be found in meta-analysis and hierarchical models. Finally, statistics can even be found involved in the generation of ideas and hypotheses through the application of techniques such as exploratory data analysis, graphical visualization, and data mining.

Einstein (1950, 59) reached a similar conclusion about the distinction between scientific thinking and everyday thinking.

> The whole of science is nothing more than a refinement of everyday thinking. It is for this reason that the critical thinking of the physicist cannot possibly be restricted to the examination of the concepts of his own specific field. He cannot proceed without considering critically a much more difficult problem, the problem of analyzing the nature of everyday thinking.

FALSIFICATIONIST PRACTICE FROM AN EVIDENTIAL STANDPOINT

Scientific evidence is generally taken to be anything tending to refute or confirm a hypothesis. Statistical evidence generally does not address hypotheses directly, but addresses their translations (Pickett, Kolasa, and Jones, 1994) into statistical models. As we have already argued, models can be neither proved nor disproved. Except possibly in a few degenerate cases, models cannot possibly be true. We have argued strongly for a comparative concept of evidence. For us, statistical evidence is a comparison of data-based estimates of the relative distance of models to truth. This differs considerably from the falsificationist logic espoused in Popper (1935, 1959) or in the "strong inference" advocated to the scientific community by Platt (1964). The difference is greater in theory than in practical application. Scientists tend not to reject a theory that has some (even if not spectacular) explanatory or predictive power if no alternative theory is available. As Salmon has said, "An old theory is never completely abandoned unless there is currently available a rival to take its place. Given that circumstance, it is a matter of choosing between the old and the new" (Salmon, 1990, 192). This conditional nature of theory rejection has been a focus of the intense philosophical criticism of the Popperian/falsificationist paradigm. We believe the falsificationism vogue in science since the 1960s is cryptically comparative, and that this is one of the causes of its great success. Because at least one hypothesis is always maintained, the hypothesis least rejected is in practice most confirmed. The real strength of the "strong inference" protocol advocated by Platt is that it embraces the benefits of multiple models.

VERISIMILITUDE AND SCIENTIFIC EVIDENCE

Popper was the arch falsificationist who influenced the philosophical discourse and scientific practice of the twentieth century. In his early work, he was very uneasy with the concept of truth and did not use it at all in constructing his philosophy of science.

In his later work, beginning with *Conjectures and Refutations* (1963), Popper recognized the need for truth as a goal for science. He conceived that science was a process of continually replacing old theories and models with new theories and models of greater truthlikeness or, as he termed it, verisimilitude. Popper struggled unsuccessfully to create a consistent quantitative definition of verisimilitude. In the end, he had to simply say that something like verisimilitude must be involved.

Verisimilitude has been critiqued severely (Miller, 1974; Tichy, 1974), but also recognized as a key concept for scientific realism (Curd and Cover, 1998). The philosophical community has pursued verisimilitude vigorously for three decades, but a satisfactory formulation has not yet been attained (Niiniluoto, 1998).

Niiniluoto ends his hopeful review of verisimilitude with

> I should recommend the following problems and tasks. What are the minimum assumptions about similarity relations that allow the construction of a comparative notion of truthlikeness? What invariance properties should measures of verisimilitude have? Which are the best applications of these notions to case studies from the history of science? (Niiniluoto, 1998, 25)

The comparative nature of statistical evidence as discussed by Royall (chapter 5), Lele (chapter 7), and Taper (chapter 15) functions in effect to select models on the basis of greater verisimilitude. Empirical comparative verisimilitude might be an appealing description of the evidence functions developed by Lele (chapter 7). Perhaps the conditions developed by Lele will prove helpful to philosophers in their work on the philosophical concept of verisimilitude.

This is not to say that evidence functions and verisimilitude are the same. Verisimilitude applies to theories. Theories are more complex and less tractable than hypotheses, which are themselves more complex and less tractable than models. Evidence functions can only compare the relative distance of models to "truth." Nevertheless, there is usually a translation (Pickett, Kolasa, and Jones, 1994) explicit or implicit from scientific models to scientific hypotheses. Further, Popper himself did not believe that theories could be tested as complete entities, but only piece by piece (Thorton, 2001).

BACK TO THE FUTURE

Throughout this volume, we have struggled to understand and define the nature of scientific evidence. Clearly, this nature is elusive; but perhaps it is not entirely beyond our grasp. In 1727, Stephen Hales, in his *Vegetative Statistik,* described the state of knowledge:

> The reasonings about the wonderful and intricate operations of nature are so full of uncertainty, that, as the wise-man truly observes, hardly do we guess aright at the things that are upon earth, and with labour do we find the things that are before us. (cited in Thompson, 1942)

REFERENCES

Berger, J. O., and R. L. Wolpert. 1984. *The Likelihood Principle.* Hayward, CA: Institute of Mathematical Statistics.

Browne, R. H. 1995. Bayesian Analysis and the GUSTO Trial. Global Utilization of Streptokinase and Tissue Plasminogen Activator in Occluded Coronary Arteries. *JAMA* 274:873.

Burnham, K. P., and D. R. Anderson. 1998. *Model Selection and Inference: A Practical Information-Theoretic Approach.* Berlin: Springer.

Butler, R. W. 1986. Predictive Likelihood Inference with Applications. *J. Roy. Stat. Soc.,* ser. B, 48:1–38.

Casella, G., and R. L. Berger. 1990. *Statistical Inference.* Belmont, CA: Duxbury.

Chatfield, C. 1995. Model Uncertainty, Data Mining, and Statistical Inference (with discussion). *J. Roy. Stat. Soc.,* ser. A, 158:419–466.

Cox, D. R. 1977. The Role of Significance Tests. *Scand. J. Statist.* 4:49–70.

Cox, D. R., and D. V. Hinkley. 1974. *Theoretical Statistics.* London: Chapman and Hall.

Curd, M., and J. A. Cover, eds. 1998. *Philosophy of Science: The Central Issues.* New York: Norton.

Dennis, B. 1996. Discussion: Should Ecologists Become Bayesians? *Ecol. Appl.* 6: 1095–1103.

Dryden, I., and K. V. Mardia. 1998. *Statistical Analysis of Shapes.* New York: Wiley.

Efron, B. 1986. Why Isn't Everyone a Bayesian? (with discussion). *Am. Stat.* 40:1–11.

Einstein, A. 1950. *Out of My Later Years.* Philosophical Library. New York.

Freedman, D. 1999. From Association to Causation: Some Remarks on the History of Statistics. *Stat. Sci.* 14:243–258.

Giere, R. N. 1999. *Science without Laws.* Chicago: University of Chicago Press.

Godambe, V. P., ed. 1991. *Estimating Functions.* Oxford: Oxford University Press.

Greenland, S., J. M. Robins, and J. Pearl. 1999. Confounding and Collapsibility in Causal Inference. *Stat. Sci.* 14:29–46.

Hacking, I. 1965. *Logic of Statistical Inference,* New York: Cambridge University Press.

Hempel, C. G. 1965a. The Thesis of Structural Identity. In *Aspects of Scientific Explanation and Other Essays.* New York: Free Press.

Hempel, C. G. 1965b. Inductive-Statistical Explanation. In *Aspects of Scientific Explanation and Other Essays.* New York: Free Press.

Hempel, C. G. 1966. Criteria of Confirmation and Acceptability. In *Philosophy of Natural Science.* Englewood Cliffs: Prentice-Hall.

Heyde, C. C. 1997. *Quasi-likelihood and Its Application: A General Approach to Optimal Parameter Estimation.* New York: Springer.

Hinkley, D. V. 1979. Predictive Likelihood. *Ann. Stat.* 7:718–728.

Johnson, D. H. 1999. The Insignificance of Statistical Significance Testing. *J. Wildlife Mgmt.* 63:763–772.

Lehmann, E. L. 1983. *Theory of Point Estimation.* New York: Wiley.

Lele, S. R., and J. T. Richtsmeier. 2000. *An Invariant Approach to Statistical Analysis of Shapes.* London: Chapman and Hall/CRC Press.

Mayo, D. G. 1996. *Error and the Growth of Experimental Knowledge.* Chicago: University of Chicago Press.

McCullogh, P., and J. A. Nelder. 1989. *Generalized Linear Models.* 2nd ed. London: Chapman and Hall.

Miller, D. 1974. Popper's Qualitative Theory of Verisimilitude. *Br. J. Phil. Sci.* 25:166–177.

Niiniluoto, I. 1998. Verisimilitude: The Third Period. *Br. J. Phil. Sci.* 49:1–29.

Pickett, S. T. A., J. Kolasa, and C. G. Jones. 1994. *Ecological Understanding.* New York: Academic Press.

Platt, J. R. 1964. Strong Inference. *Science* 146:347–353.

Popper, K. R. 1935. *Logik der Forschung.* Vienna, Julius Springer Verlag.

Popper, K. R. 1959. *The Logic of Scientific Discovery.* London: Hutchinson. (Translation of *Logik der Forschung*).

Popper, K. R. 1963. *Conjectures and Refutations: The Growth of Scientific Knowledge.* London: Routledge.

Royall, R. M. 1997. *Statistical Evidence: A Likelihood Paradigm.* London: Chapman and Hall.

Royall, R. M. 2000. On the Probability of Observing Misleading Statistical Evidence (with discussion). *J. Am. Stat. Assn* 95:760–780.

Ruben, D. H. 1990. *Explaining Explanation.* Routledge: New York

Salmon, W. C. 1990. Rationality and Objectivity in Science, or Tom Kuhn Meets Tom Bayes. In Savage, C. W., ed., *Scientific Theories.* Minneapolis: University of Minnesota Press.

Scriven, M. 1959. Explanation and Prediction in Evolutionary Theory. *Science* 130: 477–482.

Stone, R. 1993. The Assumptions on Which Causal Inference Rest. *J. Roy. Stat. Soc.* ser. B, 55:455–466.

Taper, M. L., and P. Gogan. 2002. Population Dynamics in the Northern Yellowstone Elk: Density Dependence and Climatic Conditions. *J. Wildlife Mgmt.* 66:106–122.

Thompson, D. W. 1942. *On Growth and Form.* 2nd ed. New York: Macmillan.

Thorton, S. 2001. Karl Popper. In Zalta, E. N., ed., *The Stanford Encyclopedia of Philosophy* (summer 2001 ed/). http://plato.stanford.edu/archives/ fall1999/entries/ popper/

Tichy, P. 1974. On Popper's Definitions of Verisimilitude. *Br. J. Phil. Sci.* 25:155–160.

Whewell, W. 1858. *Novum Organon Renovatum.* London.

Wilson, E. O. 1998. *Consilience: The Unity of Knowledge.* New York: Knopf.

Contributors

Prasanta S. Bandyopadhyay: Department of History and Philosophy, Montana State University, Bozeman, MT 59717, psb@montana.edu

Edward J. Bedrick: Department of Mathematics and Statistics, University of New Mexico, Albuquerque, NM 87131, bedrick@math.unm.edu

John G. Bennett: Department of Philosophy, University of Rochester, Lattimore Hall, Rochester, NY 14627-0078, jbennett@philosophy.rochester.edu

Robert J. Boik: Department of Mathematical Sciences, Montana State University, Bozeman, MT 59717-2400, rjboik@math.montana.edu

Mark S. Boyce: 321 Wood Dale Drive, Site 21, Comp 32, Mayne Island, BC V0N 2J0, Canada, boyce@ualberta.ca

Hamparsum Bozdogan: Department of Statistics, University of Tennessee, Knoxville, TN 37996, bozdogan@utk.edu

James H. Brown: Biology Department, University of New Mexico, Albuquerque, NM 87131-0001, jhbrown@unm.edu

Jean-Luc E. Cartron: Department of Biology, University of New Mexico, Albuquerque, NM 87131, jlec@unm.edu

George Casella: Department of Statistics, University of Florida, Gainesville, FL 32611-8545, casella@stat.ufl.edu

Steve Cherry: Department of Mathematical Sciences, Montana State University, Bozeman, MT 59717, cherry@math.montana.edu

D. R. Cox: Nuffield College, Oxford OX1 1NF, UK, david.cox@nuffield.oxford.ac.uk

553

Martin Curd: Department of Philosophy, Purdue University, West Lafayette, IN 47907-2098, curd@purdue.edu

Brian Dennis: Department of Fish and Wildlife Resources, University of Idaho, Moscow, ID 83844-1136, brian@uidaho.edu

Philip M. Dixon: Department of Statistics, Iowa State University, Ames, IA 50011-1210, pdixon@iastate.edu

Aaron M. Ellison: Harvard Forest, Harvard University, Petersham, MA 01366, aellison@fas.harvard.edu

Stephen P. Ellner: Department of Ecology and Evolutionary Biology, Cornell University, Ithaca, NY 14853, spe2@cornell.edu

S. K. Morgan Ernest: Biology Department, University of New Mexico, Albuquerque, NM 87131-0001, mernest@unm.edu

Malcolm Forster: Philosophy Department, University of Wisconsin, Madison WI 53706, mforster@wisc.edu

Marie-Josée Fortin: Department of Zoology, University of Toronto, Toronto, Ontario, M5S 3G5, Canada, mjfortin@zoo.utoronto.ca

V. P. Godambe: Statistics and Actuarial Science, University of Waterloo, Waterloo, Ontario, N2L 3G1, Canada, vpgodamb@uwaterloo.ca

Daniel Goodman: Department of Ecology, Montana State University, Bozeman, MT 59717-0346. goodman@rapid.msu.montana.edu

Christopher C. Heyde: Mathematical Sciences Institute, Australian National University, Canberra, ACT 0200, Australia, chris@maths.anu.edu.au; and Department of Statistics, Columbia University, New York, NY 10027, chris@stat.columbia.edu

Robert D. Holt: Department of Zoology, University of Florida, Gainesville, FL 32611-8525, rdholt@zoo.ufl.edu

Jeffrey F. Kelly: Oklahoma Biological Survey, University of Oklahoma, Norman, OK 73019, jkelly@ou.edu

Michael Kruse: 1946 Calvert Street, NW, Apt. 1, Washington, DC 20009, mbkruse@aol.com

Subhash R. Lele: Department of Mathematical and Statistical Sciences, University of Alberta, Edmonton, Alberta T6G 2G1, Canada, slele@stat.ualberta.ca

Nicholas Lewin-Koh: Block 628 Bukuit Batok Central #05-650, Singapore 650628, nikko@hailmail.net

Bruce G. Lindsay: Department of Statistics, Pennsylvania State University, University Park, PA 16802, bgl@psu.edu

William A. Link: USGS Patuxent Wildlife Research Center, Laurel, MD 20708-4017, William_Link@usgs.gov

Brian A. Maurer: Department of Fisheries and Wildlife, Michigan State University, East Lansing, MI 48824, maurerb@msu.edu

Deborah G. Mayo: Department of Philosophy, Virginia Polytechnic Institute and State University, Blacksburg, VA 24061, mayod@vt.edu

Earl D. McCoy: Department of Biology, University of South Florida, Tampa, FL 33620, mccoy@chuma.cas.usf.edu

Charles E. McCulloch: Department of Epidemiology and Biostatistics, University of California at San Francisco, San Francisco, CA 94143-0560, chuck@biostat.ucsf.edu

Jean A. Miller: Department of Philosophy, Virginia Polytechnic Institute and State University, Blacksburg, VA 24061, jemille6@vt.edu

Manuel C. Molles, Jr.: Department of Biology, University of New Mexico, Albuquerque, NM 87131-1091, molles@sevilleta.unm.edu

Paul I. Nelson: Department of Statistics, Dickens Hall, Kansas State University, Manhattan, KS 66506, nels@stat.ksu.edu

Steven Hecht Orzack: Fresh Pond Research Institute, 173 Harvey Street, Cambridge, MA 02140, orzack@freshpond.org

Richard Royall: Department of Biostatistics, Johns Hopkins University, Baltimore, MD 21205, rmroyall@earthlink.net

Samuel M. Scheiner: Division of Environmental Biology, National Science Foundation, Arlington, VA 22230, sscheine@nsf.gov

Nozer D. Singpurwalla: Department of Statistics, George Washington University, Washington, DC 20052, nozer@gwu.edu

Norman A. Slade: Natural History Museum, University of Kansas, Lawrence, KS 66045, slade@ku.edu

Elliott Sober: Department of Philosophy, Stanford University, Stanford, CA 94305, esober@stanford.edu

Mark L. Taper: Department of Ecology, Montana State University, Bozeman, MT 59717, taper@rapid.msu.montana.edu

R. Cary Tuckfield: Savannah River Technology Center, Aiken, SC 29808, cary .tuckfield@srs.gov

Isabella Verdinelli: Department of Statistics, University of Rome; and Department of Statistics, Carnegie Mellon University, Pittsburgh, PA 15213, isabella@ stat.cmu.edu

Lance A. Waller: Department of Biostatistics, Emory University, Atlanta, GA 30322, lwaller@sph.emory.edu

Larry Wasserman: Department of Statistics, Carnegie Mellon University, Pittsburgh, PA 15213, larry@stat.cmu.edu

Mark L. Wilson: Department of Epidemiology, University of Michigan, Ann Arbor, MI 48109-2029, wilsonml@umich.edu

Index

a priori, 17, 20, 21, 52, 222, 250, 253, 282, 283, 287, 383, 535, 536, 544, 545
abstraction, 62, 277, 320
adequacy, 5, 23, 27, 139, 293, 342, 437–440, 445, 448, 450, 452–455, 464, 470, 471, 476–478, 481, 485, 486, 528, 530, 533, 537, 539, 540
agriculture, 73, 371
AIC. *See* Akaike information criterion
Akaike, 76, 77, 153, 160–164, 166–170, 177, 180–182, 184–186, 188–190, 335, 373, 438, 491, 495, 509, 513, 514, 520, 521, 524
Akaike information criterion, 76, 153, 161, 163–167, 169–176, 180–187, 438, 491–495, 498, 499, 503, 507, 509–512, 514, 517, 518, 520, 521, 542
allometry, 302, 303, 321
American Naturalist, The, 330
analysis: of causation, 28; of covariance, 58; of variance, 28, 164, 342, 358
anarchism, 102
antisymmetric function, 197, 198
approximation, 174, 175, 192, 193, 201, 213, 251, 290, 342, 353, 366, 388, 439, 442, 443, 444, 445, 466, 478, 482, 492, 493, 494, 503, 511, 514
ARIMA. *See* autoregressive integrated moving average
Asimov, Isaac, 65
association, 275, 277, 280, 285, 309, 317, 421
assumptions, 6, 19–21, 24, 29, 52, 79, 84, 89,

91, 93, 95, 96, 101, 105, 107, 140, 160, 164, 183, 186, 201, 221, 223, 227–229, 236, 240, 241, 246, 247, 248, 251, 252, 256, 259, 269, 270, 276, 283, 284, 285, 293, 313, 320, 333, 366, 427, 429, 443, 453, 479, 483, 518, 545, 548
astronomy, 217
asymptotic, 169, 172, 178, 180, 186, 197, 200, 338, 341, 347, 459, 468, 469, 518; bias, 170, 187; efficiency, 336; probability, 194; properties, 201; theory, 174; variance, 336
autocorrelation, 242, 243, 247, 259, 359, 513
autoregressive integrated moving average, 241

BACI design, 223, 225, 233, 234
Bacon, Francis, 3
Bayes' theorem, 80, 105, 121, 128, 146, 154, 166, 344, 348, 365, 403
Bayesian: analysis, 85, 128, 143, 144, 150, 254, 256, 327, 347, 353, 380, 381, 392, 400, 422, 430, 536, 537; ego, 85, 86, 116; estimate, 346; estimation methods, 27; Id, 86; Superego, 85, 86, 116
belief, 12, 13, 80–82, 96, 101–105, 128, 129, 141, 171, 206, 288, 310, 328, 346, 354, 356, 363, 379, 381, 404, 411, 422, 429, 432, 446, 520, 536, 545; degree of belief, 13, 140, 343, 345
bell-shaped curve, 333
Bernoulli trials, 6, 9, 207
best model element, 441

557